Bioremediation of Petroleum and Petroleum Products

Scrivener Publishing
100 Cummings Center, Suite 541J
Beverly, MA 01915-6106

Publishers at Scrivener
Martin Scrivener (martin@scrivenerpublishing.com)
Phillip Carmical (pcarmical@scrivenerpublishing.com)

Bioremediation of Petroleum and Petroleum Products

James G. Speight

CD&W Inc., Laramie,
Wyoming, USA

Karuna K. Arjoon

KKA Associates,
California, Trinidad and Tobago

Scrivener

WILEY

Co-published by John Wiley & Sons, Inc. Hoboken, New Jersey, and Scrivener Publishing LLC, Salem, Massachusetts.
Published simultaneously in Canada.

Cover design by Kris Hackerott

Library of Congress Cataloging-in-Publication Data:

ISBN 978-0-470-93849-2

Printed in the United States of America

10 9 8 7 6 5 4 3 2 1

Contents

Preface **xv**

1 Introduction to Bioremediation **1**

1 Introduction 1
2 Principles of Bioremediation 7
3 Bioremediation and Biodegradation 10
3.1 Natural Bioremediation 10
3.2 Traditional Bioremediation Methods 11
3.3 Enhanced Bioremediation Treatment 12
3.4 Biostimulation and Bioaugmentation 12
3.5 In Situ and Ex Situ Bioremediation Techniques 13
4 Mechanism of Biodegradation 15
4.1 Chemical Reactions 15
4.2 Kinetic Aspects 19
4.3 Effect of Salt 20
5 Bioremediation Methods 22
5.1 Method Parameters 23
5.2 In Situ and Ex Situ Bioremediation 24
5.3 Biostimulation and Bioaugmentation of Contaminated Sites 25
5.4 Monitored Natural Attenuation 25
5.5 Soil Vapor Extraction, Air Sparging, and Bioventing 26
5.6 Use of Biosurfactants 28
5.7 Rhizosphere Bioremediation 28
5.8 Bioengineering in Bioremediation 29
6 Test Methods for Biodegradation 30
7 References 31

2 Petroleum Composition and Properties 39
1 Introduction 39
2 Composition 46
2.1 Elemental Composition 47
2.2 Chemical Composition 50
2.3 Composition by Volatility 52
2.4 Composition by Fractionation 55
2.5 Composition by Spectroscopy 57
2.5.1 Infrared Spectroscopy 58
2.5.2 Nuclear Magnetic Resonance 58
2.5.3 Mass Spectrometry 59
2.5.4 Other Techniques 61
3 Properties 62
3.1 Density and Specific Gravity 63
3.2 Elemental (Ultimate) Analysis 64
3.3 Chromatographic Fractionation 65
3.4 Liquefaction and Solidification 67
3.5 Metals Content 68
3.6 Surface and Interfacial Tension 69
3.7 Viscosity 71
3.8 Volatility 74
4 References 76

3 Refinery Products and By-Products 79
1 Introduction 79
2 Refinery Products 81
2.1 Liquefied Petroleum Gas 90
2.2 Naphtha, Gasoline, and Solvents 92
2.3 Kerosene and Diesel Fuel 94
2.4 Fuel Oil 95
2.5 Lubricating Oil 98
2.6 White Oil, Insulating Oil, Insecticides 100
2.7 Grease 101
2.8 Wax 102
2.9 Asphalt 104
2.10 Coke 105
3 Refinery Chemicals 106
3.1 Spent Caustic 106

3.2 Spent Acids 107
3.3 Spent Catalysts 109
3.3.1 Demet 113
3.3.2 Met-X 113
4 References 113

4 Composition and Properties of Gaseous Products 115
1 Introduction 115
2 Gaseous Products 118
2.1 Liquefied Petroleum Gas 118
2.2 Natural Gas 121
2.3 Refinery Gas 125
3 Environmental Effects 129
4 Analysis 134
4.1 Calorific Value (Heat of Combustion) 136
4.2 Composition 136
4.3 Density 141
4.4 Relative density 141
4.5 Sulfur 142
4.6 Volatility and Vapor Pressure 143
5 References 143

5 Composition and Properties of Liquid Products 147
1 Introduction 147
2 Naphtha 148
2.1 Composition 153
2.2 Density (Specific Gravity) 158
2.3 Evaporation Rate 158
2.4 Flash Point 159
2.5 Odor and Color 159
2.6 Volatility 160
2.7 Environmental Impact 161
3 Fuel Oil 162
3.1 Asphaltene Content 164
3.2 Composition 165
3.3 Density (Specific Gravity) 167
3.4 Elemental Analysis 168
3.5 Flash Point 172
3.6 Metals Content 172

3.7 Pour Point and Viscosity 173
3.8 Stability 174
3.9 Environmental Impact 175
4 Wastewaters 177
5 References 180

6 Composition and Properties of Solid Products 183
1 Introduction 183
2 Residua and Asphalt 184
2.1 Acid Number 187
2.2 Asphaltene Content 188
3.3 Carbon Disulfide Insoluble Constituents 189
2.4 Composition 190
2.5 Density (Specific Gravity) 192
2.6 Elemental Analysis 195
2.7 Float Test 196
2.8 Softening Point 196
2.9 Viscosity 197
2.10 Weathering 197
2.11 Environmental Impact 197
3 Coke 199
3.1 Ash 201
3.2 Composition 202
3.3 Density 204
3.4 Dust Control Material 205
3.5 Hardness 205
3.6 Metals 206
3.7 Sulfur 207
3.8 Environmental Impact 207
4 Sludge 208
5 References 210

7 Sample Collection and Preparation 213
1 Introduction 213
2 Petroleum Chemicals 215
3 Sample Collection and Preparation 218
3.1 Soil Sampling 220
3.2 Sampling from an Aqueous Medium 221

3.3 Sample Collection 222
3.3.1 Volatile Compounds 225
3.3.2 Condensate Releases 227
3.3.3 Semi-Volatile and Non-Volatile Compounds 228
3.3.4 Solids 233
3.3.5 Water Samples 236
3.4 Extract Concentration 237
3.5 Sample Cleanup 240
4 Measurement 240
5 Accuracy 242
6 Precision 243
7 Method Validation 245
8 Quality Control and Quality Assurance 252
8.1 Quality Control 253
8.2 Quality Assurance 255
9 Method Detection Limit 256
10 References 256

8 Analytical Methods 259
1 Introduction 259
2 Chemical And Physical Properties 262
2.1 Adhesion 263
2.2 Biological Oxygen Demand 263
2.3 Boiling Point Distribution 264
2.4 Chemical Dispersibility 265
2.5 Density, Specific Gravity, and API Gravity 266
2.6 Emulsion Formation 267
2.7 Evaporation 268
2.8 Fire Point and Flash Point 269
2.9 Fractionation 270
2.10 Leachability and Toxicity 270
2.11 Metals Content 271
2.12 Pour Point 272
2.13 Solubility in Aqueous Media 272
2.14 Sulfur Content 273
2.15 Surface Tension and Interfacial Tension 273
2.16 Total Petroleum Hydrocarbons 274

2.17 Viscosity 277
2.18 Volatile Organic Compounds 278
2.19 Water Content 279
3 Petroleum Group Analyses 279
3.1 Thin Layer Chromatography 283
3.2 Immunoassay 284
3.3 Gas Chromatography 285
3.4 High Performance Liquid Chromatography 287
3.5 Gas Chromatography-Mass Spectrometry 288
4 Other Analytical Methods 289
4.1 Total Petroleum Hydrocarbons 289
4.2 Gas Chromatography 290
4.3 Infrared Spectroscopy 294
4.4 Gravimetry 296
4.5 Immunoassay 297
5 References 298

9 Biodegradation of Petroleum 305
1 Introduction 305
2 Biodegradation of Specific Constituents 307
2.1 Alkanes 307
2.2 Aromatic hydrocarbons 309
2.3 Polynuclear Aromatic Hydrocarbons 310
2.4 Phenolic Compounds 316
2.5 Spent Caustic 317
2.6 Wastewater 318
2.7 Chlorinated Compounds 319
3 Petroleum Biodegradation 319
3.1 Effects of Biodegradation 322
3.2 Conditions for Biodegradation 323
3.3 Effect of Nutrients 325
3.4 Effect of Temperature 327
3.5 Effect of Dispersants 329
3.6 Rates of Oil Biodegradation 329
3.7 Effect of Weathering 332
4 Application to Spills 333
4.1 General Application 336
5 References 339

10 Biodegradation of Naphtha and Gasoline **361**
1 Introduction 361
2 Identity and Origin 361
3 Remediation 363
3.1 Site Remediation 364
3.2 In-Situ Air Stripping 369
3.3 Pump-and-Treat Systems 370
3.4 Vacuum Enhanced Recovery 370
3.5 Single-Phase Vacuum Extraction 371
3.6 Dual-Phase Vacuum Extraction 371
3.7 Surfactant Enhanced Aquifer Redemption 371
3.8 In Situ Chemical Oxidation 371
3.8.1 Hydrogen Peroxide/Fenton's Reagent 371
3.8.2 Permanganates 372
3.9 In Situ-Enhanced Bioremediation 372
3.10 Monitored Natural Attenuation 376
4 BTEX and MTBE 377
4.1 BTEX 377
4.2 Methyl Tert-Butyl Ether 380
5 References 382

11 Biodegradation of Kerosene and Diesel **385**
1 Introduction 385
2 Identity and Origin of Kerosene and Diesel 387
3 Bioremediation 389
3.1 Kerosene 389
3.2 Diesel Fuel 391
3.3 Diesel Fuel Additives 401
4 Jet Fuel 402
5 References 404

12 Biodegradation of Fuel Oil **413**
1 Introduction 413
2 Identity and Origin of Fuel Oil 416
2.1 Production 417
2.2 Properties 419
3 Biodegradation 421
4 References 427

13 Biodegradation of Lubricating Oil **431**
1 Introduction 431
2 Identity and Origin of Lubricating Oil 434
3 Composition and Properties of Lubricating Oil 436
3.1 Thermal Stability of Lubricating Oils 437
3.2 Environmental Aspects of Lubricating Oil Use 438
4 Biodegradation of Lubricating Oils 438
4.1 Petroleum-Based Oils 440
4.2 Synthetic Lubricating Oil 442
4.3 Used Lubricating Oil 443
4.4 Vegetable Oils 448
5 Bioremediation – The Challenge 452
6 Conclusion 453
7 References 453

14 Biodegradation of Residua and Asphalt **463**
1 Introduction 463
2 Identity and Origin of Residua and Asphalt 467
2.1 Residua 467
2.2 Asphalt 467
3 Biodegradation of Residua and Asphalt 469
3.1 Residua 470
3.2 Asphalt 471
3.3 Asphaltene Constituents 473
4 References 477

15 Bioremediation Methods **483**
1 Land Ecosystems 483
1.1 Site Evaluation 486
1.2 Soil Evaluation 489
1.2.1 Physical Properties 489
1.2.2 Chemical Properties 492
1.2.3 Biological Properties 493
1.3 Effect of Temperature 496
1.4 Effect of pH 497
1.5 Effect of Salinity 498

2 Water Ecosystems 498
2.1 Biodegradation 501
2.2 Bioremediation 502
2.2.1 Effect of Temperature 503
2.2.2 Effect of Oxygen 505
2.2.3 Nutrients 505
2.2.4 Effect of Petroleum Characteristics 506
2.2.5 Effect of Prior Exposure 506
2.2.6 Effect of Dispersants 506
2.2.7 Effect of Flowing Water 506
2.2.8 Effect of Deep-Sea Environments 507
3 References 507

16 The Future of Bioremediation 515
1 Introduction 515
2 Status 517
2.1 Conventional Bioremediation 517
2.2 Enhanced Bioremediation 519
2.3 Bioremediation in Extreme Environments 522
3 Advantages and Disadvantages 524
4 Conclusion 526
5 References 528

Glossary 535

Conversion Factors 555

Index 559

Preface

Biodegradation is a natural process, and with enough time, microorganisms can eliminate many components of petroleum oil from the environment. The concern is whether or not bioremediation technologies can accelerate this natural process enough to be considered practical, and, if so, whether they might find a niche as replacements for, or adjuncts to, other petroleum spill response technologies.

Petroleum (crude oil) is a complex mixture of thousands of different chemical compounds. In addition, the composition of each accumulation of oil is unique, varying in different producing regions, and even in different unconnected zones of the same formation. The composition of petroleum also varies with the amount of refining. Significantly, the many constituents of petroleum differ markedly in volatility, solubility, and susceptibility to biodegradation: some constituents are susceptible to microbial biodegradation, while others are non-biodegradable. Furthermore, the biodegradation of different petroleum constituents occurs simultaneously, but at very different rates. This leads to the sequential disappearance of individual components of petroleum over time and, because different species of microbes preferentially attack different compounds, to successional changes in the degrading microbial community. Thus, to evaluate the effectiveness of biodegradation through the application of bioremediation technologies, it is necessary to know the molecular effects of the process, starting with the molecular composition of the contaminants.

This book introduces the reader to the science and technology of biodegradation, a key process in the bioremediation of petroleum and petroleum based contaminants at spill sites. The contaminants of concern in the molecularly-variable petroleum and petroleum products can be degraded under appropriate conditions. But the success of the process depends on the ability to determine the necessary conditions and establish them in the contaminated environment.

Although the prime focus of the book is to determine the mechanism, extent, and efficiency of biodegradation, it is necessary to know the composition of the original petroleum or petroleum product. The laws of science dictate what can or cannot be done with petroleum and petroleum products to insure that biodegradation (hence, bioremediation) processes are effective. The science of the composition of petroleum and petroleum products is at the core of understanding the chemistry of biodegradation and bioremediation processes. Hence, inclusion of petroleum analyses and properties, along with petroleum product analyses and properties, is a necessary part of this text.

As a result, the book is divided into chapters that guide the reader through the composition of petroleum and petroleum products, as well as processes involved in the biodegradation/bioremediation of petroleum: Chapter 1: Introduction to Bioremediation; Chapter 2: Petroleum Composition and Properties; Chapter 3: Refinery Products and By-Products; Chapter 4: Composition and Properties of Gaseous Fractions; Chapter 5: Composition and Properties of Liquid Fractions; Chapter 6: Composition and Properties of Solid Fractions; Chapter 7: Sample Collection and Preparation; Chapter 8: Analytical Methods; Chapter 9: Biodegradation of Petroleum; Chapter 10: Biodegradation of Naphtha and Gasoline; Chapter 11: Biodegradation of Kerosene and Diesel; Chapter 12: Biodegradation of Fuel Oil; Chapter 13: Biodegradation of Lubricating Oil; Chapter 14: Biodegradation of Residua and Asphalt; Chapter 15: Bioremediation Methods for Oil Spills; and Chapter 16: The Future of Bioremediation.

Each chapter includes a copious reference section, and the book is further improved by the inclusion of an extensive Glossary.

James G. Speight PhD, DSc
Laramie, Wyoming, USA

Karuna K. Arjoon, MPhil
California, Trinidad and Tobago

1

Introduction to Bioremediation

1 Introduction

One of the major and continuing environmental problems is hydrocarbon contamination resulting from activities related to petroleum and petroleum products. Soil contamination with hydrocarbons causes extensive damage of local systems, since accumulation of pollutants in animals and plant tissue may cause death or mutations.

However, not all petroleum products are harmful to health and the environment. There are records of the use of *petroleum spirit* for medicinal purposes. This was probably a higher boiling fraction than *naphtha* or a low boiling fraction of *gas oil* that closely resembled the modern-day *liquid paraffin*, for medicinal purposes. In fact, the so-called liquid paraffin has continued to be prescribed up to modern times, as a means for miners to take in prescribed doses to lubricate the alimentary tract and assist coal dust, taken in during the working hours, in passing though the body.

There are, however, those constituents of petroleum that are extremely harmful to health and the environment. Indeed, petroleum constituents, either in the pure form or as the components of a fraction, have been known to belong to the various families of carcinogens and neurotoxins. Whatever the name given to these compounds, they are extremely toxic.

As a result, once a spill has occurred, every effort must be made to rid the environment of the toxins. The chemicals of known toxicity range in degree of toxicity from low to high, and represent considerable danger to human health, and must be removed (Frenzel et al., 2009). Many of these chemicals substances come in contact with, and are sequestered by, soil or water systems. While conventional methods to remove, reduce, or mitigate the effects of toxic chemical in nature are available, including (1) pump and treat systems, (2) soil vapor extraction, (3) incineration, and (4) containment, each of these conventional methods of treatment of contaminated soil and/or water suffers from recognizable drawbacks, and may involve some level of risk. In short, these methods, depending upon the chemical constituents of the spilled material, may have limited effectiveness and can be expensive (Speight, 1996; Speight and Lee, 2000; Speight, 2005).

Although the effects of bacteria (microbes) on hydrocarbons have been known for decades, this technology (now known as *bioremediation*) has shown promise and, in some cases, high degrees of effectiveness for the treatment of these contaminated sites, since it is cost-effective and will lead to complete mineralization. Bioremediation functions basically on biodegradation, which may refer to complete *mineralization* of the organic contaminants into carbon dioxide, water, inorganic compounds, and cell protein, or transformation of complex organic contaminants to other simpler organic compounds that are not detrimental to the environment. In fact, unless they are overwhelmed by the amount of the spilled material or the material is toxic, many indigenous microorganisms in soil and/or water are capable of degrading hydrocarbon contaminants.

The United States Environmental Protection Agency (US EPA) uses bioremediation because it takes advantage of natural processes and relies on microbes that occur naturally or can be laboratory cultivated; these consist of bacteria, fungi, actinomycetes, cyanobacteria, and, to a lesser extent, plants (US EPA, 2006). These microorganisms either consume and convert the contaminants, or assimilate within them all harmful compounds from the surrounding area, thereby rendering the region virtually contaminant-free. Generally, the substances that are consumed as an energy source are organic compounds, while those that are assimilated within the organism are heavy metals. Bioremediation harnesses this natural process by promoting the growth and/or rapid multiplication of these organisms that can effectively degrade specific contaminants and convert them to non-toxic by-products.

The capabilities of micro-organisms and plants to degrade and transform contaminants provide benefits in the cleanup of pollutants from spills and storage sites. These remediation ideas have provided the foundation for many ex situ waste treatment processes (including sewage treatment), and a host of in situ bioremediation methods that are currently in practice.

Thus, bioremediation – *the use of living organisms to reduce or eliminate environmental hazards resulting from accumulations of toxic chemicals and other hazardous wastes* – is an option that offers the possibility to destroy or render harmless various contaminants using natural biological activity (Gibson and Sayler, 1992). In addition, bioremediation can also be used in conjunction with a wide range of traditional physical and chemical technology to enhance their effectiveness (Vidali, 2001).

In the current context, bioremediation of petroleum and petroleum fractions (or products) is the cleanup of petroleum spills or petroleum product spills by the use of microbes to breakdown the petroleum constituents (or other organic contaminants) into less harmful (usually lower molecular weight) and easier-to-remove products (biodegradation). The microbes transform the contaminants through metabolic or enzymatic processes, which vary greatly, but the final product is usually harmless and includes carbon dioxide, water, and cell biomass. Thus, the emerging science and technology of bioremediation offers an alternative method to detoxify petroleum-related soil and water contaminants.

Briefly and by means of clarification, *biodegradation (biotic degradation, biotic decomposition) is the chemical degradation of contaminants by bacteria or other biological means*. Organic material can be degraded aerobically (in the presence of oxygen) or anaerobically (in the absence of oxygen). Most bioremediation systems operate under aerobic conditions, but a system under anaerobic conditions may permit microbial organisms to degrade chemical species that are otherwise non-responsive to aerobic treatment, and vice versa.

Thus, biodegradation is a natural process (or a series of processes) by which spilled petroleum hydrocarbons, or other organic waste material, are broken down (degraded) into nutrients that can be used by other organisms. As a result, the ability of a chemical to be biodegraded is an indispensable element in understanding the risk posed by that chemical on the environment.

Biodegradation is a key process in the natural attenuation (reduction or disposal) of chemical compounds at hazardous waste sites. The contaminants of concern in crude oil are able to degrade under appropriate conditions, but the success of the process depends on the ability to determine these conditions and establish them in the contaminated environment. Thus, important site factors required for success include (1) the presence of metabolically capable and sustainable microbial populations, (2) suitable environmental growth conditions, such as the presence of oxygen, (3) temperature, which is an important variable: keeping a substance frozen or below the optimal operating temperature for microbial species can prevent biodegradation, and most biodegradation occurs at temperatures between 10 and 35°C (50 and 95°F), (4) the presence of water, (5) appropriate levels of nutrients and contaminants, and (6) favorable acidity or alkalinity (Table 1.1).

Table 1.1 Essential factors for microbial bioremediation.

Factor	Optimal Conditions
Microbial population	Suitable kinds of organisms that can biodegrade all of the contaminants
Oxygen	Enough to support aerobic biodegradation (about 2% oxygen in the gas phase or 0.4 mg/liter in the soil water)
Water	Soil moisture should be from 50–70% of the water holding capacity of the soil
Nutrients	Nitrogen, phosphorus, sulfur, and other nutrients to support good microbial growth
Temperature	Appropriate temperatures for microbial growth (0–40°C)
pH	Best range is from 6.5 to 7.5

In regard to the last parameter, soil pH is extremely important because most microbial species can survive only within a certain pH range: generally the biodegradation of petroleum hydrocarbons is optimal at a pH 7 (neutral), and the *acceptable* (or optimal) pH range is on the order of 6 to 8. Furthermore, soil (or water) pH can affect availability of nutrients.

Thus, through biodegradation processes, living microorganisms (primarily bacteria, but also yeasts, molds, and filamentous fungi) can alter and/or metabolize various classes of compounds present in petroleum. Furthermore, biodegradation also alters subsurface oil accumulations of petroleum (Winters and Williams, 1969; Speight, 2007). Shallow oil accumulations, such as heavy oil reservoirs and tar sand deposits, where the reservoir temperature is low-to-moderate (<80°C, <176°F) are commonly found to have undergone some degree of biodegradation (Winters and Williams, 1969; Speight, 2007).

Temperature influences the rate of biodegradation by controlling the rate of enzymatic reactions within microorganisms. Generally, the rate of an enzymatic reaction approximately doubles for each 10°C (18°F) rise in temperature (Nester et al., 2001). However, there is an upper limit to the temperature that microorganisms can withstand. Most bacteria found in soil, including many bacteria that degrade petroleum hydrocarbons, are mesophile organisms, which have an optimum working temperature range on the order of 25 to 45°C (77 to 113°F) (Nester et al., 2001). Thermophilic bacteria (those that survive and thrive at relatively high temperatures) that are normally found in hot springs and compost heaps exist indigenously in cool soil environments, and can be activated to degrade hydrocarbons with an increase in temperature to 60°C (140°F). This indicates the potential for natural attenuation in cool soils through thermally enhanced bioremediation techniques (Perfumo et al., 2007).

In order to enhance and make favorable the parameters presented above to ensure microbial activity, there are two other bioremediation technologies that offer useful options for cleanup of spills of petroleum and petroleum products: (1) fertilization and (2) seeding.

Fertilization (*nutrient enrichment*) is the method of adding nutrients, such as phosphorus and nitrogen, to a contaminated environment to stimulate the growth of the microorganisms capable of biodegradation. Limited supplies of these nutrients in nature usually control the growth of native microorganism populations. When more nutrients are added, the native microorganism population can grow rapidly, potentially increasing the rate of biodegradation.

Seeding is the addition of microorganisms to the existing native oil-degrading population. Some species of bacteria that do not naturally exist in an area will be added to the native population. As with fertilization, the purpose of seeding is to increase the population of microorganisms that can biodegrade the spilled oil.

Thus, biodegradation is an environmentally acceptable, naturally occurring process that takes place when all of the nutrients and physical conditions involved are suitable for growth. The process allows for the breakdown of a compound to either fully oxidized or reduced simple molecules, such as carbon dioxide/methane, nitrate/ammonium, and water. However, in some cases, where the process is not complete, the products of biodegradation can be more harmful than the substance degraded.

Intrinsic bioremediation is the combined effect of natural destructive and non-destructive processes to reduce the mobility, mass, and associated risk of a contaminant. Non-destructive mechanisms include sorption, dilution, and volatilization. Destructive processes are aerobic and anaerobic biodegradation.

Intrinsic aerobic biodegradation is well-documented as a means of remediating soil and groundwater contaminated with fuel hydrocarbons. In fact, intrinsic aerobic degradation should be considered an integral part of the remediation process (McAllister et al., 1995; Barker et al., 1995). There is growing evidence that natural processes influence the immobilization and biodegradation of chemicals such as aromatic hydrocarbons, mixed hydrocarbons, and chlorinated organic compounds (Ginn et al., 1995; King et al., 1995).

Phytoremediation is the use of living green plants for the removal of contaminants and metals from soil, and is essentially an in situ treatment of pollutant contaminated soils, sediments, and water: terrestrial, aquatic and wetland plants, and algae can be used for the phytoremediation process under specific cases and conditions of hydrocarbon contamination (Brown, 1995; Nedunuri et al., 2000; Radwan et al., 2000; Siciliano et al., 2000; Magdalene et al., 2009). It is best applied at sites with relatively shallow contamination of pollutants that are amenable to the various subcategories of phytoremediation: (1) phytotransformation, the breakdown of organic contaminants sequestered by plants, (2) rhizosphere bioremediation, the use of rhizosphere microorganisms to degrade organic pollutants, (3) phytostabilization, a containment process using plants, often in combination with soil additives, to assist plant installation, mechanically stabilize the site, and reduce pollutant transfer to other ecosystem compartments and the food chain, (4) phytoextraction, the ability of some plants to accumulate metals/metalloids in their shoots, (5) rhizofiltration, and/or (6) phytovolatilization/rhizovolatilization, processes employing metabolic capabilities of plants and associated rhizosphere microorganisms to transform pollutants into volatile compounds that are released to the atmosphere (Korade and Fulekar, 2009).

These technologies are especially valuable where the contaminated soils are fragile, and prone to erosion. The establishment of a stable vegetation community stabilizes the soil system, and prevents erosion. This aspect is especially relevant to certain types of soil where removal of large volumes of soil destabilizes the soil system, which leads to extensive erosion. However, when the above parameters are not conducive to bacterial activity, the bacteria (1) grow too slowly, (2) die, or (3) create more harmful chemicals.

Phytotransformation and *rhizosphere bioremediation* are applicable to sites that have been contaminated with organic pollutants, including pesticides. It is a technology that should be considered for remediation of contaminated sites because of its cost effectiveness, aesthetic advantages, and long-term applicability (Brown, 1995).

Plants have shown the capacity to withstand relatively high concentrations of organic chemicals without toxic effects, and they can uptake and convert chemicals quickly to less toxic metabolites in some cases. In addition, they stimulate the degradation of organic chemicals in the rhizosphere by the release of root exudates, enzymes, and the buildup of organic carbon in the soil.

Micro-organisms degrade or transform contaminants by a variety of mechanisms. Petroleum hydrocarbons (particularly alkanes), for example, are converted to carbon dioxide and water:

$$2C_{12}H_{26} + 37O_2 \rightarrow 24CO_2 + 26H_2O$$

Or the hydrocarbon may be used as a primary food source by the bacteria, which use the energy to generate new cells.

Some contaminants, such as chlorinated organic or high aromatic hydrocarbons, are generally resistant to microbial attack. They are degraded either slowly or not at all, and hence, it is not easy to predict the rates of cleanup for a bioremediation exercise; there are no rules to predict if a contaminant can be degraded.

When the hydrocarbons are chlorinated (as might occur in several additives to improve the performance of petroleum products) the degradation takes place as a secondary or co-metabolic process, rather than a primary metabolic process. In such a case, enzymes, which are produced during aerobic utilization of carbon sources such as methane, degrade the chlorinated compounds. Under aerobic conditions, a chlorinated solvent, such as trichloroethylene ($ClCH{=}CCl_2$), which may have been mixed with the petroleum product during processing or during use, can be degraded through a sequence of metabolic steps, where some of the intermediary by-products may be more hazardous than the parent compound (e.g., vinyl chloride, $CH_2{=}CHCl$).

Since many of the contaminants of concern in petroleum and petroleum-related products oil are readily biodegradable under the appropriate conditions, the success of oil-spill bioremediation depends mainly on the ability to establish these conditions in the contaminated environment, using the technology to optimize the total efficiency of the microorganisms.

Over the past two decades, opportunities for applying bioremediation to a much broader set of contaminants have been identified. Indigenous and enhanced organisms have been shown to degrade industrial solvents, polychlorinated biphenyls (PCBs), explosives, and many different agricultural chemicals. Pilot, demonstration, and full-scale applications of bioremediation have been carried out on a limited basis. However, the full benefits of bioremediation have not been realized, because processes and organisms that are effective in controlled laboratory tests are not always equally effective in full-scale applications. The failure to perform optimally in the field setting stems from a lack of predictability, due in part to inadequacies in the fundamental scientific understanding of how and why these processes work.

2 Principles of Bioremediation

As already stated, bioremediation is an environmentally friendly technique used to restore soil and water to its original state by using indigenous microbes to break down and eliminate contaminants. Biological technologies are often used as a substitute to chemical or physical cleanup of oil spills, because bioremediation does not require as much equipment or labor as other methods: therefore, it is usually cheaper. It also allows cleanup workers to avoid contact with polluted soil and water.

The microorganisms used for bioremediation may be indigenous to a contaminated area, or they may be isolated from elsewhere and brought to the contaminated site. Contaminants are transformed by living organisms through reactions that take place as a part of their metabolic processes. Biodegradation of a compound is often a result of the actions of multiple organisms. When microorganisms are imported to a contaminated site to enhance degradation, it is a process known as bioaugmentation.

For bioremediation to be effective, microorganisms must convert the pollutants to harmless products. As bioremediation can be effective only where environmental conditions permit microbial growth and activity, its application often involves the manipulation of environmental parameters to allow microbial growth and degradation to proceed at a faster rate. However, as is the case with other technologies, bioremediation has its limitations, and there are several disadvantages that must be recognized (Table 1.2).

The control and optimization of bioremediation processes are a complex system of many factors. These factors include: the existence of a microbial population capable of degrading the pollutants; the availability of contaminants to the microbial population; and the environment factors (type of soil, temperature, pH, the presence of oxygen or other electron acceptors, and nutrients).

One of the important factors in biological removal of hydrocarbons from a contaminated environment is their bioavailability to an active microbial population, which is *the degree of interaction of chemicals with living organisms* or the degree to which a contaminant can be readily taken up and metabolized by a bacterium (Harms et al., 2010). Moreover, the bioavailability of a contaminant is controlled by factors such as the physical state of the hydrocarbon in situ, its hydrophobicity, water solubility, sorption to environmental matrices such as soil, and diffusion out of the soil matrix. When contaminants have very low solubility in water, as in the case of n-alkanes and polynuclear aromatic hydrocarbons, the organic phase components will not partition efficiently into the aqueous phase supporting the microbes.

In the case of soil, the contaminants will also partition to the soil organic matter, and become even less bioavailable. Two-phase bioreactors containing an aqueous phase and a non-aqueous phase liquid (NAPL) have been developed and used for bioremediation of hydrocarbon-contaminated soil to address this very problem, but the adherence of microbes to the NAPL-water interface can still be an important factor in reaction kinetics. Similarly, two-phase bioreactors, sometimes with silicone oil as the non-aqueous phase, have been proposed for biocatalytic conversion of hydrocarbons like styrene (Osswald et al., 1996) to make the substrate more bioavailable to microbes in the aqueous phase. When the carbon source is in limited supply, its availability will control the rate of metabolism, and hence, biodegradation, rather than catabolic capacity of the cells or availability of oxygen or other nutrients.

Table 1.2 Advantages and disadvantages of bioremediation.

Advantages	**Disadvantages**
Remediates contaminants that are adsorbed onto or trapped within the geologic materials of which the aquifer is composed along with contaminants dissolved in groundwater.	Injection wells and/or infiltration galleries may become plugged by microbial growth or mineral precipitation.
Application involves equipment that is widely available and easy to install.	High concentrations (TPH greater than 50,000 ppm) of low solubility constituents may be toxic and/or not bioavailable
Creates minimal disruption and/or disturbance to ongoing site activities.	Difficult to implement in low-permeability aquifers.
Time required for subsurface remediation may be shorter than other approaches (e.g., pump-and-treat).	Re-injection wells or infiltration galleries may require permits or may be prohibited. Some states require permit for air injection.
Generally recognized as being less costly than other remedial options.	May require continuous monitoring and maintenance.
Can be combined with other technologies (e.g., bioventing, SVE) to enhance site remediation.	Remediation may only occur in more permeable layer or channels within the aquifer.
In many cases, this technique does not produce waste products that must be disposed.	

In the case of the biomediation of waterways, similar principles apply. Under enhanced conditions, (1) certain fuel hydrocarbons can be removed preferentially over others, but the order of preference is dependent upon the geochemical conditions, and (2) augmentation and enhancement via electron acceptors accelerate the biodegradation process. For example, with regard to the aromatic benzene-toluene-ethylbenzene-xylenes (BTEX): (1) toluene can be preferentially removed under intrinsic bioremediation conditions,

(2) biodegradation of benzene is relatively slow, (3) augmentation with sulfate can preferentially stimulate biodegradation of o-xylene, and (4) ethylbenzene may be recalcitrant under sulfate-reducing conditions, but readily degradable under denitrifying conditions (Cunningham et al., 2000).

3 Bioremediation and Biodegradation

In the current context, bioremediation (biodegradation) is a method for dealing with contamination by petroleum, petroleum products, and petroleum waste streams. The process typically occurs through the degradation of petroleum or a petroleum product through the action of microorganisms (biodegradation).

The method utilizes indigenous bacteria (microbes), compared to the customary (physical and chemical) remediation methods. Also, the microorganisms engaged are capable of performing almost any detoxification reaction. Biodegradation studies provide information on the fate of a chemical or mixture of petroleum-derived chemicals (such as oil spills and process wastes) in the environment, thereby opening the scientific doorway to develop further methods of cleanup by (1) analyzing the contaminated sites, (2) determining the best method suited for the environment, and (3) optimizing the cleanup techniques which lead to the emergence of new processes.

3.1 Natural Bioremediation

Natural biodegradation/bioremediation typically involves the use of molecular oxygen (O_2), where oxygen (the *terminal electron acceptor*) receives electrons transferred from an organic contaminant:

$$\text{Organic substrate} + O_2 \rightarrow \text{biomass} + CO_2 + H_2O + \text{other products}$$

In the absence of oxygen, some microorganisms obtain energy from fermentation and anaerobic oxidation of organic carbon. Many anaerobic organisms (*anaerobes*) use nitrate, sulfate, and salts of iron (III) as practical alternates to oxygen acceptor, as, for example, in the anaerobic reduction process of nitrates, sulfates, and salts of iron (III):

$$2NO_3^- + 10e^- + 12H^+ \rightarrow N_2 + 6H_2O$$

$$SO_4^{2-} + 8e^- + 10H^+ \rightarrow H_2S + 4H_2O$$

$$Fe(OH)_3 + e^- + 3H^+ \rightarrow Fe^{2+} + 3H_2O$$

3.2 Traditional Bioremediation Methods

Methods for the cleanup of pollutants have usually involved removal of the polluted materials and their subsequent disposal by land filling or incineration (so-called *dig, haul, bury, or burn* methods) (Speight, 1996; Speight and Lee, 2000; Speight, 2005). Furthermore, available space for landfills and incinerators is declining. Perhaps one of the greatest limitations to traditional cleanup methods is the fact that, in spite of their high costs, they do not always ensure that contaminants are completely destroyed.

Conventional bioremediation methods that have been, and are still, used are (1) composting, (2) land farming, (3) biopiling, and (4) use of a bioslurry reactor (Speight, 1996; Speight and Lee, 2000; Semple et al., 2001).

Composting is a technique that involves combining contaminated soil with nonhazardous organic materials, such as manure or agricultural wastes; the presence of the organic materials allows the development of a rich microbial population and elevated temperature characteristic of composting. *Land farming* is a simple technique, in which contaminated soil is excavated and spread over a prepared bed and periodically tilled until pollutants are degraded. *Biopiling* is a hybrid of land farming and composting: it is essentially engineered cells that are constructed as aerated composted piles. A *bioslurry reactor* can provide rapid biodegradation of contaminants, due to enhanced mass transfer rates and increased contaminant-to-microorganism contact. These units are capable of aerobically biodegrading aqueous slurries created through the mixing of soils or sludge with water. The most common state of bioslurry treatment is batch; however, continuous-flow operation is also possible.

The technology selected for a particular site will depend on the limiting factors present at the location. For example, where there is insufficient dissolved oxygen, bioventing or sparging is applied, and biostimulation or bioaugmentation are suitable for instances where the biological count are low. On the other hand, application of the composting technique, if the operation is unsuccessful, will result in a greater quantity of contaminated materials. Land farming is only effective if the contamination is near the soil surface, or else bed preparation is required. The main drawback with slurry bioreactors is that high-energy input is required to maintain suspension and the potential needed for volatilization.

Other techniques are also being developed to improve the microbe-contaminant interactions at treatment sites, so as to use bioremediation technologies at their fullest potential. These bioremediation technologies consist of monitored natural attenuation, bioaugmentation, biosimulation, surfactant addition, anaerobic bioventing, sequential anaerobic/aerobic treatment, soil vapor extraction, air sparging, enhanced anaerobic dechlorination, and bioengineering (Speight, 1996; Speight and Lee, 2000).

The use of traditional methods of bioremediation continues, but there is also method evolution, which may involve the following steps:

1. Isolating and characterizing naturally-occurring microorganisms with bioremediation potential.
2. Laboratory cultivation to develop viable populations.
3. Studying the catabolic activity of these microorganisms in contaminated material through bench scale experiments.
4. Monitoring and measuring the progress of bioremediation through chemical analysis and toxicity testing in chemically-contaminated media.
5. Field applications of bioremediation techniques using either/both steps: (1) *in-situ* stimulation of microbial activity by the addition of microorganisms and nutrients and the optimization of environmental factors at the contaminated site itself, and/or (2) *ex-situ* restoration of contaminated material in specifically designated areas by land-farming and composting method.

3.3 Enhanced Bioremediation Treatment

Enhanced bioremediation is a process in which indigenous or inoculated microorganisms (e.g., fungi, bacteria, and other microbes) degrade (metabolize) organic contaminants found in soil and/or ground water, and convert the contaminants to innocuous end products. The process relies on general availability of naturally occurring microbes to consume contaminants as a food source (petroleum hydrocarbons in aerobic processes) or as an electron acceptor (chlorinated solvents, which may be waste materials from petroleum processing). In addition to microbes being present, in order to be successful, these processes require nutrients, such as carbon, nitrogen, and phosphorus.

Enhanced bioremediation involves the addition of microorganisms (e.g., fungi, bacteria, and other microbes) or nutrients (e.g. oxygen, nitrates) to the subsurface environment to accelerate the natural biodegradation process.

3.4 Biostimulation and Bioaugmentation

Biostimulation is the method of adding nutrients, such as phosphorus and nitrogen, to a contaminated environment to stimulate the growth of the microorganisms that break down oil. Additives are usually added to the subsurface through injection wells, although injection well technology for biostimulation purposes is still emerging. Limited supplies of these necessary nutrients usually control the growth of native microorganism populations. Thus, addition of nutrients causes rapid growth of the indigenous microorganism population, thereby increasing the rate of biodegradation.

It is to be anticipated that the success of biostimulation is case-specific and site-specific, depending on oil properties, the nature of the nutrient products, and the characteristics of the contaminated environments. When oxygen is not a limiting factor, one of keys for the success of oil biostimulation is to maintain an optimal nutrient level in the interstitial pore water. Several types of commercial biostimulation agents are available for use in bioremediation (Zhu et al., 2004).

Bioaugmentation is the addition of pre-grown microbial cultures to enhance microbial populations at a site to improve contaminant clean up and reduce clean up time and cost. Indigenous or native microbes are usually present in very small quantities and may not be able to prevent the spread of the contaminant. In some cases, native microbes do not have the ability to degrade a particular contaminant. Therefore, bioaugmentation offers a way to provide specific microbes in sufficient numbers to complete the biodegradation (Atlas, 1991).

Mixed cultures have been most commonly used as inocula for seeding because of the relative ease with which microorganisms with different and complementary biodegradative capabilities can be isolated (Atlas, 1977). Different commercial cultures were reported to degrade petroleum hydrocarbons (Compeau et al., 1991; Leavitt and Brown, 1994; Chhatre et al., 1996; Mangan, 1990; Mishra et al., 2001; Vasudevan and Rajaram, 2001).

Microbial inocula (the microbial materials used in an inoculation) are prepared in the laboratory from soil or groundwater, either from the site where they are to be used, or from another site where the biodegradation of the chemicals of interest is known to be occurring. Microbes from the soil or groundwater are isolated, and are added to media containing the chemicals to be degraded. Only microbes capable of metabolizing the chemicals will grow on the media. This process isolates the microbial population of interest. One of the main environmental applications for bioaugmentation is at sites with chlorinated solvents. Microbes called *Dehalococcoides nethenogenes* usually perform reductive dechlorination of solvents such as perchloroethylene and trichloroethylene.

Bioaugmentation adds highly concentrated and specialized populations of specific microbes to the contaminated area, while biostimulation is dependent on appropriate indigenous microbial population and organic material being present at the site.

3.5 In Situ and Ex Situ Bioremediation Techniques

Bioremediation can be used as a cleanup method for both contaminated soil and water. Its applications fall into two broad categories: *in situ* or *ex situ*. In situ bioremediation treats the contaminated soil or groundwater in the location in which it was found, while ex situ bioremediation processes require excavation of contaminated soil or pumping of groundwater before they can be treated.

In situ technologies do not require excavation of the contaminated soils, so they may be less expensive, create less dust, and cause less release of contaminants than ex situ techniques. Also, it is possible to treat a large volume of soil at once. *In situ* techniques, however, may be slower than ex situ techniques, may be difficult to manage, and are only most effective at sites with permeable soil.

The most effective means of implementing *in situ* bioremediation depends on the hydrology of the subsurface area, the extent of the contaminated area, and the nature (type) of the contamination. In general, this method is effective only when the subsurface soils are highly permeable, the soil horizon to be treated falls within a depth of 8–10 m, and shallow groundwater is present at 10 m or less below ground surface. The depth of contamination plays an important role in determining whether or not an *in situ* bioremediation project should be employed. If the contamination is near the groundwater, but the groundwater is not yet contaminated, then it would be unwise to set up a hydrostatic system. It would be safer to excavate the contaminated soil and apply an on-site method of treatment away from the groundwater.

The typical time frame for an *in situ* bioremediation project can be in the order of 12 to 24 months, depending on the levels of contamination and depth of contaminated soil. Due to the poor mixing in this system, it becomes necessary to treat for long periods of time to ensure that all the pockets of contamination have been treated.

In situ bioremediation is a very site-specific technology that involves establishing a hydrostatic gradient through the contaminated area by flooding it with water carrying nutrients and possibly organisms adapted to the contaminants. Water is continuously circulated through the site until it is determined to be clean.

In situ bioremediation of groundwater speeds the natural biodegradation processes that take place in the water-soaked underground region that lies below the water table. One limitation of this technology is that differences in underground soil layering and density may cause re-injected conditioned groundwater to follow certain preferred flow paths. On the other hand, *ex situ* techniques can be faster, easier to control, and used to treat a wider range of contaminants and soil types than *in situ* techniques. However, they require excavation and treatment of the contaminated soil before, and sometimes after, the actual bioremediation step.

In situ bioremediation is the preferred method for large sites, and is used when physical and chemical methods of remediation may not completely remove the contaminants, leaving residual concentrations that are above regulatory guidelines. This method has the potential to provide advantages such as complete destruction of the contaminant(s), lower risk to site workers, and lower equipment/operating costs. *In situ* bioremediation can be used as a cost-effective secondary treatment scheme to decrease the concentration of contaminants to acceptable levels, or as a primary treatment method, which is followed by physical or chemical methods for final site closure.

Finally, evidence for the effectiveness of petroleum bioremediation and petroleum product bioremediation should include: (1) faster disappearance of oil in treated areas than in untreated areas, and (2) a demonstration that biodegradation was the main reason for the increased rate of oil disappearance. To obtain such evidence, the analytical procedures must be chosen carefully, and careful data interpretation is essential, but there are disadvantages and errors when the method is not applied correctly (Chapter 9) (Speight, 2005).

4 Mechanism of Biodegradation

Biodegradation involves chemical transformations mediated by microorganisms that: (1) satisfy nutritional requirements, (2) satisfy energy requirements, (3) detoxify the immediate environment, or (4) occur fortuitously, such that the organism receives no nutritional or energy benefit (Stoner, 1994).

Mineralization is the complete biodegradation of organic materials to inorganic products, and often occurs through the combined activities of microbial consortia rather than through a single microorganism (Shelton and Tiedje, 1984).

Co-metabolism is the partial biodegradation of organic compounds that occurs fortuitously and that does not provide energy or cell biomass to the microorganisms. Co-metabolism can result in partial transformation to an intermediate that can serve as a carbon and energy substrate for microorganisms, as with some hydrocarbons, or can result in an intermediate that is toxic to the transforming microbial cell, as with trichloroethylene and methanotrophs.

4.1 Chemical Reactions

Biodegradation of petroleum constituents can occur under both aerobic (oxic) and anaerobic (anoxic) conditions (Zengler et al., 1999), albeit by the action of different consortia of organisms. In the subsurface, oil biodegradation occurs primarily under anoxic conditions, mediated by sulfate reducing bacteria (e.g., Holba et al., 1996) or other anaerobes, using a variety of other electron acceptors as the oxidant. Thus, two classes of biodegradation reactions are (1) aerobic biodegradation and (2) anaerobic biodegradation.

Aerobic biodegradation involves the use of molecular oxygen (O_2), where oxygen (the "terminal electron acceptor") receives electrons transferred from an organic contaminant:

$$\text{Organic substrate} + O_2 \rightarrow \text{biomass} + CO_2 + H_2O + \text{other inorganic products}$$

Thus, the organic substrate is oxidized (addition of oxygen), and the oxygen is reduced (addition of electrons and hydrogen) to water (H_2O). In this case, the organic substrate serves as the sources of energy (electrons) and the source of cell carbon used to build microbial cells (biomass). Some microorganisms (chemo-autotrophic aerobes or litho-trophic aerobes) oxidize reduced inorganic compounds (NH_3, Fe^{2+}, or H_2S) to gain energy and fix carbon dioxide to build cell carbon:

$$NH_3 \text{ (or } Fe^{2+} \text{ or } H_2S) + CO_2 + H_2 + O_2 \rightarrow \text{biomass} + NO_3 \\ \text{(or Fe or } SO_4) + H_2O$$

At some contaminated sites, as a result of consumption of oxygen by aerobic microorganisms and slow recharge of oxygen, the environment becomes anaerobic (lacking oxygen), and mineralization, transformation, and co-metabolism depend upon microbial utilization of electron acceptors other than oxygen (anaerobic biodegradation). Nitrate (NO_3), iron (Fe^{3+}), manganese (Mn^{4+}), sulfate (SO_4), and carbon dioxide (CO_2) can act as electron acceptors if the organisms present have the appropriate enzymes (Sims, 1990).

Anaerobic biodegradation is the microbial degradation of organic substances in the absence of free oxygen. While oxygen serves as the electron acceptor in aerobic biodegradation processes forming water as the final product, degradation processes in anaerobic systems depend on alternative acceptors, such as sulfate, nitrate or carbonate, yielding, in the end, hydrogen sulfide, molecular nitrogen, and/or ammonia and methane (CH_4), respectively.

In the absence of oxygen, some microorganisms obtain energy from fermentation and anaerobic oxidation of organic carbon. Many anaerobes use nitrate, sulfate, and salts of iron (III) as practical alternates to oxygen acceptor. The anaerobic reduction process of nitrates, sulfates and salts of iron is an example:

$$2NO_3^- + 10e^- + 12H^+ \rightarrow N_2 + 6H_2O$$
$$SO_42^- + 8e^- + 10H^+ \rightarrow H_2S + 4H_2O$$
$$Fe(OH)_3 + e^- + 3H^+ \rightarrow Fe^{2+} + 3H_2O$$

Anaerobic biodegradation is a multistep process performed by different bacterial groups. It involves hydrolysis of polymeric substances like proteins or carbohydrates to monomers and the subsequent decomposition to soluble acids, alcohols, molecular hydrogen, and carbon dioxide, Depending on the prevailing environmental conditions, the final steps of ultimate anaerobic biodegradation are performed by denitrifying, sulfate-reducing or methanogenic bacteria.

In contrast to the strictly anaerobic sulfate-reducing and methanogenic bacteria, the nitrate-reducing microorganisms, as well as many other decomposing bacteria, are mostly facultative anaerobic, insofar as these microorganisms are able to grow and degrade organic substances under aerobic as well as anaerobic conditions. Thus, aerobic and anaerobic environments represent the two extremes of a continuous spectrum of environmental habitats which are populated by a wide variety of microorganisms with specific biodegradation abilities.

Anaerobic conditions occur where vigorous decomposition of organic matter and restricted aeration result in the depletion of oxygen. Anoxic conditions may represent an intermediate stage where oxygen supply is limited, while still allowing a slow (aerobic) degradation of organic compounds.

In a digester, the various bacteria also have different requirements to the surrounding environment. For example, acidogenic bacteria need pH values from 4 to 6, while methanogenic bacteria need values from 7 to 7.5. In batch tests, the dynamic equilibrium is often interrupted because of an enrichment of acidogenic bacteria as a consequence of lacking substrate in- and outflow.

On a structural basis, the hydrocarbons in crude oil are classified as alkanes (*normal* or *iso*), cycloalkanes, and aromatics. Alkenes are rare in petroleum, but occur in many refined petroleum products as a consequence of the cracking process (Speight, 2007). Increasing carbon numbers of alkanes (homology), variations in carbon chain branching (*iso*-alkanes), ring condensations, and interclass combinations, such as phenyl alkanes, account for the high numbers of hydrocarbons that occur in crude oil.

In addition, smaller amounts of oxygen-containing compounds (phenol derivatives, naphthenic acids), nitrogen-containing compounds (pyridine derivatives, pyrrole derivatives, indole derivatives), sulfur-containing compounds (thiophene derivatives), and the high molecular weight polar asphalt fraction also occur in petroleum, but not in refined petroleum products (Speight, 2007).

The inherent biodegradability of these individual components is a reflection of their chemical structure, but is also strongly influenced by the physical state and toxicity of the compounds. As an example, while n-alkanes as a structural group are the most biodegradable petroleum hydrocarbons, the C_5 to C_{10} homologs have been shown to be inhibitory to the majority of hydrocarbon degraders. As solvents, these homologs tend to disrupt lipid membrane structures of microorganisms. Similarly, alkanes in the C_{20} to C_{40} range are hydrophobic solids at physiological temperatures. Apparently, it is this physical state that strongly influences their biodegradation (Bartha and Atlas, 1977).

Primary attack on intact hydrocarbons requires the action of oxygenases, and, therefore, requires the presence of free oxygen. In the case of alkanes, mono-oxygenase attack results in the production of alcohol. Most microorganisms attack alkanes terminally, whereas some perform sub-terminal oxidation. The alcohol product is oxidized finally into an aldehyde. Extensive methyl branching interferes with the beta-oxidation process, and necessitates terminal attack or other bypass mechanisms. Therefore, n-alkanes are degraded more readily than iso alkanes.

Cycloalkanes are transformed by an oxidase system to a corresponding cyclic alcohol, which is dehydrated to ketone, after which a mono-oxygenase system lactonizes the ring, which is subsequently opened by a lactone hydrolase. These two oxygenase systems usually never occur in the same organisms and hence, result in the frustrated attempts to isolate pure cultures that grow on cycloalkanes (Bartha, 1986b). However, synergistic actions of microbial communities are capable of dealing with degradation of various cycloalkanes quite effectively.

As in the case of alkanes, the monocyclic compounds, cyclopentane, cyclohexane, and cycloheptane have a strong solvent effect on lipid membranes, and are toxic to the majority of hydrocarbon degrading microorganisms. Highly condensed cycloalkane compounds resist biodegradation, due to their relatively complex structure and physical state (Bartha, 1986a).

Condensed polycyclic aromatics are degraded, one ring at a time, by a similar mechanism, but biodegradability tend to decline with the increasing number of rings and degree of condensation (Atlas and Bartha, 1998). Aromatics with more than four condensed rings are generally not suitable as substrates for microbial growth, though they may undergo metabolic transformations. The biodegradation process also declines with the increasing number of alkyl substituents on the aromatic nucleus.

Asphaltic constituents of petroleum (Speight, 2007) tend to increase during biodegradation in relative and sometimes absolute amounts. This would suggest that they not only tend to resist biodegradation, but may also be formed *de novo* by condensation reactions of biodegradation and photo-degradation intermediates.

In crude petroleum as well as in refined products, petroleum hydrocarbons occur in complex mixtures and influence each others' biodegradation. The effects may go in negative as well as positive directions. Some iso-alkanes are apparently spared as long as n-alkanes are available as substrates, while some condensed aromatics are metabolized only in the presence of more easily utilizable petroleum hydrocarbons, a process referred to as co-metabolism (Wackett, 1996).

Finally, a word on the issue of adhesion as it affects biodegradation and, hence, bioremediation.

Adhesion to hydrophobic surfaces is a common strategy used by microorganisms to overcome limited bioavailability of hydrocarbons (Bouchez-Naïtali et al., 1999). Intuitively, it may be assumed that adherence of cells to a hydrocarbon would correlate with the ability to utilize it as a growth substrate, and conversely, that cells able to utilize hydrocarbons would be expected to be able to adhere to them. However, species like *Staphylococcus aureus* and *Serratia marcescens*, which are unable to grow on hydrocarbons, adhere to them (Rosenberg et al., 1980). Thus, adherence to hydrocarbons does not necessarily predict utilization (Abbasnezhad et al., 2011).

Biodegradation of poorly water-soluble liquid hydrocarbons is often limited by low availability of the substrate to microbes. Adhesion of microorganisms to an oil–water interface can enhance this availability, whereas detaching cells from the interface can reduce the rate of biodegradation. The capability of microbes to adhere to the interface is not limited to hydrocarbon degraders, nor is it the only mechanism to enable rapid uptake of hydrocarbons, but it represents a common strategy. The general indications are that microbial adhesion can benefit growth on and biodegradation of very poorly water-soluble hydrocarbons, such as n-alkanes and large polycyclic aromatic hydrocarbons dissolved in a non-aqueous phase. Adhesion is particularly important when the hydrocarbons are not emulsified, thereby giving limited interfacial area between the two liquid phases.

When mixed communities are involved in biodegradation, the ability of cells to adhere to the interface can enable selective growth and enhance bioremediation with time. The critical challenge in understanding the relationship between growth rate and biodegradation rate for adherent bacteria is to accurately measure and observe the population resides at the interface of the hydrocarbon phase.

4.2 Kinetic Aspects

The kinetics for modeling the bioremediation of contaminated soils can be extremely complicated. This is largely due to the fact that the primary function of microbial metabolism is not for the remediation of environmental contaminants. Instead, the primary metabolic function, whether bacterial or fungal in nature, is to grow and sustain more of the microorganism. Therefore, the formulation of a kinetic model must start with the active biomass and factors, such as supplemental nutrients and oxygen source, that are necessary for subsequent biomass growth (Cutright, 1995; Rončević et al., 2005; Pala et al., 2006).

Studies of the kinetics of the bioremediation process proceed in two directions: (1) the first is concerned with the factors influencing the amount of transformed compounds with time, and (2) the other approach seeks the types of curves describing the transformation, and determines which of them fits the degradation of the given compounds by the microbiologic culture in the laboratory microcosm, and sometimes in the field. However, studies of biodegradation kinetics in the natural environment are often empiric, reflecting only a basic level of knowledge about the microbiologic population and its activity in a given environment (Maletić et al., 2009).

One such example of the empirical approach is the simple (perhaps over-simplified) model:

$$dC/dt = kC^n$$

C is the concentration of the substrate, t is time, k is the degradation rate constant of the compound, and n is a fitting parameter (most often taken to be unity) (Hamaker, 1972; Wethasinghe et al., 2006). Using this model, one can fit the curve of substrate removal by varying n and k until a satisfactory fit is obtained. It is evident from this equation that the rate is proportional to the exponent of substrate concentration. First order kinetics are the most often used equation for representation of the degradation kinetics (Heitkamp et al., 1987; Heitkamp and Cerniglia, 1987; Venosa et al., 1996; Seabra et al., 1999; Holder et al., 1999; Winningham et al., 1999; Namkoonga et al., 2002; Grossi et al., 2002; Hohener et al., 2003; Collina et al., 2005; Rončević et al., 2005; Pala et al., 2006).

However, researchers involved in kinetic studies do not always report whether the model they used was based on theory or experience, and whether the constants in the equation have a physical meaning or if they just serve as fitting parameters (Bazin et al., 1976; Rončević et al., 2005).

4.3 Effect of Salt

Salt is a common co-contaminant that can adversely affect the bioremediation potential at sites such as flare pits and drilling sites (*upstream sites*) contaminated with saline produced formation water, or at oil and gas processing facilities contaminated by refinery wastes containing potassium chloride (KCl) and sodium chloride (NaCl) salts (Pollard et al., 1994). Because of increasing emphasis and interest in the viability of intrinsic bioremediation as a remedial alternative, the impact of salt on these processes is of interest.

The effect of salinity on microbial cells varies from disrupted tertiary protein structures and denatured enzymes to cell dehydration (Pollard et al., 1994), with different species having different sensitivities to salt (Tibbett et al., 2011). A range of organic pollutants, including hydrocarbons, has been shown to be mineralized by marine or salt-adapted terrestrial microorganisms that are able to grow in the presence of salt (Margesin and Schinner, 2001; Oren et al., 1992; Nicholson and Fathepure, 2004). In naturally saline soils, it has been shown that bioremediation of diesel fuel is possible at salinities up to 17.5% w/v (Riis et al., 2003; Kleinsteuber et al. (2006).

However, an inverse relationship between salinity and the biodegradation of petroleum hydrocarbons by halophilic enrichment cultures from the Great Salt Lake (Utah) has been observed (Ward and Brock, 1978). These cultures were unable to metabolize petroleum hydrocarbons at salt concentrations above 20% (w/v) in this hyper-saline environment. An inhibitory effect of salinity at concentrations above 2.4% (w/v) NaCl was found to be greater for the biodegradation of aromatic and polar fractions than for the saturated fraction of petroleum hydrocarbons in crude oil incubated with marine sediment (Mille et al., 1991). This represents ex-situ petroleum hydrocarbon degradation by salt-adapted terrestrial microorganisms.

Furthermore, the effects of salt as a co-contaminant on hydrocarbon degradation in naturally non-saline systems has been described (deCarvalho and daFonseca, 2005). The results showed that in the degradation of C_5 to C_{16} hydrocarbons at 28°C (82°F) in the presence of 1.0, 2.0 or 2.5% (w/v) NaCl by the isolate *Rhodococcus erythropolis* DCL14, the lag phase of the cultures increased and growth rates decreased with increasing concentrations of sodium chloride. In a similar study (Rhykerd et al., 1995), soils were fertilized with inorganic nitrogen and phosphorus, and amended with sodium chloride at 0.4, 1.2, or 2% (w/w). After 80 days at 25°C (77°F), the highest salt concentration had inhibited motor oil mineralization.

However, investigation of the combinations of factors limiting biodegradation of petroleum hydrocarbon contamination at upstream oil and gas production facilities has received relatively little attention. A laboratory solid-phase bioremediation study reported that high salinity levels reduced the degradation rate of flare pit hydrocarbons (Amatya et al., 2002), and more recently it has been observed that addition of sodium chloride to a petroleum-contaminated Arctic soil decreased hexadecane mineralization rates in the initial stages of bioremediation and increased lag times, but the final extent of mineralization was comparable over a narrow range of salinity from 0 to 0.4% w/w (Børresen and Rike, 2007).

Continuing investigations are necessary to determine whether the effects observed in the laboratory are site-specific or contaminant-specific, or are applicable more broadly to sub-surface hydrocarbon bioremediation. Further research using more sites, including those previously having been impacted by sodium chloride, may allow inference of salt tolerance at upstream oil and gas sites. Particularly important is the impact of sodium chloride on anaerobic hydrocarbon degradation, which should be investigated.

Field evidence is sparse with respect to anaerobic biodegradation at salt contaminated upstream oil and gas sites. Before embarking on anaerobic microcosm tests, field evidence of indicators of anaerobic biodegradation, including changes in terminal electron acceptors, presence of metabolites, and isotopic analysis, would be a reasonable way to initiate the research (Ulrich et al., 2009).

5 Bioremediation Methods

Bioremediation technology exploits various naturally occurring mitigation processes: (1) *natural attenuation*, (2) *biostimulation*, and (3) *bioaugmentation*. Bioremediation which occurs without human intervention other than monitoring is often called *natural attenuation*. This natural attenuation relies on natural conditions and behavior of soil microorganisms that are indigenous to soil. *Biostimulation* also utilizes indigenous microbial populations to remediate contaminated soils, and consists of adding nutrients and other substances to soil to catalyze natural attenuation processes. *Bioaugmentation* involves introduction of exogenic microorganisms (sourced from outside the soil environment) capable of detoxifying a particular contaminant, sometimes employing genetically altered microorganisms.

In recent years, *in situ* bioremediation concepts have been applied in treating contaminated soil and ground water. Removal rates and extent vary based on the contaminant of concern and site-specific characteristics. Removal rates also are affected by variables such as contaminant distribution and concentration; co-contaminant concentrations; indigenous microbial populations and reaction kinetics; and parameters such as pH, moisture content, nutrient supply, and temperature. Many of these factors are a function of the site and the indigenous microbial community and, thus, are difficult to manipulate. Specific technologies may have the capacity to manipulate some variables, and may be affected by other variables as well (US EPA, 2006).

During bioremediation, microbes utilize chemical contaminants in the soil as an energy source and, through oxidation-reduction reactions, metabolize the target contaminant into usable energy for microbes. By-products (metabolites) released back into the environment are typically in a less toxic form than the parent contaminants. For example, petroleum hydrocarbons can be degraded by microorganisms in the presence of oxygen through aerobic respiration. The hydrocarbon loses electrons and is oxidized, while oxygen gains electrons and is reduced. The result is formation of carbon dioxide and water (Nester et al., 2001).

When oxygen is limited in supply or absent, as in saturated or anaerobic soils or lake sediment, anaerobic (without oxygen) respiration prevails. Generally, inorganic compounds, such as nitrate, sulfate, ferric iron, manganese, or carbon dioxide, serve as terminal electron acceptors to facilitate biodegradation.

Generally, a contaminant is more easily and quickly degraded if it is a naturally occurring compound in the environment, or chemically similar to a naturally occurring compound, because microorganisms capable of its biodegradation are more likely to have evolved. Petroleum hydrocarbons are naturally occurring chemicals; therefore, microorganisms which are capable of attenuating or degrading hydrocarbons exist in the environment. Development of biodegradation technologies of synthetic chemicals, such chlorocarbons or chlorohydrocarbons, is dependent on outcomes of research that searches for natural or genetically improved strains of microorganisms to degrade such contaminants into less toxic forms.

In summary, bioremediation is increasingly viewed as an appropriate remediation technology for hydrocarbon-contaminated polar soils. As for all soils, the successful application of bioremediation depends on appropriate biodegradative microbes and environmental conditions in situ. Laboratory studies have confirmed that hydrocarbon-degrading bacteria typically assigned to the genera *Rhodococcus*, *Sphingomonas*, or *Pseudomonas* are present in contaminated polar soils. However, as indicated by the persistence of spilled hydrocarbons, environmental conditions in situ are suboptimal for biodegradation in polar soils.

Therefore, it is likely that ex situ bioremediation will be the method of choice for ameliorating and controlling the factors limiting microbial activity, i.e., low and fluctuating soil temperatures, low levels of nutrients, and possible alkalinity and low moisture. Care must be taken when adding nutrients to the coarse-textured, low-moisture soils prevalent in continental Antarctica and the high Arctic, because excess levels can inhibit hydrocarbon biodegradation by decreasing soil water potentials. Bioremediation experiments conducted on site in the Arctic indicate that land farming and biopiles may be useful approaches for bioremediation of polar soils (Aislabie et al., 2006; Nugroho et al., 2010).

5.1 Method Parameters

Several factors that affect the decision of which method is chosen are (1) the nature of the contaminants, (2) the location of contaminated site, (3) the time allotted to the cleanup, (4) effects on humans, animals and plants, and last but by no means least (5) the cost of the cleanup. Sometimes when one method is no longer effective and efficient, another remediation method can be introduced into the contaminated soil.

Oil spills introduce large amounts of toxic compounds into the environment, and though different methods of bioremediation have been successful in remediating soils and water contaminated with the lighter organics, the high-viscosity crude oils (heavier oils) are still less susceptible to these techniques.

Conventional bioremediation methods used are biopiling, composting, land farming, and bioslurry reactors, but there are limitations affecting the applicability and effectiveness of these methods (Speight and Lee, 2000). With the application of the composting technique, if the operation is unsuccessful, it will result in a greater quantity of contaminated materials. Land farming is only effective if the contamination is near the soil surface, or else bed preparation needs to take place. The main drawback with slurry bioreactors is that high-energy inputs are required to maintain suspension and the potential needed for volatilization.

For a bioremediation method to be successful in soil and water cleanup, the physical, chemical, and biological environment must be feasible. Parameters that affect the bioremediation process are (1) low temperatures, (2) preferential growth of microbes obstructive to bioremediation, (3) high concentrations of chlorinated organics, heavy metals, and heavy oils poisoning the microorganisms, (4) preferential flow paths severely decreasing contact between injected fluids and contaminants throughout the contaminated zones, and (5) the soil matrix prohibiting contaminant-microorganism contact.

Since most of the contaminants of concern in crude oil are readily biodegradable under the appropriate conditions, the success of oil spill bioremediation depends mainly on the ability to establish these conditions in the contaminated environment using the above new developing technologies to optimize the microorganisms' total efficiency. The technologies used at various polluted sites depend on the limiting factor present at the location. For example, where there is insufficient dissolved oxygen, bioventing or sparging is applied, while biostimulation or bioaugmentation is suitable for instances where the biological count is low.

The bioventing process combines an increased oxygen supply with vapor extraction. A vacuum is applied at some depth in the contaminated soil, which draws air down into the soil from holes drilled around the site, and sweeps out any volatile organic compounds. The development and application of venting and bioventing for *in situ* removal of petroleum from soil have been shown to remediate hydrocarbons by venting and biodegradation (van Eyk, 1994).

Even though a particular technology may have reports of improving biodegradation efficiency (for example, surfactant addition), this may not be the case at times, depending on the sample.

5.2 In Situ and Ex Situ Bioremediation

Bioremediation applications fall into two broad categories: (1) *in situ* or (2) *ex situ*. In situ bioremediation is used when physical and chemical methods of remediation may not completely remove the contaminants, leaving residual concentrations that are above regulatory guidelines. Bioremediation can be used as a cost-effective secondary treatment scheme to decrease the concentration of contaminants to acceptable levels. In other cases, bioremediation can be the primary treatment method, and followed by physical or chemical methods for final site closure. Also, it is the preferred method for very large sites.

5.3 Biostimulation and Bioaugmentation of Contaminated Sites

The primary advantage of biostimulation is that bioremediation will be undertaken by already present native microorganisms that are well suited to the subsurface environment, and are well distributed spatially within the subsurface, but the main disadvantage is that the delivery of additives in a manner that allows the additives to be readily available to subsurface microorganisms is based on the local geology of the subsurface.

Bioaugmentation is the addition of pre-grown microbial cultures to enhance microbial populations at a site to improve contaminant cleanup and reduce cleanup time and cost. Indigenous or native microbes are usually present in very small quantities, and may not be able to prevent the spread of the contaminant. In some cases, native microbes do not have the ability to degrade a particular contaminant. Therefore, bioaugmentation offers a way to provide specific microbes in sufficient numbers to complete the biodegradation.

Bioaugmentation adds highly concentrated and specialized populations of specific microbes to the contaminated area, while biostimulation is dependent on appropriate indigenous microbial population and organic material being present at the site. Therefore, one might be led to believe that bioaugmentation is more effective than biostimulation.

5.4 Monitored Natural Attenuation

The term *monitored natural attenuation* refers to the reliance on natural attenuation to achieve site-specific remedial objectives within a time frame that is reasonable compared to that offered by other more active methods.

The *natural attenuation processes* that are at work in such a remediation approach include a variety of physical, chemical, or biological processes that, under favorable conditions, act without human intervention to reduce the mass, toxicity, mobility, volume, or concentration of contaminants in soil or groundwater. These in situ processes include biodegradation, dispersion, dilution, sorption, volatilization, and chemical or biological stabilization, transformation, or destruction of contaminants. A study of any contaminated site must first be performed to decide whether natural attenuation would make a positive input, and though it has degraded lighter chain hydrocarbons quite extensively, the heavier chain hydrocarbons are less susceptible.

As with any technique, there are disadvantages: the disadvantages of *monitored natural attenuation* method are the need for longer time frames to achieve remediation objectives compared to active remediation, the site characterization may be more complex and costly, and long-term monitoring will generally be necessary.

5.5 Soil Vapor Extraction, Air Sparging, and Bioventing

Soil vapor extraction removes harmful chemicals, in the form of vapors, from the soil above the water table. The vapors are extracted from the ground by applying a vacuum to pull it out.

Like soil vapor extraction, *air sparging* uses a vacuum to extract the vapors. Air sparging uses air to help remove harmful vapors like the lighter gasoline constituents (i.e., benzene, ethylbenzene, toluene, and xylene [BTEX]), because they readily transfer from the dissolved to the gaseous phase, but is less applicable to diesel fuel and kerosene. When air is pumped underground, the chemicals evaporate faster, which makes them easier to remove. Methane can be used as an amendment to the sparged air to enhance co-metabolism of chlorinated organics. Soil vapor extraction and air sparging are often used at the same time to clean up both soil and groundwater.

Biosparging is used to increase the biological activity in soil by increasing the oxygen supply via sparging air or oxygen into the soil. In some instances, air injections are replaced by pure oxygen to increase the degradation rates. However, in view of the high costs of this treatment, in addition to the limitations in the amount of dissolved oxygen available for microorganisms, hydrogen peroxide (H_2O_2) was introduced as an alternative, and it was used on a number of sites to supply more oxygen (Schlegel, 1977), and is more efficient in enhancing microbial activity during the bioremediation of contaminated soil and ground water (Brown and Norris, 1994; Flathman et al., 1991; Lee et al., 1988; Lu, 1994; Lu and Hwang, 1992; Pardieck et al., 1992), but it can be a disadvantage if the toxicity is sufficient to microorganisms even at low concentrations (Brown and Norris, 1994; Scragg, 1999).

Soil vapor extraction requires drilling extraction wells within the polluted area. The necessary equipment to create a vacuum is attached to the well, which pulls air and vapors through the soil and up to the surface. Once the extraction wells pull the air and vapors out of the ground, special air pollution control equipment collects them. The equipment separates the harmful vapors from the clean air.

Air sparging works very much like soil vapor extraction. However, the wells that pump air into the ground are drilled into water-soaked soil below the water table. Air pumped into the wells disturbs the groundwater. This helps the pollution change into vapors. The vapors rise into the drier soil above the groundwater, and are pulled out of the ground by extraction wells. The harmful vapors are removed in the same way as soil vapor extraction. The air used in soil vapor extraction and air sparging also helps clean up pollution by encouraging the growth of microorganisms. In general, the wells and equipment are simple to install and maintain, and can reach greater depths than other methods that involve digging up soil. Soil vapor extraction and air sparging are effective at removing many types of pollution that can evaporate.

Air sparging should not be used if free products are present. Air sparging can create groundwater mounding, which could potentially cause free products to migrate and contamination to spread. Also, it is not suitable around basements, sewers, or other subsurface confined spaces that are present at the site. Potentially dangerous constituent concentrations could accumulate in basements unless a vapor extraction system is used to control vapor migration. If the contaminated groundwater is located in a confined aquifer system, air sparging is not advisable, because the injected air would be trapped by the saturated confining layer and could not escape to the unsaturated zone.

Anaerobic sparging, an innovative technique in bioremediation, depends on the delivery of an inert gas (nitrogen or argon) with low (<2%) levels of hydrogen. Co-metabolic air sparging is the delivery of oxygen containing gas with enzyme-inducing growth substrate (such as methane or propane).

Bioventing is a technology that stimulates the natural in situ biodegradation of any aerobically degradable compounds in soil by providing oxygen to existing soil microorganisms. In contrast to soil vapor vacuum extraction, bioventing uses low airflow rates to provide only enough oxygen to sustain microbial activity. Two basic criteria have to be satisfied for successful bioventing: (1) the air must be able to pass through the soil in sufficient quantities to maintain aerobic conditions, and (2) natural hydrocarbon-degrading microorganisms must be present in concentrations large enough to obtain reasonable biodegradation rates.

Bioventing is a medium to long-term technology: cleanup ranges from a few months to several years, and it is applicable to any chemical that can be aerobically biodegraded. The technique has been successfully used to remediate soils contaminated by petroleum hydrocarbons, non-chlorinated solvents, some pesticides, wood preservatives, and other organic chemicals.

Though there are limitations, this technology does not require expensive equipment, and relatively few personnel are involved in operation and maintenance; therefore, bioventing is receiving increased exposure in the remediation consulting community. Potential improvements on the current bioventing methods that have been taking place are the use of electrochemical oxygen gas sensors, detailed characterization of *non-aqueous phase liquid* (NAPL) distribution, and neutron probe logging. Another bioventing enhancement is the use of this technique with bio-slurping and soil vapor extraction (Baker, 1999).

Bio-slurping is an in situ remediation technology that combines the two remedial approaches of bioventing and vacuum-enhanced free-product recovery. It is faster than the conventional remedy of product recovery followed by bioventing. The system is made to minimize groundwater recovery and drawdown in the aquifer. Bio-slurping was designed and is being tested to address contamination by petroleum products with a floating lighter non aqueous phase liquids (LNAPL) layer.

Bio-slurping efficiently recovers free product and extracts less groundwater for treatment, which speeds up remediation and reduces water handling and treatment costs. It enhances natural in-situ bioremediation of vadose zone (which extends from the top of the ground surface to the water table) soils, and may be the only feasible remediation technology at low-permeability sites. But for bio-slurping to even be considered at a contamination site, free product must be present, and the product must be biodegradable; also, the soil must respond to bioventing.

5.6 Use of Biosurfactants

Another common emerging technology is the use of *biosurfactants*, which are microbially-produced surface-active compounds. They are amphiphilic molecules with both hydrophilic and hydrophobic regions, causing them to aggregate at interfaces between fluids with different polarities found in oil spills. Many of the known biosurfactant producers are hydrocarbon-degrading organisms.

Biosurfactants have comparable solubilization properties to synthetic surfactants, but have several additional advantages that make them superior candidates in bioremediation schemes. First, biosurfactants are biodegradable, and are not a pollution threat. Furthermore, most studies indicate that they are non-toxic to microorganisms, and therefore are unlikely to inhibit biodegradation of nonpolar organic contaminants. Biosurfactant production is less expensive, can be easily achieved ex situ at the contaminated site, and has the potential of occurring in situ.

Biosurfactants are also effective in many diverse geologic formations, and are compatible with many existing remedial technologies (such as pump and treat rehabilitation, air sparging, and soil flushing), and significantly accelerate innovative approaches, including microbial, natural attenuation enhanced soil flushing, and bio-slurping.

5.7 Rhizosphere Bioremediation

Rhizosphere bioremediation is the interaction between plants and microorganisms, and is also known as phytostimulation or plant-assisted bioremediation.

The plant root zone (the rhizosphere) has significantly larger numbers of microorganisms than soils that do not have plants growing in them, which appears to enhance the biodegradation of organic compounds

(Wenzel, 2009). In the rhizosphere bioremediation process, plants provide oxygen, bacteria, and organic carbon to encourage the degradation of organics in the soil. The microorganisms in the environment created by the plants, together with the roots of the plants, can degrade more contaminants than could occur in a purely microbial system.

Plants release stimulants into the soil environment that help to motivate the degradation of organic chemicals by inducing enzyme systems of existing bacterial populations, stimulating growth of new species that are able to degrade the wastes, and/or increasing soluble substrate concentrations for all microorganisms. Plants help with microbial conversions where certain bacteria that metabolize pollutants are able to encourage degradation of chemicals in the soil, allowing bioremediation to occur with less retardation.

Evaluation of the current efforts (Wenzel, 2008) suggests that pollutant bioavailability in the rhizosphere of phytoremediation crops is decisive for designing phytoremediation technologies with improved, predictable remedial success. For phytoextraction, emphasis should be put on improved characterization of the bioavailable metal pools and the kinetics of resupply from less available fractions to support decision making on the applicability of this technology to a given site. Limited pollutant bioavailability may be overcome by the design of plant-microbial consortia that are capable of mobilizing metals/metalloids by modification of rhizosphere pH and ligand exudation, or by enhancing bioavailability of organic pollutants by the release of biosurfactants.

The complexity and heterogeneity of polluted soils will require the design of integrated approaches of rhizosphere management, such as (1) combining co-cropping of phytoextraction and rhizodegradation crops, (2) inoculation of microorganisms, and (3) soil management.

5.8 Bioengineering in Bioremediation

In many cases, after an oil spill, the natural microbial systems for degrading the oil are overwhelmed. Therefore, molecular engineers are constructing starvation promoters to express heterologous genes needed in the field for survival, and adding additional bioremediation genes that code for enzymes able to degrade a broader range of compounds present in the contaminated environments. Various bacterial strains are also being developed where each strain is specific for a certain organic compound present in oil spills. This will help increase the speed of bioremediation, and allow detailed cleanup to take place where no organic contamination remains in the environment.

Thus, the decision to bioremediate a site is dependent on cleanup, restoration, and habitat protection objectives and the factors that are present that would have an impact on success. If the circumstances are such that no amount of nutrients will accelerate biodegradation, then the decision should be made on the need to accelerate oil disappearance to protect a vital living resource or simply to speed up restoration of the ecosystem. These decisions are clearly influenced by the circumstances of the spill.

6 Test Methods for Biodegradation

Various methods exist for the testing of biodegradability of substances. Biodegradability is assessed by following certain parameters which are considered to be indicative of the consumption of the test substance by microorganisms, or the production of simple basic compounds which indicate the mineralization of the test substance.

Hence, there are various biodegradability testing methods that measure the amount of carbon dioxide (or methane, for anaerobic cases) produced during a specified period; there are those that measure the loss of dissolved organic carbon for substances which are water soluble, those that measure the loss of hydrocarbon infrared bands, and there are yet others that measure the uptake of oxygen by the activities of microorganisms (biochemical oxygen demand, BOD).

However, when the reference is specifically to lubricants, there are two major methods of biodegradability testing, and these are outlined in brief below:

1. ASTM D5864 determines lubricant biodegradation. This test determines the rate and extent of aerobic aquatic biodegradation of lubricants when exposed to an inoculum under laboratory conditions. The inoculum may be the activated sewage-sludge from a domestic sewage-treatment plant, or it may be derived from soil or natural surface waters, or any combination of the three sources. The degree of biodegradability is measured by calculating the rate of conversion of the lubricant to carbon dioxide. A lubricant, hydraulic fluid or grease is classified as readily biodegradable when 60% or more of the test material carbon is converted to carbon dioxide in 28 days, as determined using this test method.

2. The most established test methods used by the lubricant industry for evaluating the biodegradability of their products are as follows: Method CEC-L-33-A-94, developed by the Coordinating European Council (CEC); Method OEC D 301B, the Modified Sturm Test, developed by the Organization for Economic Cooperation and Development (OECD); and Method EPA 560/6-82-003, number CG-2000, the Shake Flask Test, adapted by the U.S. Environmental Protection Agency (EPA). These tests also determine the rate and extent of aerobic aquatic biodegradation under laboratory conditions. The Modified Sturm Test and Shake Flask Test also calculate the rate of conversion of the lubricant to CO_2. The CEC

test measures the disappearance of the lubricant by analyzing test material at various incubation times through infrared spectroscopy. Laboratory tests have shown that the degradation rates may vary widely among the various test methods indicated above (US Army Corps of Engineers, 1999).

Biodegradability tests based on the CEC method described above have certain trends which indicate that mineral oils, along with alkylated benzenes and polyalkeleneglycols among others, generally have poor biodegradability: between 0–40%. Specifically, mineral oil biodegradability varies from 15–35% as conducted by the CEC biodegradability test.

7 References

Abbasnezhad, H., Gray, M., and Foght, J.M. 2011. "Influence of Adhesion on Aerobic Biodegradation and Bioremediation of Liquid Hydrocarbons." *Appl. Microbiol. Biotechnol.*, 92: 653–675.

Aislabie, J., Saul, D.J., and Foght, J.M. 2006. "Bioremediation of Hydrocarbon-Contaminated Polar Soils." *Extremophiles*, 10: 171–179.

Amatya, P.L., Hettiaratchi, J.P.A., and Joshi, R.C. 2002. "Biotreatment of Flare Pit Waste." *J. Can. Petrol. Technol.*, 41: 30–36.

Atlas, R.M. 1977. "Stimulated Petroleum Biodegradation." *Crit. Rev. Microbiol.*, 5: 371–386.

Atlas, R.M. 1991. "Bioremediation: Using Nature's Helpers-Microbes and Enzymes to Remedy Mankind's Pollutants." *Proceedings of Biotechnology in the Feed Industry*. Alltech's Thirteenth Annual Symposium. T.P. Lyons and K.A. Jacques (Editors), Alltech Technical Publications, Nicholasville, Kentucky. pp. 255–264.

Atlas, R.M., and Bartha, R. 1998. "Fundamentals and Applications." In *Microbial Ecology* 4th Edition. Benjamin/Cummings Publishing Company Inc., California. pp. 523–530.

Baker, R.S. 1999. "Bioventing Systems: A Critical Review. In: Bioremediation of Contaminated Soils." In D.C. Adriana, J.M. Bollag, W.T. Frankenberger, and R.C. Sims, R.C. (Editors). *Agronomy Monograph No. 37*. Amer. Soc. Agron., Crop Sci. Soc. Amer., Soil Sci. Soc. Amer. Madison, Wisconsin. pp. 595–630.

Barker, G.W., Raterman, K.T., Fisher, J.B., Corgan, J.M., Trent, G.L., Brown, D.R., and Sublette, G.L. 1995. "Assessment of Natural Hydrocarbon Bioremediation at Two Gas Condensate Production Sites." In R.E. Hinchee, J.T. Wilson, and D.C. Downey (Editors), *Intrinsic Bioremediation*, Battelle Press, Columbus, Ohio. pp. 181–188.

Bartha, R. 1986a. *Microbial Ecology: Fundamentals and Applications*. Addisson-Wesley Publishers, Reading, Massachusetts.

Bartha, R. 1986b. "Biotechnology of Petroleum Pollutant Biodegradation." *Microb. Ecol.*, 12: 155–172.

Bartha, R., and Atlas, R.M. 1977. "The Microbiology of Aquatic Oil Spills." *Adv. Appl. Microb.*, 22: 225–266.

Børresen, M.H., and Rike, A.G. 2007. "Effects of Nutrient Content, Moisture Content and Salinity on Mineralization of Hexadecane in an Arctic Soil." *Cold Regions Sci. Technol.*, 48: 129–138.

Bouchez-Naïtali, M., Rakatozafy, H., Marchal, R., Leveau, J.Y., and Vandecasteele, J.P. 1999. "Diversity of Bacterial Strains Degrading Hexadecane in Relation to the Mode of Substrate Uptake." *J. Appl. Microbiol.*, 86: 421–428.

Brown, K.S. 1995. "The Green Clean: The Emerging Field of Phytoremediation Takes Root." *BioScience*, 45: 579–582.

Brown, R.A., and Norris, R.D. 1994. "The Evolution of a Technology: Hydrogen Peroxide in In Situ Bioremediation." In R.E. Hinchee, B.C. Alleman, R.E. Hoeppel, and R.N. Miller (Editors). *Hydrocarbon Bioremediation*. CRC Press, Boca Raton, Florida. pp. 148–162.

Chhatre, S., Purohit, H., Shanker, R., and Khanna, P. 1996. "Bacterial consortia for crude oil spill remediation." *Water. Sci. Technol.*, 34: 187–193.

Collina, E., Bestetti, G., Di Gennaro, P., Franzetti, A., Gugliersi, F., Lasagni, M., and Pitea, D. 2005. "Naphthalene Biodegradation Kinetics in an Aerobic Slurry-Phase Bioreactor." *Environment International*, 31(2): 167–171.

Compeau, G.C., Mahaffey, W.D., and Patras, L. 1991. "Full-Scale Bioremediation of a Contaminated Soil and Water Site." In G.S. Sayler, R. Fox and J.W. Blackburn (Editors). *Environmental Biotechnology for Waste Treatment*. Plenum Press, New York. pp. 91–110.

Cunningham, J.A., Hopkins, G.D., Lebron, C.A., and Reinhard, M. 2000. "Enhanced Anaerobic Bioremediation of Groundwater Contaminated by Fuel Hydrocarbons at Seal Beach, California." *Biodegradation*, 11: 159–170.

Cutright, T.J. 1995. "Polycyclic Aromatic Hydrocarbon Biodegradation and Kinetics using *Cunninghamella echinulatu var. elegans*." *International Biodeterioration & Biodegradation*, 35(4): 397–408.

deCarvalho, C.C.C.R., and daFonseca, M.M.R. 2005. "Degradation of Hydrocarbons and Alcohols at Different Temperatures and Salinities by *Rhodococcus erythropolis* DCL14." *FEMS Microbiol. Ecol.*, 51: 389–399.

Flathman, P.E., Carson, J.H. Jr., Whitenhead, S.J., Khan, K.A., Barnes, D.M., and Evans, J.S. 1991. "Laboratory Evaluation of the Utilization of Hydrogen Peroxide for Enhanced Biological Treatment of Petroleum Hydrocarbon Contaminants in Soil." In R.E. Hinchee and R.F. Olfenbuttel (Editors), *In Situ Bioreclamation: Applications and Investigations for Hydrocarbon and Contaminated Site Remediation*. Butterworth-Heinemann, Stoneham, Massachusetts. pp. 125–142.

Frenzel, M., James, P., Burton, S.K., Rowland, S.J., and Lappin-Scott, H.M. "Towards Bioremediation of Toxic Unresolved Complex Mixtures of

Hydrocarbons: Identification of Bacteria Capable of Rapid Degradation Of Alkyltetralins." *J. Soils Sediments*, 9:129–136.

Gibson, D.T. and Sayler, G.S. 1992. *Scientific Foundation for Bioremediation: Current Status and Future Needs*. American Academy of Microbiology, Washington, DC.

Ginn, J.S., Sims, R.C., and Murarka, I.P. 1995. "In Situ Bioremediation (Natural Attenuation) at a Gas Plant Waste Site." In R.E. Hinchee, J.T. Wilson, and D.C. Downey (Editors), *Intrinsic Bioremediation*. Battelle Press, Columbus, Ohio. pp. 153–162.

Grossi, V., Massias, D., Stora, G., and Bertrand Burial, J.C. 2002. "Exportation and Degradation of Acyclic Petroleum Hydrocarbons Following Simulated Oil Spill in Bioturbated Mediterranean Coastal Sediments." *Chemosphere*, 48(9): 947–954.

Hamaker, W. 1972. "Decomposition: Quantitative aspects." In C.A.I. Goring, J.W. Hamaker, and J. Thomson (Editors), *Organic Chemicals in the Soil Environment*. Marcel Dekker Inc., New York.

Harms, H., Smith, K.E.C., and Wick, L.Y. 2010. "Problems of Hydrophobicity / Bioavailability." In K.N. Timmis (Editor), *Handbook of Hydrocarbon and Lipid Microbiology*. Springer, Berlin. Chapter 42, pp. 1439–1450.

Heitkamp, M.A., and Cerniglia, C.E. 1987. "Effects of Chemical Structure and Exposure on the Microbial Degradation of Polycyclic Aromatic Hydrocarbons in Freshwater and Estuarine Ecosystems." *Environmental Toxicology and Chemistry*, 6(7): 535–546.

Heitkamp, M. A., Freeman, J. P., & Cerniglia, C. E. (1987). "Naphthalene Biodegradation in Environmental Microcosms: Estimates of Degradation Rates and Characterization of Metabolites." *Applied and Environmental Microbiology*, 53(1): 129–136.

Hohener, P., Duwig, C., Pasteris, G., Kaufmann, K., Dakhel, N., and Harms, H. 2003. "Biodegradation of Petroleum Hydrocarbon Vapors: Laboratory Studies on Rates and Kinetics in Unsaturated Alluvial Sand." *Journal of Contaminant Hydrology*, 66(1–2): 93–115.

Holba, A.G., Dzou, I.L., Hickey, J.J., Franks, S.G., May, S.J., and Lenney, T. 1996. "Reservoir Geochemistry of South Pass 61 Field, Gulf of Mexico: Compositional Heterogeneities Reflecting Filling History and Biodegradation." *Org. Geochem.*, (24): 1179–1198.

Holder, E.L., Miller, K.M., and Haines, J.R. 1999. "Crude Oil Component Biodegradation Kinetics by Marine And Freshwater Consortia." In B.C. Alleman and A. Leeson (Editors), *In Situ Bioremediation of Polycyclic Hydrocarbons and Other Organic Compounds*. Battelle, Columbus, Ohio. pp. 245–250.

King, M.W.G., Barker, J.F., and Hamilton, L.K. 1995. "Natural Attenuation of Coal Tar Organics in Groundwater." In R.E. Hinchee, J.T. Wilson, and D.C. Downey (Editors), *Intrinsic Bioremediation*. Battelle Press, Columbus, Ohio. pp. 171–180.

Kleinsteuber, S., Riis, V., Fetzer, I., Harms, H., and Müller, S. 2006. "Population dynamics within a microbial consortium during growth on diesel fuel in saline environments." *Appl. Environ. Microbiol.*, 72: 3531–3542.

Korade, D.L., and Fulekar, M.H. 2009. "Development and Evaluation of Mycorrhiza for Rhizosphere Bioremediation." *Journal of Applied Biosciences* 17: 922–929.

Leavitt, M.E. and Brown, K.L. 1994. "Bioremediation versus Bioaugmentation – Three Case Studies." In Hinchee, R.E., B.C. Alleman, R.E. Hoeppel and R.N. Miller (Editors), *Hydrocarbon Bioremediation*. CRC Press, Inc., Boca Raton, Florida. pp. 72–79.

Lee, M.D., Thomas, J.M., Borden, R.C., Bedient, P.B., and Ward, C.H. 1988. "Biorestoration of Aquifers Contaminated with Organic Compounds." *CRC Crit. Rev. Environ. Control*, 18: 29–89.

Lu, C.J. and Hwang, M.C. 1992. "Effects of Hydrogen Peroxide on the In Situ Biodegradation of Chlorinated Phenols in Groundwater." *Proceedings of Water Environ. Federation 65th Annual Conference*, New Orleans, Louisiana. September 20–24.

Lu, C.J. 1994. "Effects of Hydrogen Peroxide on the In Situ Biodegradation of Organic Chemicals in a Simulated Groundwater System." In Hinchee, R.E., B.C. Alleman, R.E. Hoeppel and R.N. Miller (Editors), *Hydrocarbon Bioremediation*. CRC Press, Inc., Boca Raton, Florida. pp. 140–147.

Magdalene, O.E., Ufuoma, A., and Gloria, O. 2009. "Screening of Four Common Nigerian Weeds for use in Phytoremediation of Soil Contaminated with Spent Lubricating Oil." *African Journal of Plant Science* Vol. 3 (5), pp. 102–106.

Maletić, S., Dalmacija, B., Rončević, S., Agbaba, J., and Petrović, O. 2009. "Degradation Kinetics of an Aged Hydrocarbon-Contaminated Soil." *Water Air Soil Pollut.*, 202: 149–159.

Margesin, R., and Schinner, F. 2001. "Biodegradation and Bioremediation of Hydrocarbons in Extreme Environments." *Appl. Microbiol. Biotechnol.*, 56: 650–663.

McAllister, P.M., Chiang, C.Y., Salanitro, J.P., Dortch, I.J., and Williams, P. 1995. "Enhanced Aerobic Bioremediation of Residual Hydrocarbon Sources." In R.E. Hinchee, J.T. Wilson, and D.C. Downey (Editors), *Intrinsic Bioremediation*. Battelle Press, Columbus, Ohio. pp. 67–76.

Mille, G., Almallah, M., Bianchi, M., Van Wambeke, F., and Bertrand, J.C. 1991. "Effect of Salinity on Petroleum Biodegradation." *Fresenius J. Anal. Chem.*, 339: 788–791.

Mishra, S., Jyot, J., Kuhad, R.C., and Lal, B. 2001. "In Situ Bioremediation Potential of an Oily Sludge-Degrading Bacterial Consortium." *Current Microbiol.*, 43: 328–335.

Namkoonga, W., Hwangb, E.Y., Parka, J.S., and Choic, J.Y. 2002. "Bioremediation of Diesel-Contaminated Soil With Composting." *Environmental Pollution*, 119(1): 23–31.

Nedunuri, K.V., Govundaraju, R.S., Banks, M.K., Schwab, A.P., and Chen, Z. 2000. "Evaluation of Phytoremediation for Field Scale Degradation of Total Petroleum Hydrocarbons." *J. Environ. Eng.*, 126: 483–490.

Nicholson, C.A., and Fathepure, B.Z. 2004. "Biodegradation of Benzene by Halophilic and Halotolerant Bacteria under Aerobic Conditions." *Appl. Environ. Microbiol.*, 70: 1222–1225.

Nester, E.W., Anderson, D.G., Roberts, C.E. Jr., Pearsall, N.N., and Nester, M.T. 2001. *Microbiology: A Human Perspective* 3rd Edition. McGraw-Hill, New York.

Nugroho, A., Effendi, E., and Karonta, Y. 2010. "Petroleum Degradation in Soil by Thermophilic Bacteria with Biopile Reactor." *Makara, Teknologi*, 14(1): 43–46.

Obire, O., and Nwaubeta, O. 2001. "Biodegradation of Refined Petroleum Hydrocarbons in Soil." *Journal of Applied Sciences & Environmental Management*, Vol. 5(1): 43–46.

Obire, O., and Anyanwu, E.C. 2009. "Impact of Various Concentrations of Crude Oil on Fungal Populations of Soil." *Int. J. Environ. Sci. Tech.*, 6(2): 211–218.

Oren, A., Gurevich, P., Azachi, M., and Henis, Y. 1992. "Microbial Degradation of Pollutants at High Salt Concentrations." *Biodegradation*, 3: 387–398.

Osswald, P., Baveye. P., and Block, J.C. 1996. "Bacterial Influence on Partitioning Rate during the Biodegradation of Styrene In A Biphasic Aqueous-Organic System." *Biodegradation*, 7: 297–302.

Pala, D.M., de Carvalho, D.D., Pinto, J.C., and & Sant'Anna, G.L. Jr. 2006. "A Suitable Model to Describe Bioremediation of a Petroleum-Contaminated Soil." *International Biodeterioration & Biodegradation*, 58(3–4): 254–260.

Pardieck, D.L., Bouwer, E.J., and Stone, A.T. 1992. "Hydrogen Peroxide Use to Increase Oxidant Capacity for In Situ Bioremediation of Contaminated Soils and Aquifers: A Review." *J. Contaminant Hydrol.*, 9: 221–242.

Perfumo, A., Banat, I.M., Marchant, R., and Vezzulli, L. 2007. "Thermally Enhanced Approaches for Bioremediation of Hydrocarbon-Contaminated Soils." *Chemosphere*, 66: 179–184.

Pollard, S.J.T., Hrudey, S.E., and Fedorak, P.M. 1994. "Bioremediation of Petroleum- and Creosote-Contaminated Soils: A Review Of Constraints." *Waste Management Res.*, 12: 173–194.

Radwan, S.S., Al-Mailem, D., El-Nemr, I., and Salamah, S. 2000. "Enhanced Remediation of Hydrocarbon Contaminated Desert Soil Fertilized with Organic Carbons." *Int. Biodet. Biodeg.*, 46: 129–132.

Rhykerd, R.L., Weaver, R.W., and McInnes, K.J. 1995. "Influence of Salinity on Bioremediation of Oil in Soil." *Environ. Pollut.*, 90: 127–130.

Riis, V., Kleinsteuber, S., and Babel, W. 2003. "Influence of High Salinities on the Degradation of Diesel Fuel by Bacterial Consortia." *Can. J. Microbiol.*, 49: 713–721.

Rončević, S., Dalmacija, B., Ivančev-Tumbas, I., Petrović, O., Klašnja, M., and Agbaba, J. 2005. "Kinetics of Degradation of Hydrocarbons in the Contaminated Soil Layer." *Archives of Environmental Contamination and Toxicology*, 49(1): 27–36.

Rosenberg, M., Gutnick, D., and Rosenberg, E. 1980. "Adherence of Bacteria to Hydrocarbons: a Simple Method for Measuring Cell-Surface Hydrophobicity." *FEMS Microbiol. Lett.*, 9: 29–33.

Schlegel, H.G. 1977. "Aeration without air: oxygen supply by hydrogen peroxide." *Biotechnol. Bioeng.*, 19: 413.

Scragg, A. 1999. *Environmental Biotechnology*. Pearson Education Limited, Harlow, Essex, England.

Seabra, P.N., Linhares, M.M., and Santa Anna, L.M. (999. "Laboratory Study of Crude Oil Remediation by Bioaugmentation." In B.C. Alleman, and A. Leeson (Editors), *In Situ Bioremediation of Polycyclic Hydrocarbons and Other Organic Compounds*. Battelle, Columbis, Ohio. pp. 421–426.

Semple, K.T., Reid, B.J., and Fermor, T.R. 2001. "Impact of Composting Strategies on the Treatment of Soils Contaminated with Organic Pollutants." *Environmental Pollution*, 112: 269–283.

Shelton, D.R., and Tiedje, J.M. 1984. "Isolation and Partial Characterization of Bacteria in an Anaerobic Consortium That Mineralizes 3-Chlorobenzoic Acid." *Appl. Environ. Microbiol.*, 48: 840–848.

Sims, R.C. 1990. "Soil Remediation Techniques at Uncontrolled Hazardous Waste Sites." *J. Air Waste Mgmt. Assoc.*, 40(5): 703–732.

Speight, J.G. 1996. *Environmental Technology Handbook*. Taylor & Francis, Washington, DC.

Speight, J.G., and Lee, S. 2000. *Environmental Technology Handbook*. 2nd Edition. Taylor & Francis, New York.

Speight, J.G. 2005. *Environmental Analysis and Technology for the Refining Industry*. John Wiley & Sons Inc., Hoboken, New Jersey.

Speight, J.G. 2007. *The Chemistry and Technology of Petroleum* 4th Edition. CRC Press, Taylor & Francis Group, Boca Raton, Florida.

Stoner, D.L. 1994. *Biotechnology for the Treatment of Hazardous Waste*. CRC Press, Boca Raton, Florida.

Suflita, J.M. 1989. "Microbiological Principles Influencing the Biorestoration of Aquifers." In *Transport and Fate of Contaminants in the Subsurface*. EPA/625/4-89/019, Robert S. Kerr Environmental Research Laboratory, US Environmental Protection Agency, Ada, Oklahoma.

Tibbett, M., George, S.J., Davie, A., Barron, A., Milton, N., and Greenwood, P.F. 2011. "Just Add Water and Salt: the Optimisation of Petrogenic Hydrocarbon Biodegradation in Soils from Semi-arid Barrow Island, Western Australia." *Water Air Soil Pollut.*, 216:513–525.

Ulrich, A.C., Guigard, S.E., Foght, J.M., Semple, K.M., Pooley, K., Armstrong, J.E., and Biggar, K.W. 2009. "Effect of Salt on Aerobic Biodegradation of Petroleum Hydrocarbons in Contaminated Groundwater." *Biodegradation*, 20: 27–38.

US Army Corps of Engineers. 1999. *US Army Manual EM1110-2-1424*. Chapter 8. www.usace.army.mil/usace-docs/engmanuals/em1110-2-1424/c-8.pdf.

US EPA. 2006. "In Situ and Ex Situ Biodegradation Technologies for Remediation of Contaminated Sites." Report No. EPA/625/R-06/015. Office of Research and Development National Risk Management Research Laboratory, United States Environmental Protection Agency, Cincinnati, Ohio.

Van Eyk, J. 1994. "Venting and Bioventing for the In Situ Removal of Petroleum From Soil." In R.E. Hinchee, B.C. Alleman, R.E. Hoeppel, and R.N. Miller (Editors), *Hydrocarbon Bioremediation*. CRC Press, Boca Raton, Florida. pp. 234–251.

Vasudevan, N., and Rajaram, P. 2001. "Bioremediation of Oil Sludge-Contaminated Soil." *Environ. Int.*, 26: 409–411.

Venosa, A., Suidan, M., Wrenn, B., Strohmeier, K., Haines, J., and Eberhart, B. 1996. "Bioremediation of an Experimental Oil Spill on the Shoreline of Delaware Bay." *Environmental Science & Technology*, 30(5): 1764–1775.

Vidali, M. 2001. "Bioremediation: An Overview." *Pure and Applied Chemistry*, 73(7): 1163–1172.

Wackett, L.P. 1996. "Co-metabolism: Is the Emperor Wearing any Clothes?" *Curr. Opin. Biotechn.*, 7: 321–325.

Ward, D.M., and Brock, T.D. 1978. "Hydrocarbon Biodegradation in Hypersaline Environments." *Appl. Environ. Microbiol.*, 35: 353–359.

Wenzel, W. 2009. "Rhizosphere Processes and Management in Plant-Assisted Bioremediation (Phytoremediation) of Soils." *Plant and Soil*, 321 (1-2): 385–408.

Wethasinghe, C., Yuen, S.T.S., Kaluarachchi, J.J., and Hughes, R. 2006. "Uncertainty in Biokinetic Parameters on Bioremediation: Health Risks and Economic Implications." *Environment International*, 32(3): 312–323.

Winningham, J., Britto, R., Patel, M., and McInturff, F. 1999. "A Land Farming Field Study of Creosote-Contaminated Soil." In B.C. Alleman, and A. Leeson (Editors), *Bioremediation Technologies for PAH Compounds*. Battelle, Columbus, Ohio. pp. 421–426.

Winters, J.C., and Williams, J.A. 1969. *Microbiological Alteration of Crude Oil in the Reservoir*. Preprints. Division of Petroleum Chemistry, American Chemical Society. Volume 14(4): E22–E31.

Zengler, K., Richnow, H.H., Rossello-Mora, R., Michaelis, W., and Widdel, F. 1999. "Methane Formation from Long-Chain Alkanes by Anaerobic Microorganisms." *Natr.*, 401: 266–269.

Zhu, X., Venosa, A.D., and Suidan, M.T. 2004. "Literature Review on the Use of Commercial Bioremediation Agents for Clean-up of Oil Contaminated Estuarine Environments." Report No. EPA/600/R-04/075. National Risk Management Research Laboratory, Environmental Protection Agency, Cincinnati, Ohio.

2

Petroleum Composition and Properties

1 Introduction

Spills from tankers, pipelines, and oil wells are examples of *point sources* of pollution, where the origin of the contaminants is a single identifiable point (Speight, 1996; Speight and Lee, 2000; Speight, 2005). They also represent catastrophic releases of a large volume of pollutants in a short period of time. But the majority of pollution from oil is from nonpoint sources, where small amounts coming from many different places over a long period of time add up to large-scale effects. For example, the majority of the petroleum products released by human activity into oceans worldwide are a result of small spills during petroleum consumption. These minor, unreported spills can include routine discharges of fuel from commercial vessels or leakage from recreational boats.

Spills of petroleum and petroleum products tend to collect in hazardous concentrations in the soil or wastewater coming out of cities and other populated areas. Runoff from asphalt-covered roads and parking lots enters storm drains, streams, and lakes, and eventually travels to the ocean, affecting all of the ecosystems through which it passes. As cities grow, more and more petroleum products – gasoline, solvents, and lubricants – are often improperly disposed into drains and sewage pipes.

In order to apply bioremediation technologies to petroleum or petroleum products (or, for that matter, any collection of organic compounds), it is essential to know the origin of the spilled material, as well as the methods of manufacture and the composition of the material, and the role played by the composition in biodegradation (Chapter 10). Without such background knowledge and information, the cleanup technology could be more difficult to apply, or even register as a failed technology.

Petroleum and petroleum products are no exception to this, and even require, on the basis of their complexity, additional efforts to assure cleanup in the form of complete removal of any such contaminants from a spill area. The fate of spilled of petroleum or petroleum products in water systems is exceptionally difficult to predict: petroleum and petroleum products oxidize on the surface, after which they form oil-in-water emulsions.

The inclusion of polar functions, such as hydroxyl groups (-OH) or carbonyl groups (>C=O) (a result of the oxidation process), causes an increase in the density of the emulsion (relative to the original unoxidized petroleum), and with an increased propensity to form emulsions. As a result, the emulsion forms and sinks to various depths, or even to the seabed, depending on the extent of the oxidation and the resulting density. This may give the erroneous appearance (leading to erroneous deductions with catastrophic consequences) that the petroleum spill (as evidenced from the petroleum remaining on the surface of the water) is less than it actually was. The so-called *missing* oil will undergo further chemical changes, and eventually reappear on the water surface or on a distant beach.

Such behavior can confuse (and has confused) many cleanup (or oil spill) *experts* to the detriment of the environment. Thus, understanding the chemistry of petroleum and its products is not merely a chemist's pipe dream, but is an essential part of understanding the behavior and subsequent cleanup of petroleum (and petroleum product) spills.

Historically, petroleum fractions (in various forms) have been known and used for millenia (Table 2.1) (Speight, 2007). However, just as the chemical and physical properties of petroleum have offered challenges in selecting and designing optimal upgrading schemes, these same properties also introduce challenges when determining the effects of petroleum and its product on the environment.

In particular, predicting the fate of the polynuclear aromatic systems, the heteroatom systems (principally compounds containing nitrogen and sulfur), and the metal-containing systems (principally compounds of vanadium, nickel, and iron) in the feedstocks is the subject of many studies and migration models. These constituents generally cause processing problems, and knowledge of the behavior of these elements is essential for process improvements, process flexibility, and environmental compliance.

Table 2.1 Petroleum and derivatives such as asphalt have been known and used for six thousand years.

3800 BC	First documented use of asphalt for caulking reed boats.
3500 BC	Asphalt used as cement for jewelry and for ornamental applications.
3000 BC	Use of asphalt as a construction cement by Sumerians; also believed to be used as a road material; asphalt used to seal bathing pool or water tank at Mohenjo Daro.
2500 BC	Use of asphalt and other petroleum liquids (oils) in the embalming process; asphalt believed to be widely used for caulking boats.
1500 BC	Use of asphalt for medicinal purposes and (when mixed with beer) as a sedative for the stomach; continued reference to use of asphalt liquids (oil) as illuminant in lamps.
1000 BC	Use of asphalt as a waterproofing agent by lake dwellers in Switzerland.
500 BC	Use of asphalt mixed with sulfur as an incendiary device in Greek wars; also use of asphalt liquid (oil) in warfare.
350 BC	Occurrence of flammable oils in wells in Persia.
300 BC	Use of asphalt and liquid asphalt as incendiary device (Greek fire) in warfare.
250 BC	Occurrences of asphalt and oil seepage in several areas of the Fertile Crescent (Mesopotamia); repeated use of liquid asphalt (oil) as an illuminant in lamps.
750 AD	Use in Italy of asphalt as a color in paintings.
950 AD	Report of destructive distillation of asphalt to produce distillate; reference to distillate as nafta (naft, naphtha).
1500 AD	Discovery of asphalt deposits in the Americas; first attempted documentation of the relationship of asphalt and naphtha (petroleum).
1600 AD	Asphalt used for a variety of tasks; relationship of asphalt to coal and wood tar studied; asphalt studied; used for paving; continued documentation of the use of naphtha as an illuminant and the production of naphtha from asphalt; importance of naphtha as fuel realized.
1859 AD	Discovery of petroleum in North America; birth of modern-day petroleum science and refining.

Because of the variation in the amounts of chemical types and bulk fractions (Table 2.2 and Table 2.3), it should not be surprising that petroleum exhibits a wide range of physical properties, and several relationships can be made between various physical properties (Table 2.4, Figure 2.1). Whereas the properties such as viscosity, density, boiling point, and color of petroleum may vary widely, the ultimate or

Table 2.2 Boiling fractions from petroleum (Speight, 2002).

Product	Lower Carbon Limit	Upper Carbon Limit	Lower Boiling Point C	Upper Boiling Point C	Lower Boiling Point F	Upper Boiling Point F
Refinery gas	C1	C4	–161	–1	–259	31
Liquefied petroleum gas	C3	C4	–42	–1	–44	31
Naphtha	C5	C17	36	302	97	575
Gasoline	C4	C12	–1	216	31	421
Kerosene/ diesel fuel	C8	C18	126	258	302	575
Aviation turbine fuel	C8	C16	126	287	302	548
Fuel oil	C12	>C20	216	421	>343	>649
Lubricating oil	>C20		>343		>649	
Wax	C17	>C20	302	>343	575	>649
Asphalt	>C20		>343		>649	
Coke	>C50*		>1000*		>1832*	

*Carbon number and boiling point difficult to assess; inserted for illustrative purposes only.

Table 2.3 Compound types in petroleum and petroleum fractions.

Class	Compound Types
Saturated hydrocarbons	n-Paraffins iso-Paraffins and other branched paraffins Cycloparaffins (naphthenes) Condensed cycloparaffins (including steranes, hopanes) Alkyl side chains on ring systems
Unsaturated hydrocarbons	Olefins not indigenous to petroleum; present in products of thermal reactions
Aromatic hydrocarbons	Benzene systems Condensed aromatic systems Condensed aromatic-cycloalkyl systems Alkyl side chains on ring systems
Saturated heteroatomic systems	Alkyl sulfides Cycloalkyl sulfides Alkyl side chains on ring systems
Aromatic heteroatomic systems	Furans (single-ring and multi-ring systems) Thiophenes (single-ring and multi-ring systems) Pyrroles (single-ring and multi-ring systems) Pyridines (single-ring and multi-ring systems) Mixed heteroatomic systems Amphoteric (acid-base) systems Alkyl side chains on ring systems

elemental analysis varies, as already noted, over a narrow range for a large number of petroleum samples. The carbon content is relatively constant, while the hydrogen and heteroatom contents are responsible for the major differences between petroleum samples (Speight, 2007). The nitrogen, oxygen, and sulfur can be present in only trace amounts in

Table 2.4 Variation of density and API gravity, and residuum content for selected crude oils.

Crude oil	Specific Gravity	API Gravity	Residuum >1000 F
US Domestic			
California	0.858	33.4	23.0
Oklahoma	0.816	41.9	20.0
Pennsylvania	0.800	45.4	2.0
Texas	0.827	39.6	15.0
Texas	0.864	32.3	27.9
Foreign			
Bahrain	0.861	32.8	26.4
Iran	0.836	37.8	20.8
Iraq	0.844	36.2	23.8
Kuwait	0.860	33.0	31.9
Saudi Arabia	0.840	37.0	27.5
Venezuela	0.950	17.4	33.6

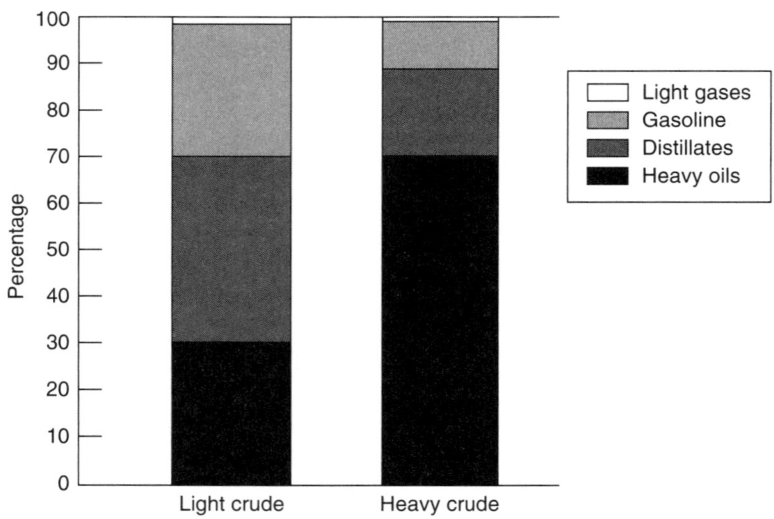

Figure 2.1 Variation in the distribution of the various fractions from light crude oil and from heavy crude oil.

some petroleum, which, as a result consists primarily of hydrocarbons. On the other hand, petroleum containing 9.5% heteroatoms may contain essentially no true hydrocarbon constituents, insofar as the constituents contain *at least one or more* nitrogen, oxygen, and/or sulfur atoms within the molecular structures. Coupled with the changes brought about to the feedstock constituents by refinery operations, it is not surprising that petroleum characterization is a monumental task (Speight, 2007).

Thus, initial inspection of the nature of the petroleum will provide deductions about the most logical means of cleanup and any subsequent environmental effects. Indeed, careful evaluation of petroleum from physical property data is a major part of the initial study of any petroleum that has been released to the environment. Proper interpretation of the data resulting from the inspection of crude oil requires an understanding of their significance.

Consequently, various standards organizations, such as the American Society for Testing and Materials in North America, the Institute of Petroleum in Britain, and the Deutsches Institut fur Normung EV, as well as many other standards organizations (Table 2.5), have devoted considerable time and effort to the correlation and standardization of methods for the inspection and evaluation of petroleum and petroleum products.

The data derived from any one or more of the evaluation techniques present an indication of the nature of petroleum and its products, and can (or should) be employed to give the environmental scientist or engineer an

Table 2.5 Various standards organization.

AENOR	Asociacion Espanola de de Normalizacion y Certificacion
AFNOR	Assocaition Francaise de Normalisation
ANSI	American National Standards Institute
BSI	British Standards Institution
CEN	European Committee for Standardization
CSA	CSA Group
DS	Dansk Standardiseringsrad
FSA	Fluid Sealing Association
GOST	Gosudarstvennye Standarty Stae Standard Gost
ISO	International Organization for Standardization

(Continued)

Table 2.5 (cont.) Various standards organization.

JSA	Japanese Standards Association
KSA	Korean Standards Association
NEN	Netherlands Normalisatie-Instituut
SAI	Standards Australia International Ltd....
SIS	Standardiseringkommissio
SN	Standards Norway
SNV	Schweizerische Normen-Vereinigung
SNZ	Standards New Zealand
UNI	Ente Nazionale Italiano di Unificazione

indication of the means by which the spilled material can be, or should be, recovered. Other properties (Speight, 2001, 2002, 2005, 2007) may also be required for further evaluation, or, more likely, for comparison on before and after scenarios, even though they may not play any role in dictating which cleanup operations are necessary.

To take this one step further, it is then possible to consider the nature of the spill, followed by development of preferred cleanup methods from one (but preferably more) of the physical properties, as determined by the evaluation test methods.

2 Composition

An important step in assessing the effects of petroleum products that have been released into the environment is to evaluate the nature of the particular mixture and eventually select an optimum remediation technology for that mixture. As a general rule, crude oil (unrefined petroleum) is a complex mixture composed of the same compounds, but the quantities of the individual compounds differ in crude oils from different locations. This rule of thumb implies that the quantities of some compounds can be zero in a given mixture of compounds that comprise a crude oil from a specific location (Speight, 2007).

In very general terms, petroleum is a mixture of: (a) hydrocarbon types, (b) nitrogen compounds, (c) oxygen compounds, (d) sulfur compounds, and (e) metallic constituents. Petroleum products are less well-defined in terms of heteroatom compounds, and are better defined in terms of the hydrocarbon types present. However, this general definition is not adequate to describe the true composition as it relates to the behavior of the petroleum, and its products, in the environment. For example, the occurrence of amphoteric species (i.e., compounds having a mixed acid/base nature) is not always addressed, nor is the phenomenon of molecular size, or the occurrence of specific functional types that can play a major role.

In the present context, petroleum composition is defined in terms of (1) the elemental composition, (2) the chemical composition, and (3) the fractional composition. All three parameters are interrelated, although the closeness of the elemental composition makes it difficult to relate precisely to the chemical composition and the fractional composition. The chemical composition and the fractional composition are somewhat easier to relate because of the quantities of heteroatoms (nitrogen, oxygen, sulfur, and metals) that occur in the higher boiling fractions and contribute to the separation methods (Speight, 2001, 2002, 2005, 2007).

2.1 Elemental Composition

The elemental composition of petroleum varies greatly from crude oil to crude oil. Most compounds in petroleum (usually more than 75%) are types of hydrocarbons, and the majority of the chemical components in petroleum are made up of five main elements: carbon 82 to 87% w/w, hydrogen 11 to 15% w/w/, sulfur 0 to 8% w/w, nitrogen 0 to 1% w/w, and oxygen 0.0 to 0.5% w/w.

The analysis for oxygen is subject to interpretation, because petroleum readily absorbs oxygen from the air, followed by reaction to produce oxygen-containing species. There has been considerable discussion about including oxygen in petroleum analysis (Speight, 2001, 2007). However, no matter how the oxygen arises or the origin of the oxygen, it is present in spilled petroleum and petroleum products, and must be taken into account. The aforementioned presence (inclusion) of oxygen arising from interaction of spilled petroleum or spilled petroleum products to form emulsions (above) is a case in point.

The elements are combined to form a complex mixture of organic compounds that range in molecular weight from 16 (methane; CH_4) to several hundred, perhaps even to several thousand, when the constituents of the high-boiling residua are considered (Table 2.6) (Speight, 2007). A wide range of metals is also found in trace amounts in crude oil. All metals through the atomic number 42 (molybdenum) have been found, with the exception of rubidium and niobium; a few heavier elements also have been detected. Nickel and vanadium are the most important, because they are present in all crude oil, usually at concentrations far higher than any other metal.

Table 2.6 Properties of atmospheric and vacuum residua.

Feedstock	Gravity API	Sulfur wt. %	Nitrogen wt. %	Nickel ppm	Vanadium ppm	Asphaltene Constituents (Heptane) wt. %	Carbon Residue (Conradson) wt. %
Arabian Light, >650 F	17.7	3.0	0.2	10.0	26.0	1.8	7.5
Arabian Light, >1050 F	8.5	4.4	0.5	24.0	66.0	4.3	14.2
Arabian Heavy, >650 F	11.9	4.4	0.3	27.0	103.0	8.0	14.0
Arabian Heavy, >1050 F	7.3	5.1	0.3	40.0	174.0	10.0	19.0
Alaska, North Slope, >650 F	15.2	1.6	0.4	18.0	30.0	2.0	8.5
Alaska, North Slope, >1050 F	8.2	2.2	0.6	47.0	82.0	4.0	18.0
Lloydminster (Canada), >650 F	10.3	4.1	0.3	65.0	141.0	14.0	12.1

(Continued)

Table 2.6 Properties of atmospheric and vacuum residua.

Feedstock	Gravity API	Sulfur wt. %	Nitrogen wt. %	Nickel ppm	Vanadium ppm	Asphaltene Constituents (Heptane) wt. %	Carbon Residue (Conradson) wt. %
Lloydminster (Canada), >1050 F	8.5	4.4	0.6	115.0	252.0	18.0	21.4
Kuwait, >650 F	13.9	4.4	0.3	14.0	50.0	2.4	12.2
Kuwait, >1050 F	5.5	5.5	0.4	32.0	102.0	7.1	23.1
Tia Juana, >650 F	17.3	1.8	0.3	25.0	185.0		9.3
Tia Juana, >1050 F	7.1	2.6	0.6	64.0	450.0		21.6
Taching, >650 F	27.3	0.2	0.2	5.0	1.0	4.4	3.8
Taching, >1050 F	21.5	0.3	0.4	9.0	2.0	7.6	7.9
Maya, >650 F	10.5	4.4	0.5	70.0	370.0	16.0	15.0

2.2 Chemical Composition

The chemical composition of petroleum also varies over wide ranges of molecular types with a varying range of weights. A broad functional definition of petroleum hydrocarbons is that hydrocarbons are primarily composed of many organic compounds of natural origin and low water solubility.

Remembering that hydrocarbons are (by definition) compounds containing carbon and hydrogen *only*, the hydrocarbon content may be as high as 97% by weight in a conventional (lighter) paraffinic crude oil, or about 50% by weight in heavy crude oil, and less than 30% by weight in tar sand bitumen. However, within the hydrocarbon constituents, there are also considerable variations of chemical type and molecular weight. Petroleum hydrocarbons may be paraffinic, alicyclic, or aromatic, and occur in varying concentrations within the different fraction of a single crude oil. Thus, the constituents of petroleum that occur in varying amounts depend on the source and character of the oil and are (Table 2.3) (Speight, 2001):

1. *Alkanes* (also called *normal paraffins* or *n-paraffins*).
   These constituents are characterized by unbranched (linear) or branched (non-linear) chains of carbon atoms with attached hydrogen atoms; alkanes contain no carbon-carbon double bonds (hence the designation *saturated*), and are generally insoluble in cold water – examples of alkanes are pentane (C_5H_{12}) and heptane (C_7H_{16}).
2. *Cycloalkanes* or *cycloparaffins* (also called *naphthenes*).
   These constituents are characterized by the presence of simple closed rings of carbon atoms (such as the cyclopentane ring or the cyclohexane ring); naphthenes can also contain alky moieties on the ring, and are generally stable and relatively insoluble in water – examples are cyclohexane and methyl cyclohexane.
3. *Alkenes* (also called *olefins*).
   These constituents are characterized by the presence of a carbon-carbon double bond (>C=C<), and can be unbranched (linear) or branched (non-linear) chains of carbon atoms; typically, alkenes are not generally found in crude oil (having reacted over the millenia during which petroleum was formed in the ground), but are common in thermally-produced products, such as naphtha (a precursor to gasoline) – common gaseous alkenes include ethylene ($CH_2{=}CH_2$) and propene (also called propylene, $CH_3CH{=}CH_2$).
4. *Single-ring Aromatics.*
   Aromatic constituents are characterized by the presence of an aromatic ring with six carbon atoms, and are considered to be the

most acutely toxic component of crude oil constituents because of their association with chronic and carcinogenic effects; the rings can carry alkyl of naphthene substituents, further distinguishing the aromatic constituents: low molecular weight aromatic constituents may have a noticeable solubility in water, increasing the potential for exposure to aquatic resources.

5. *Multi-ring Aromatics.*
   Aromatic constituents with two or more condensed rings are referred to as polynuclear aromatic hydrocarbons (PNAs, sometime PAHs); the most abundant aromatic hydrocarbon families in petroleum and petroleum products have two and three fused rings with one to four carbon atom alkyl group substitutions: condensed aromatic constituents with more than two condensed rings (three-to-five) are also present in the higher boiling fractions of petroleum, but higher condensed systems are unlikely (Speight, 2007).

A typical crude oil contains, generally speaking, approximately 1% polynuclear aromatic hydrocarbons. Concentrations of total carcinogenic polynuclear aromatic hydrocarbons [such as benzo(a)pyrene] range from 12 ppm to <100 ppm. Fresh crude oil will contain a fraction of volatile hydrocarbons, some of which may pose a threat to human health, such as benzene, toluene, xylenes, and other aromatics. However, the relative mass fraction of volatile hydrocarbons in crude oil is significantly less than that found in crude oil distillate products, such as gasoline. The percent by volume (% v/v) of benzene in gasoline may range up to 3% (30,000 ppm), while the benzene content of crude oil is usually on the order of 0.2% (2000 ppm) or less. As a result of the lower percentage of volatile aromatic constituents in a product such as gasoline, vapor emissions from crude oil-contaminated soils are expected to be (but not always, depending upon the crude petroleum) much less than potential emissions from gasoline-contaminated soils.

If the only constituents of petroleum were the hydrocarbons, the complexity is further illustrated by the number of potential isomers, i.e., molecules having the same atomic formula, that can exist for a given number of paraffinic carbon atoms, and that increases rapidly as molecular weight increases:

Carbon atoms per hydrocarbon	Number of isomers
4	2
8	18
12	355
18	60,523

This same increase in number of isomers with molecular weight also applies to the other molecular types present. Since the molecular weights of the molecules found in petroleum can vary from that of methane (CH_4; molecular weight: 16) to several thousand (Speight, 2001, 2002, 2005, 2007), it is clear that the heavier nonvolatile fractions can contain virtually unlimited numbers of molecules. However, in reality, the number of molecules in any specified fraction is limited by the nature of the precursors of petroleum, their chemical structures, and the physical conditions that are prevalent during the maturation (conversion of the precursors) processes (Speight, 2007).

In addition to hydrocarbons, petroleum also contains compounds that contain nitrogen, oxygen, and sulfur (in the minority), as well as trace amounts of metals, such as vanadium, nickel, iron, and copper. Porphyrins, the major organometallic compounds present in petroleum, are complex large cyclic carbon structures derived from chlorophyll, and are characterized by the ability to contain a central metal atom (trace metals are commonly found within these compounds).

2.3 Composition by Volatility

Distillation (the separation of the constituents of a mixture based on relative volatility) is a common method for the fractionation of petroleum that is used in the laboratory as well as in refineries. The technique of distillation has been practiced for many centuries, and the stills that have been employed have taken many forms (Speight, 2001, 2002, 2005, 2007).

Distillation is the first and the most fundamental step in the refining process (after the crude oil has been cleaned and any remnants of brine removed) (Speight, 2001, 2002; Speight and Ozum, 2002; Speight, 2005, 2007) and references cited therein), which is often referred to as the *primary refining process*. Distillation involves the separation of the different hydrocarbon compounds that occur naturally in a crude oil into a number of different fractions (a fraction is often referred to as a *cut*).

In the atmospheric distillation process (Figure 2.2), heated crude oil is separated in a distillation column (*distillation tower, fractionating tower, atmospheric pipe still*) into streams that are then purified, transformed, adapted, and treated in a number of subsequent refining processes, into products for the refinery's market. The lower boiling products separate out higher up the column, whereas the heavier, less volatile, products settle out toward the bottom of the distillation column. The fractions produced in this manner are known as *straight-run fractions*, ranging from (atmospheric tower) gas, naphtha, kerosene, and gas oil, to atmospheric residuum.

The atmospheric residuum is then fed to the vacuum distillation unit at the pressure of 10 mmHg, where light vacuum gas oil, heavy vacuum gas oil, and vacuum residue are the products (Figure 2.3). The fractions obtained by vacuum distillation of the reduced crude (atmospheric residuum) from an atmospheric distillation unit depend on whether or not the

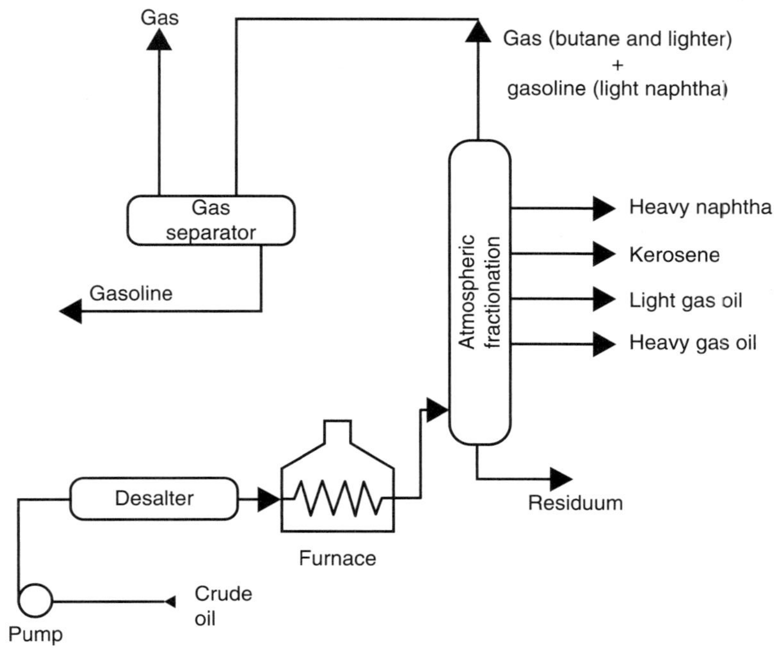

Figure 2.2 An atmospheric distillation unit.

unit is designed to produce lubricating or vacuum gas oils. In the former case, the fractions include (1) heavy gas oil, which is an overhead product and is used as catalytic cracking stock or, after suitable treatment, a light lubricating oil, (2) lubricating oil (usually three fractions: light, intermediate, and heavy), which is obtained as a side-stream product, and (3) asphalt (or residuum), which is the non-volatile (bottom) product and may be used directly as, or to produce, asphalt, and which may also be blended with gas oils to produce a heavy fuel oil. Operating conditions for vacuum distillation are usually 350 to 410°C (660 to 770°F) and 50 to 100 mm Hg (atmospheric pressure = 760 mm Hg). Although the temperature range is above the cracking temperature (350°C, 660°F) of many petroleum constituents, the residence time is adjusted to maintain cracking at a minimum.

Fractions from the atmospheric and vacuum towers (Figure 2.4) are often used as feedstocks to these second stage refinery processes that break down the fractions, or cause sufficient chemical changes in the nature of a particular hydrocarbon compound to produce specific products.

However, fractionation of petroleum by volatility, informative as it might be, does not give any indication of the physical nature of petroleum. This is more often achieved by subdivision of the petroleum into bulk fractions that are separated by a variety of solvent and adsorption methods.

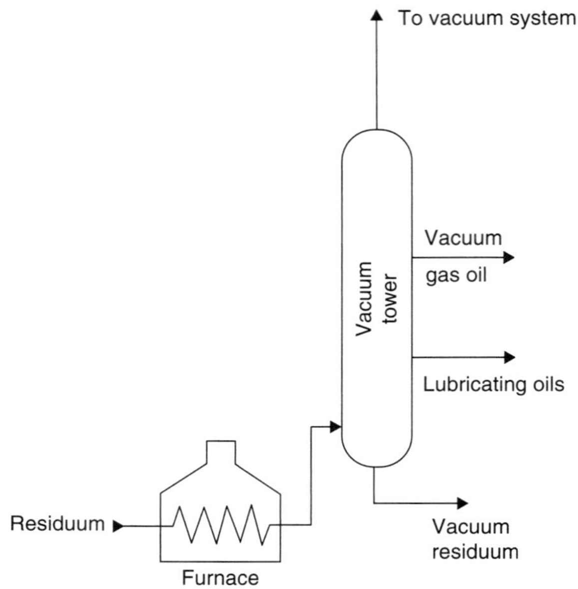

Figure 2.3 A vacuum distillation unit.

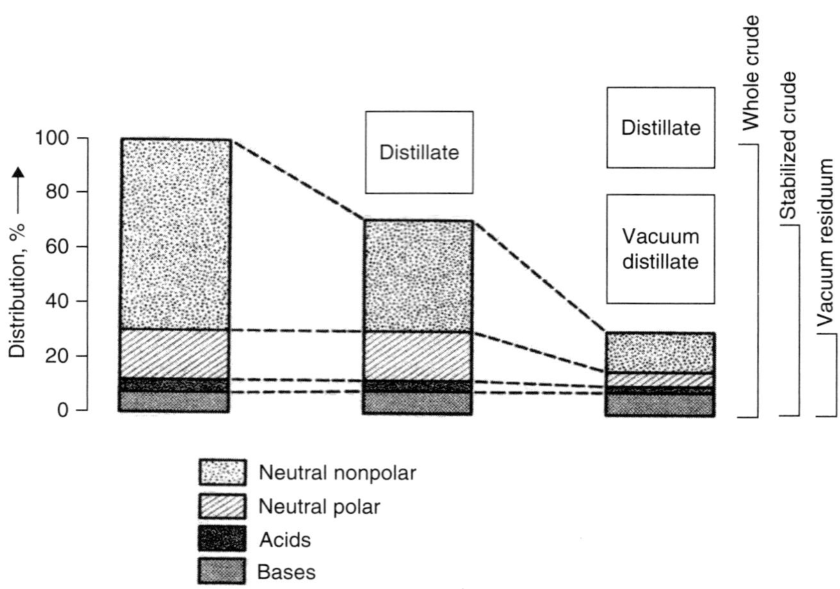

Figure 2.4 Accumulation of polar functions in residua during distillation.

2.4 Composition by Fractionation

An understanding of the chemical types (or composition) in petroleum can lead to an understanding of the chemical aspects of petroleum behavior. Indeed, this is not only a matter of knowing the elemental composition of a feedstock; it is also a matter of understanding the bulk properties as they relate to the chemical or physical composition of the material. For example, it is difficult to understand, *a priori*, the behavior of petroleum and petroleum products from the elemental composition alone, and more information is necessary to understand environmental behavior.

Thus, in the simplest sense, petroleum and petroleum products can be considered to be composites of four major fractions (saturates, aromatics, resins, and asphaltene constituents) (SARA analysis) (Figure 2.5) in varying amounts, depending upon the type of petroleum and the type of petroleum product (Speight, 2001, 2002). In fact, the lower boiling product may only be compositions of one or two fractions (saturates and aromatics). Whatever the case, it must never be forgotten that the nomenclature of these fractions lies within the historical development of petroleum science, and that the fraction names are operational, and are related more to the general characteristics of the fraction rather than to the identification of specific compound types.

Asphaltene constituents and resin constituents can comprise a large fraction of crude oil and heavy fuel oil (formerly called *residual fuel oil*), making those oils very dense and viscous. Composition is dependent upon source (these structures have the highest individual molecular weight of all crude oil components, and are basically colloidal aggregates).

Asphaltene constituents are substances in petroleum that are insoluble in solvents of low molecular weight, such as pentane or heptane. These

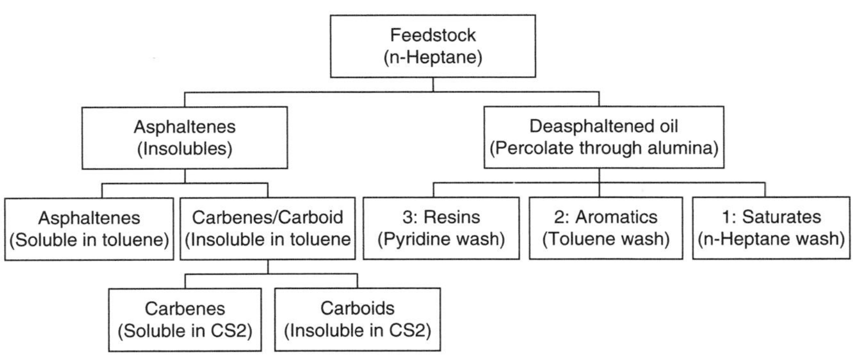

Figure 2.5 A SARA-type analysis (showing two additional fractions – carbenes and carboids – that are generally recognized as the results of thermal processes).

compounds are composed of very polynuclear aromatic and heterocyclic molecules, and are solids at normal temperatures. Consequently, oils that have high asphaltene contents are very viscous, have a high pour point, and are generally non-volatile in nature. During distillation, the porphyrin constituents, asphaltene constituents, and resin constituents are concentrated in the non-volatile residuum. During the weathering process, this fraction is the last to degrade, and its persistence over years has been noted (Chapter 15).

There are also two other operational definitions that should be noted at this point, and these are the terms *carbenes* and *carboids* (Figure 2.5). Both such fractions are, by definition, insoluble in benzene (or toluene), but the carbenes are soluble in carbon disulfide, whereas the carboids are insoluble in carbon disulfide (Speight, 2001, 2002, 2005, 2007). Typically, only traces of these materials occur in crude petroleum, and none occur in the products, unless the product is a high-boiling product of thermal treatment (such as visbroken feedstocks). On the other hand, oxidized petroleum and oxidized high-boiling product that have been susceptible to oxidation through a spill may contain such fractions (as defined by solubility or insolubility in solvents). But again, it must be remembered that the fractions separated by the various techniques are based on solubility or adsorption properties, and not on specific chemical types.

Thus, the separation of crude oil into two fractions, asphaltene constituents (the asphaltene fraction) and maltene constituents (the maltene fraction), is conveniently brought about by means of low-molecular-weight paraffinic hydrocarbons (ASTM D2006, ASTM D2007, ASTM D3279, ASTM D4124, IP 143) that were recognized to have selective solvency for hydrocarbons, and simple relatively low molecular weight heteroatom derivatives. The more complex, higher molecular weight compounds are precipitated by addition of 40 volumes of *n*-pentane or *n*-heptane in the methods that are generally preferred at present (Speight, 2001, 2002, 2005, 2007 and references cited therein). The insoluble fraction, the asphaltene fraction, should correctly be called *n-pentane asphaltene fraction* or *n-heptane asphaltene fraction*: the method is qualitatively and quantitatively reproducible.

After removal of the asphaltene fraction, further fractionation of petroleum is also possible, and there are three standard (ASTM) methods that provide for the separation of a feedstock into four or five constituent fractions (Speight, 2001, 2002, 2005, 2007 and references cited therein). It is interesting to note that, as the methods have evolved, there has been a change from the use of pentane (ASTM D2006, ASTM D2007) to heptane (ASTM D4124) to separate asphaltene constituents. This is, in fact, in keeping with the production of a more consistent fraction that represents the higher molecular weight constituents of petroleum.

Two of the methods (ASTM D2007, ASTM D4124) use adsorbents to fractionate the deasphaltened oil, but the third method (ASTM D2006) advocates the use of various grades of sulfuric acid to separate the material into compound types. Caution is advised in the application of this method, since the method does not work well with all feedstocks. For example, when the *sulfuric acid* method (ASTM D2006) is applied to the separation of heavy feedstocks, complex, difficult-to-break emulsions can be produced.

There are precautions that must be taken when attempting to separate heavy feedstocks or polar feedstocks into constituent fractions. The disadvantages in using ill-defined adsorbents are that adsorbent performance differs with the same feed and, in certain instances, may even cause chemical and physical modification of the feed constituents. The use of a chemical reactant, like sulfuric acid, should only be advocated with caution, since feeds react differently, and may even cause irreversible chemical changes and/or emulsion formation. These advantages may be of little consequence when it is not, for various reasons, the intention to recover the various product fractions *in toto* or in the original state, but in terms of the compositional evaluation of different feedstocks, the disadvantages are very real.

The terminology used for the identification of the various methods might differ. However, in general terms, group-type analysis of petroleum is often identified by the acronyms for the names: PONA (paraffins, olefins, naphthenes, and aromatics), PIONA (paraffins, *iso*-paraffins, olefins, naphthenes, and aromatics), PNA (paraffins, naphthenes, and aromatics), PINA (paraffins, *iso*-paraffins, naphthenes, and aromatics), or SARA (saturates, aromatics, resins, and asphaltene constituents). However, it must be recognized that the fractions produced by the use of different adsorbents will differ in content, and will also be different from fractions produced by solvent separation techniques.

The variety of fractions isolated by these methods and the potential for the differences in composition of the fractions makes it even more essential that the method is described accurately, and that it is reproducible not only in any one laboratory, but also among various laboratories (Speight, 2001, 2007).

2.5 Composition by Spectroscopy

Analytical spectroscopy-based methods can be used to investigate the reactivity, structures, and associations of petroleum constituents in the environment. Because analytical environmental chemistry deals with very complex naturally occurring mixtures, new spectroscopic approaches must be utilized or further developed to fully understand global environmental processes involved in contamination by petroleum constituents and their subsequent remediation.

Moreover, continually evolving environmental legislation and the introduction of stringent environmental quality standards relating to environmental contaminants continues to provide significant analytical challenges to traditional methods of analysis. Advancements in spectroscopic techniques and associated chromatographic instrumentation have become indispensable in the environmental analysis laboratory. Without some of the recent advancements in mass spectrometry, many of the applications (discussed below) would have not been possible without significant additional effort (Kanda and Glendinning, 2011).

Thus, spectroscopic studies continue to play an important role in the evaluation of petroleum and of petroleum products, and many of the methods are now used as standard methods of analysis of petroleum and its products before and after a spill. Application of these methods to petroleum and its and products is a natural consequence for the environmental scientist and engineer.

2.5.1 Infrared Spectroscopy

Conventional infrared spectroscopy yields information about the functional features of various petroleum constituents. For example, infrared spectroscopy will aid in the identification of N-H and O-H functions, the nature of polymethylene chains, the C-H out-of-place bending frequencies, and the nature of any polynuclear aromatic systems (Speight, 2001, 2002, 2005, 2007 and references cited therein).

Infrared spectroscopy is used for the determination of benzene in motor and/or aviation gasoline (ASTM D4053), while ultraviolet spectroscopy is employed for the evaluation of mineral oils (ASTM D2269) and for determining the naphthalene content of aviation turbine fuels (ASTM D1840). With the recent progress of *Fourier transform infrared (FTIR) spectroscopy*, which allows the analysis of a relevant amount of compositional and structural information concerning environmental samples, quantitative estimates of the various functional groups can also be made (Mecozzi et al., 2009). This is particularly important for application to the higher molecular weight solid constituents of petroleum (i.e., the asphaltene fraction).

2.5.2 Nuclear Magnetic Resonance

Because nuclear magnetic resonance is one of the premiere techniques for studying intermolecular interactions, it has also been used extensively to investigate the environmental chemistry of humic substances through their interactions with organic contaminants and toxic metals (Nanny et al., 1997; Cardoza et al., 2004).

Nuclear magnetic resonance has frequently been employed for general studies and for the structural studies of petroleum constituents (Bouquet and Bailleul, 1982; Hasan et al., 1989). In fact, *proton magnetic resonance* (PMR) studies (along with infrared spectroscopic studies) were perhaps the first studies of the modern era that allowed structural inferences to be made about the polynuclear aromatic systems that occur in the high molecular weight constituents of petroleum.

Nuclear magnetic resonance spectroscopy has been developed as a standard method for the determination of hydrogen types in aviation turbine fuels (ASTM D3701). *X-ray fluorescence spectrometry* has been applied to the determination of lead in gasoline (ASTM D2599) as well as to the determination of sulfur in various petroleum products (ASTM D2622 and ASTM D4294).

Carbon-13 magnetic resonance (CMR) can play a useful role. Since carbon magnetic resonance deals with analyzing the carbon distribution types, the obvious structural parameter to be determined is the aromaticity, f_a. A direct determination from the various carbon type environments is one of the better methods for the determination of aromaticity. Thus, through a combination of proton and carbon magnetic resonance techniques, refinements can be made on the structural parameters and, for the solid-state high-resolution carbon magnetic resonance technique, additional structural parameters can be obtained.

2.5.3 Mass Spectrometry

Mass spectrometry can play a key role in the identification of the constituents of feedstocks and products (Aczel, 1989). The principal advantages of mass spectrometric methods are: (1) high reproducibility of quantitative analyses; (2) the potential for obtaining detailed data on the individual components and/or carbon number homologues in complex mixtures; and (3) a minimal sample size is required for analysis. The ability of mass spectrometry to identify individual components in complex mixtures is unmatched by any modern analytical technique, but perhaps the exception is gas chromatography.

The methods include the use of *mass spectrometry* to determine the (1) hydrocarbon types in middle distillates (ASTM D2425); (2) hydrocarbon types of gas oil saturate fractions (ASTM D2786); (3) hydrocarbon types in low-olefin gasoline (ASTM D2789); and (4) aromatic types of gas oil aromatic fractions (ASTM D3239).

However, there are disadvantages arising from the use of mass spectrometry, and these are: (1) the limitation of the method to organic materials that are volatile and stable at temperatures up to 300°C (570°F); and (2) the difficulty of separating isomers for absolute identification. The sample is usually destroyed, but this is seldom a disadvantage.

Mass spectrometry coupled to gas chromatography and liquid chromatography are ideal techniques for the determination of organic contaminants in wastewater, due to the separation capability of gas chromatography and liquid chromatography, coupled to the selectivity and sensitivity of mass spectrometry. This coupling of chromatography with mass selective detection allows the routine determination of many contaminants in parts-per-trillion (ppt) or lower concentrations.

In recent years, developments in chromatography, such as the introduction of ultra-high-performance liquid chromatography, which utilizes the same separation methodology as conventional high-performance liquid chromatography, but uses columns packed with smaller particles (generally about 2 μm particle size or less), has further enhanced the applicability of liquid chromatography-mass spectrometry in wastewater analysis.

The introduction of hybrid mass analyzers, such as the triple-stage quadrupole mass spectrometers, allowed the use of tandem mass spectrometry for quantitative low level analysis of analytes in complex mixtures or in the presence of a high sample matrix background. In the triple-stage quadrupole mass spectrometer, the first quadrupole (Q1) is connected with a collision cell (Q2) to another quadrupole (Q3). Both quadrupoles (Q1 and Q3) can be used in scanning or static mode, depending on the type of tandem mass spectrometry analysis being performed. Types of analysis that can be performed include product ion scans, precursor ion scans, neutral loss scan, and selected reaction monitoring, also referred to as multiple reaction monitoring.

In the multiple reaction monitoring mode, Q1 is in static mode (precursor ion selected), and Q3 is also in static mode (product ion selected for the chosen precursor ion). The multiple reaction monitoring mode is highly specific, and virtually eliminates matrix background, resulting in very selective analysis.

Gas chromatography-mass spectrometry has become the technique of choice for volatile and semi-volatile, non-polar compounds, including polycyclic aromatic hydrocarbons (PAH), polychlorinated biphenyls (PCB), pesticides, for example, organo-chlorine insecticides, dioxins, and other volatile priority pollutants. Compounds containing polar functional groups are difficult to analyze by gas chromatography, either because they are not sufficiently volatile, tail badly even on polar analytical columns, or are thermally unstable and decompose in the gas chromatography or mass spectrometry interface source.

Chemical derivatization of samples prior to analysis can be used to overcome some of these problems. Derivatization allows polar compounds to be modified to produce new compounds that have properties which are suitable for analysis using chromatography-mass spectrometry. Common chemical derivatization methods include: (1) silylation: used to volatilize compounds, (2) alkylation: used as the first step to further derivatization or as a method of protection of certain active hydrogen atoms, and (3) acylation: used to add fluorinated groups.

Gas chromatography or mass spectrometry is most often used in positive ion electron ionization mode, in either full scan (useful in determining unknown compounds in a sample or confirming compounds present) or in selected ion monitoring mode, in which only certain ion fragments specific to the analytes are monitored, allowing more scans to take place each second. Ion fragmentation patterns in electron ionization are dependent upon the electron energy applied: typically 70 eV is used to facilitate comparison of generated spectra with library spectra.

Chemical ionization is sometimes used for compounds chemically similar to those analyzed by electron ionization to enhance the abundance of the molecular ion, resulting in less fragmentation. In positive ion chemical ionization, a chemical reagent gas (for example, methane, ammonia, or isobutene) is used. The reagent gas is present at a much higher concentration than the analyte, and is ionized by electron ionization to produce primary reagent ions, and the collision of these ions with neutral reagent molecules leads to the formation of stable secondary ions, which are the reactant species that then ionize analyte molecules by ion-molecule reactions.

Negative chemical ionization can be performed when an analyte contains electron-capturing moieties (for example, halogenated compounds). In negative ion chemical ionization, there are two primary mechanisms whereby negative ions are produced: electron capture and reactant ion chemical ionization. Under chemical ionization conditions, electronegative molecules can capture electrons to generate negative ions. True negative ion chemical ionization occurs by reaction of an analyte compound with negatively charged reactant ions. Several types of ion-molecule reactions can occur, with the most common being proton abstraction of the analyte.

2.5.4 Other Techniques

Other techniques include the use of *flame emission spectroscopy* for determining trace metals in gas turbine fuels (ASTM D3605) and the use of *absorption spectrophotometry* for the determination of the alkyl nitrate content of diesel fuel (ASTM D4046). *Atomic absorption* has been employed as a means of measuring the lead content of gasoline (ASTM D3237), and also for the manganese content of gasoline (ASTM D3831), as well as for determining the barium, calcium, magnesium, and zinc contents of lubricating oils (ASTM D4628). *Flame photometry* has been employed as a means of measuring the lithium/sodium content of lubricating greases (ASTM D3340) and the sodium content of residual fuel oil (ASTM D1318).

Liquid chromatography is a fundamental analytical technique in environmental analysis used for the determination of a wide range of compounds, including large molecular weight compounds and thermally labile compounds not suitable for gas chromatography. Liquid chromatography-mass spectrometry is a rapidly growing technique for routine analysis of organic compounds in environmental analysis. For most compounds, a mass spectrometer is more sensitive, and has greater specifity than other liquid chromatographic detectors.

Many of the advancements in liquid chromatography-mass spectrometry have been in the development of ionization techniques (interface/ ion source). Earlier liquid chromatography-mass spectrometry systems used interfaces that either did not separate the mobile phase from analyte molecules (direct liquid inlet, thermospray), or did so before ionization (particle beam). The analyte molecules were then ionized in the mass spectrometer under vacuum, often by traditional electron ionization.

The introduction of atmospheric pressure ionization techniques greatly expanded the applicability of liquid chromatography-mass spectrometry. In atmospheric pressure ionization liquid chromatography-mass spectrometry, the analyte molecules are ionized at atmospheric pressure. The analyte ions are then mechanically and electrostatically separated from neutral molecules.

Atmospheric pressure ionization techniques are: (1) electrospray ionization, (2) atmospheric pressure chemical ionization, and (3) atmospheric pressure photoionization.

Electrospray ionization is used for the analysis of a wide range of compounds, including large biomolecules. Atmospheric pressure chemical ionization is used for molecules less than 1500 microns, and is less well-suited than electrospray for analysis of large biomolecules and thermally unstable compounds. Atmospheric pressure chemical ionization is used with normal-phase chromatography more often than electrospray, because the analytes are usually non-polar. Atmospheric pressure photoionization is applicable to many of the compounds that are suitable for analysis by atmospheric pressure chemical ionization. Atmospheric pressure photoionization is particularly suitable for the analysis of highly nonpolar compounds.

3 Properties

The chemical composition of petroleum products is complex, and will change over time following release into the environment. These factors make it difficult to select the most appropriate analytical methods for evaluating environmental samples. This brings about the need for an understanding of the properties of petroleum and petroleum products.

Petroleum is typically are described in terms of their physical properties (such as density and pour point) and chemical composition (such as percent composition of various petroleum hydrocarbons, asphaltene constituents, and sulfur). Although very complex in makeup, crude can be broken down into four basic classes of petroleum hydrocarbons. Each class is distinguished on the basis of molecular composition.

In addition, properties important for characterizing the behavior of petroleum and petroleum products when spilled into the waterways or on to land and/or released into the air include flash point, density (read: specific gravity and/or API gravity), viscosity, emulsion formation in waterways, and adhesion to soil.

Generally, the properties of petroleum constituents (excluding the volatile gases and the non-volatile residuum) vary over the boiling range 0 to >565°C (32 to 1050°F). Variations in density also occur over the range 0.6 to 1.3, and pour points can vary from <0°C to >100°C. Although these properties may seem to be of lower consequence in the grand scheme of environmental cleanup, they are important insofar as these properties influence (1) evaporation rate, (2) ability of the petroleum constituents or petroleum product to float on water, and (3) fluidity or mobility of the petroleum or petroleum product at various temperatures.

3.1 Density and Specific Gravity

Density is the mass of a unit volume of material at a specified temperature, and has the dimensions of grams per cubic centimeter (a close approximation to grams per milliliter). *Specific gravity* is the ratio of the mass of a volume of the substance to the mass of the same volume of water, and is dependent on two temperatures, those at which the masses of the sample and the water are measured. When the water temperature is 4C (39F), the specific gravity is equal to the density in the centimeter-gram-second (cgs) system, since the volume of 1 g of water at that temperature is, by definition, 1 ml. Thus, the density of water, for example, varies with temperature, and its specific gravity at equal temperatures is always unity. The standard temperatures for a specific gravity in the petroleum industry in North America are 60/60F (15.6/15.6C).

The *density* and *specific gravity* of crude oil (ASTM D287, D1298, D941, D1217, and D1555) are two properties that have found wide use in the industry for preliminary assessment of the character of the crude oil (Table 2.4). Density also is a determinant as to whether or not the crude petroleum or petroleum product will float on water, and therefore remain susceptible to aerial oxidation with subsequent emulsion formation.

Although density and specific gravity are used extensively, the API (American Petroleum Institute) gravity is the preferred property by which petroleum is often referenced.

$$\text{Degrees API} = 141.5/\text{sp gr} \ @\ 60/60\text{F} - 131.5$$

The specific gravity of petroleum usually ranges from about 0.8 (45.3 API) for the light crude oil and heavy crude oil, to over 1.0 (les than 10 API) for tar sand bitumen. This is of lesser importance in determining the behavior of petroleum or petroleum products after a spill.

Density or specific gravity or API gravity may be measured by means of a hydrometer (ASTM D287 and D1298) or by means of a pycnometer (ASTM D941 and D1217). The variation of density with temperature, effectively the coefficient of expansion, is a property of great technical importance, since most petroleum products are sold by volume, and specific gravity is usually determined at the prevailing temperature (21C, 70F) rather than at the standard temperature (60F, 15.6C). The tables of gravity corrections (ASTM D1555) are based on an assumption that the coefficient of expansion of all petroleum products is a function (at fixed temperatures) of density only. Recent work has focused on the calculation and predictability of density using new mathematical relationships (Gomez, 1989, 1992).

3.2 Elemental (Ultimate) Analysis

The analysis of petroleum for the percentages of carbon, hydrogen, nitrogen, oxygen, and sulfur is perhaps the first method used to examine the general nature, and perform an evaluation, of a feedstock. The atomic ratios of the various elements to carbon (i.e., H/C, N/C, O/C, and S/C) are frequently used for indications of the overall character of the feedstock. It is also of value to determine the amounts of trace elements, such as vanadium and nickel, in a feedstock, since these materials can have serious deleterious effects on catalyst performance during refining by catalytic processes.

For example, *carbon content* can be determined by the method designated for coal and coke (ASTM D3178) or by the method designated for municipal solid waste (ASTM E-777). There are also methods designated for:

1. *hydrogen content* (ASTM D1018, ASTM D3178, ASTM D3343, ASTM D3701, and ASTM E-777);
2. *nitrogen content* (ASTM D3179, ASTM D3228, ASTM D3431, ASTM E-148, ASTM E-258, and ASTM E-778);
3. *oxygen content* (ASTM E-385); and

4. *sulfur content* (ASTM D124, ASTM D1266, ASTM D1552, ASTM D1757, ASTM D2662, ASTM D3177, ASTM D4045 and ASTM D4294).

Of the data that are available, the proportions of the elements in petroleum vary only slightly over narrow limits:

Carbon	83.0–87.0%
Hydrogen	10.0–14.0%
Nitrogen	0.1–2.0%
Oxygen	0.1–1.5%
Sulfur	0.1–6.0%

Of the ultimate analytical data, more has been made of the sulfur content than any other property. For example, the sulfur content (ASTM D124, D1552, and D4294) and the API gravity represent the two properties that have, in the past, had the greatest influence on determining the value of petroleum as a feedstock.

The sulfur content varies from about 0.1 wt. % to about 3 wt. % for the more conventional crude oils, to as much as 5–6% for heavy oil and bitumen. Residua, depending on the sulfur content of the crude oil feedstock, may be of the same order, or even have substantially higher sulfur content (Table 2.5).

The role of sulfur in producing obnoxious emission during refining and product use has been a major determinant in attempts to control sulfur emissions. There are similar cautions about nitrogen, although the amount of nitrogen in crude petroleum is significantly lower than the amount of sulfur. Considered alone as a petroleum constituent, these compounds are dangerous, but they can react with the environment to produce secondary poisonous chemicals (specifically the respective oxides), thereby contributing to the formation of acid rain (Speight, 1996; Speight and Lee, 2000; Speight, 2005, 2007).

3.3 Chromatographic Fractionation

The evaluation of petroleum must of necessity involve a study of composition because of the interrelationship of the physical properties and composition as part of the overall evaluation of different feedstocks. Fractionation also informs the environmental scientist or engineer where the most environmentally recalcitrant constituents of the crude oil are located. The data will also present indications of which constituents can be removed from coil by steam stripping, and which constituents (or how much of the spilled material) will remain in the contaminated soil.

There are several ASTM procedures for feedstock/product evaluation. These are:

1. Determination of aromatic content of olefin-free gasoline by silica gel adsorption (ASTM D936).
2. Separation of aromatic and non-aromatic fractions from high-boiling oils (ASTM D2549).
3. Determination of hydrocarbon groups in rubber extender oils by clay-gel adsorption (ASTM D2007).
4. Determination of hydrocarbon types in liquid petroleum products by a fluorescent indicator adsorption test (ASTM D1319).

Gel permeation chromatography is an attractive technique for the determination of the number average molecular weight (Mn) distribution of petroleum fractions, especially the heavier constituents, and petroleum products (Altgelt, 1968; Oelert, 1969; Altgelt, 1970; Baltus and Anderson, 1984; Hausler and Carlson, 1985; Reynolds and Biggs, 1988).

Ion exchange chromatography is also widely used in the characterization of petroleum constituents and products. For example, cation exchange chromatography can be used primarily to isolate the nitrogen constituents in petroleum (Snyder and Buell, 1965; Drushel and Sommers, 1966; McKay et al., 1974), thereby giving an indication of how the feedstock might behave during refining, and also an indication of any potential deleterious effects on catalysts.

Liquid chromatography (also called *adsorption chromatography*) has helped to characterize the group composition of crude oils and hydrocarbon products since the beginning of this century. The type and relative amount of certain hydrocarbon classes in the matrix can have a profound effect on the quality and performance of the hydrocarbon product. The *fluorescent indicator adsorption* (FIA) method (ASTM D1319) has been used to measure the paraffinic, olefinic, and aromatic content of gasoline, jet fuel, and liquid products in general (Suatoni and Garber, 1975; Miller et al., 1983; Norris and Rawdon, 1984).

High-performance liquid chromatography (*HPLC*) has found great utility in separating different hydrocarbon group types and identifying specific constituent types (Colin and Vion, 1983; Drushel, 1983; Miller et al., 1983; Chartier et al., 1986). Of particular interest is the application of the HPLC technique to the identification of the molecular types in the heavier feedstocks, especially the molecular types in the asphaltene fraction. This technique is especially useful for studying such materials on a *before-processing* and *after-processing* basis (Chmielowiec et al., 1980; Alfredson, 1981; Bollet et al., 1981; Colin and Vion, 1983; George and Beshai, 1983; Felix et al., 1985; Coulombe and Sawatzky, 1986; Speight, 1986).

Several recent high performance liquid chromatographic separation schemes are applicable, since they also incorporate detectors not usually associated with conventional hydrocarbon group-type analyses (Matsushita et al., 1981; Miller et al., 1983; Rawdon, 1984; Lundanes and Greibokk, 1985; Schwartz and Brownlee, 1986; Hayes and Anderson, 1987).

The general advantages of high performance liquid chromatography method are: (1) each sample may be analyzed *as received*, even though the boiling range may vary over a considerable range; (2) the total time per analysis is usually on the order of minutes; and, perhaps most important, (3) the method can be adapted for any recoverable petroleum sample or product.

In recent years, *supercritical fluid chromatography* has found use in the characterization and identification of petroleum constituents and products. A supercritical fluid is defined as a substance above its critical temperature. A primary advantage of chromatography using supercritical mobile phases results from the mass transfer characteristics of the solute. The increased diffusion coefficients of supercritical fluids compared with liquids can lead to greater speed in separations or greater resolution in complex mixture analyses. Another advantage of supercritical fluids compared with gases is that they can solubilize thermally labile and non-volatile solutes and, upon expansion (decompression) of this solution, introduce the solute into the vapor phase for detection (Lundanes et al., 1986).

Currently, supercritical fluid chromatography is leaving the stages of infancy (Smith et al., 1988). The indications are that it will find wide applicability to the problems of characterization and identification of the higher molecular weight species in petroleum (Schwartz et al., 1988), thereby adding an extra dimension to our understanding of refining chemistry. It will still retain the option as a means of product characterization, although the use may be somewhat limited because of the ready availability of other characterization techniques.

3.4 Liquefaction and Solidification

Petroleum (with the exception of some heavy oils and tar sand bitumen) and the majority of petroleum products are liquids at ambient temperature, and problems that may arise from solidification during normal use are not common. Nevertheless, the *melting point* is a test (ASTM D87 and D127) that often serves to determine the state of the petroleum or the product under various weather conditions or under applied conditions, such as steam stripping of the material from the soil. The reverse process, *solidification*, has received attention, again to determine the behavior of the material in nature. However, solidification of petroleum and petroleum products has been differentiated into four categories, namely, freezing point, congealing point, cloud point, and pour point.

Petroleum becomes more or less a plastic solid when cooled to sufficiently low temperatures. This is due to the congealing of the various hydrocarbons that constitute the oil. The cloud point of petroleum (or a product) is the temperature at which paraffin wax or other solidifiable compounds present in the oil appear as a haze when the oil is chilled under definitely prescribed conditions (ASTM D2500 and ASTM D3117). As cooling is continued, petroleum becomes more viscous and the pour point is the lowest temperature at which the oil pours or flows under definitely prescribed conditions when it is chilled without disturbance at a standard rate (ASTM D97).

The solidification characteristics of petroleum and its products depend on its grade or kind. For pure or essentially pure hydrocarbons, the solidification temperature is the freezing point, the temperature at which a hydrocarbon passes from a liquid to a solid state (ASTM D1015 and ASTM D1016). For grease and residua, the temperature of interest is that at which fluidity occurs, commonly known as the dropping point. The dropping point of grease is the temperature at which the grease passes from a plastic solid to a liquid state and begins to flow under the conditions of the test (ASTM D566 and ASTM D2265). For another type of plastic solid, including petrolatum and microcrystalline wax, both melting point and congealing point are of interest.

3.5 Metals Content

Heteroatoms (*nitrogen*, *oxygen*, *sulfur*, and *metals*) are found in every crude oil, and the concentrations have to be reduced to convert the oil to transportation fuel. The reason is that if nitrogen and sulfur are present in the final fuel during combustion, nitrogen oxides (NO_X) and sulfur oxides (SO_X) form, respectively. In addition, metals affect many upgrading processes adversely, poisoning catalysts in refining and causing deposits in combustion. In addition, heteroatoms also play a major role in environmental issues, and can cause the petroleum to adhere to the soil, ensuring long-term contamination.

A variety of tests (ASTM D1026, ASTM D1262, ASTM D1318, ASTM D1368, ASTM D1548, ASTM D1549, ASTM D2547, ASTM D2599, ASTM D2788, ASTM D3340, ASTM D3341, and ASTM D3605) have been designated for the determination of metals in petroleum and petroleum products. At the time of writing, the specific test for the determination of metals in whole feeds has not been designated. However, this task can be accomplished by combustion of the sample so that only inorganic ash remains. The ash can then be digested with an acid and the solution examined for metal species by atomic absorption (AA) spectroscopy or by inductively coupled argon plasma (ICP,) spectrometry.

3.6 Surface and Interfacial Tension

Surface tension is a measure of the force acting at a boundary between two phases. If the boundary is between a liquid and a solid or between a liquid and a gas (air) the attractive forces are referred to as surface tension, but the attractive forces between two immiscible liquids are referred to as *interfacial tension*.

Temperature and molecular weight have a significant effect on surface tension (Table 2.7) (Speight, 2001). For example, in the normal hydrocarbon series, a rise in temperature leads to a decrease in the surface tension, but an increase in molecular weight increases the surface tension. A similar trend, that is, an increase in molecular weight causing an increase in surface tension, also occurs in the acrylic series and, to a lesser extent, in the alkylbenzene series.

The surface tension of petroleum and petroleum products has been studied for many years. The narrow range of values (24–38 dyne/cm) for such widely diverse materials as gasoline (26 dyne/cm), kerosene (30 dyne/cm), and the lubricating fractions (34 dyne/cm), has rendered the surface tension of little value for any attempted characterization. However, it is generally acknowledged that non-hydrocarbon materials dissolved in oil reduce the surface tension: polar compounds, such as soaps and fatty acids, are particularly active. The effect is marked at low concentrations up to a critical value, beyond which further additions cause little change; the critical value corresponds closely with that required for a monomolecular layer on the exposed surface, where it is adsorbed and accounts for the lowering. Recent work has focused on the predictability of surface tension using mathematical relationships (Gomez, 1987):

$$\text{Dynamic surface tension} = 681.3/K(1-T/13.488^{1.7654} \times sg^{2.1250})^{1.2056}$$

K is the Watson characterization factor, sg is the specific gravity, and T is the temperature in °K.

A high proportion of the complex phenomena shown by emulsions and foams that are common when petroleum enters the environment can be traced to these induced surface tension effects. Dissolved gases, even hydrocarbon gases, lower the surface tension of oils, but the effects are less dramatic, and the changes probably result from dilution. The matter is of some importance in environmental issues, because the viscosity and surface tension of the petroleum govern the amount of oil that migrates or can be recovered under certain conditions.

On the other hand, although petroleum products show little variation in surface tension, within a narrow range the *interfacial tension* of petroleum, especially of petroleum products, against aqueous solutions provides valuable information (ASTM D971) that can be used to predict the

Table 2.7 Surface tension of selected hydrocarbons.

Hydrocarbon		Surface Tension			
	C	20	38	93	
	F	68	100	200	
n-Pentane		16.0	14.0	8.0	dyn/cm
		16.0	14.0	8.0	mN/m
n-Hexane		18.4	16.5	10.9	dyn/cm
		18.4	16.5	10.9	mN/m
n-Heptane		20.3	18.6	13.1	dyn/cm
		20.3	18.6	13.1	mN/m
n-Octane		21.8	20.2	14.9	dyn/cm
		21.8	20.2	14.9	mN/m
Cyclopentane		22.4			dyn/cm
		22.4			mN/m
Cyclohexane		25.0			dyn/cm
		25.0			mN/m
Tetralin		35.2			dyn/cm
		35.2			mN/m
Decalin		29.9			dyn/cm
		29.9			mN/m
Benzene		28.8			dyn/cm
		28.8			mN/m
Toluene		28.5			dyn/cm
		28.5			mN/m
Ethylbenzene		29.0			dyn/cm
		29.0			mN/m
n-Butylbenzene		29.2			dyn/cm
		29.2			mN/m

potential for emulsion formation. Thus, the interfacial tension of petroleum is subject to the same constraints as surface tension, that is, differences in composition, molecular weight, and so on. When oil-water systems are involved, the pH of the aqueous phase influences the tension at the interface; the change is small for highly refined oils, but increasing pH causes a rapid decrease for poorly refined, contaminated, or slightly oxidized oils.

A change in interfacial tension between oil and alkaline water has been proposed as an index for following the refining or deterioration of certain products, such as turbine and insulating oils. When surface or interfacial tensions are lowered by the presence of solutes, which tend to concentrate on the surface, some time is required to obtain the final concentration, and hence, the final value of the tension. In such systems, dynamic and static tension must be distinguished; the first concerns the freshly exposed surface having nearly the same composition as the body of the liquid; it usually has a value only slightly less than that of the pure solvent. The static tension is that existing after equilibrium concentration has been reached at the surface.

The interfacial tension between oil and distilled water provides an indication of compounds in the oil that have an affinity for water. The measurement of interfacial tension has received special attention because of its possible use in predicting when oil in constant use will reach the limit of its serviceability. This interest is based on the fact that oxidation decreases the interfacial tension of the oil. Furthermore, the interfacial tension of turbine oil against water is lowered by the presence of oxidation products, impurities from the air or rust particles, and certain antirust compounds intentionally blended in the oil. Thus, a depletion of the antirust additive may cause an increase in interfacial tension, whereas the formation of oxidation products or contamination with dust and rust lowers the interfacial tension.

3.7 Viscosity

Viscosity is the force in dynes required to move a plane of 1 cm^2 area at a distance of 1 cm from another plane of 1 cm^2 area through a distance of 1 cm in 1 s. In the centimeter-gram-second (cgs) system, the unit of viscosity is the poise (P) or centipoise (1 cP = 0.01 P). Two other terms in common use are *kinematic viscosity* and *fluidity*. The kinematic viscosity is the viscosity in centipoises divided by the specific gravity, and the unit is the stoke (cm^2/s), although centistokes (0.01 cSt) is in more common usage; fluidity is simply the reciprocal of viscosity.

The viscosity (ASTM D445, D88, D2161, D341, and D2270) of petroleum oils varies markedly over a very wide range (less than 10 cP at room temperature to many thousands of centipoises at the same temperature).

Many types of instruments have been proposed for the determination of viscosity. The simplest and most widely used are capillary types (ASTM D445), and not only are such capillary instruments the most simple, but when designed in accordance with known principles and used with known necessary correction factors, they are probably the most accurate viscometers available. It is usually more convenient, however, to use relative measurements, and for this purpose, the instrument is calibrated with an appropriate standard liquid of known viscosity.

Batch flow times are generally used; in other words, the time required for a fixed amount of sample to flow from a reservoir through a capillary is the datum actually observed. Any features of technique that contribute to longer flow times are usually desirable. Some of the principal capillary viscometers in use are those of Cannon-Fenske, Ubbelohde, Fitzsimmons, and Zeitfuchs.

The Saybolt universal viscosity (SUS) (ASTM D88) is the time in seconds required for the flow of 60 ml of petroleum from a container, at constant temperature, through a calibrated orifice. The Saybolt furol viscosity (SFS) (ASTM D88) is determined in a similar manner, except that a larger orifice is employed.

As a result of the various methods for viscosity determination, it is not surprising that much effort has been spent on interconversion of the several scales, especially converting Saybolt to kinematic viscosity (ASTM D2161),

$$\text{Kinematic viscosity} = a \times \text{Saybolt s} + b/\text{Saybolt s}$$

Where a and b are constants.

The Saybolt universal viscosity equivalent to a given kinematic viscosity varies slightly with the temperature at which the determination is made, because the temperature of the calibrated receiving flask used in the Saybolt method is not the same as that of the oil. Conversion factors are used to convert kinematic viscosity from 2 to 70 cSt at 38°C (100°F) and 99°C (210F) to equivalent Saybolt universal viscosity in seconds. Appropriate multipliers are listed to convert kinematic viscosity over 70 cSt. For a kinematic viscosity determined at any other temperature, the equivalent Saybolt universal value is calculated by use of the Saybolt equivalent at 38°C (100F) and a multiplier that varies with the temperature:

$$\text{Saybolt s at 100°F (38C)} = \text{cSt} \times 4.635$$

$$\text{Saybolt s at 210°F (99C)} = \text{cSt} \times 4.667$$

Viscosity decreases as the temperature increases, and the rate of change appears to depend primarily on the nature or composition of the petroleum, but other factors, such as volatility, may also have an effect. The effect of temperature on viscosity is generally represented by the equation:

$$\log \log (n + c) = A + B \log T$$

Where n is absolute viscosity, T is temperature, and A and B are constants. This equation has been sufficient for most purposes, and has come into very general use. The constants A and B vary widely with different oils, but c remains fixed at 0.6 for all oils having a viscosity over 1.5 cSt; it increases only slightly at lower viscosity (0.75 at 0.5 cSt). The viscosity-temperature characteristics of any oil, so plotted, thus create a straight line, and the parameters A and B are equivalent to the intercept and slope of the line. To express the viscosity and viscosity-temperature characteristics of an oil, the slope and the viscosity at one temperature must be known; the usual practice is to select 38°C (100F) and 99°C (210F) as the observation temperatures.

Suitable conversion tables are available (ASTM D341), and each table or chart is constructed in such a way that, for any given petroleum or petroleum product, the viscosity-temperature points result in a straight line over the applicable temperature range. Thus, only two viscosity measurements need be made at temperatures far enough apart to determine a line on the appropriate chart from which the approximate viscosity at any other temperature can be read.

Since the viscosity-temperature coefficient of high-boiling fractions, such as lubricating oil, is an important expression of its suitability, a convenient number to express this property is very useful, and hence, a viscosity index (ASTM D2270) was derived. Thus:

$$\text{Viscosity index} = \text{L} - \text{U} /\ \text{L} - \text{H}(100)$$

where L and H are the viscosities of the zero and 100 index reference oils, both having the same viscosity at 99°C (210F), and U is that of the unknown, all at 38°C (100°F). Originally the viscosity index was calculated from Saybolt viscosity data, but subsequently figures were provided for kinematic viscosity.

The viscosity of petroleum fractions increases on the application of pressure, and this increase may be very large. The pressure coefficient of viscosity correlates with the temperature coefficient, even when oils of widely different types are compared. At higher pressures, the viscosity decreases with increasing temperature, as at atmospheric pressure; in fact, viscosity changes of small magnitude are usually proportional to density changes, whether these are caused by pressure or by temperature.

Because of the importance of viscosity in determining the transport properties of petroleum, and this is particularly important in the migration of petroleum and petroleum products through soil, recent work has focused on the development of an empirical equation for predicting the dynamic viscosity of low molecular weight and high molecular weight hydrocarbon vapors at atmospheric pressure (Gomez, 1995). The equation uses molar mass and specific temperature as the input parameters, and offers a means of estimation of the viscosity of a wide range of petroleum fractions. Other work has focused on the prediction of the viscosity of blends of lubricating oils as a means of accurately predicting the viscosity of the blend from the viscosity of the base oil components (Al-Besharah et al., 1989).

3.8 Volatility

The volatility of a liquid or liquefied gas may be defined as its tendency to vaporize, that is, to change from the liquid to the vapor or gaseous state. Because one of the three essentials for combustion in a flame is that the fuel be in the gaseous state, volatility is a primary characteristic of liquid fuels.

The vaporizing tendencies of petroleum and petroleum products are the basis for the general characterization of liquid petroleum fuels, such as liquefied petroleum gas, natural gasoline, motor and aviation gasoline, naphtha, kerosene, gas oil, diesel fuel, and fuel oil (ASTM D2715). A test (ASTM D6) also exists for determining the loss of material when crude oil and asphaltic compounds are heated. Another test (ASTM D20) is a method for the distillation of road tars that might also be applied to estimating the volatility of high molecular weight unknown residues.

For many environmental purposes, it is necessary to have information on the initial stage of vaporization. To supply this need, flash and fire, vapor pressure, and evaporation methods are available. The data from the early stages of the several distillation methods are also useful. For other uses, it is important to know the tendency of a product to partially vaporize or to completely vaporize, and in some cases to know if small quantities of high-boiling components are present. For such purposes, chief reliance is placed on the distillation methods.

The *flash point* of petroleum or a petroleum product is the temperature to which the product must be heated under specified conditions to give off sufficient vapor to form a mixture with air that can be ignited momentarily by a specified flame (ASTM D56, D92, and D93). The *fire point* is the temperature to which the product must be heated under the prescribed conditions of the method to burn continuously when the mixture of vapor and air is ignited by a specified flame (ASTM D92).

From the viewpoint of safety, information about the flash point is of most significance at or slightly above the maximum temperatures (30–60°C, 86–140°F) that may be encountered in storage, transportation, and use of liquid petroleum products, in either closed or open containers. In this temperature range, the relative fire and explosion hazard can be estimated from the flash point. For products with flash point below 40°C (104°F), special precautions are necessary for safe handling. Flash points above 60°C (140°F) gradually lose their safety significance, until they become indirect measures of some other quality.

The flash point of a petroleum product is also used to detect contamination. A substantially lower flash point than expected for a product is a reliable indicator that a product has become contaminated with a more volatile product, such as gasoline. The flash point is also an aid in establishing the identity of a particular petroleum product.

A further aspect of volatility that receives considerable attention is the vapor pressure of petroleum and its constituent fractions. The *vapor pressure* is the force exerted on the walls of a closed container by the vaporized portion of a liquid. Conversely, it is the force that must be exerted on the liquid to prevent it from vaporizing further (ASTM D323). The vapor pressure increases with temperature for any given gasoline, liquefied petroleum gas, or other product. The temperature at which the vapor pressure of a liquid, either a pure compound of a mixture of many compounds, equals 1 atmosphere pressure (14.7 psi, absolute) is designated as the boiling point of the liquid.

In each homologous series of hydrocarbons, the boiling points increase with molecular weight; structure also has a marked influence: it is a general rule that branched paraffin isomers have lower boiling points than the corresponding *n*-alkane. Furthermore, in any specific aromatic series, steric effects notwithstanding, there is an increase in boiling point with an increase in carbon number of the alkyl side chain. This particularly applies to alkyl aromatic compounds where alkyl-substituted aromatic compounds can have higher boiling points than polycondensed aromatic systems. This fact is very meaningful when attempts are made to develop hypothetical structures for asphaltene constituents.

One of the main properties of petroleum that serves to indicate the comparative ease with which the material evaporates after a spill is the volatility. Investigation of the volatility of petroleum is usually carried out under standard conditions in which petroleum or the product is subdivided by distillation into a variety of different *fractions*, sometimes referred to as *cut points* (Table 2.2) (Speight, 2002). In fact, distillation involves the general procedure of vaporizing the petroleum liquid in a suitable flask either at *atmospheric pressure* (ASTM D86, D216, D285, D447, and D2892) or at *reduced pressure* (ASTM D1160).

Simulated distillation (*simdis*) by gas chromatography is often applied to petroleum to obtain true boiling point data for distillates and crude oils, and may also be useful in predicting the amount of material that can be (or needs to be) recovered after a spill. Two standardized methods (ASTM D2887 and ASTM D3710) are available for the boiling point determination of petroleum fractions and gasoline, respectively. The ASTM D2887 method utilizes non-polar, packed gas chromatographic columns in conjunction with flame ionization detection. The upper limit of the boiling range covered by this method is to approximately 540°C (1000°F) atmospheric equivalent boiling point. Recent efforts in which high temperature gas chromatography was used have focused on extending the scope of the ASTM D2887 method for higher boiling petroleum materials to 800°C (1470°F) atmospheric equivalent boiling point (Schwartz et al., 1988).

4 References

Aczel, T. 1989. "Preprints." *Div. Petrol. Chem. Am. Chem. Soc.* 34(2): 318.

Al-Besharah, J.M., Mumford, C.J., Akashah, S.A., and Salman, O. 1989. "Prediction of the Viscosity of Lubricating Oil Blends." *Fuel*. 68: 609–811.

Alfredson, T.V. 1981. *J. Chromatography*. 218: 715.

Altgelt, K.H. 1968. "Preprints." *Div. Petrol. Chem. Am. Chem. Soc.* 13(3): 37.

Altgelt, K.H. 1970. *Bitumen, Teere, Asphalte, Peche*. 21: 475.

ASTM. 2011. *Annual Book of Standards*. American Society for Testing and Materials, West Conshohocken, Pennsylvania.

Baltus, R.E., and Anderson, J.L. 1984. *Fuel*. 63: 530.

Bollet, C., Escalier, J.C., Souteyrand, C., Caude, M., and Rosset, R. 1981. *J. Chromatography*. 206: 289.

Bouquet, M., and Bailleul, A. 1982. In *Petroananlysis '81. Advances in Analytical Chemistry in the Petroleum Industry 1975–1982*. G.B. Crump (Editor). John Wiley & Sons, Chichester, England.

Cardoza, L., Korir, A., Otto, W., Wurrey, C., and Larive, C. 2004. "Applications of NMR Spectroscopy in Environmental Science." Progress in Nuclear Magnetic Resonance Spectroscopy, 45(3–4): 209–238.

Chartier, P., Gareil, P. Caude, M., Rosset, R., Neff, B., Bourgognon, H.F., and Husson, J.F. 1986. *J. Chromatography*. 357: 381.

Chmielowiec, J., Beshai, J.E., and George, A.E. 1980. *Fuel*. 59: 838.

Colin, J.M., and Vion, G. 1983. *J. Chromatography*. 280: 152.

Coulombe, S., and Sawatzky, H. 1986. *Fuel*. 65: 552.

Drushel, H.V., and Sommers, A.L. 1966. *Anal. Chem.* 38: 19.

Drushel, H.V. 1983. *J. Chromatographic Sci.* 21: 375.

Felix, G., Bertrand, C., and Van Gastel, F. 1985. *Chromatographia*. 20(3): 155.

George, A.E., and Beshai, J.E. 1983. *Fuel*. 62: 345.

Gomez, J.V. 1987. *Oil & Gas Journal*. December 7. p. 68.

Gomez, J.V. 1989. *Oil & Gas Journal*. March 27. p. 66.

Gomez, J.V. 1992. *Oil & Gas Journal*. July 13. p. 49.
Gomez, J.V. 1995. *Oil & Gas Journal*. February 6. p. 60.
Hasan, M., Ali, M.F., and Arab, M. 1989. *Fuel*. 68: 801.
Hausler, D.W., and Carlson, R.S. 1985. "Preprints." *Div. Petrol. Chem. Am. Chem. Soc.* 30(1): 28.
Hayes, P.C., and Anderson, S.D. 1987. *J. Chromatography*. 387: 333.
IP 143. *Standard Methods for Analysis and Testing of Petroleum and Related Products 2011*. Institute of Petroleum, London, England.
Kanda, R., and Glendinning, R. 2011. "Mass Spectrometry for Environmental and Wastewater Monitoring." *Spectroscopy Europe*, 23(5): 14–27.
Lundanes, E., and Greibokk, T. 1985. *J. Chromatography*. 349: 439.
Lundanes, E., Iversen, B., and Greibokk, T. 1986. *J. Chromatography*. 366: 391.
Matsushita, S., Tada, Y., and Ikushige. 1981. *J. Chromatography*. 208: 429.
McKay, J.F., Cogswell, T.E., and Latham, D.R. 1974. "Preprints." *Div. Petrol. Chem. Am. Chem. Soc.* 19(1): 25.
Mecozzi, M., Moscato, F., Pietroletti, M., Quarto, F., Oteri, F., and Cicero, A.M. 2009. "Applications of FTIR Spectroscopy in Environmental Studies Supported by Two Dimensional Correlation Analysis." *Global NEST Journal*, Vol 11, No 4, pp. 593–600.
Miller, R.L., Ettre, L.S., and Johansen, N.G. 1983. *J. Chromatography*. 259: 393.
Nanny, M.A., Minear, R.A., and Leenheer, J.A. (Editors). 1997. *Nuclear Magnetic Resonance Spectroscopy in Environmental Chemistry*. Oxford University Press, Oxford, United Kingdom.
Norris, T.A., and Rawdon, M.G. 1984. *Anal. Chem.* 56: 1767.
Oelert, H.H. 1969. *Erdoel und Kohle*. 22: 536.
Rawdon, M. 1984. *Anal. Chem.* 56: 831.
Reynolds, J.G., and Biggs, W.R. 1988. *Fuel Sci. and Technology Intl.* 6: 329.
Schwartz, H.E., and Brownlee, R.G. 1986. *J. Chromatography*. 353: 77.
Schwartz, H.E., Brownlee, R.G., Boduszynski, M.M., and Su, F. 1988. *Anal. Chem.* 59: 1393.
Speight, J.G. 1986. "Polynuclear Aromatic Systems in Petroleum." Preprints, *Am. Chem. Soc., Div. Petrol. Chem.*, 1986, 31(4), 818.
Speight, J.G. 1996. *Environmental Technology Handbook*. Taylor & Francis, Washington, DC.
Speight, J.G., and Lee, S. 2000. *Environmental Technology Handbook*. 2nd Edition. Taylor & Francis, New York.
Speight, J.G. 2001. *Handbook of Petroleum Analysis*. John Wiley & Sons Inc., Hoboken, New Jersey.
Speight, J.G. 2002. *Handbook of Petroleum Product Analysis*. John Wiley & Sons Inc., Hoboken, New Jersey.
Speight, J.G., and Ozum, B. 2002. *Petroleum Refining Processes*. Marcel Dekker Inc., New York.

Speight, J.G. 2005. *Environmental Analysis and Technology for the Refining Industry*. John Wiley & Sons Inc., Hoboken, New Jersey.
Speight, J.G. 2007. *The Chemistry and Technology of Petroleum*. 4th Edition. CRC Press, Taylor & Francis Group, Boca Raton, Florida.
Suatoni, J.C., and Garber, H.R. 1975. *J. Chromatographic Sci*. 13: 367.

3

Refinery Products and By-Products

1 Introduction

The constituents of petroleum and petroleum products, like many organic chemical entities, have the *potential* to be biodegradable into the raw materials of nature and disappear into the environment. However, sustainable biodegradation of petroleum and petroleum products is not as straightforward as it may seem. Petroleum will biodegrade in its natural state, but once it is converted into saleable products, unsustainable pollution problems can arise: instead of returning to the natural cycle, these products pollute and litter the land, water, and air.

While the basic concept of petroleum biodegradation seems straightforward, there are five factors to consider in determining the biodegradability of petroleum and petroleum products.

1. The first factor relates to the inherent biodegradability of the constituents of petroleum and its products. A natural product, such as petroleum, has the potential to return to nature as long as it remains in its relatively natural form. Any plant-based or animal-based natural product has the capability to biodegrade, but petroleum products, having undergone chemical changes, may require advanced

biodegradation chemistry, since they may not biodegrade readily.

2. The second factor relates to the time it takes for petroleum constituents to biodegrade. In nature, different materials biodegrade at different rates. The rate for petroleum and petroleum products to biodegrade depends on the composition of the material.
3. The third factor relates to the products of biodegradation, and whether or not toxic substances are formed during biodegradation, or as the end result of biodegradation. To be truly biodegradable, petroleum constituents should undergo complete mineralization (i.e., complete degradation to carbon dioxide and water), but this is not always the case, even though petroleum is a natural product.
4. The fourth factor relates to the characteristics of the environment that the substance or material is in, which can also affect its ability to biodegrade. Petroleum constituents and the constituents of petroleum-based products might biodegrade in aerobic environments, but not in anaerobic environments, such as swamps, flooded soils, or water-bearing sediments.
5. The fifth factor relates to the capacity of the ecosystem to accommodate the invasive chemical. Once it is determined that petroleum constituents or a petroleum product will biodegrade under specific environmental conditions, there is the issue of the amount of the contaminant that can be sustained by the ecosystem that is receiving it, and the sustainable rate of biodegradation, which is dependent upon the amount of the contaminant that an ecosystem can absorb as a nutrient, and, if necessary, render harmless. Spills of petroleum and petroleum products are devastating, not because some of the constituents do not biodegrade at an acceptable rate, but because the amount of petroleum is much greater than the size of the microbial colony available for biodegradation. Indeed, as much as it is necessary to consider the biodegradability of petroleum and its products, it is necessary to consider the capacity of the system the biodegradable substance or material is being placed into.

Consideration of the above factors leads to many issues which can, however, be alleviated somewhat by knowing the character of petroleum and the history of the processes by which petroleum is refined into saleable products. There are also those chemicals that are not petroleum products, but are used in the production of products, and are often designated as refinery chemicals or refinery wastes (Chapter 10).

Because a critical aspect of assessing the toxic effects of the release of petroleum and petroleum products is the measurement of the compounds in the environment, the approach is not only to understand the composition and properties of petroleum itself (Chapter 2), but also the processes by which petroleum is refined to products, as well as the properties of the various fractions and products derived from petroleum (Chapter 4, Chapter 5, Chapter 6).

Therefore, it is the purpose of this chapter to describe major refinery operations and the products therefrom, and the chapter focuses on their composition and properties and uses. This presents to the reader the essence of petroleum processes, the types of feedstocks employed, and the product produced.

By so doing, the reader is also presented with a forewarning of the types of the chemicals that can be released to the environment whenever an accident occurs. Being forewarned offers the environmental analyst the ability to design the necessary test methods to examine the released chemical(s). The chapter offers environmental scientists and engineers the ability to start forming opinions and predictions about the nature of the released chemical(s), the potential effects of the chemical(s) on the environment, and the possible methods of cleanup.

2 Refinery Products

Petroleum, in the unrefined or crude form, like many industrial feedstocks has little or no direct use, and its value as an industrial commodity is only realized after the production of salable products by a series of processing steps as performed in a refinery (Figure 3.1). Each processing step is, in fact, a separate process, and thus, a refinery is a series of integrated processes that generate the desired products according to the market demand (Meyers, 1997; Speight, 2007). Therefore, the value of petroleum is directly related to the yield of saleable products, and is subject to the market pull. In general, crude oil, once refined, yields four basic groupings of distillation products that are produced when it is broken down into fractions: naphtha, middle distillates (kerosene and light gas oil), heavy gas oil/vacuum gas oil, and the residuum (Table 3.1). It is from these unrefined fractions that saleable petroleum products are produced by a series of refinery processes (Speight, 2007).

Naphtha, a precursor to gasoline and solvents, is extracted from both the light and middle range of distillate cuts, and is also used as a feedstock for the petrochemical industry. The term *middle distillates* refers to products from the middle boiling range of petroleum, and includes kerosene, diesel fuel, distillate fuel oil, and light gas oil; waxy distillate and lower boiling lubricating oils are sometimes included in the middle distillates. The remainder of the crude oil includes the higher boiling lubricating oils,

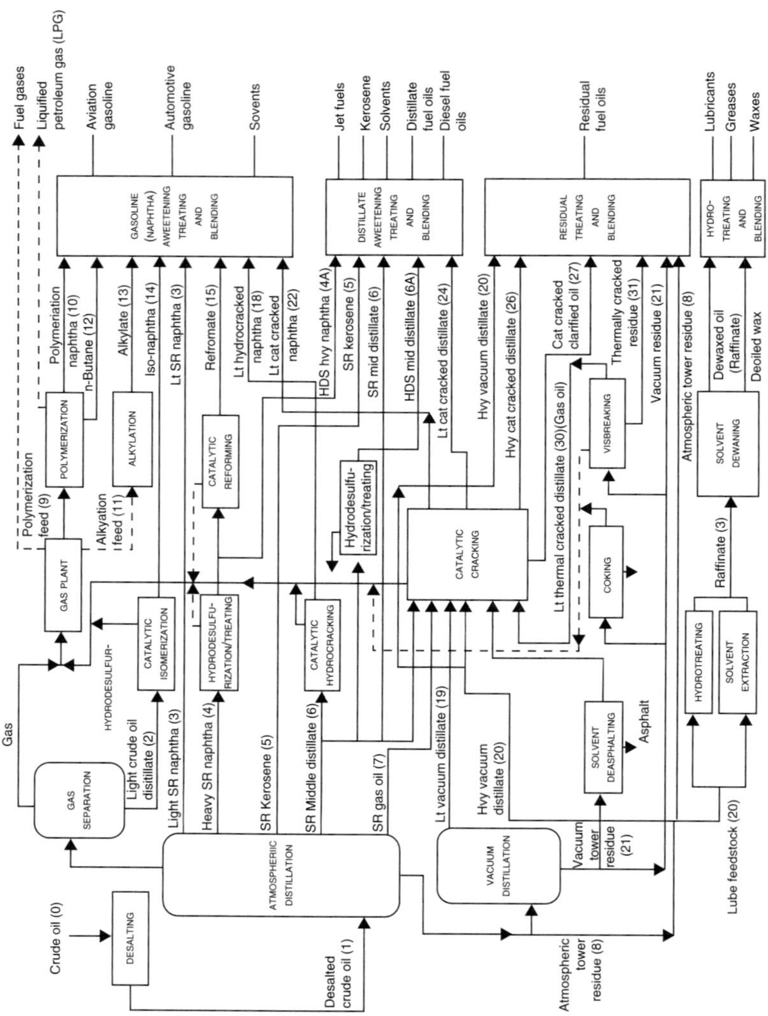

Figure 3.1 General layout of a petroleum refinery. (Source: http://www.osha.gov/dts/osta/otm/otm_iv/otm_iv_2.html)

Table 3.1 Crude petroleum is a mixture of compounds that can be separated into different generic boiling fractions.

	Boiling Range*	
Fraction	**°C**	**°F**
Light naphtha	–1–150	30–300
Medium naphtha	–1–180	30–355
Heavy naphtha	150–205	300–400
Kerosene	205–260	400–500**
Light gas oil	260–315	400–600
Heavy gas oil	315–425	600–800
Vacuum gas oil	425–600	800–1100
Residuum	>510	>950

*For convenience, boiling ranges are converted to the nearest 5°.

**Kerosene destined for conversion to diesel fuel; crude kerosene is often quoted as having a wider boiling range (150–350°C, 302–660°F) before separation into various fuel and/or solvent products.

Source: Speight, J.G. 2007. *The Chemistry and Technology of Petroleum*. 4th Edition. CRC Press, Taylor & Francis Group, Boca Raton, Florida.

gas oil, and residuum (the non-volatile fraction of the crude oil). The residuum can also produce heavy lubricating oils and waxes, but is more often used for asphalt production. The complexity of petroleum is emphasized insofar as the actual proportions of light, medium, and heavy fractions vary significantly from one crude oil to another.

Refinery processes for crude oil are generally divided into two categories: (1) separation processes (Table 3.2), of which atmospheric distillation (Figure 3.2) and vacuum distillation (Figure 3.3) are the prime examples, and (2) conversion processes (Table 3.3), of which visbreaking (Figure 3.4), delayed coking (Figure 3.5), fluid coking (Figure 3.6), catalytic cracking (Figure 3.7), and hydrocracking (Figure 3.8) are prime examples. Distillates and other fractions are usually treated further by hydrotreating (Figure 3.9) to meet product specifications before sale.

The current trend throughout the refining industry is to produce more fuel products from each barrel of petroleum, and to process those products in different ways to meet the product specifications for use in various engines (automobile, diesel, aircraft, and marine). Overall, the demand for

Table 3.2 Separation processes and conversion processes:

(1) Separation Processes:					
Process Name	Action	Method	Purpose	Feedstock(s)	Product(s)
Atmospheric distillation	Separation	Thermal	Separate fractions without cracking	Desalted crude oil	Gas, gas oil, distil-late, residual
Vacuum distillation	Separation	Thermal	Separate fraction without cracking	Atmospheric tower residua	Gas oil, lube stock, residual
(2) Conversion Processes:					
Process Name	Action	Method	Purpose	Feedstock(s)	Product(s)
Catalytic cracking	Alteration	Catalytic	Upgrade gasoline	Gas oil, coke, distillate	Gasoline, petrochemical feedstock
Coking	Polymerize	Thermal	Convert vacuum residuals	Gas oil, coke distillate	Gasoline, petrochemical feedstock
Hydrocracking	Hydrogenate	Catalytic	Convert to lighter hydrocarbons	Gas oil, cracked oil, residual	Lighter, higher-quality products
Visbreaking	Decompose	Thermal	Reduce viscosity	Atmospheric tower resid	Distillate, tar

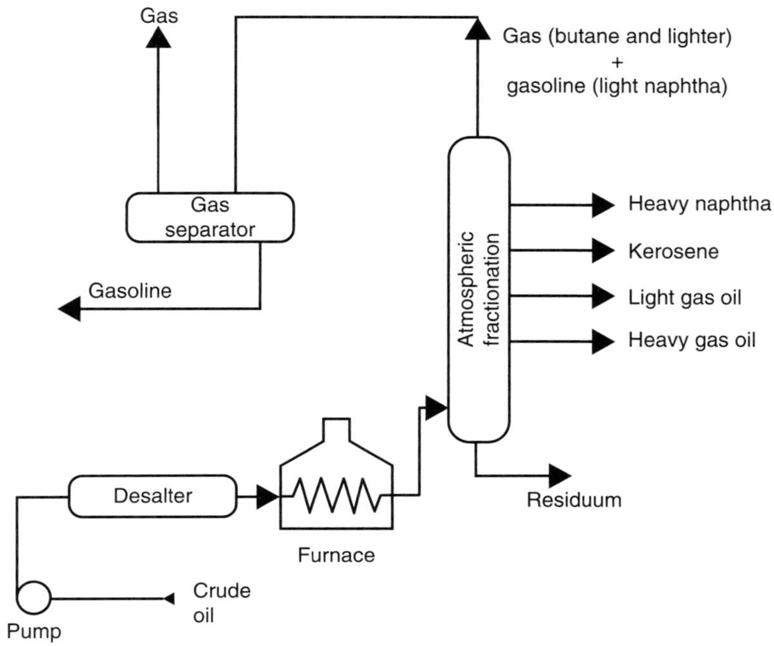

Figure 3.2 An atmospheric distillation unit. (Source: http://www.osha.gov/dts/osta/otm/otm_iv/otm_iv_2.html)

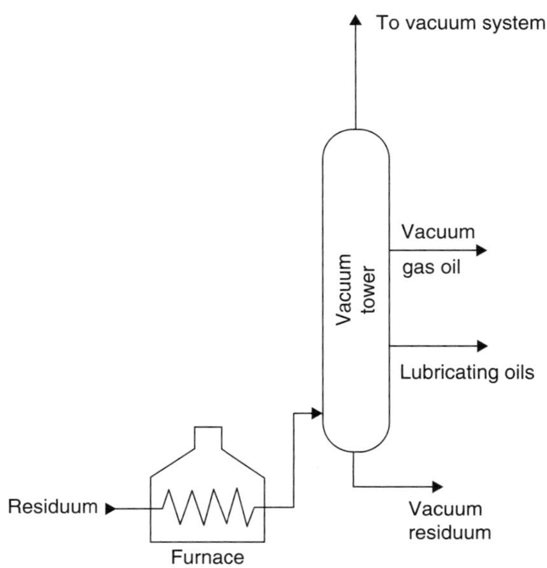

Figure 3.3 A vacuum distillation unit. (Source: http://www.osha.gov/dts/osta/otm/otm_iv/otm_iv_2.html)

Table 3.3 Component streams for gasoline.

Stream	Producing Process	Boiling Range	
		°C	°F
Paraffinic			
Butane	Distillation	0	32
Conversion			
Iso-pentane	Distillation	27	81
Conversion			
Isomerization (1)			
Alkylate	Alkylation (2)	40–150	105–300
Isomerate	Isomerization	40–70	105–160
Naphtha	Distillation	30–100	85–212
Hydrocrackate	Hydrocracking	40–200	105–390
Olefinic			
Catalytic naphtha	Catalytic cracking	40–200	105–390
Cracked naphtha	Steam cracking	40–200	105–390
Polymer	Polymerization (3)	60–200	140–390
Aromatic			
Catalytic reformate	Catalytic reforming (4)	40–200	105–390

Source: Speight, J.G. 2007. *The Chemistry and Technology of Petroleum*. 4th Edition. CRC Press, Taylor & Francis Group, Boca Raton, Florida.

liquid fuels has rapidly expanded, and demand has also developed for gas oils and fuels for domestic central heating, and fuel oil for power generation, as well as for light distillates and other inputs, derived from crude oil, for the petrochemical industries.

However, as the need for the lower boiling products developed, petroleum yielding the desired quantities of the lower boiling products became less available, and refineries had to introduce conversion processes to produce greater quantities of lighter products from the higher boiling fractions. The means by which a refinery operates in terms of

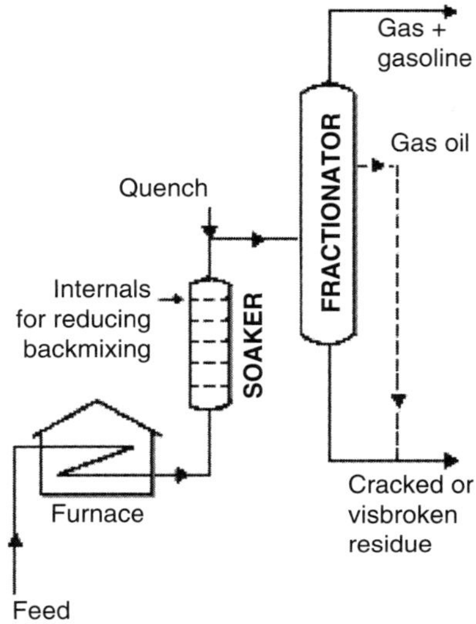

Figure 3.4 Visbreaking unit. (Source: http://www.osha.gov/dts/osta/otm/otm_iv/otm_iv_2.html)

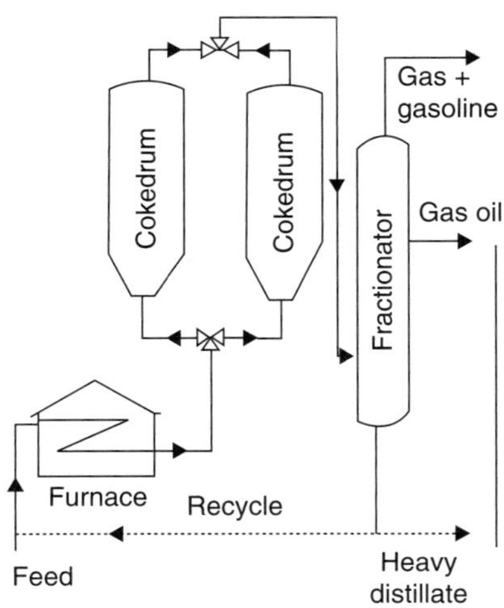

Figure 3.5 A delaying coking unit. (Source: http://www.osha.gov/dts/osta/otm/otm_iv/otm_iv_2.html)

producing the relevant products depends not only on the nature of the petroleum feedstock, but also on its configuration (i.e., the number of types of the processes that are employed to produce the desired product slate), and the refinery configuration is, therefore, influenced by the specific demands of a market. Therefore, refineries need to be constantly

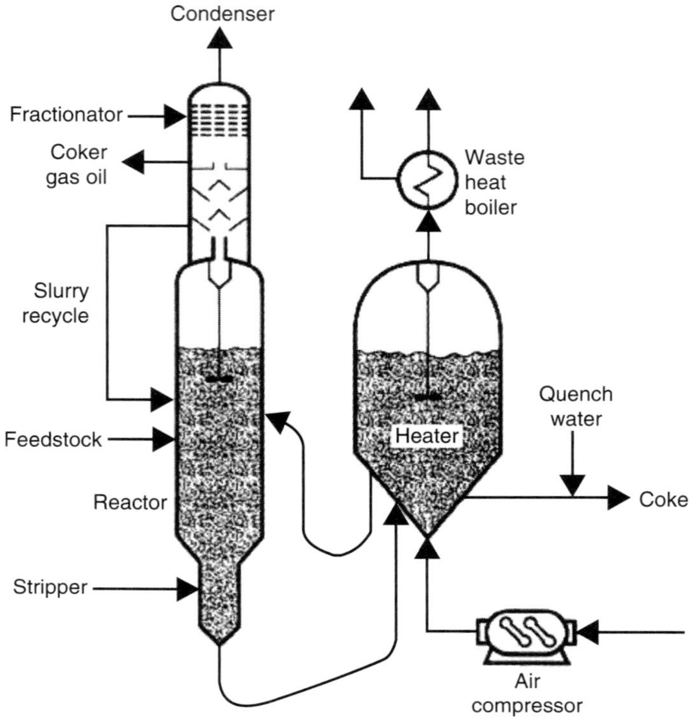

Figure 3.6 A fluid coking unit. Speight, J.G. 2007. *The chemistry and technology of petroleum*. 4th edition. CRC press, taylor & francis group, Boca Raton, Florida.

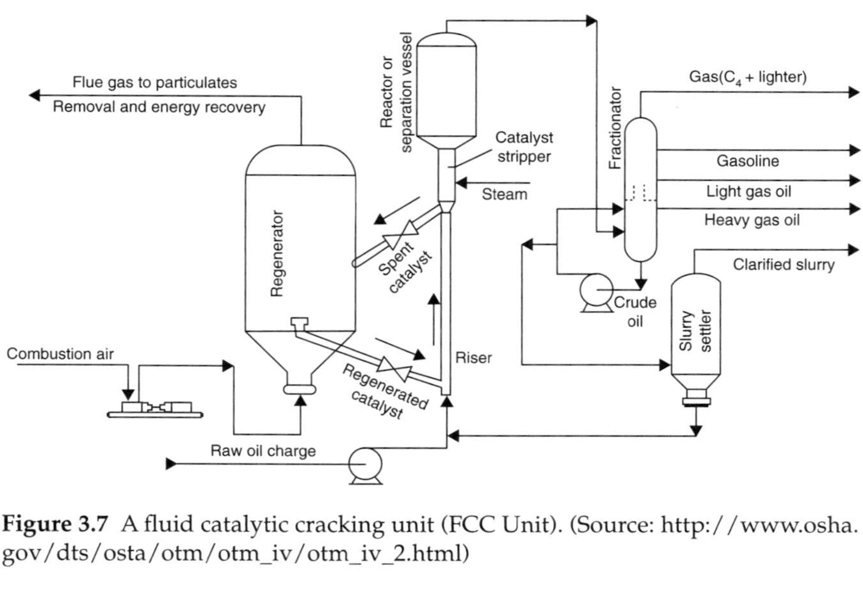

Figure 3.7 A fluid catalytic cracking unit (FCC Unit). (Source: http://www.osha.gov/dts/osta/otm/otm_iv/otm_iv_2.html)

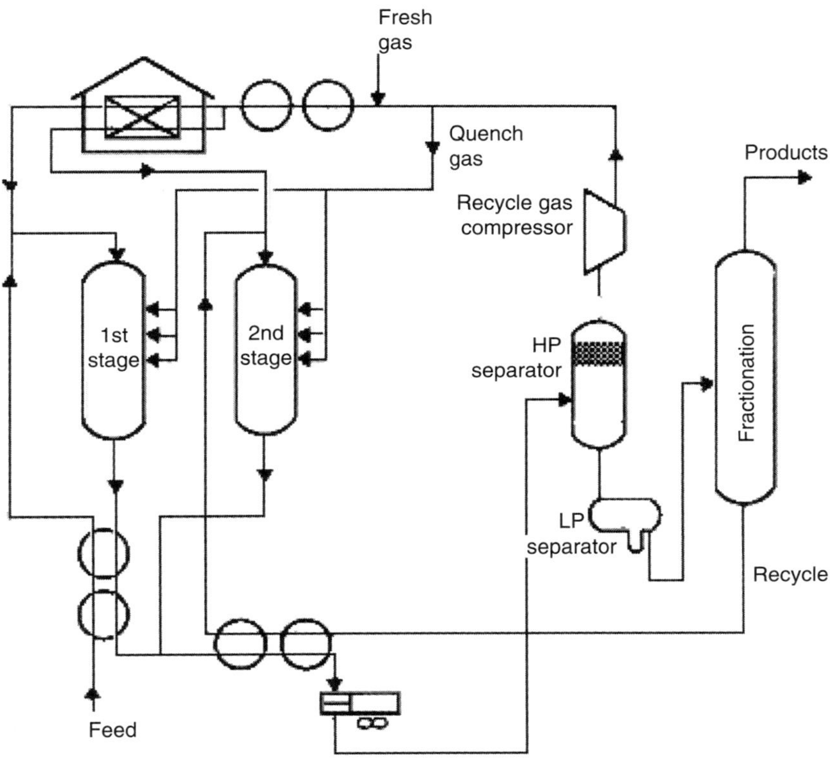

Figure 3.8 A two-stage hydrocracking unit. (Source: http://www.osha.gov/dts/osta/otm/otm_iv/otm_iv_2.html)

upgraded (in terms of process options, i.e., modernized) to adapt and remain viable and responsive to ever changing patterns of crude supply and product market demands. As a result, refineries have been introducing increasingly complex (and expensive) processes to gain more and more low-boiling products from the high-boiling and residual portions of crude petroleum.

Petroleum products (in contrast to *petrochemicals*) are those bulk fractions (Table 3.1) that are derived from petroleum and have commercial value as a bulk product (Mushrush and Speight, 1995; Speight, 2007). Petroleum products are, generally speaking, hydrocarbon compounds that have various combinations of hydrogen and carbon, and can take many molecular forms. Many of the combinations exist naturally in the original raw materials, but other combinations are created by an ever-growing number of commercial processes for altering one combination to another.

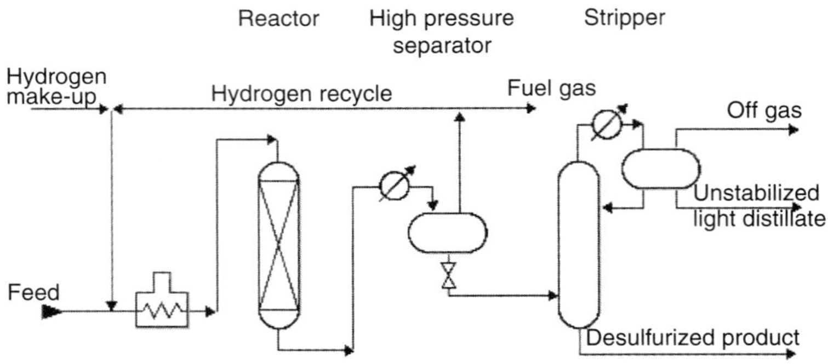

Figure 3.9 A distillate hydrotreater for hydrodesulfurization. (Source: http://www.osha.gov/dts/osta/otm/otm_iv/otm_iv_2.html)

The specifications for petroleum products (which define whether or not a product is suitable for sale, and to assure that the product can perform the task for which it is intended) are based on properties such as density and boiling range (to mention only two such properties). In addition, each product has its own unique set of chemical and physical properties. As a consequence, each product can, and does, have a different effect on the environment (Speight, 2005).

In the strictest sense, *petrochemicals* are also petroleum products but, typically, they are individual chemicals that are used as the basic building blocks of the chemical industry. Many of the petrochemicals are substitutes for earlier products from non-petroleum sources. Each new use often imposes additional specifications on the new product, and product specifications evolve to stay abreast of advances in both product application and manufacturing methods. Like petroleum products, the petrochemicals cover such a vast range of chemical and physical properties that they also have a wide range of effects on the environment.

2.1 Liquefied Petroleum Gas

Fuels with four or less carbon atoms in the hydrogen-carbon combination have boiling points that are lower than room temperature; hence, these products are gases at ambient temperature and pressure, and are often referred to as fuel gases.

Liquefied petroleum gas (LPG, LP Gas, liquid propane gas) is a flammable mixture of gaseous hydrocarbons that is used as a fuel in heating appliances, and sometimes in vehicles. Liquefied petroleum gas (LPG) is frequently used as domestic bottled gas for cooking and heating, and forms an important feedstock for the petrochemical industry. It is also used in industry for cutting metals.

Liquefied petroleum gas is composed of propane (C_3H_8) and/or butane (C_4H_{10}), and is stored under pressure in order to keep these hydrocarbons liquefied at normal atmospheric temperatures. Before liquefied petroleum gas is burned, it passes through a pressure relief valve that causes a reduction in pressure, and the liquid vaporizes (gasifies).

Varieties of liquefied petroleum gas include mixtures that are primarily propane (C_3H_8), primarily butane (C_4H_{10}), and, most commonly, mixes including both propane and butane, depending on the season: in winter, the mixture will contain more propane (because of the increased volatility necessary for ignition), while in summer, the mixture will contain more of the relatively less volatile butane. Propylene (C_3H_6, $CH_3CH=CH_2$) and butylenes (C_4H_8, $CH_3CH_2CH=CH_2$ and/or $CH_3CH=CHCH_3$) are usually also present in small concentration. An odorant, such as ethanethiol (C_2H_5SH) or amyl mercaptan ($C_5H_{11}SH$), is added to facilitate detection of gas leaks.

Natural gas (predominantly *methane*) denoted by the chemical structure CH_4 is the lowest boiling and least complex of all hydrocarbons. Natural gas from an underground reservoir, when brought to the surface, can contain other higher boiling hydrocarbons, and is often referred to as *wet gas*. Wet gas is usually processed to remove the entrained hydrocarbons that are higher boiling than methane and, when isolated, the higher boiling hydrocarbons sometimes liquefy, and are called *natural gas condensate*. As with liquefied petroleum gas, natural gas for sales and use by the consumer also has an odorant, such as ethanethiol (C_2H_5SH) or amyl mercaptan ($C_5H_{11}SH$) added to facilitate detection of gas leaks.

Pipestill gas (*still gas*) is broad terminology for low-boiling hydrocarbon mixtures, and is the lowest boiling fraction isolated from a distillation unit (*still*) unit in the refinery (Speight, 2007). If the distillation unit is separating light hydrocarbon fractions, the still gas will be almost entirely methane, with only traces of ethane (CH_3CH_3). If the distillation unit is handling higher boiling fractions, the still gas might also contain propane ($CH_3CH_2CH_3$), butane ($CH_3CH_2CH_2CH_3$), and their respective isomers. *Fuel gas* and still gas are terms that are often used interchangeably, but the term *fuel gas* is intended to denote the product's destination: to be used as a fuel for boilers, furnaces, or heaters in industrial or domestic markets.

Fuel gas or *refinery gas* is produced in considerable quantities during the different refining processes, and is used as fuel for the refinery itself, and as an important feedstock for the petrochemical industry.

2.2 Naphtha, Gasoline, and Solvents

Naphtha is the general term that is applied to refined, partly refined, or unrefined low-boiling petroleum products. Naphtha is prepared by any one of several methods, including (1) fractionation of distillates or even crude petroleum; (2) solvent extraction; (3) hydrogenation of distillates; (4) polymerization of unsaturated (olefinic) compounds; and (5) alkylation processes. Naphtha may also be a combination of product streams from more than one of these processes (Speight, 2007).

The main uses of naphtha fall into the general areas of (1) precursor to gasoline and other liquid fuels, (2) solvents/diluents for paints, (3) dry-cleaning solvents, (4) solvents for cutback asphalt, (5) solvents in rubber industry, and (6) solvents for industrial extraction processes. Turpentine, the older and more conventional solvent for paints, has now been almost completely replaced by the cheaper, and more abundant, naphtha.

The term aliphatic naphtha refers to naphtha containing less than 0.1% benzene and with carbon numbers from C_3 through C_{16}. Aromatic naphtha also spans the range of carbon numbers from C_6 through C_{16}, but contains significant quantities of aromatic hydrocarbons, such as benzene (>0.1% v/v), toluene, xylene, and (on occasion) ethyl benzene. The final gasoline product as a transport fuel is a carefully blended mixture having a predetermined octane value (Table 3.3). Thus, gasoline is a complex mixture of hydrocarbons that typically boil below 200°C (390°F). The hydrocarbon constituents in this boiling range are those that have four to twelve carbon atoms in their molecular structure.

Gasoline varies widely in composition, and even those with the same octane number may be compositionally very different. The variation in aromatics content as well as the variation in the content of normal paraffins, branched paraffins, cyclopentane derivatives, and cyclohexane derivatives all involve characteristics of any one individual crude oil, and influence the octane number of the gasoline.

Automotive gasoline is a mixture of low-boiling hydrocarbon compounds suitable for use in spark-ignited internal combustion engines and having an octane rating of at least 60. Additives that have been used in gasoline include alkyl tertiary butyl ethers, such as methyl t-butyl ether (e.g., MTBE), and ethanol (ethyl alcohol). Other categories of compounds that may be added to gasoline include anti-knock agents, antioxidants, metal deactivators, lead scavengers, anti-rust agents, anti-icing agents, upper-cylinder lubricants, detergents, and dyes (Speight, 2007).

Automotive gasoline contains one hundred fifty or more different chemical compounds, and the relative concentrations of the compounds vary considerably, depending on the source of crude oil, refinery process, and product specifications. Typical hydrocarbon constituents are: alkanes (4 to 8% v/v), alkenes (2 to 5% v/v), iso-alkanes (25 to 40% v/v), cycloalkanes (3 to 7% v/v), cycloalkenes (1 to 4% v/v), and aromatics (20 to 50% v/v).

Reformulated gasoline (RFG) is gasoline that is blended such that hydrocarbon and air toxic emissions are significantly reduced. It is not considered to be an alternative fuel, gasohol, or part of the oxygenated fuel program. Oxygenated fuel contains higher amounts of oxygen, and is required during the winter months in areas with the most serious carbon monoxide problems.

The requirements of the reformulated gasoline program, as outlined in the Clean Air Act Amendments, are as follows: (1) an aromatic hydrocarbon content (benzene and its derivatives) of no more than 25% v/v, (2) a maximum benzene content of 1% v/v, (3) an oxygen content of 2% w/w, (4) no heavy metals, including lead and manganese, (5) detergent additives to prevent the accumulation of deposits in engines or vehicle fuel supply systems, and (6) a reduced Reid Vapor Pressure (a measure of how quickly fuel evaporates) during the summer.

Aviation fuel comes in two types: aviation gasoline and jet fuel. Aviation gasoline, now usually found in use in light aircraft and older civil aircraft, typically has a narrower boiling range than conventional (automobile) gasoline, i.e., 38 to 170°C (100 to 340°F), compared to −1 to 200°C (30 to 390°F) for automobile gasoline. The narrower boiling range ensures better distribution of the vaporized fuel through the more complicated induction systems of aircraft engines. Since aircraft operate at altitudes where the prevailing pressure is less than the pressure at the surface of the earth (pressure at 17,500 ft is 7.5 psi (0.5 atmosphere) compared to 14.8 psi (1.0 atmosphere) at the surface of the earth), the vapor pressure of aviation gasoline must be limited to reduce boiling in the tanks, fuel lines, and carburetors.

Aviation gasoline consists primarily of straight and branched alkanes and cycloalkanes. Aromatic hydrocarbons are limited to 20 to 25% v/v of the total mixture, because they produce smoke when burned. A maximum of 5% v/v alkenes is allowed in JP-4. The approximate distribution by chemical class is: 32% v/v straight alkanes, 31% v/v branched alkanes, 16% v/v cycloalkanes, and 21% v/v aromatic hydrocarbons.

Jet fuel is classified as *aviation turbine fuel*, and in the specifications, ratings relative to octane number are replaced with properties concerned with the ability of the fuel to burn cleanly. Jet fuel is a light petroleum distillate that is available in several forms suitable for use in various types of jet engines. The exact composition of jet fuel is established by the U.S. Air Force using specifications that yield maximum performance by the aircraft. The major jet fuels used by the military are JP-4, JP-5, JP-6, JP-7, and JP-8.

Briefly, JP-4 is a wide-cut fuel, developed for broad availability in times of need. JP-6 is a higher cut than JP-4, and is characterized by fewer impurities. JP-5 is specially blended kerosene, and JP-7 is high flash point kerosene used in advanced supersonic aircraft. JP-8 is a kerosene fraction that is modeled on Jet A-l fuel (used in civilian aircraft). For this profile, JP-4 will be used as the prototype jet fuel, due to its broad availability and extensive use.

The quality of gasoline is determined by means of the octane number (octane rating), which is a standard measure of the performance of automobile gasoline or aviation fuel. The higher the octane number, the more compression the fuel can withstand before detonating. Generally, gasoline-type fuels with a higher octane rating are used in high-compression engines (spark ignition engines) that generally have higher performance. In contrast, fuels with low octane numbers (but high cetane numbers) are ideal for diesel engines (compression ignition engines).

Petroleum solvents (also called *naphtha*) are valuable because of their good dissolving power. The wide range of naphtha available and the varying degree of volatility possible offer products suitable for many uses.

Stoddard solvent is a petroleum distillate widely used as a dry cleaning solvent and as a general cleaner and degreaser. It may also be used as paint thinner, as a solvent in some types of photocopier toners, in some types of printing inks, and in some adhesives. Stoddard solvent is considered to be a form of mineral spirits, white spirits, and naphtha; however, not all forms of mineral spirits, white spirits, and naphtha are considered to be Stoddard solvent.

Stoddard solvent consists (volume basis) of linear and branched alkanes (30–50%), cycloalkanes (30–40%), and aromatic hydrocarbons (10–20%). Alcohols, glycols, and ketones are not included in the composition, as few, if any, of these types of compounds are expected to be present in Stoddard solvent. Possible contaminants may include lead (<1 ppm) and sulfur (3.5 ppm).

2.3 Kerosene and Diesel Fuel

Kerosene was the major refinery product before the onset of the *automobile age*, but now kerosene might be termed as one of several other petroleum products after gasoline. Kerosene originated as a straight-run (distilled) petroleum fraction that boiled between approximately 150 to 350°C (300 to 660°F). In the early days of petroleum refining, some crude oils contained kerosene fractions of very high quality, but other crude oils, such as those having a high proportion of asphaltic materials, must be thoroughly refined to remove aromatics and sulfur compounds before a satisfactory kerosene fraction can be obtained. Apart from the removal of excessive quantities of aromatics, kerosene fractions may need only a lye (alkali) wash if hydrogen sulfide is present.

Chemically, kerosene is a mixture of hydrocarbons; the chemical composition depends on its source, but it usually consists of hydrocarbons each containing 10 to 16 carbon atoms per molecule. The constituents include *n*-dodecane (n-$C_{12}H_{26}$), alkyl benzenes, and naphthalene and its derivatives.

Diesel fuel also forms part of the kerosene boiling range (or middle distillate group of products). Diesel fuels come in two broad groups, one for high speed engines in cars and trucks requiring a high quality product, and lower quality heavier diesel fuel for slower engines, such as in marine engines or for stationary power plants. An important property of diesel is cetane number (analogous to the gasoline octane number), and the cetane number determines the ease of ignition under compression.

The quality of diesel fuel is measured using the cetane number that is a measure of the tendency of a diesel fuel to knock in a diesel engine, and the scale from which the cetane number is derived is based upon the ignition characteristics of two hydrocarbons: (1) *n*-hexadecane (cetane) and (2) 2,3,4,5,6,7,8-heptamethylnonane.

2.4 Fuel Oil

Fuel oil is classified in several ways, but generally may be divided into two main types: (1) *distillate fuel oil* and (2) *residual fuel oil*. These classifications are still employed but, of late, the terms distillate fuel oil and residual fuel oil have lost some of their significance, because fuel oils are now made for specific uses, and may be distillates, residuals, or mixtures of the two. The terms domestic fuel oils, diesel fuel oils, and heavy fuel oils are more indicative of the uses of fuel oils. More often than not, fuel oil is prepared by using a visbreaker unit (Figure 3.4) to perform mild thermal cracking (insofar as the reactions are not allowed to proceed to completion) on a residuum or on a high boiling distillate so that the product meets specifications.

Distillate fuel oil is vaporized and condensed during a distillation process; it has a definite boiling range and does not contain high-boiling oils or asphaltic components. A fuel oil that contains any amount of the residue from crude distillation or thermal cracking is a residual fuel oil. *Domestic fuel oil* is fuel oil that is used primarily in the home, and includes kerosene, stove oil, and furnace fuel oil. *Diesel fuel oil* is also a distillate fuel oil, but residual oil has been successfully used to power marine diesel engines, and mixtures of distillates and residuals have been used on locomotive diesels. *Furnace fuel oil* is similar to diesel fuel, but the proportion of cracked gas oil in diesel fuel is usually less, since the high aromatic content of the cracked gas oil reduces the cetane number of the diesel fuel.

Stove oil is a straight-run (distilled) fraction from crude oil, whereas other fuel oils are usually blends of two or more fractions. The straight-run fractions available for blending into fuel oils are heavy naphtha, light and heavy gas oils, and residua. Cracked fractions, such as light and heavy gas oils from catalytic cracking, cracking coal tar, and fractionator bottoms from catalytic cracking, may also be used as blends to meet the specifications of the different fuel oils.

Heavy fuel oil includes a variety of oils ranging from distillates to residual oils that must be heated to 260°C (500°F) or higher before they can be used. In general, heavy fuel oils consist of residual oils blended with distillates to suit specific needs. Included among heavy fuel oils are various industrial oils; when used to fuel ships, heavy fuel oil is called bunker oil.

Fuel oil that is used for heating is graded from No. 1 Fuel Oil to No. 6 Fuel Oil, and covers light distillate oils, medium distillate, heavy distillate, a blend of distillate and residue, and residue oil. For instance, No. 2 and No. 3 Fuel oils refer to medium-to-light distillate grades used in domestic central heating (Table 3.4).

No. 1 fuel oil is a petroleum distillate that is one of the most widely used of the fuel oil types. It is used in atomizing burners that spray fuel into a combustion chamber, where the tiny droplets burn while in suspension. It is also used as a carrier for pesticides, as a weed killer, as a mold release agent in the ceramic and pottery industry, and in the cleaning industry. It is found in asphalt coatings, enamels, paints, thinners, and varnishes. No. 1 fuel oil is a light petroleum distillate (straight-run kerosene) consisting primarily of hydrocarbons in the range C9-C16, and is similar in composition to diesel fuel; the primary difference is in the additives.

Table 3.4 Properties of various fuel oils.

No. 1 Fuel Oil	Similar to kerosene or range oil (fuel used in stoves for cooking). A distillate intended for vaporizing in pot-type burners and other burners where a clean flame is required.
No. 2 Fuel Oil	Often called domestic heating oil with properties similar to diesel and higher-boiling jet fuel. Defined as a distillate for general purpose heating in which the burners do not require the fuel to be completely vaporized before burning.

(Continued)

Table 3.4 (cont.) Properties of various fuel oils.

No. 4 fuel oil	A light industrial heating oil that is intended for use where preheating is not required for handling or burning Two grades that differ primarily in safety (flash) and flow (viscosity) properties. .
No. 5 fuel oil	A heavy industrial oil that often requires preheating for burning and, in cold climates, for handling.
No. 6 Fuel Oil	A more viscous oil usually containing residuum. Commonly referred to as *Bunker C oil* when it is used to fuel ocean-going vessels. Preheating is required for both handling and burning this grade oil.

No. 2 fuel oil is a petroleum distillate that may be referred to as domestic fuel oil or industrial fuel oil. The domestic fuel oil is usually lighter and straight-run refined; it is used primarily for home heating and to produce diesel fuel. Industrial distillate is the cracked type, or a blend of both. It is used in smelting furnaces, ceramic kilns, and packaged boilers.

No. 2 fuel oil is characterized by hydrocarbons in the C11-C20 range, whereas diesel fuels predominantly contain a mixture of C10-C19 hydrocarbons. The composition consists of approximately 64% aliphatic hydrocarbons (straight chain alkanes and cycloalkanes), 1-2% unsaturated hydrocarbons (alkenes), and 35% aromatic hydrocarbons (including alkylbenzenes and 2-ring and 3-ring aromatics). No. 2 fuel oil contains less than 5% polycyclic aromatic hydrocarbons.

No. 6 fuel oil (also called Bunker C or residual fuel oil) is the residual from crude oil after the light oil, gasoline, naphtha, No. 1 fuel oil, and No. 2 fuel oil have been distilled. No. 6 fuel oil can be blended directly to heavy fuel oil or made into asphalt. It is limited to commercial and industrial uses where sufficient heat is available to fluidize the oil for pumping and combustion.

Residual fuel oils are residues remaining after distillation or cracking, or blends of such residues with distillates. Residual fuel oil is generally more complex in composition and impurities than distillate fuels. Limited data are available, but there are indications that the composition of No. 6 fuel oil includes (volume basis) aromatics (25%), paraffins (15%), naphthenes (45%), and non-hydrocarbon compounds (15%). Polynuclear aromatic hydrocarbons and their alkyl derivatives and metals are important hazardous and persistent components of No. 6 fuel oil.

In summary, all fuel oils consist of complex mixtures of aliphatic and aromatic hydrocarbons. The aliphatic alkanes (paraffins) and cycloalkanes (naphthenes) are hydrogen saturated, and compose approximately 80 to 90% v/v of the fuel oils. Aromatics (e.g., benzene) and olefins (e.g., styrene and indene) compose 10 to 20% v/v and 1% v/v, respectively, of the fuel oils. Fuel oil No. 1 (straight-run kerosene) is a light distillate that consists primarily of hydrocarbons in the C_9 to C_{16} range. Fuel oil No. 2 is a higher boiling, usually blended, distillate with hydrocarbons in the C_{11} to C_20 range. Straight-run distillates may also be used to produce fuel oil No. 1 and diesel fuel oil No. 1.

Diesel fuel No. 1 and No. 2 are similar in chemical composition to fuel oil No. 1 and fuel oil No. 2, respectively, with the exception of the additives. Diesel fuels predominantly contain a mixture of C_{10} through C_{19} hydrocarbons, which include approximately 64% v/v aliphatic hydrocarbons, 1 to 2% v/v olefin hydrocarbons, and 35% v/v aromatic hydrocarbons.

Jet fuels are based primarily on straight-run kerosene, as well as additives. All of the above fuel oils contain less than 5% v/v polycyclic aromatic hydrocarbons. Fuel oil No. 4 (marine diesel fuel) is less volatile than diesel fuel oil No. 2, and may contain up to 15% v/v residual process streams, in addition to more than 5% v/v polycyclic aromatic hydrocarbons. Residual fuel oils are generally more complex in composition and impurities than distillate fuel oils; therefore, a specific composition is difficult to determine and is, at best, speculative. The sulfur content in residual fuel oils has been reported to vary from 0.18% to 4.36% w/w.

2.5 Lubricating Oil

Lubricating oil is distinguished from other fractions of crude oil by a usually high (>400°C, >750°F) boiling point, as well as high viscosity.

The development of vacuum distillation provided the means of separating lubricating oil fractions with predetermined viscosity ranges, and removed the limit on the maximum viscosity that might be obtained in distillate oil. Vacuum distillation prevented residual asphaltic material from contaminating lubricating oils, but did not remove other undesirable materials, such as acidic components, or components that caused the oil to thicken excessively when cold, and become very thin when hot.

Lubricating oil may be divided into many categories according to the types of service it is intended to perform. However, there are two main groups: (1) oils used in intermittent service, such as motor and aviation oils, and (2) oils designed for continuous service, such as turbine oils.

The mineral-based oils are produced from heavy-end crude oil distillates. Distillate streams may be treated in several ways, such as vacuum-, solvent-, acid-, or hydrotreated, to produce oils with commercial properties. Hydrocarbon types ranging from C_{15} to C_{50} are found in the various types of oils, with the heavier distillates having higher percentages of the higher carbon number compounds.

Crankcase oil or motor oil may be either petroleum-based or synthetic. The petroleum-based oils are more widely used than the synthetic oils, and may be used in automotive engines, railroad and truck diesel engines, marine equipment, jet and other aircraft engines, and most small 2- and 4-stroke engines.

The petroleum-based oils contain hundreds to thousands of hydrocarbon compounds, including a substantial fraction of nitrogen- and sulfur-containing compounds. The hydrocarbons are mainly mixtures of straight and branched chain hydrocarbons (alkanes), cycloalkanes, and aromatic hydrocarbons. Polynuclear aromatic hydrocarbons, alkyl polynuclear aromatic hydrocarbons, and metals are important components of motor oils and crankcase oils, with the used oils typically having higher concentrations than the new unused oils. Typical carbon numbers range from C_{15} to C_{50}.

Because of the wide range of uses and the potential for close contact with the engine to alter oil composition, the exact composition of crankcase oil/motor oil has not been specifically defined.

Viscosity is the basic property by which lubricating oils are classified. The requirements vary, from very thin oil needed for the high-speed spindles of textile machinery, to the viscous, tacky materials applied to open gears or wire ropes. Between these extremes is a wide range of products with special characteristics. Automotive oils represent the largest product segment in the market. In the United States, specifications for these products are defined by the Society of Automotive Engineers (SAE), which issues viscosity ratings with numbers that range from 5 to 50. In the United Kingdom, standards are set by the Institute of Petroleum, which conducts tests that are virtually identical to those of the SAE.

When ordinary mineral oils having satisfactory lubricity at low temperatures are used over an extended temperature range, excessive thinning occurs, and the lubricating properties are found to be inadequate at higher temperatures. To correct this, multigrade oils have been developed using long-chain polymers. Thus, oil designated SAE 10W40 has the viscosity of an SAE 10W oil at –18°C (0°F) and of an SAE 40 oil at 99°C (210° F). Such an oil performs well under cold starting conditions in winter (hence, the W designation), yet will lubricate under high-temperature running conditions in the summer as well. Other additives that improve the performance of lubricating oils are antioxidants and detergents, which maintain engine cleanliness and keep fine carbon particles suspended in the circulating oil.

Used lubricating oil (waste crankcase oil) is the used lubricating oils removed from the crankcase of internal combustion engines. Before they are used, lubricating oils consist of a base lubricating oil (a complex mixture of hydrocarbons, 80 to 90% v/v) and performance-enhancing additives (10 to 20% v/v). Lubricating oils are altered during service because of the breakdown of additives, contamination with the products of combustion, and the addition of metals from the wear and tear of the engine. Therefore, the composition of used lubricating oil is difficult to generalize in exact chemical terms. Generally, the major components of used lubricating oil consist of aliphatic and aromatic hydrocarbons (such as phenol, naphthalene, benz(a)anthracene, benzo(a)pyrene, and fluoranthene).

2.6 White Oil, Insulating Oil, Insecticides

The term *white oil* (*white distillate*) is applied to all the refinery streams with a distillation range between approximately 80 and 360°C (175 to 680°F) at atmospheric pressure, and with properties similar to the corresponding straight-run distillate from atmospheric crude distillation. Light distillate products, i.e., naphtha, kerosene, jet fuel, diesel fuel, and heating oil, are all manufactured by appropriate blending of white distillate streams.

There is also a category of petroleum products known as *white oil* that generally falls into two classes: (1) *technical white oil* that is employed for cosmetics, textile lubrication, insecticide vehicles, and paper impregnation, and (2) *pharmaceutical white oil* that is employed medicinally (e.g., as a laxative) or for the lubrication of food-handling machinery.

White oil is manufactured with a mixture of paraffinic and naphthenic hydrocarbons of the highest purity. It is available in various grades and is colorless, odorless, and tasteless during normal storage and service life. It is nontoxic and has outstanding storage stability. White oil is hydrophobic, colorless, tasteless, odorless, and does not change color over time. Many white oils are chemically and biologically stable, and do not support pathogenic bacterial growth.

Insulating oil (*transformer oil, dielectric oil*) falls into two general classes: (1) oil used in transformers, circuit breakers, and oil-filled cables, and (2) oil employed for impregnating the paper covering of wrapped cables. The first are highly refined fractions of low viscosity and comparatively high boiling range, and resemble heavy burning oils, such as mineral seal oil, or the very light lubricating fractions known as non-viscous neutral oils. The second are usually highly viscous products, and are often naphthenic distillate that is not usually highly refined.

The main function of insulating oil (transformer oil) is insulating and cooling of the transformer. Accordingly, the oil should have the following properties: (1) high dielectric strength, (2) low viscosity, (3) no inorganic acids, alkali, and corrosive sulfur present, (4) resistant to emulsification,

(5) does not form sludge under normal operation, (6) low pour point, and (7) high flash point.

Oxidation is the most common cause of oil deterioration: careful and routine vacuum dehydration to remove air and water is essential to maintaining good oil. Moisture is the main source of contamination, and tends to lower the dielectric strength of the oil, and promotes acid formation when combined with air and sulfur. Finally, excessive heat breaks down the oil and will increase the rate of oxidation: this usually occurs when the transformer is overloaded.

Insecticides are not always derived from petroleum (Ware and Whitacre, 2004). However, insecticides are usually applied in water-emulsion form and have marked killing power for certain species of insects. For many applications for which their own effectiveness is too slight, the oils serve as carriers for active poisons, as in the household and livestock sprays.

Nearly all insecticides have the potential to significantly alter ecosystems; many are toxic to humans, and others are concentrated in the food chain.

2.7 Grease

The term *grease* is applied to many different chemicals, and hence, has no specific formula. Generally, there are fatty or oily chemicals making up grease.

In gear lubrication, the oil separates metal surfaces, reducing friction and wear. Extreme pressures develop in some gears, notably those in the rear axles of cars, and special additives must be employed to prevent the seizing of the metal surfaces. These oils contain sulfur compounds that form a resistant film on the surfaces, preventing actual metal-to-metal contact.

In the current context, grease is lubricating oil to which a thickening agent has been added for the purpose of holding the oil to surfaces that must be lubricated. The most widely used thickening agents are soaps of various kinds, and grease manufacture is essentially the mixing of soaps with lubricating oils.

Soap is made by chemically combining a metal hydroxide with a fat or fatty acid:

$$\underset{\text{Fatty acid}}{RCO_2H} + NaOH \rightarrow \underset{\text{soap}}{RCO^-_2Na^+} + H_2O$$

The most common metal hydroxides used for this purpose are calcium hydroxide, lye, lithium hydroxide, and barium hydroxide. Fats are chemical combinations of fatty acids and glycerin. If a metal hydroxide is reacted with a fat, a soap containing glycerin is formed. Frequently, a fat is separated into its fatty acid and glycerin components, and only the fatty acid portion is used to make soap. Commonly used fats for grease-making soaps are cottonseed oil, tallow, and lard. Among the fatty acids used are stearic acid (from tallow), oleic acid (from cottonseed oil), and animal fatty acids (from lard).

A wide range of lubricant base fluids is used in grease technology. However, the largest segment consists of a variety of products derived from the refining of crude oil and downstream petroleum raw materials. These mineral oils can contain a very wide spectrum of chemical components, depending on the origin and composition of the crude oil, as well as the refining processes to which they have been submitted.

There are three basic groups of mineral oils: (1) aromatic, (2) naphthenic, and (3) paraffinic. Historically, the aromatic and naphthenic mineral oils have represented the principle volumes used in grease formulation, largely due to availability, but also due to their solubility characteristics. However, concerns about the carcinogenic aspects of base oil constituents containing aromatic and polyaromatic ring structures have led to their replacement by paraffinic oils as the mineral fluids of choice.

The decline in the use of oils containing aromatic and polyaromatic components has reduced the range of solubility characteristics available to the grease technologist. However, naphthenic oils contain not only polyaromatics, but also similar, less toxic, polycyclic compounds without a benzene ring structure. Using modern selective refining techniques, it is possible to remove only the unwanted components and produce oils with low toxicity. Thus, these severely refined oils provide a range of beneficial solubility properties, similar to conventional types of naphthenic oils, but without the toxicity problems.

Modern base oils in lubricating greases are, therefore, often a blend of severely refined paraffinic and naphthenic oils, designed to provide the final product with the appropriate characteristics of mechanical stability, lubricity and dropping point.

2.8 Wax

Wax is of two general types: (1) paraffin wax in petroleum distillates, and (2) microcrystalline wax in petroleum residua.

Paraffin wax is a solid crystalline mixture of straight-chain (normal) hydrocarbons ranging from twenty to thirty carbon atoms per molecular, and even higher. Wax constituents are solid at ordinary temperatures (25°C, 77°F) whereas *petrolatum* (*petroleum jelly*) does contain both solid and liquid hydrocarbons. The melting point of wax is not always directly related to its boiling point, because wax contains hydrocarbons of different chemical structure.

There are three main grades of paraffin wax: fully refined, semi-refined, and scale. These differ mainly by the degree to which entrapped oil has been removed during refining, and by color. Fully refined paraffins are hard, brittle, white, odorless materials with less than 0.5% w/w oil, melting points from 46 to 68°C (115 to 155°F). On the other hand, semi-refined paraffin waxes contain more oil, 0.5% to 1% w/w, and the additional oil detracts from gloss, and the waxes are softer, light-colored, and have a slight odor and taste. Scale waxes are softer materials, with 1 to 3% w/w oil content. Color varies from white to yellow.

Wax recrystallization, like wax sweating, separates wax into fractions, but instead of relying upon differences in melting points, the process makes use of the different solubility of the wax fractions in a solvent such as a ketone.

More generally, there are three main methods are used in modern refinery technology: (1) *solvent dewaxing*, in which the feedstock is mixed with one or more solvents, then the mixture is cooled down to allow the formation of wax crystals, and the solid phase is separated from the liquid phase by filtration, (2) *urea dewaxing*, in which urea forms an adduct with straight chain paraffins that separated by filtration from the dewaxed oil, and (3) *catalytic dewaxing*, in which straight-chain paraffin hydrocarbons are selectively cracked on zeolite-type catalysts, and the lower-boiling reaction products are separated from the dewaxed lubricating oil by fractionation.

Individual test methods are determined by molecular size and structure, chemical composition, and oil content. Paraffin wax consists mostly of straight chain hydrocarbons with 80 to 90% normal paraffin content, and the balance consists of branched paraffins (iso-paraffins) and cycloparaffins.

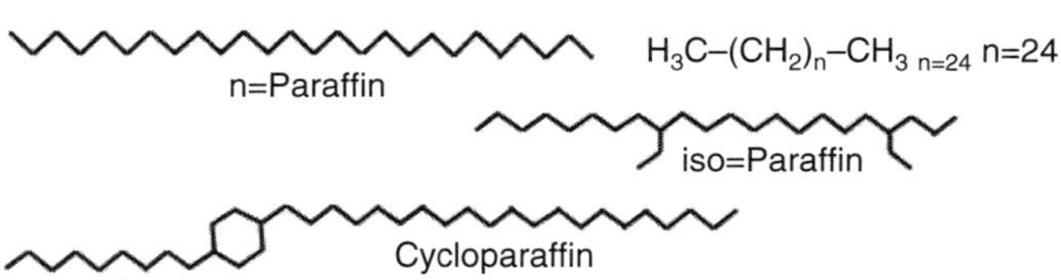

Microcrystalline waxes are produced from a combination of heavy lube distillates and residual oils. They differ from paraffin waxes in that they have poorly defined crystalline structure, darker color, and generally higher viscosity and melting points. Microcrystalline waxes (sometimes also called micro wax) tend to vary much more widely than paraffin waxes with regard to physical characteristics. Microcrystalline waxes can range from being soft and tacky to being hard and brittle, depending on the compositional balance.

The last category of petroleum wax is referred to as petrolatum, which is derived from heavy residual oils, and is separated by a dilution and filtering (or centrifuging) process. Petrolatum is microcrystalline in nature and semi-solid at room temperature.

Typical test methods that can be used to determine the properties of wax include (1) melting point, (2) congealing point, (3) oil content, and (4) viscosity. However, these physical properties alone do not completely define the suitability of a wax for a particular application: the functional properties of wax should be considered as well.

2.9 Asphalt

Asphalt (sometimes referred to as bitumen in some parts of the world) is produced from the distillation residuum. In addition to road asphalt, a variety of asphalt grades for roofing and waterproofing are also produced. Asphalt has complex chemical and physical compositions that usually vary with the source of the crude oil, and it is produced to certain standards of hardness or softness in controlled vacuum distillation processes (Barth, 1962; Bland and Davidson, 1967; Speight, 2007 and references cited therein; Speight and Ozum, 2002).

There are wide variations in refinery operations and in the types of crude oils, so different asphalts will be produced that have different environmental effects (EPA, 1996). Blending with higher and lower softening point asphalts may make asphalts of intermediate softening points. If lubricating oils are not required, the reduced crude may be distilled in a flash drum that is similar to a distillation tower, but has few, if any, trays. Asphalt descends to the base of the flasher as the volatile components pass out of the top. Asphalt is also produced by propane deasphalting, and can be made softer by blending the hard asphalt with the extract obtained in the solvent treatment of lubricating oils. On the other hand, soft asphalts can be converted into harder asphalts by oxidation (air blowing).

Road oil is liquid asphalt material intended for easy application to earth roads, and provides a strong base or a hard surface that will maintain a satisfactory passage for light traffic. *Cutback asphalt* is a mixture in which hard asphalt has been diluted with a lower boiling oil to permit application as a liquid without drastic heating. *Asphalt emulsions* are usually the oil-in-water type emulsion that breaks on application to a stone or earth surface, so that the oil clings to the stone and the water disappears.

In addition to their usefulness in road and soil stabilization, asphalt emulsions are also used for paper impregnation and waterproofing.

The exact chemical composition of asphalt is dependent on the chemical complexity of the original crude petroleum and the manufacturing process. In fact, because of the precursors (residua) from which asphalt is manufactured, attempts to identify individual constituents of asphalt have only led to wide speculation.

Crude petroleum consists mainly of aliphatic compounds, cyclic alkanes, aromatic hydrocarbons, polycyclic aromatic compounds, and metals (nickel and vanadium, with smaller amounts of iron and copper). The proportions of these chemicals can vary greatly because of significant differences in crude petroleum from oil field to oil field, or even at different locations in the same oil field. While the manufacturing process may change the physical properties of asphalt dramatically, the chemical nature of the asphalt does not change unless thermal cracking occurs. Although no two asphalts are chemically identical, and chemical analysis cannot be used to define the exact chemical structure or chemical composition of asphalt, elemental analyses indicate that most asphalts contain 79 to 88% w/w carbon, 7 to 13% w/w hydrogen, traces to 8% w/w sulfur, 2 to 8% w/w oxygen, and traces to 3% w/w nitrogen.

2.10 Coke

Petroleum coke is the residue left by the destructive distillation (thermal cracking or coking) of petroleum residua. The coke formed in catalytic cracking operations is usually non-recoverable because of adherence to the catalyst, as it is often employed as fuel for the process.

Petroleum coke is characterized by its chemical composition and physical characteristics. The chemical composition of petroleum coke is dependent upon the composition of the feedstocks that are used in the coking process, which in turn are dependent upon the composition of the crude oil from which they are derived. The metals and sulfur composition of calcined coke is directly dependent upon the composition of the green coke from which it was produced.

Petroleum coke is insoluble on organic solvents, has a honeycomb-type appearance, and can be categorized generally as either (1) green coke or (2) calcined coke. The initial product of the coking process, green coke, is used as a solid fuel. Further processing of green coke at higher temperatures and higher pressures results in the production of calcined coke, which is used in: (1) the manufacture of electrodes, (2) smelting applications, (3) graphite electrode production, or (4) for the carbonization of steel.

Typical parameters measured to define the chemical composition of petroleum coke are: (1) ash % w/w, (2) sulfur % w/w, (3) extractable material % w/w, and (4) nickel ppm and vanadium ppm. Because of the lower temperature used in its production, green or fuel-grade coke contains higher levels of residual hydrocarbon than other grades of coke. The calcining process removes essentially the entire extractable material.

The use of coke as a fuel must proceed with some caution with the acceptance by refiners of the heavier crude oils as refinery feedstocks. The higher contents of sulfur and nitrogen in these oils mean a product coke containing substantial amounts of sulfur and nitrogen. Both of these elements will produce unacceptable pollutants, sulfur oxides and nitrogen oxides, during combustion. These elements must also be regarded with caution in any coke that is scheduled for electrode manufacture and removal procedures, for these elements are continually being developed.

Because petroleum coke is a solid, fairly inert material, there has not been much concern about its environmental effects. The typical battery of tests used to measure a chemical's impact on the environment, such as breakdown by sunlight, stability in water, breakdown in the soil, and volatility, cannot be measured for petroleum coke. In addition, there are suggestions that any residual oil associated with petroleum coke either does not come off, or is present at amounts too low to cause harmful effects.

Although the potential for coke to cause environmental damage is often thought to be very low, green coke, with higher oil content and the potential for leaching of metals from the coke during periods of snow melt, rain, or acid rain, might be expected to have a lower hazard than calcined coke.

3 Refinery Chemicals

Typical refinery products include (1) natural gas and liquefied petroleum gas (LPG), (2) solvent naphtha, (3) kerosene, (4) diesel fuel, (5) jet fuel, (6) lubricating oil, (7) various fuel oils, (8) wax, (9) residua, and (10) asphalt (Chapter 3). A single refinery does not necessarily produce all of these products, but any refinery may also produce a variety of waste chemicals that must be disposed in an environmentally acceptable manner. Example of such products are (1) spent caustic, (2) spent acids, and (3) spent catalysts.

3.1 Spent Caustic

Spent caustic solutions is an alkaline solution that has become weakened, diluted, or exhausted by constant use, and the solution retains very little of the original alkaline nature.

Several process use the principle of an alkali (caustic) wash, which is a process in which distillate is treated with sodium hydroxide to remove acidic contaminants that contribute to poor odor and stability. The alkaline wastewater from oil refinery is a high concentration alkaline wastewater, which contains a great deal of sulfide, phenol, and oil pollutants.

Used caustic (spent caustic) in a refinery comes from multiple sources (Speight, 2007) in which sulfides and organic acids are removed from the product streams into the caustic phase. The sodium hydroxide is consumed, and the resulting wastewaters are often mixed and called refinery spent caustic. This spent caustic is contaminated with sulfides, carbonates, and in many cases, a high fraction of heavy organic acids.

Thus, refinery wastewater may be highly alkaline and contain oil, sulfides, ammonia, and/or phenol. The potential exists in the coking process for exposure to burns when handling hot coke, or in the event of a steam-line leak, or from steam, hot water, hot coke, or hot slurry that may be expelled when opening cokers. Safe work practices and/or the use of appropriate personal protective equipment may be needed for exposures to chemicals and other hazards, such as heat and noise, and during process sampling, inspection, maintenance, and turnaround activities.

The caustic streams used in various refining, petrochemical, and chemical operations often end up in a waste caustic tank for disposal. This waste caustic can produce a unique disposal challenge, depending upon the components to be removed. A common method of remediation is a wet air oxidation process that is designed to convert the reactive sulfide components, such as sodium sulfide and sodium mercaptide, to sodium sulfate. Organic contaminants, such as phenols, can also be significantly reduced. Wet air oxidation treatment allows spent caustic to be routed to a standard biological wastewater treatment plant without creating operational or odor problems.

Light and heavy hydrocarbons (oils) will adversely affect the operation of a wet air oxidation process by one or more of the following: (1) foul the system heat exchangers, causing frequent heat exchanger cleaning, system shutdown, and localized corrosion problems, (2) exceed the supply for of available oxygen, reducing system capacity or resulting in system shutdown due to oxygen deficiency, and (3) cause the system to be oversized in design that increases the capital and operating cost.

3.2 Spent Acids

In addition to spent caustic solutions, spent acid solutions also occur in a refinery. By analogy to spent caustic, spent acid is an acidic solution that has become weakened, diluted, or exhausted by constant use, and the solution retains very little of the original acidic nature.

Naphthenic acids, complex carboxylic acids that are believed to have a cyclopentane ring or cyclohexane ring in the molecule, occur in petroleum.

They seem to be of little consequence environmentally, since thermal decarboxylation can occur during the distillation process. During this process, the temperature of the crude oil in the distillation column can reach as high as 395°C (740°F). Hence, decarboxylation is possible (Speight and Francisco, 1990).

$$RCO_2H \rightarrow RH + CO_2$$

However, inorganic acids are used in various processes to treat unfinished petroleum products, such as gasoline, kerosene, and lubricating oil stocks, which are treated with sulfuric acid for improvement of color, odor, and other properties.

For example, treating of petroleum distillates with sulfuric acid is generally applied to dissolve unstable or colored substances and sulfur compounds, as well as to precipitate asphaltic materials. When drastic conditions are employed, as in the treatment of lubricating fractions with large amounts of concentrated acid, or when fuming acid is used in the manufacture of white oils, considerable quantities of petroleum sulfonic acids are formed.

$$\begin{array}{ll} RH + H_2SO_4 \rightarrow & RSO_3H + H_2O \\ \text{Paraffin} & \text{sulfonic acid} \end{array}$$

Two general methods are applied for the recovery of sulfonic acids from sulfonated oils and their sludge. (1) In one case, the acids are selectively removed by adsorbents or by solvents (generally low-molecular-weight alcohols), and (2) in the other case, the acids are obtained by salting out with organic salts or bases.

Petroleum sulfonic acids may be roughly divided into those soluble in hydrocarbons, and those soluble in water. Because of their color, hydrocarbon-soluble acids are referred to as mahogany acids, and the water-soluble acids are referred to as green acids. The composition of each type varies with the nature of the oil sulfonated and the concentration of the acids produced. In general, those formed during light acid treatment are water-soluble; oil-soluble acids result from more drastic sulfonation.

Sulfonic acids are used as detergents made by the sulfonation of alkylated benzene. The number, size, and structure of the alkyl side chains are important in determining the performance of the finished detergent. The salts of mixed petroleum sulfonic acids have many other commercial applications. They find use as anticorrosion agents, leather softeners, and flotation agents, and have been used in place of red oil (sulfonated castor oil) in the textile industry. Lead salts of the acids have been employed in greases as extreme pressure agents, and alkyl esters have been used as alkylating agents. The alkaline earth metal (magnesium, Mg, calcium, Ca, and barium, Ba) salts are used in detergent compositions for motor oils, and the alkali metal (K and Na) salts are used as detergents in aqueous systems.

The *sulfuric acid sludge* from sulfuric acid treatment is frequently used as a source (through thermal decomposition) to produce sulfur dioxide (SO_2) that is returned to the sulfuric acid plant and *sludge acid coke*. The coke, in the form of small pellets, is used as a substitute for charcoal in the manufacture of carbon disulfide. Sulfuric acid coke is different from other petroleum coke, in that it is pyrophoric in air, and also reacts directly with sulfur vapor to form carbon disulfide.

In recent years, the petroleum refining industry has placed an increasing emphasis on the safety of the use of hydrogen fluoride (HF) in petroleum refineries. Refineries use the acid in a manufacturing process called "alkylation," which is increasingly important in producing a high-quality gasoline. Hydrofluoric acid is hazardous and corrosive and, if accidentally released, can form a vapor cloud. If the vapor cloud is concentrated enough, it can be toxic until sufficiently dispersed. In the past five years, there have been a number of accidental releases of this acid from alkylation units at major petroleum refineries in the United States.

Pure hydrogen fluoride is a clear, colorless, corrosive liquid that has roughly the same weight as water (comparing equal volumes). It boils at 67 degrees Fahrenheit and, depending on the release conditions, can form a vapor cloud if released to the atmosphere. It has a sharp, penetrating odor that human beings can detect at very low concentrations (0.04–0.13 ppm(1)) in the air. It is completely soluble in water, in which it forms HF, which, in concentrated solutions, vaporizes in air to form a noticeable cloud.

To protect against adverse effects from exposure to hydrofluoric acid in the workplace, the Occupational Safety and Health Administration has established a permissible exposure limit (PEL) of 3 ppm averaged over an eight-hour work shift. The National Institute for Occupational Safety and Health immediately dangerous to life or health (IDLH) concentration for HF is 30 ppm(2).

3.3 Spent Catalysts

Many processes involve the use of a catalyst, a material that aids or promotes a chemical reaction between other substances, but does not react itself. Catalysts increase reaction speeds, and can provide control by increasing desirable reactions and decreasing undesirable reactions.

The cracking of crude oil fractions occurs over many types of catalytic materials, and cracking catalysts can differ markedly, in both activity to promote the cracking reaction, and in the quality of the products obtained from cracking the feedstocks (Gates et al., 1979; Wojciechowski and Corma, 1986; Stiles and Koch, 1995; Cybulski and Moulijn, 1998; Occelli and O'Connor, 1998). Activity can be related directly to the total number of active (acid) sites per unit weight of catalyst, and also to the acidic strength of these sites. Differences in activity and acidity regulate the extent of various secondary reactions occurring, and thus, the product quality differences. The acidic sites are considered to be Lewis- or Brønsted-type acid sites, but there is much controversy as to which type of site predominates

The first cracking catalysts were acid leached *montmorillonite clays*. The acid leach was to remove various metal impurities, principally iron, copper, and nickel that could exert adverse effects on the cracking performance of the catalyst. The catalysts were first used in fixed-bed and moving-bed reactor systems in the form of shaped pellets. Later, with the development of the fluid catalytic cracking process, the clay catalysts were made in the form of a ground, sized powder. Clay catalysts are relatively inexpensive, and have been used extensively for many years.

The desire to have catalysts that were uniform in composition and catalytic performance led to the development of *synthetic catalysts*. The first synthetic cracking catalyst, consisting of 87% silica (SiO_2) and 13% alumina (Al_2O_3), was used in pellet form and used in fixed-bed units in 1940. Catalysts of this composition were ground and sized for use in fluid catalytic cracking units. In 1944, catalysts in the form of beads about 2.5 to 5.0 mm diameter were introduced, comprised about 90% silica and 10% alumina, and were extremely durable. One version of these catalysts contained a minor amount of chromia (Cr_2O_3) to act as an oxidation promoter.

Neither silica (SiO_2) nor alumina (Al_2O_3) alone is effective in promoting catalytic cracking reactions. In fact, they (and also activated carbon) promote hydrocarbon decompositions of the thermal type. Mixtures of anhydrous silica and alumina ($SiO_2{\cdot}Al_2O_3$) or anhydrous silica with hydrated alumina ($2SiO_2{\cdot}2Al_2O_3{\cdot}6H_2O$) are also essentially non-effective. A catalyst having appreciable cracking activity is obtained only when prepared from hydrous oxides followed by partial dehydration (*calcining*). The small amount of water remaining is necessary for proper functioning.

Commercial synthetic catalysts are amorphous, and contain more silica than is called for by the preceding formulae; they are generally composed of 10–15% alumina (Al_2O_3) and 85–90% silica (SiO_2). The natural materials, montmorillonite, a non-swelling bentonite, and halloysite, are hydrosilicates of aluminum, with a well-defined crystal structure and approximate composition of Al_2O_3 $4Si_2O.xH_2O$. Some of the newer catalysts contain up to 25% alumina, and are reputed to have a longer active life.

The catalysts are porous and highly adsorptive, and their performance is affected markedly by the method of preparation. Two catalysts that are chemically identical but have pores of different size and distribution may have different activity, selectivity, temperature coefficient of reaction rate, and response to poisons. The intrinsic chemistry and catalytic action of a surface may be independent of pore size, but small pores appear to produce different effects, because of the manner and time in which hydrocarbon vapors are transported into and out of the interstices.

In addition to synthetic catalysts comprising silica-alumina, other combinations of *mixed oxides* were found to be catalytically active, and were developed during the 1940s. These systems included silica (SiO_2), magnesia (MgO), silica-zirconia (SiO_2-ZrO), silica-alumina-magnesia, silica-alumina-zirconia, and alumina-boria (Al_2O_3-B_2O_3). Of these, only silica-magnesia was used in commercial units, but operating difficulties developed with the regeneration of the catalyst, which at the time demanded a switch to another catalyst. Further improvements in silica-magnesia catalysts have since been made. High yields of desirable products are obtained with hydrated aluminum silicates. These may be either activated (acid-treated natural clays of the bentonite type) or synthesized silica-alumina or silica-magnesia preparations. Both the natural and the synthetic catalysts can be used as pellets or beads, and also in the form of powder; in either case, replacements are necessary because of attrition and gradual loss of efficiency (DeCroocq, 1984; Le Page et al., 1987; Thakur, 1985).

During the period of 1940 to 1962, the cracking catalysts used most widely commercially were the aforementioned acid-leached clays and silica alumina. The latter was made in two versions: *low alumina* (about 13% Al_2O_3) and *high alumina* (about 25% Al_2O_3) contents. High alumina content catalysts showed a higher equilibrium activity level and surface area.

During the 1958 to 1960 period, *semi-synthetic catalysts* of silica-alumina catalyst were used, in which approximately 25 to 35% kaolin was dispersed throughout the silica-alumina gel. These catalysts could be offered at a lower price and therefore, were disposable, but they were marked by a lower catalytic activity and greater stack losses because of increased attrition rates. One virtue of the semi-synthetic catalysts was that a lesser amount of adsorbed, unconverted, high molecular weight products on the catalyst were carried over to the stripper zone and regenerator. This resulted in a higher yield of more valuable products, and also smoother operation of the regenerator, as local hot spots were minimized.

The catalysts must be stable to physical impact loading and thermal shocks, and must withstand the action of carbon dioxide, air, nitrogen compounds, and steam. They should also be resistant to sulfur compounds; the synthetic catalysts and certain selected clays appear to be better in this regard than average untreated natural catalysts.

Commercially used cracking catalysts are *insulator catalysts* possessing strong protonic (acidic) properties. They function as catalyst by altering the cracking process mechanisms through an alternative mechanism involving *chemisorption* by *proton donation* and *desorption*, resulting in cracked oil and, theoretically, restored catalyst. Thus, it is not surprising that all cracking catalysts are poisoned by proton-accepting vanadium.

The catalyst-oil volume ratios range from 5:1 to 30:1 for the different processes, although most processes are operated to 10:1. However, for moving-bed processes, the catalyst-oil volume ratios may be substantially lower than 10:1.

Crystalline *zeolite catalysts* having molecular sieve properties were introduced as selective adsorbents in the 1955–1959 period. In a relatively short time period, all of the cracking catalyst manufacturers were offering their versions of zeolite catalysts to refiners. The intrinsically higher activity of the crystalline zeolites vis-à-vis conventional amorphous silica-alumina catalysts, coupled with the much higher yields of gasoline and decreased coke and light ends yields, served to revitalize research and development in the mature refinery process of catalytic cracking.

A number of *zeolite catalysts* have been mentioned as having catalytic cracking properties, such as synthetic faujasite (X- and Y-types), offretite, mordenite, and erionite. Of these, the faujasites have been most widely used commercially. While faujasite is synthesized in the sodium form, base exchange removes the sodium with other metal ions that, for cracking catalysts, include magnesium, calcium, rare earths (mixed or individual), and ammonium. In particular, mixed rare earths alone, or in combination with ammonium ions, have been the most commonly used forms of faujasite in cracking catalyst formulations. Empirically, X-type faujasite has a stoichiometric formula of $Na_2O\ Al_2O_3\ 2{\cdot}5SiO_2$ and Y-type faujasite $Na_2O\ Al_2O_3\ 4.8SiO_2$. Slight variations in the silica/alumina ($SiO_2/A1_2O_3$) ratio exist for each of the types. Rare earth exchanged Y-type faujasite retains much of its crystallinity after steaming at 825°C (1520°F) with steam for 12 hr; rare earth form X-faujasite, while thermally stable in dry air, will lose its crystallinity at these temperatures in the presence of steam.

The latest technique developed by the refining industry to increase gasoline yield and quality is to treat the catalysts from the cracking units to remove metal poisons that accumulate on the catalyst. Nickel, vanadium, iron, and copper compounds contained in catalytic cracking feedstocks are deposited on the catalyst during the cracking operation, thereby adversely affecting both catalyst activity and selectivity. Increased catalyst metal contents affect catalytic cracking yields by increasing coke formation, decreasing gasoline and butane and butylene production, and increasing hydrogen production.

The recent commercial development and adoption of cracking catalyst-treating processes definitely improve the overall catalytic cracking process economics.

3.3.1 Demet

A cracking catalyst is subjected to two pretreatment steps. The first step effects vanadium removal; the second, nickel removal, is to prepare the metals on the catalyst for chemical conversion to compounds (chemical treatment step) that can be readily removed through water washing (catalyst wash step). The treatment steps include use of a sulfurous compound followed by chlorination with an anhydrous chlorinating agent (e.g., chlorine gas), and washing with an aqueous solution of a chelating agent (e.g., citric acid). The catalyst is then dried and further treated before returning to the cracking unit.

3.3.2. Met-X

This process consists of cooling, mixing, and ion-exchange separation, filtration, and resin regeneration. Moist catalyst from the filter is dispersed in oil and returned to the cracking reactor as a slurry. On a continuous basis, the catalyst from a cracking unit is cooled and then transported to a stirred reactor and mixed with an ion-exchange resin (introduced as slurry). The catalyst-resin slurry then flows to an elutriator for separation. The catalyst slurry is taken overhead to a filter, and the wet filter cake is slurried with oil and pumped into the catalytic cracked feed system. The resin leaves the bottom of the elutriator, and is regenerated before returning to the reactor.

4 References

Barth, E.J. 1962. *Asphalt Science and Technology*, Gordon and Breach, New York.

Bland, W.F., and Davidson, R.L. 1967. *Petroleum Processing Handbook*, McGraw-Hill, New York.

EPA. 1996. *Study of Selected Petroleum Refining Residuals*, Office of solid Waste Management, Environmetnal Protection Agency, Washington, DC.

EPA. 2004. Environmental Protection Agency, Washington, DC. Web site: http://www.epa.gov

Meyers, R.A. 1997. *Handbook of Petroleum Refiing Processes*, 2nd Edition. McGraw-Hill, New York.

Mushrush, G.W., and Speight, J.G. 1995. *Petroleum Products: Instability and Incompatibility*, Taylor & Francis, Washington, D.C., 1995.

Speight, J.G. 2007. *The Chemistry and Technology of Petroleum*, 4th Edition. CRC Press, Taylor & Francis Group, Boca Raton, Florida.

Speight, J.G., and Francisco, M.A. 1990. *Rev. de l'Institut Francais du Petrole*, 1990, 45: 733.

Speight, J.G., and Ozum, B. 2002. *Petroleum Refining Processes*, Marcel Dekker Inc., New York.

Speight, J.G. 2007. *The Chemistry and Technology of Petroleum*, 4th Edition. CRC Press, Taylor & Francis Group, Boca Raton, Florida.

Ware, G.W., and Whitacre, D.M. 2004. "An Introduction to Insecticides." In *The Pesticide Book*, 6th Edition. MeisterPro Information Resources, Meister Media Worldwide, Willoughby, Ohio.

4

Composition and Properties of Gaseous Products

1 Introduction

Petroleum product gases and refinery gases (process gases), produced from petroleum refining, upgrading, or natural gas processing facilities, are a category of saturated and unsaturated light hydrocarbons, predominantly C_1 to C_6. Some gases may also contain inorganic compounds, such as hydrogen, nitrogen, hydrogen sulfide, carbon monoxide, and carbon dioxide. As such, petroleum and refinery gases (unless produced as a salable product that must meet specifications prior to sale) are considered to be of unknown or variable composition. The site-restricted petroleum and refinery gases (i.e., those not produced for sale) often serve as fuels consumed on-site, as intermediates for purification and recovery of various gaseous products, or as feedstocks for isomerization and alkylation processes within a facility.

In terms of waste definition, there are three basic approaches to defining petroleum or a petroleum product as hazardous: (1) a qualitative description of the waste by origin, type, and constituents; (2) classification by characteristics based upon testing procedures; and (3) classification as a result of the concentration of specific chemical substances (Speight, 2005). In addition, there are recommended protocols that must occur as a prelude to cleanup of emissions and mitigating future releases (Table 4.1).

Table 4.1 General properties of liquefied petroleum gas.

		Butane	**Propane**
Relative density (to water) of liquid at 15.6°C		0.57–0.580	.50–0.51
Liters/Ton of liquid at 15.6°C		1723–1760	1957–2091
Relative density (to air) of vapor at 15.6°C and 1015.9 mbar		1.90–2.10	1.40–1.55
Ratio of gas to liquid volume at 15.6°C and 1015.0 mbar		.233	.274
Vapor pressure at 20°C	bar	2.5	9
	psig	40	130
Vapor pressure at 50°C	bar	7	19.6
	psig	100	283
Lower limit of flammability	%	1.8	2.2
Upper limit of flammability	%	9.0	10.0

Petroleum is capable of producing gaseous pollutants (Guthrie, 1967; Rawlinson and Ward, 1973; Francis and Peters, 1980; Hoffman, 1983; Loehr, 1992; Olschewsky and Megna, 1992; Moustafa, 1996; Speight, 1996, 1999, 2005, 2007) which are often characterized by chemical species identification, e.g., *inorganic gases*, such as sulfur dioxide (SO_2), nitrogen oxides (NO_x), and carbon monoxide (CO), or *organic gases*, such as chloroform ($CHCl_3$) and formaldehyde (HCHO). The rate of release or concentrating in the exhaust air stream (in parts per million or comparable units) along with the type of gaseous emission greatly predetermines the applicable control technology.

The three main greenhouse gases that are products of refining are carbon dioxide, nitrous oxide, and methane (Fogg and Sangster, 2003). Carbon dioxide is the main contributor to climate change. Methane is generally not as abundant as carbon dioxide, but is produced during refining, and, if emitted into the atmosphere, it is a powerful greenhouse gas, and more effective at trapping heat. However, gaseous emissions associated with petroleum refining are more extensive that carbon dioxide and methane, and typically include process gases, petrochemical gases, volatile organic compounds (VOCs), carbon monoxide (CO), sulfur oxides (SO_x), nitrogen oxides (NO_x), particulates, ammonia (NH_3), and hydrogen sulfide (H_2S). These effluents may be discharged as air emissions, and must be treated. However, gaseous emissions are more difficult to capture than wastewater or solid waste, and thus, are the largest source of untreated wastes released to the environment.

In the refining industry, as in other industries, air emissions include point and non-point sources. Point sources are emissions that exit stacks and flares, and thus, can be monitored and treated. Non-point sources are *fugitive emissions* that are difficult to locate and capture. Fugitive emissions occur throughout refineries and arise from, for example, the thousands of valves, pumps, tanks, pressure relief valves, and flanges. While individual leaks are typically small, the sum of all fugitive leaks at a refinery can be one of its largest emission sources.

The numerous process heaters used in refineries to heat process streams or to generate steam (boilers) for heating or steam stripping can be potential sources of sulfur oxides (SO_2, and SO_3), nitrogen oxides (NO and NO_2), carbon monoxide (CO), particulates, and hydrocarbons emissions. When operating properly, and when burning cleaner fuels, such as refinery fuel gas, fuel oil, or natural gas, these emissions are relatively low. If, however, combustion is not complete, or heaters are fired with refinery fuel pitch or residuals, emissions can be significant.

In addition to the corrosion of equipment of acid gases, the escape into the atmosphere of sulfur-containing gases can eventually lead to the formation of the constituents of acid rain, i.e., the oxides of sulfur (SO_2 and SO_3). Similarly, the nitrogen-containing gases can also lead to nitrous and nitric acids (through the formation of the oxides NO_x, where x = 1 or 2), which are the other major contributors to acid rain. The release of carbon dioxide and hydrocarbons as constituents of refinery effluents can also influence the behavior and integrity of the ozone layer.

The processes that have been developed to accomplish gas purification vary from a simple once-through wash operation to complex multi-step recycling systems. In many cases, the process complexities arise because of the need for recovery of the materials used to remove the contaminants, or even recovery of the contaminants in the original, or altered, form (Kohl and Riesenfeld, 1979; Newman, 1985; Mokhatab et al., 2006; Speight, 2007).

The majority of gas streams exiting each refinery process are collected and sent to the gas treatment and sulfur recovery units to recover the refinery fuel gas and sulfur. Emissions from the sulfur recovery unit typically contain some hydrogen sulfide (H_2S), sulfur oxides, and nitrogen oxides. Other emissions sources from refinery processes arise from periodic regeneration of catalysts. These processes generate streams that may contain relatively high levels of carbon monoxide, particulates, and volatile organic compounds (VOCs). Before being discharged to the atmosphere, such off-gas streams may be treated first through a carbon monoxide boiler to burn carbon monoxide and any volatile organic compounds, and then through an electrostatic precipitator or cyclone separator to remove particulates.

Analysts need consistent, reliable, and credible methodologies to produce analytical data about gaseous emissions (Patnaik, 2004). To fulfill this need in this text, this chapter is devoted to descriptions and the various analytical methods that can be applied to identify gaseous emissions from a refinery. Each gas is, in turn, referenced by its name rather than the generic term *petroleum gas* (ASTM D4150). However, the composition of each gas varies, and recognition of this is essential before testing protocols are applied.

2 Gaseous Products

2.1 Liquefied Petroleum Gas

Liquefied petroleum gas (LPG) is a generic term used to describe a mixture of the gaseous hydrocarbons propane ($CH_3CH_2CH_3$, boiling point: –42°C, –44°F) and butane ($CH_3CH_2CH_2CH_3$, boiling point: 0°C, 32°F) that are produced during natural gas refining, petroleum stabilization, and petroleum refining (Table 4.1) (Austin, 1984; Speight, 1999; Ritter, 2000). The propane and butane can be derived from natural gas or from refinery operations, but in this latter case, substantial proportions of the corresponding olefins will be present, and need to be separated. The hydrocarbons are normally liquefied under pressure for transportation and storage.

Recovery of higher molecular weight hydrocarbons (C_{3+}) or liquefied petroleum gas from refinery purge and fuel gas streams is more profitable than sending these high-value components to fuel. Liquefied petroleum gas components are produced in many refinery operations. Traditionally, absorption and cryogenic systems have been used for the recovery of liquefied petroleum gas (Mokhatab et al., 2006; Speight, 2007).

The presence of propylene and butylenes in liquefied petroleum gas used as fuel gas is not critical. The vapor pressures of these olefins are slightly higher than those of propane and butane, and the flame speed is substantially higher, but this may be an advantage, since the flame speeds of propane and butane are slow. However, one issue that often limits the amount of olefins in liquefied petroleum gas is the propensity of the olefins to form soot, as well as the presence of mechanically entrained water (that may be further limited by specifications) (ASTM D1835). The presence of water in liquefied petroleum gas (or in natural gas) is undesirable, since it can produce hydrates that will cause, for example, line blockage due to the formation of hydrates under conditions where the water *dew point* is attained (ASTM D1142). If the amount of water is above acceptable levels, the addition of a small methanol will counteract any such effect. Another component of liquefied petroleum gas is propylene ($CH_3CH=CH_2$), which has a significantly lower octane number (ASTM D2623) than propane, so there is a limit to the amount of this component that can be tolerated in the mixture. Analysis by gas chromatography is possible (ASTM D5504, ASTM D6228, IP 405).

Liquefied petroleum gas and liquefied natural gas can share the facility of being stored and transported as a liquid and then vaporized and used as a gas. In order to achieve this, liquefied petroleum gas must be maintained at a moderate pressure, but at ambient temperature. The liquefied natural gas can be at ambient pressure, but must be maintained at a temperature of roughly –1 to 60°C (30 to 140°F). In fact, in some applications, it is actually economical and convenient to use liquefied petroleum gas in the liquid phase. In such cases, certain aspects of gas composition (or quality, such as the ratio of propane to butane, and the presence of traces of heavier hydrocarbons, water, and other extraneous materials) may be of lesser importance compared to the use of the gas in the vapor phase.

For normal (gaseous) use, the contaminants of liquefied petroleum gas are controlled at a level at which they do not corrode fittings and appliances or impede the flow of the gas. For example, hydrogen sulfide (H_2S) and carbonyl sulfide (COS) should be absent, although organic sulfur compounds, to the level required for adequate odorization, are permissible (ASTM D5305). In fact, *stenching* is a normal requirement in liquefied petroleum gas, and dimethyl sulfide (CH_3SCH_3) and ethyl mercaptan (C_2H_5SH) are commonly used at a concentration of up to 50 ppm. Natural gas is similarly treated, possibly with a wider range of volatile sulfur compounds.

The boiling point of liquefied petroleum gas ranges from –42 to 0°C (–44 to 32°F), depending on the amount of propane and butane in the mixture. The combustion of liquefied petroleum gas produces carbon dioxide (CO_2) and water vapor, but sufficient air must be available. Inadequate appliance ventilation or flue construction can result in the production of carbon monoxide, which can be toxic.

Liquefied petroleum gas is stored as a liquid under pressure. It is colorless, and its weight is approximately half that of an equivalent volume of water. The pressure inside a closed container in which liquefied petroleum gas is stored is equal to the vapor pressure of the liquid and gaseous liquefied petroleum gas in the container, and corresponds to its temperature.

The vapor of liquefied petroleum gas is denser than air: butane is about twice as heavy as air, and propane is about one and half times as heavy as air. Consequently, the vapor may flow along the ground and into drains, sinking to the lowest level of the surroundings, and be ignited at a considerable distance from the source of leakage.

In the liquid form, liquefied petroleum gas can cause severe cold burns to the skin, owing to its rapid vaporization, which can cool equipment so that it may be cold enough to cause cold burns.

Liquefied petroleum gas forms a flammable mixture with air in concentrations of between 2% and 10%. It can, therefore, be a fire and explosion hazard if stored or used incorrectly. Vapor/air mixtures arising from leakages may be ignited some distance from the point of escape, and the flame can travel back to the source of the leak. At very high concentrations when mixed with air, liquefied petroleum gas vapor is an anesthetic, and subsequently an asphyxiant by diluting the available oxygen.

The main disadvantage associated with the usage of liquefied petroleum gas has to do with the storage and safety. To store liquefied petroleum gas, sturdy tanks and cylinders are required. The gas has to be kept pressurized to accommodate it in 274 times less space, which is often emphasized by the number of cases LPG cylinders have exploded, and resulted in serious damages to lives and property.

In colder climates or conditions, there is a known problem related to starting, due to the low vapor pressure of propane. This is known to happen in conditions with sub-zero (0°C, 32°F) 32 degrees Fahrenheit temperatures.

Boilers and gas stoves that use liquefied petroleum gas need regular maintenance to ensure that they are running efficiently. Also, there has to be increased awareness yet to be created with regard to safe storage of liquefied petroleum gas cylinders in domestic properties.

Finally, since liquefied petroleum gas is highly inflammable, it is potentially very hazardous.

The major harmful emissions from engines running on liquefied petroleum gas are similar to those from other internal combustion engines: (1) carbon monoxide, (2) hydrocarbons, and (3) nitrogen oxides. Unlike diesel engines, there are practically no particulate emissions from liquefied petroleum gas engines.

Carbon monoxide is generated in the exhaust as the result of incomplete combustion of fuel. Carbon monoxide is a very toxic, colorless, and odorless gas. Liquefied petroleum gas emissions may contain considerable amounts of carbon monoxide. When engines operate in enclosed spaces, such as warehouses, buildings under construction, or tunnels, carbon monoxide can accumulate quickly, and reach concentrations which are dangerous for humans. It causes headaches, dizziness, lethargy, and death. Carbon monoxide is usually the major concern whenever liquefied petroleum gas engines are used indoors.

Hydrocarbons are also a product of incomplete combustion of fuel. Liquefied petroleum gas emissions, because of the composition of fuel, contain only short chain hydrocarbons. They are not likely to contain toxic components that are found in gasoline hydrocarbon emissions. Also, the environmental impact of liquefied petroleum gas hydrocarbon emissions (ozone reactivity contributing to smog) is much smaller than that of gasoline. However, hydrocarbon derivatives are responsible for the characteristic smell, which is often a nuisance when liquefied petroleum gas engines operate in an enclosed (inadequately ventilated) area.

Nitrogen oxides are generated from nitrogen and oxygen under the high temperature and pressure conditions in the engine cylinder. Nitrogen oxides consist mostly of nitric oxide (NO) and some nitrogen dioxide (NO_2). Nitrogen dioxide is a reactive gas, and is very toxic for humans. Nitrogen oxide emissions are also a serious environmental concern, because of their ozone reactivity and important role in smog and acid rain formation.

2.2 Natural Gas

Natural gas is found in petroleum reservoirs as free gas (*associated gas*), in solution with petroleum in the reservoir (*dissolved gas*), or in reservoirs that contain only gaseous constituents and no (or little) petroleum (*unassociated gas*) (Austin, 1984; Speight, 1999; Cranmore and Stanton, 2000). The hydrocarbon content varies from mixtures of methane and ethane with very few other constituents (*dry* gas) to mixtures containing all of the hydrocarbons, from methane to pentane, and even hexane (C_6H_{14}) and heptane (C_7H_{16}) (*wet* gas) (Speight, 1999). In both cases, some carbon dioxide (CO_2) and inert gases, including helium (He), are present, together with hydrogen sulfide (H_2S) and a small quantity of organic sulfur.

While the major constituent of natural gas is methane, there are components such as carbon dioxide (CO), hydrogen sulfide (H_2S), and mercaptans (thiols; R-SH), as well as trace amounts of sundry other emissions. The fact that methane has a foreseen and valuable end-use makes it a desirable product, but in several other situations, it is considered a pollutant, having been identified as one of several greenhouse gases (Graedel and Crutzen, 1989; Smith et al., 1994).

Carbon dioxide (ASTM D1137, ASTM D1945, ASTM D4984) in excess of 3 per cent is normally removed for reasons of corrosion prevention (ASTM D1838). Hydrogen sulfide (ASTM D2420, ASTM D2385, ASTM D2725, ASTM D4084, ASTM D4810, IP 103, IP 272) is also removed, and the odor of the gas must not be objectionable (ASTM D6273), so mercaptan content (ASTM D1988, ASTM D2385, IP 272) is important. A simple lead acetate test (ASTM D2420, ASTM D4084) is available for detecting the presence of hydrogen sulfide, and is an additional safeguard that hydrogen sulfide not be present (ASTM D1835). The odor of the gases must not be objectionable. Methyl mercaptan, if present, produces a transitory yellow stain on the lead acetate paper that fades completely in less than 5 min. Other sulfur compounds (ASTM D5504, ASTM D6228) present in liquefied petroleum gas do not interfere.

In the lead acetate test (ASTM D2420), the vaporized gas is passed over moist lead acetate paper under controlled conditions. Hydrogen sulfide reacts with lead acetate to form lead sulfide, resulting in a stain on the paper varying in color from yellow to black, depending on the amount of hydrogen sulfide present. Other pollutants can be determined by gas chromatography (ASTM D5504, ASTM D6228, IP 318).

The total sulfur content (ASTM D1072, ASTM D2784, ASTM D3031) is normally acceptably low, and frequently so low that it needs augmenting, by means of alkyl sulfides, mercaptans, or thiophenes, in order to maintain an acceptable safe level of odor.

The hydrocarbon dew point is reduced to such a level that retrograde condensation, i.e., condensation resulting from pressure drop, cannot occur under the worst conditions likely to be experienced in the gas transmission system. Similarly, the water dew point is reduced to a level sufficient to preclude formation of C_1 to C_4 hydrates in the system.

The natural gas, after appropriate treatment for acid gas reduction, odorization, and hydrocarbon and moisture dew point adjustment (ASTM D1142), would then be sold within prescribed limits of pressure, calorific value, and possibly Wobbe Index (cv/(sp. gr.)).

One issue that has arisen with respect to natural gas and the greenhouse effect is the fact that methane, the principle component of natural gas, is itself a potent greenhouse gas. Methane has an ability to trap heat almost 21 times more effectively than carbon dioxide.

In addition to natural gas, sources of methane emissions in the U.S. include the waste management and operations industry and the agricultural industry, as well as leaks and emissions from the oil and gas industry itself.

Smog and poor air quality is a pressing environmental problem, particularly for large metropolitan cities. Smog, the primary constituent of which is ground level ozone, is formed by a chemical reaction of carbon monoxide, nitrogen oxides, volatile organic compounds, and heat from sunlight. Pollutants contributing to smog come from a variety of sources, including vehicle emissions, smokestack emissions, paints, and solvents. Because the reaction to create smog requires heat, smog problems are the worst in the summertime.

The use of natural gas does not contribute significantly to smog formation, as it emits low levels of nitrogen oxides, and virtually no particulate matter. For this reason, it can be used to help combat smog formation in those areas where ground level air quality is poor.

The main sources of nitrogen oxides are electric utilities, motor vehicles, and industrial plants. Increased natural gas use in the electric generation sector, a shift to cleaner natural gas vehicles, or increased industrial natural gas use could all serve to combat smog production, especially in urban centers, where it is needed the most. Particularly in the summertime, when natural gas demand is lowest and smog problems are the greatest, industrial plants and electric generators could use natural gas to fuel their operations instead of other, more polluting fossil fuels. This would effectively reduce the emissions of smog-causing chemicals, and result in clearer, healthier air around urban centers.

Acid rain is another environmental problem that affects much of the eastern United States, damaging crops, forests, wildlife populations, and causing respiratory and other illnesses in humans. Acid rain is formed when sulfur dioxide and nitrogen oxides react with water vapor and other chemicals in the presence of sunlight to form various acidic compounds in the air (John and Gordon, 1987; Speight, 1996; Speight and Lee, 2000; Speight, 2005, 2007).

$$CO_2 + H_2O \rightarrow H_2CO_3 \text{ (carbonic acid)}$$
$$SO_2 + H_2O \rightarrow H_2SO_3 \text{ (sulfurous acid)}$$
$$2SO_2 + O_2 \rightarrow 2SO_3$$
$$SO_3 + H_2O \rightarrow H_2SO_4 \text{ (sulfuric acid)}$$
$$NO + H_2O \rightarrow HNO_2 \text{ (nitrous acid)}$$
$$2NO + O_2 \rightarrow NO_2$$
$$NO_2 + H_2O \rightarrow HNO_3 \text{ (nitric acid)}$$

The principle source of acid rain-causing pollutants, sulfur dioxide and nitrogen oxides, are coal-fired power plants. Since natural gas emits virtually no sulfur dioxide, and up to 80 percent less nitrogen oxides than the combustion of coal, increased use of natural gas could provide for fewer acid rain-causing emissions.

As a result, natural gas is becoming an increasingly important fuel in the generation of electricity. As well as providing an efficient, competitively priced fuel for the generation of electricity, the increased use of natural gas allows for the improvement in the emissions profile of the electric generation industry.

Natural gas-fired electric generation and natural gas-powered industrial applications offer a variety of environmental benefits and environmentally friendly uses, including: (1) fewer emissions, (2) reduced sludge, (3) reburning and cogeneration, and (4) combined cycle generation.

In terms of *fewer emissions,* combustion of natural gas, used in the generation of electricity, industrial boilers, and other applications, emits lower levels of nitrogen oxides, carbon dioxide, much lesser amounts (if not no) sulfur oxide emissions, and no mercury emissions. Natural gas can be used in place of, or in addition to, other fossil fuels, including coal, oil, or petroleum coke, which emit significantly higher levels of these pollutants.

Use of natural gas also reduced the amounts of *process sludge.* Coal-fired power plants and industrial boilers that use scrubbers to reduce sulfur dioxide emissions generate significant amounts (thousands of tons) of sludge. Combustion of natural gas emits extremely low levels of sulfur dioxide, eliminating the need for scrubbers, and reducing the amounts of sludge associated with power plants and industrial processes.

In addition, the reburning process involves injecting natural gas into coal or oil fired boilers. The addition of natural gas to the fuel mix can result in reduced nitrogen oxide and sulfur dioxide emissions. Along similar lines, the use of the concept of *cogeneration* for the production and use of both heat and electricity can increase the energy efficiency of electric generation systems and industrial boilers, which translates to the combustion of less fuel and the emission of fewer pollutants. Natural gas is often the preferred choice for new cogeneration applications.

Combined-cycle generation units generate electricity and capture normally wasted heat energy, using it to generate more electricity. In a manner similar to *cogeneration* applications, this increases energy efficiency, uses less fuel, and thus, produces fewer emissions. Natural gas-fired combined-cycle generation units can be up to 60 percent energy efficient, whereas coal and oil generation units may reach an energy efficiency on the order of 30 to 40 percent.

Essentially, electric generation and industrial applications that require energy, particularly for heating, use the combustion of fossil fuels for that energy. Because of its clean burning nature, the use of natural gas wherever possible, either in conjunction with other fossil fuels, or instead of them, can help to reduce the emission of harmful pollutants.

Nevertheless, in spite of the apparent virtues of using natural gas in place of coal and petroleum, there is still the need to be ever watchful. Relative to other fuels, natural gas is clean-burning, but combustion of natural gas is a significant contributor to global warming.

In a manner similar to petroleum, gas enters the environment due to both natural and anthropogenic processes. Among the major mechanisms of methane natural production in the biosphere, the decomposition of organic matter by methane-producing bacteria (such as *Methanococcus* and *Methanosarica*) generate internal energy by reducing carbon dioxide:

$$CO_2 + 4H_2 \rightarrow CH_4 + 2H_2O$$

These processes are typical for the silt deposits of lakes and marshes, and for marine sediments that are lacking in oxygen and rich in organic matter.

Microbial methane formation in the oceans is usually accompanied by sulfur reduction and the release of hydrogen sulfide. These take place inside the upper part of sediments from the seafloor surface to tens and even hundreds of meters deep. In regions with a cold and moderate climate at depths of over 1625 feet (500 meters), methane can accumulate in a form of crystal gas hydrates. In areas with a warmer climate, some methane from shallow formations is often released from the sediments into the water column, and then into the atmosphere.

Natural gas (specifically methane) can appear in the marine environment not only due to microbial and biochemical decomposition of the organic substance in bottom sediments. It can also occur as a result of the natural bottom seepage of combustible gases from shallow oil- and gas-bearing structures. This process can lead to intensive vertical flows of hydrocarbon gases from the bottom to the sea surface. Sometimes it is accompanied by gas hydrate decomposition, which can be exceedingly dangerous when not controlled.

Another component of natural gas, hydrogen sulfide (H_2S), is water soluble (in contrast to methane, that has little or no solubility in water under ambient conditions), and can cause hazardous pollution situations in both the atmosphere and the water environment. The proportion of hydrogen sulfide in natural gas and gas condensate, as previously mentioned, sometimes reaches more than 20% v/v. Pollution by hydrogen sulfide can lead to disturbances in the chemical composition of surface waters.

Other sources of atmospheric pollution also include flaring of natural gas on the offshore platforms and on land terminals. The behavior and distribution of the products of natural gas flaring in the atmosphere, their removal by precipitation, and the impact on the water environment can be severe: the same is true for gas emissions at different stages of natural gas production, transportation, and processing.

2.3 Refinery Gas

Most refinery processes produce gaseous by-products (*process gas*, known collectively as *refinery gas*), which is then used for other purposes or cleaned from the salable products. The production rate refinery gas generation is approximately 5% of the raw materials. In addition to the refinery furnace fuels, refinery gas can be used as catalysts congruency, alkylation of the materials used in the production of composite gasoline, alkylation gasoline, industrial iso-octane, iso-pentane high-octane gasoline components, or for the production of petrochemical products (Speight, 2007).

Still gas is a specific term applied to a mixture of low-boiling hydrocarbons, and is the lowest-boiling fraction isolated from a distillation (still) unit in the refinery. If the distillation unit is separating only low-boiling hydrocarbon fractions, the still gas will be almost entirely methane, with only trace amounts of ethane (CH_3CH_3) and ethylene ($CH_2{=}CH_2$). If the distillation unit is handling higher-boiling fractions, the still gas may also contain propane ($CH_3CH_2CH_3$), butane ($CH_3CH_2CH_2CH_3$), and their respective isomers. Fuel gas and still gas are terms that are often used interchangeably. However, fuel gas is intended to denote the product's future use as a fuel for boilers, furnaces, or heaters.

Refinery gas is a collective term used to denote the non-condensable gas that is obtained during distillation of crude oil or treatment (cracking, thermal decomposition) of petroleum (Gary and Handwerk; 1975; Austin, 1984; Speight, 1999; Robinson and Faulkner, 2000; Speight and Ozum, 2002). There are also components of the gaseous products that must be removed prior to release of the gases to the atmosphere, or prior to use of the gas in another part of the refinery, i.e., as a fuel gas or as a process feedstock.

Unlike products such as liquefied petroleum gas, there are no recognized product specifications for the chemical constituents that fall under the general description of *refinery gas*. Typically, these constituents are processed to recover the hydrocarbons and hydrogen, or burned as fuel gas within the plant. Some inorganic constituents are recovered as commercial products (liquid ammonia, sulfur or bisulfide from hydrogen sulfide), while others are routed to flares or oxidation devices (carbon monoxide).

The composition of refinery gas varies, depending on the composition of the crude it originates from and the processes from which the gas is produced. Typically, refinery gas consists mainly of hydrogen (H_2), methane (CH_4), ethane (C_2H_6), propane (C_3H_8), butane (C_4H_{10}), and olefins ($RCH=CHR^1$, where R and R^1 can be hydrogen or a methyl group), and may also include off-gases from petrochemical processes. Olefins such as ethylene (ethene, $CH_2=CH_2$, boiling point: –104°C, –155°F), propene (propylene, $CH_3CH=CH_2$, boiling point: –47°C, –53°F), butene (butene-1, $CH_3CH_2CH=CH_2$, boiling point: –5°C, 23°F) *iso*-butylene ($(CH_3)_2C=CH_2$, –6°C, 21°F), *cis*- and *trans*-butene-2 ($CH_3CH=CHCH_3$, boiling point: ca. 1°C, 30°F), and butadiene ($CH_2=CHCH=CH_2$, boiling point: –4°C, 24°F), as well as higher boiling olefins, are produced by various refining processes.

As stated above, refinery gas varies in composition and volume, depending on crude origin and on any additions to the crude made at the loading point. It is not uncommon to reinject light hydrocarbons, such as propane and butane, into the crude before dispatch by tanker or pipeline. This results in a higher vapor pressure of the crude, but it allows one to increase the quantity of light products obtained at the refinery. Since light ends in most petroleum markets command a premium, while in the oil field itself propane and butane may have to be reinjected or flared, the practice of *spiking* crude oil with liquefied petroleum gas is becoming fairly common.

In addition to the gases obtained by distillation of crude petroleum, more highly volatile products result from the subsequent processing of naphtha and middle distillate to produce gasoline, from desulfurization processes involving hydrogen treatment of naphtha, distillate, and residual fuel, and from the coking or similar thermal treatments of vacuum gas oils and residual fuels. The most common processing step in the production of gasoline is the catalytic reforming of hydrocarbon fractions in the heptane (C_7) to decane (C_{10}) range.

In a series of processes commercialized under Platforming, Powerforming, Catforming, and Ultraforming, paraffinic and naphthenic (cyclic non-aromatic) hydrocarbons, in the presence of hydrogen and a catalyst, are converted into aromatics, or isomerized to more highly branched hydrocarbons. Catalytic reforming processes thus not only result in the formation of a liquid product of higher octane number, but also produce substantial quantities of gases. The latter are rich in hydrogen, but also contain hydrocarbons from methane to butanes, with a preponderance of propane ($CH_3CH_2CH_3$), n-butane ($CH_3CH_2CH_2CH_3$), and isobutane [$(CH_3)_3CH$]. Their composition will vary in accordance with reforming severity and reformer feedstock. Since all catalytic reforming processes require substantial recycling of a hydrogen stream, it is normal to separate reformer gas into a propane ($CH_3CH_2CH_3$) and/or a butane [$CH_3CH_2CH_2CH_3/(CH_3)_3CH$] stream, which becomes part of the refinery liquefied petroleum gas production, and a lighter gas fraction, part of which is recycled. In view of the excess of hydrogen in the gas, all products of catalytic reforming are saturated, and there are usually no olefinic gases present in either gas stream.

A second group of refining operations that contributes to gas production is that of the catalytic cracking processes. These consist of fluid-bed catalytic cracking, Thermofor catalytic cracking, and other variants, in which heavy gas oils are converted into cracked gas, liquefied petroleum gas, catalytic naphtha, fuel oil, and coke by contacting the heavy hydrocarbon with the hot catalyst. Both catalytic and thermal cracking processes, the latter being now largely used for the production of chemical raw materials, result in the formation of unsaturated hydrocarbons, particularly ethylene ($CH_2=CH_2$), but also propylene (propene, $CH_3CH=CH_2$), isobutylene [isobutene, $(CH_3)_2C=CH_2$] and the n-butenes ($CH_3CH_2CH=CH_2$, and $CH_3CH=CHCH_3$), in addition to hydrogen (H_2), methane (CH_4), and smaller quantities of ethane (CH_3CH_3), propane ($CH_3CH_2CH_3$), and butane isomers [$CH_3CH_2CH_2CH_3$, $(CH_3)_3CH$]. Diolefins, such as butadiene ($CH_2=CHCH=CH_2$), are also present.

Additional gases are produced in refineries with coking or visbreaking facilities for the processing of their heaviest crude fractions. In the visbreaking process, fuel oil is passed through externally fired tubes and undergoes liquid phase cracking reactions, which result in the formation of lighter fuel oil components. Oil viscosity is thereby reduced, and some gases, mainly hydrogen, methane, and ethane, are formed. Substantial quantities of both gas and carbon are also formed in coking (both fluid coking and delayed coking), in addition to the middle distillate and naphtha. When coking a residual fuel oil or heavy gas oil, the feedstock is preheated and contacted with hot carbon (coke), which causes extensive cracking of the feedstock constituents of higher molecular weight to produce lower molecular weight products ranging from methane, via liquefied petroleum gas(es) and naphtha, to gas oil and heating oil. Products from coking processes tend to be unsaturated, and olefinic components predominate in the tail gases from coking processes.

A further source of refinery gas is hydrocracking, a catalytic high-pressure pyrolysis process in the presence of fresh and recycled hydrogen. The feedstock is again heavy gas oil or residual fuel oil, and the process is mainly directed at the production of additional middle distillates and gasoline. Since hydrogen is to be recycled, the gases produced in this process again have to be separated into lighter and heavier streams; any surplus recycle gas and the liquefied petroleum gas from the hydrocracking process are both saturated.

Both hydrocracker and catalytic reformer tail gases are commonly used in catalytic desulfurization processes. In the latter, feedstocks ranging from light to vacuum gas oils are passed at pressures of 500–1000 psi (3.5-7.0 x 10^3 kPa) with hydrogen over a hydrofining catalyst. This results mainly in the conversion of organic sulfur compounds to hydrogen sulfide (Bland and Davidson, 1967; Speight, 1999; Speight, 2000):

$$[S]_{feedstock} + H_2 \rightarrow H_2S + \text{hydrocarbons}$$

but also produces some light hydrocarbons by hydrocracking.

Petroleum refining also produces substantial amounts of carbon dioxide that, with hydrogen sulfide, corrode refining equipment, harm catalysts, pollute the atmosphere, and prevent the use of hydrocarbon components in petrochemical manufacture. When the amount of hydrogen sulfide is high, it may be removed from a gas stream and converted to sulfur or sulfuric acid. Some natural gases contain sufficient carbon dioxide to warrant recovery as dry ice (Bartoo, 1985; Kumar, 1987).

Thus refinery streams, while ostensibly being hydrocarbon in nature, may contain large amounts of acid gases, such as hydrogen sulfide and carbon dioxide. Most commercial plants employ hydrogenation to convert organic sulfur compounds into hydrogen sulfide. Hydrogenation is effected by means of recycled hydrogen-containing gases or external hydrogen over a nickel molybdate or cobalt molybdate catalyst.

In summary, refinery process gas, in addition to hydrocarbons, may contain other contaminants, such as carbon oxides (CO_x, where x = 1 and/or 2), sulfur oxides (So_x, where x = 2 and/or 3), as well as ammonia (NH_3), mercaptans (R-SH), and carbonyl sulfide (COS).

Residual refinery gases, usually in more than one stream, which allows a degree of quality control, are treated for hydrogen sulfide removal, and gas sales are usually on a thermal content (calorific value, heating value) basis, with some adjustment for variation in the calorific value and hydrocarbon type (Rawlinson and Ward, 1973; McKetta, 1993; Speight, 1993; Johansen, 1998; Cranmore and Stanton, 2000). For fuel uses, gas as specified above presents little difficulty used as supplied. Alternatively, a gas of constant Wobbe Index, say for gas turbine use, could readily be produced by the user. Part of the combustion air would be diverted into the gas stream by a Wobbe Index controller. This would be set to supply gas at the lowest Wobbe Index of the undiluted gas.

To determine the composition of refinery gas, the most convenient method of analysis involves the use of gas chromatography, which creates a compositional profile of the gas. Chromatography profiles are used by refinery officials as they decide what they want to do with refinery gas.

Release of refinery gas is tightly controlled, not only because of the potential salability of the constituents, but also because of concerns about the environment, which restrict the types of materials which can be released by companies that process chemicals and gases that can cause environmental harm.

3 Environmental Effects

Refineries (and gas processing plants) create and process a variety of gaseous product streams and waste streams that contain inorganic constituents, such as hydrogen sulfide, hydrogen, carbon monoxide, and ammonia. Exposure to these substances is limited by exposure standards and their inherent flammability hazard.

Potential releases of petroleum and refinery gases from petroleum facilities can be characterized as either controlled or unintentional releases. Controlled releases are planned releases from pressure relief valves and venting valves, for safety purposes or maintenance, and are considered part of routine operations, and occur under controlled conditions. Unintentional releases are typically characterized as unplanned releases due to spills or leaks from various equipment, valves, piping, flanges, etc., and may result from equipment failure, poor maintenance, a lack of proper operating practices, adverse weather conditions, or other unforeseen factors. Petroleum facilities are highly regulated, and regulatory requirements established under various jurisdictions, as well as voluntary non-regulatory measures implemented by the petroleum industry, are in place to manage potential releases.

More generally, these gaseous products or waste materials rarely, if ever, leave the plant without further processing to recover the valuable hydrocarbons or inorganic constituents. Some are used within the plant for process heating (burned). Because they are piped and tanked under elevated pressure and present an extreme explosion hazard should there be a release, control technologies to prevent exposure have been in place since the earliest days of the petroleum industry.

The refinery gases are made up of predominantly one to four carbon atom hydrocarbons and inorganic components such as ammonia, hydrogen, nitrogen, hydrogen sulfide, mercaptans, carbon monoxide, and carbon dioxide. Several refinery gases (process gases) also contain benzene and/or 1,3-butadiene. These gases can be used within the refinery as fuel gases to provide energy for other refinery processes. Alternatively, they can also undergo further refining to separate components and make them into commercially salable products.

All refinery gases contain one or more inorganic compounds in addition to hydrocarbons, which are (with the exception of asphyxiant gases, such as hydrogen and nitrogen) typically more toxic than the C_1 to C_6 to both fauna and flora. Unlike other petroleum product categories (such as gasoline, diesel fuel, and lubricating oil), the inorganic and hydrocarbon constituents of refinery gases can be evaluated for hazardous properties individually. The predominant categories of refinery gases are (1) inorganic gases, (2) hydrocarbon gases, and (3) asphyxiant gases.

The *inorganic gases* are sub-categorized as (1) ammonia, (2) carbon monoxide, (3) volatile mercaptans, and (4) hydrogen sulfide. The *hydrocarbon gases* include (1) C1 to C4 hydrocarbons, (2) C5 and C6 hydrocarbons, and (3) benzene. Finally, the *asphyxiant gases* include (1) carbon dioxide, (2) hydrogen, and (3) nitrogen.

Inorganic compounds (with the exception of asphyxiant gases, such as hydrogen and nitrogen) are typically more toxic than the majority of hydrocarbons in the C_1 to C_4 range. In contrast, hydrogen sulfide, ammonia, methyl mercaptan, and carbon monoxide are acutely toxic (API, 2009).

Air pollutants are responsible for a number of adverse environmental effects, such as photochemical smog, acid rain, death of forests, or reduced atmospheric visibility. Emissions of greenhouse gases are associated with the global warming. Certain air pollutants, including black carbon, not only contribute to global warming, but are also suspected of having immediate effect on regional climates.

Sulfur oxides, nitrogen oxides, hydrogen sulfide, and carbon dioxide are commonly produced during refining operations or during use of the refined products. For example, the most common toxic gases present in diesel exhaust include carbon monoxide, sulfur dioxide, nitric oxide, and nitrogen dioxide.

These gases are also classed as primary pollutants because they are emitted directly from the source and then react to produce secondary pollutant, such as acid rain (Speight, 1996). The emissions may include a number of biologically active substances that can pose a major health concern. These gases are classed as pollutants because (1) they may not be indigenous to the location or (2) they are ejected into the atmosphere in a greater-than natural concentration and are, in the current context, the product of human activity. Thus, they can have a detrimental effect on the environment, in part or *in toto*.

For these pollutants, the atmosphere has the ability to cleanse itself within hours, especially when the effects of the pollutant are minimized by the natural constituents of the atmosphere. For example, the atmosphere might be considered to be cleaning as a result of rain. However, removal of some pollutants from the atmosphere (e.g., sulfates and nitrates) by rainfall results in acid rain that can/will cause serious environmental damage to ecosystems within the water and land systems.

Several methods have been developed to estimate the exposure to such emissions. Most methods are based on either ambient air quality surveys or emission modeling. Exposure to other components of diesel emissions, such as polynuclear aromatic hydrocarbons, is also higher in occupational settings than it is in ambient environments. The principles of the techniques most often used in exhaust gas analysis include infrared (NDIR and FTIR), chemiluminescence, flame ionization detector (FID and fast FID), and paramagnetic methods.

Sulfur is removed from a number of refinery process off-gas streams (sour gas) in order to meet the sulfur oxide emissions limits of the Clean Air Act and to recover saleable elemental sulfur. Process off-gas streams, or sour gas, from the coker, catalytic cracking unit, hydrotreating units, and hydroprocessing units, can contain high concentrations of hydrogen sulfide mixed with light refinery fuel gases. Before elemental sulfur can be recovered, the fuel gases (primarily methane and ethane) need to be separated from the hydrogen sulfide. This is typically accomplished by dissolving the hydrogen sulfide in a chemical solvent. Solvents most commonly used are amines, such as diethanolamine (DEA). Dry adsorbents, such as molecular sieves, activated carbon, iron sponge, and zinc oxide, are also used. In the amine solvent processes, DEA solution or another amine solvent is pumped to an absorption tower, where the gases are contacted, and hydrogen sulfide is dissolved in the solution. The fuel gases are removed for use as fuel in process furnaces in other refinery operations. The amine-hydrogen sulfide solution is then heated and steam stripped to remove the hydrogen sulfide gas.

Since the Claus process by itself removes only about 90 percent of the hydrogen sulfide in the gas stream, the Beavon process (Speight, 1993, p. 268), SCOT (Shell Claus Off-gas Treating) process (Speight, 1993, p. 316; Hydrocarbon Processing, 2002), or the Wellman-Lord process (Speight, 1993, p. 327) are often used to further recover sulfur. The Claus process consists of partial combustion of the hydrogen sulfide-rich gas stream (with one-third the stoichiometric quantity of air), and then reacting the resulting sulfur dioxide and unburned hydrogen sulfide in the presence of a bauxite catalyst to produce elemental sulfur.

In the Beavon process, the hydrogen sulfide in the relatively low concentration gas stream from the Claus process can be almost completely removed by absorption in a quinone solution. The dissolved hydrogen sulfide is oxidized to form a mixture of elemental sulfur and hydroquinone. The solution is injected with air or oxygen to oxidize the hydroquinone back to quinone. The solution is then filtered or centrifuged to remove the sulfur, and the quinone is then reused. The Beaven process is also effective in removing small amounts of sulfur dioxide, carbonyl sulfide, and carbon disulfide that are not affected by the Claus process. These compounds are first converted to hydrogen sulfide at elevated temperatures in a cobalt molybdate catalyst prior to being fed to the Beaven unit. Air emissions from sulfur recovery units will consist of hydrogen sulfide, sulfur oxides, and nitrogen oxides in the process tail gas, as well as fugitive emissions and releases from vents.

In the SCOT process, the sulfur compounds in the Claus tail gas are converted to hydrogen sulfide by heating and passing it through a cobalt-molybdenum catalyst with the addition of a reducing gas. The gas is then cooled and contacted with a solution of di-isopropanolamine (DIPA) that removes all but trace amounts of hydrogen sulfide. The sulfide-rich di-isopropanolamine is sent to a stripper, where hydrogen sulfide gas is removed and sent to the Claus plant. The di-isopropanolamine is returned to the absorption column.

In the Wellman-Lord process, sodium sulfite is used to capture the sulfur dioxide. The sodium bisulfite thus formed is later heated to evolve sulfur dioxide and regenerate the sulfite scrubbing material. The sulfur dioxide-rich product stream can be compressed or liquefied and oxidized to sulfuric acid, or reduced to sulfur.

Most refinery process units and equipment are manifolded into a collection unit, called the blowdown system. Blowdown systems provide for the safe handling and disposal of liquid and gases that are either automatically vented from the process units through pressure relief valves, or that are manually drawn from units. Recirculated process streams and cooling water streams are often manually purged to prevent the continued build up of contaminants in the stream. Part or all of the contents of equipment can also be purged to the blowdown system prior to shutdown before normal or emergency shutdowns. Blowdown systems utilize a series of flash drums and condensers to separate the blowdown into its vapor and liquid components. The liquid is typically composed of mixtures of water and hydrocarbons containing sulfides, ammonia, and other contaminants, which are sent to the wastewater treatment plant. The gaseous component typically contains hydrocarbons, hydrogen sulfide, ammonia, mercaptans, solvents, and other constituents, and is either discharged directly to the atmosphere, or is combusted in a flare. The major air emissions from blowdown systems are hydrocarbons in the case of direct discharge to the atmosphere, and sulfur oxides when flared.

Particulate matter is a complex emission that is classified as either *suspended particulate matter*, *total suspended particulate matter*, or simply *particulate matter*. For human health purposes, the fraction of particulate matter that has been shown to contribute to respiratory diseases is termed PM_{10} (i.e., particulate matter with sizes less than 10 microns). From a control standpoint, particulate matter can be characterized as follows: (1) particle size distribution, and (2) particulate matter concentration in the emission (mg/m^3). On occasion, physical property descriptions may also be employed when there are specific control applications.

Traditionally, regulatory and compliance testing requires gravimetric determination of, for example, fuel mass emissions. Instruments utilizing collecting or in-situ measurement techniques are used for the analysis of various particle parameters for non-regulatory purposes.

In collecting methods, particulate matter emissions are determined through gravimetric analysis of the particulates collected on a sampling filter. Alternatively, the sample can be analyzed using thermal mass analysis (e.g., coulometric analysis). A number of other properties, for instance, surface area or biological activity, can be also analyzed. Collecting methods, and especially gravimetric analysis, are well established as the most common method of particulate matter emission determination.

The health effects of particulate matter (a complex mixture of solids and liquids) emissions are not yet well understood, but are recognized as major contributors to health problems. Biological activity of particulate matter may be related to particle sizes and/or particle composition. Furthermore, it has generally been concluded that exposure to particulate matter may cause increased morbidity and mortality, such as from cardiovascular disease. Long-term exposure to particulate emissions is also associated with small increase in the relative risk of lung cancer.

Some of the non-hydrocarbon fraction of the refinery gases would not be expected to biologically degrade, as these substances do not contain the chemical linkages necessary for microbial metabolism. For this reason, hydrogen, nitrogen, and carbon dioxide would not be susceptible to biodegradation. Furthermore, carbon dioxide is the final product in the biological mineralization of organic compounds. In contrast, ammonia can be readily oxidized to nitrite under aerobic conditions by autotrophic nitrifying bacteria (API, 2009).

Carbon monoxide has been reported to be microbially oxidized to carbon dioxide in pure cultures by a number of microbial species. It was also shown to be rapidly converted to carbon dioxide by indigenous soil microbial communities. Methanethiol can be both evolved and consumed in nature. It is produced by a variety of organisms through the decay of sulfur-containing organic matter under anoxic conditions. Methanethiol is known to undergo both aerobic and anaerobic biodegradation, but hydrogen sulfide does not usually biodegrade. The reduction of sulfate to hydrogen sulfide occurs in anoxic environments by anaerobic bacteria. Conversely, hydrogen sulfide can be oxidized to elemental sulfur and sulfate by a number of bacteria. Much of this cycling of sulfur occurs in sediments at the boundary layer between oxic and anoxic conditions.

Biodegradation of the hydrocarbon components in refinery gases may occur in soil and water. Gaseous hydrocarbons are widespread in nature, and numerous types of microbes have evolved that are capable of oxidizing these substances as their sole energy source. Although volatilization is the predominant behavior for these gases, sufficient aqueous solubility and bioavailability are exhibited by these compounds. The use of gaseous carbon sources for cell growth is common among autotrophic organisms (Chapter 1). Higher chain length hydrocarbons typical of naphtha streams also are known to inherently biodegrade in the environment (Chapter 11).

Several of the constituents in refinery gases were shown to be highly hazardous to aquatic organisms in laboratory toxicity tests where exposure concentrations can be maintained over time (API, 2009). Hydrogen sulfide has been shown to be the most toxic constituent to fish and invertebrates. Given the physical-chemical characteristics of the refinery gases and the confined production and use within refineries, potential exposures to aquatic organisms would be greatest from accidental catastrophic releases. Fugitive emissions in refineries would not be expected to impact aquatic systems. Based on a simple conceptual exposure model analysis, emissions of refinery gases to the atmosphere would not likely result in acutely toxic concentrations in adjacent water bodies, because such emitted gases will tend to remain in the atmosphere.

4 Analysis

One of the more critical aspects for the analysis of gaseous (or low boiling) hydrocarbons is the question of volumetric measurement (ASTM D1071) and sampling (ASTM D1145, ASTM D1247, ASTM D1265). However, sampling liquefied petroleum gas from a liquid storage system is complicated by existence of two phases (gas and liquid), and the composition of the supernatant vapor phase will, most probably, differ from the composition of the liquid phase. Furthermore, the compositions of both phases will vary as a sample is removed from one or both phases. An accurate check of composition can only be made if samples are taken during filling of the tank or from a fully charged tank.

In general, the sampling of gaseous constituents and of liquefied gases is the subject of a variety of sampling methods (ASTM D5503), such as the manual method (ASTM D1265, ASTM D4057), the floating piston cylinder method (ASTM D3700), and the automatic sampling method (ASTM D4177, ASTM D5287). Methods for the preparation of gaseous and liquid blends are also available (ASTM D4051, ASTM D4307), including the sampling and handling of fuels for volatility measurements (ASTM D5842).

Sampling methane (CH_4) and ethane (C_2H_6) hydrocarbons is usually achieved using stainless steel cylinders, either lined or unlined. However, other containers may also be employed, dependent upon particular situations. For example, glass cylinder containers or polyvinyl fluoride (PVF) sampling bags may also be used, but, obviously, cannot be subjected to pressures that are far in excess of ambient pressure. The preferred method for sampling propane (C_3H_8) and butane (C_4H_{10}) hydrocarbons is by the use of piston cylinders (ASTM D3700), although sampling these materials as gases is also acceptable in many cases. The sampling of propane and higher boiling hydrocarbons is dependent upon the vapor pressure of the sample (IP 410). Piston cylinders or pressurized steel cylinders are recommended for high-vapor pressure sampling where significant amounts of low boiling gases are present, while atmospheric sampling may be used for samples having a low-vapor pressure.

To monitor a process, measurement of gaseous emissions is typically performed over the time of a process cycle, or over the time of use of a particular product. Emission test cycles are repeatable sequences of operating conditions. Such timely analyses allow process monitoring as well as identification of any changes that can lead to potential leakage of the gas.

Hydrocarbon gases are amenable to analytical techniques, and there has been the tendency, and it remains, for the determination of both major constituents and trace constituents than is the case with the heavier hydrocarbons. The complexity of the mixtures that are evident as the boiling point of petroleum fractions and petroleum products increases makes identification of many of the individual constituents difficult, if not impossible. In addition, methods have been developed for the determination of physical characteristics, such as calorific value, specific gravity, and enthalpy, from the analyses of mixed hydrocarbon gases, but the accuracy does suffer when compared to the data produced by methods for the direct determination of these properties.

Bulk physical property tests, such as density and heating value, as well as some compositional tests, such as the Orsat analysis and the mercuric nitrate method for the determination of unsaturation, are still used. However, the choice of a particular test is dictated by (1) the requirements of the legislation, (2) the properties of the gas under study, and (3) the selection of a suitable suite of tests by the analyst to meet the various requirements. For example, judgment by the analyst is necessary as to whether or not a test that is applied to liquefied petroleum gas is suitable for process gas or natural gas, insofar as inference from the non-hydrocarbon constituents will be minimal.

The first and most important aspect of gaseous testing is the measurement of the volume of gas (ASTM D1071). In this test method, several techniques are described and may be employed for any purpose where it is necessary to know the quantity of gaseous fuel. In addition, the thermophysical properties of methane (ASTM D3956), ethane (ASTM D3984), propane (ASTM D4362), *n*-butane (ASTM D4650), and *iso*-butane (ASTM D4651) should be available for use and consultation (see also Stephenson and Malanowski, 1987).

4.1 Calorific Value (Heat of Combustion)

Satisfactory combustion of hydrocarbon gases depends upon the matching of burner and appliance design with certain gas characteristics. Various types of test methods are available for the direct determination of calorific value (ASTM D900, ASTM D1826, ASTM D3588, ASTM D4981).

The most important of these are the Wobbe Index (or Wobbe Number = calorific value/specific gravity), and the flame speed, usually expressed as a factor or an arbitrary scale on which that of hydrogen is 100. This factor can be calculated from the gas analysis. In fact, calorific value and specific gravity can be calculated from compositional analysis (ASTM D3588).

The Wobbe number gives a measure of the heat input to an appliance through a given aperture at a given gas pressure. Using this as a vertical co-ordinate and the flame speed factor as the horizontal co-ordinate, a combustion diagram can be constructed for an appliance, or a whole range of appliances, with the aid of appropriate test gases. This diagram shows the area within which variations in the WI and fsf of gases may occur for the given range of appliances without resulting in either incomplete combustion, flame lift, or the lighting back of pre-aerated flames. This method of prediction of combustion characteristics is not sufficiently accurate to eliminate entirely the need for the practical testing of new gases.

Another important combustion criterion is the Gas Modulus, $M = P/W$ where P is the gas pressure and W is the Wobbe Number of the gas. This must remain constant if a given degree of aeration is to be maintained in a pre-aerated burner using air at atmospheric pressure.

4.2 Composition

Liquefied petroleum gas, natural gas, and refinery gas are mixtures of products or naturally occurring materials, and fortunately are relative simple mixtures and do not suffer the complexities of the isomeric variations of the higher molecular weight hydrocarbons (Speight, 1999; Drews, 1998).

Thus, because of the lower molecular weight constituents of these gases and their volatility, gas chromatography has been the technique of choice for fixed gas and hydrocarbon speciation, and mass spectrometry is also a method of choice for compositional analysis of low molecular weight hydrocarbons (ASTM D2421, ASTM D2650). More recently, piggyback methods (such as gas chromatography/mass spectrometry and other double technique methods) have been developed for the identification of gaseous and low boiling liquid constituents of mixtures. The hydrocarbon composition is limited to set to the total amount of ethane, butane, or pentane, as well as ethylene and total dienes.

By limiting the amount of hydrocarbons that are lower boiling than the main component, the vapor pressure control is reinforced. Tests are available for vapor pressure 100°F (38°C) (ASTM D1267) and at 113°F (45°C) (IP 161). The limitation on the amount of higher boiling hydrocarbons supports the volatility clause. The vapor pressure and volatility specifications will often be met automatically if the hydrocarbon composition is correct.

The amount of ethylene is limited because it is necessary to restrict the amount of unsaturated components, so as to avoid the formation of deposits caused by the polymerization of the olefin(s). In addition, ethylene (boiling point: –104°C, –155°F) is more volatile than ethane (boiling point: –88°C, –127°F), and therefore, a product with a substantial proportion of ethylene will have a higher vapor pressure and volatility than one that is predominantly ethane. Butadiene is also undesirable, because it may also produce polymeric products that form deposits and cause blockage of lines.

Currently, the preferred method for the analysis of liquefied petroleum gas, and indeed for most petroleum-related gases, is gas chromatography (ASTM D2163, IP 264). This technique can be used for the identification and measurement of both main constituents and trace constituents. However, there may be some accuracy issues that arise in the measurement of the higher boiling constituents, due to relative volatility under the conditions in which the sample is held.

Capillary column gas chromatography is an even quicker and equally accurate alternative. Mass spectrometry (ASTM D1137) is also suitable for analysis of petroleum gases. Of the other spectroscopic techniques, infrared and ultra-violet absorption may be applied to petroleum gas analysis for some specialized applications. Gas chromatography has also largely supplanted chemical absorption methods of analysis, but again, these may have some limited specialized application.

Once the composition of a mixture has been determined, it is possible to calculate various properties, such as specific gravity, vapor pressure, calorific value, and dew point.

Simple evaporation tests, in conjunction with vapor pressure measurement, give a further guide to composition. In these tests, a liquefied petroleum gas sample is allowed to evaporate naturally from an open graduated vessel. Results are recorded on the basis of volume/temperature changes, such as the temperature recorded when 95% per cent has evaporated, or volume left at a particular temperature (ASTM D1837).

Since dew point can be calculated from composition, direct determination of dew point for a particular liquefied petroleum gas sample is a measure of composition. It is, of course, of more direct practical value, and if there are small quantities of higher molecular weight material present, it is preferable to use a direct measurement.

Specific gravity again can be calculated, but if it is necessary to measure it, several pieces of apparatus are available. For determining the density or specific gravity of liquefied petroleum gas in its liquid state, there are two methods, using a metal pressure pycnometer. A pressure hydrometer may be used (ASTM D1267) for the relative density that may also be calculated from composition al analysis (ASTM D2598). Various procedures, manual and recording, for specific gravity or density in the gaseous state are given in two methods (ASTM D1070, IP 59). Calculation of the density is also possible using any one of four models, depending upon the composition of the gas (ASTM D4784).

Gases such as ethane that are destined for use as petrochemical feedstocks must adhere to stringent composition controls that are dependent upon the process. For example, moisture content (ASTM D1142), oxygen content (ASTM D1945), carbon dioxide content (ASTM D1945), and sulfur content (ASTM D1072) must be monitored, as they all interfere with catalyst performance in petrochemical processes.

The hydrocarbon composition of natural gasoline (although not specifically a gas) for petrochemical use must undergo a compositional analysis (ASTM D2427) and a test for total sulfur (ASTM D1266, IP 107, IP 191).

The presence of any component substantially less volatile than the main constituents of the liquefied petroleum gas will give rise to unsatisfactory performance. It is difficult to set limits to the amount and nature of the 'residue' that will make a product unsatisfactory. Obviously, small amounts of oily material can block regulators and valves. In liquid vaporizer feed systems, even gasoline type material could cause difficulty. The residue (ASTM D2158) is a measure of the concentration of contaminants boiling above 37.8°C (100°F) that may be present in the gas. The residue as determined by the End Point Index (EPI) endeavors to give a measure of the heavier hydrocarbons, but the relationship between EPI, hydrocarbon range, and performance is not established.

Analytical methods are available in standard form for determining volatile sulfur content and certain specific corrosive sulfur compounds that are likely to be present. Volatile sulfur determination is made by a combustion procedure (ASTM D126, IP 107) that uses a modification of the standard wick-fed lamp. Many laboratories use rapid combustion techniques with an oxy-hydrogen flame in a Wickbold or Martin-Floret burner (ASTM D2784, IP 243).

This test method (ASTM D2784, IP 243) is valid for sulfur levels of >1 microgram/g of sulfur in liquefied petroleum gas, but the samples should not contain more than 100 microgram/g of chlorine. In the test, the sample is burned in an oxy-hydrogen burner or in a lamp in a closed system in a carbon dioxide-oxygen atmosphere. The latter is not recommended for trace quantities of sulfur, due to inordinately long combustion times needed. The sulfur oxides produced are absorbed and oxidized to sulfuric acid in a hydrogen peroxide solution. The sulfate ions are then determined, by either titrating with barium perchlorate solution using a thorinmethylene blue mixed indicator, or by precipitating as barium sulfate and measuring the turbidity of the precipitate with a photometer.

Trace hydrocarbons that may be regarded as contaminants may be determined by the gas chromatographic methods already discussed. Heavier hydrocarbons in small amounts may not be completely removed from the column. If accurate information is required about the nature and amount of heavy ends, then temperature programming or a concentration procedure may be used.

Analytical methods for determining traces of various other impurities, such as chlorides (ASTM D2384), are known to be in use. The presence of acetylenes in refinery gases, although unlikely, must still be considered. Acetylenes can be determined using a chemical test method while carbonyls are determined by the classical hydroxylamine hydrochloride reaction (ASTM D1089).

The determination of traces of higher boiling hydrocarbons and oily matter involves use of a method for residue that involves a preliminary weathering. The residue after weathering is dissolved in a solvent, and the solution is applied to a filter paper. The presence of residue is indicated by the formation of an oil stain. The procedure is taken further by combining the oil stain observation with other observed values to calculate an End Point Index (ASTM D2158). The method is not very precise, and work is proceeding in several laboratories to develop a better method for the determination of residue in the form of oily matter.

In liquefied petroleum gas where the composition is such that the hydrocarbon dew point is known to be low, a dew point method will detect the presence of traces of water (ASTM D1142). Propane, iso-butane (boiling point: −12°C, 11°F), and butane generally constitute this sample type, and are used for heating, motor fuels, and as chemical feedstocks (ASTM D2597, ASTM D2504, ASTM D2505). Important constituents of gas not accounted for in these analyses are moisture (water) and hydrogen sulfide, as well as other sulfur compounds (ASTM D1142, ASTM D1988, ASTM D5454, ASTM D4888, ASTM D5504, ASTM D6228).

Procedures for the determination of hydrogen, helium, oxygen, nitrogen, carbon monoxide, carbon dioxide, methane, ethene, ethane, propane, butanes, pentanes, and hexanes-plus in natural and reformed gases by packed column gas chromatography are available (ASTM D1945, ASTM D1946). These compositional analyses are used to calculate many other properties of gases, such as density, heating value, and compressibility. The first five components listed are determined using a molecular sieve column (argon carrier gas), while the remaining components are determined using polydimethylsiloxane partition or porous polymer columns. The hexanes-plus analysis is accomplished by backflushing the column after the elution of pentane, or by the use of a backflushed pre-column.

Olefins (ethylene, propylene, butylenes, and pentylenes) that occur in refinery (process) gas have specific characteristics and require specific testing protocols (ASTM D5234, ASTM D5273, ASTM D5274). Thus, hydrocarbon analysis of ethylene is accomplished by two methods (ASTM D2505, ASTM D6159), one of which (ASTM D6159) uses wide-bore (0.57-mm) capillary columns, including a Al_2O_3/KCl PLOT column. Another method (ASTM D2504) is recommended for determination of non-condensable gases, and yet another (ASTM D2505) is used for the determination of carbon dioxide.

Hydrocarbon impurities in propylene can be determined by gas chromatographic methods (ASTM D2712, ASTM D2163), and another test is available for determination of traces of methanol in propylene (ASTM Test Method D4864). A gas chromatographic method (ASTM D5303) is available for the determination of trace amounts of carbonyl sulfide in propylene using a flame photometric detector. Also, sulfur in petroleum gas can be determined by oxidative microcoulometry (ASTM D3246).

Commercial butylenes, high-purity butylenes, and butane-butylene mixtures are analyzed for hydrocarbon constituents (ASTM D4424), and hydrocarbon impurities in 1,7-butadiene can also be determined by gas chromatography (ASTM D2593). The presence of butadiene dimer and styrene are determined in butadiene by gas chromatography (ASTM D2426).

Carbonyls in C4 hydrocarbons are determined by a titrimetric technique (ASTM D4423) and by use of a peroxide method (ASTM D5799).

In general, gas chromatography will undoubtedly continue to be the method of choice for characterization of light hydrocarbon materials. New and improved detection devices and techniques, such as chemiluminescence, atomic emission, and mass spectroscopy, will enhance selectivity, detection limits, and analytical productivity. Laboratory automation through autosampling, computer control, and data handling will provide improved precision and productivity as well as simplified method operation.

Compositional analysis can be used to calculate calorific value, specific gravity, and compressibility factor (ASTM D3588).

Mercury in natural gas is also measured by atomic fluorescence spectroscopy (ASTM D6350) and by atomic absorption spectroscopy (ASTM D5954).

4.3 Density

The density of light hydrocarbons can be determined by several methods (ASTM D1070), including a hydrometer method (ASTM D1298) or by a pressure hydrometer method (ASTM D1657, IP 235). The specific gravity (relative density) (ASTM D1070, ASTM D1657) by itself has little significance compared to its use for higher molecular weight liquid petroleum products, and can only give an indication of quality characteristics when combined with values for volatility and vapor pressure. It is important for stock quantity calculations, and is used in connection with transport and storage.

4.4 Relative density

The *relative density* (*specific gravity*) is the ratio of the density (mass of a unit volume) of a substance to the density of a given reference material. Specific gravity usually means relative density (of a liquid) with respect to water.

$$\text{Relative density} = [\rho(\text{substance})]/[\rho(\text{reference})]$$

As it pertains to gases, particularly in relation to safety considerations at commercial and industrial facilities in the United States, the relative density of a gas is usually defined with respect to air, in which air is assigned a *vapor density* of one (unity). With this definition, the vapor density indicates whether a gas is denser (greater than one) or less dense (less than one) than air.

The vapor density has implications for container storage and personnel safety: if a container can release a dense gas, its vapor could sink and, if flammable, collect until it is at a concentration sufficient for ignition. Even if not flammable, it could collect in the lower floor or level of a confined space and displace air, possibly presenting a smothering hazard to individuals entering the lower part of that space.

Gases can be divided into two groups based upon their vapor density: (1) gases which are heavier than air, and (2) gases which are as light as, or lighter than, air. Gases that have a vapor density greater than one will be found in the bottom of storage containers, and will tend to migrate downhill and accumulate in low lying areas. Gases that have a vapor density that is the same or less than the vapor density of air will disperse readily into the surrounding environment. Additionally, chemicals that have the same vapor density as air (1.0) tend to disperse uniformly into the surrounding air when contained, and, when released into the open air, chemicals that are lighter than air will travel up and away from the ground.

Table 4.2 Relative density (specific gravity) of natural gas hydrocarbons relative to air.

Gas	Specific Gravity
Air	1.000
Methane - CH_4	0.5537
Ethane - C_2H_6	1.0378
Propane - C_3H_8	1.5219
Butane - C_4H_{10}	2.0061
Pentane	2.487
Hexane	2.973

Methane is the only hydrocarbon constituent of natural gas that is lighter than air (Table 4.2). The higher molecular weight hydrocarbons have a higher vapor density than air, and are likely, after a release, to accumulate in low-lying areas and represent a danger to the investigator (of the release).

4.5 Sulfur

The manufacturing processes for liquefied petroleum gas are designed so that the majority, if not all, of the sulfur compounds are removed. The total sulfur level is, therefore, considerably lower than for other petroleum fuels, and a maximum limit for sulfur content helps to define the product more completely. The sulfur compounds that are mainly responsible for corrosion are hydrogen sulfide, carbonyl sulfide, and, sometimes, elemental sulfur. Hydrogen sulfide and mercaptans have distinctive unpleasant odors.

A control of the total sulfur content, hydrogen sulfide, and mercaptans ensures that the product is not corrosive or nauseating. Stipulating a satisfactory copper strip test further ensures the control of the corrosion.

Total sulfur in gas can be determined by combustion (ASTM D1072), by the lamp method (ASTM D1266), or by hydrogenation (ASTM D3031, ASTM D4468). Trace total organic and bound nitrogen is determined (ASTM D4629). The current test method for heavy residues in liquefied petroleum gas (ASTM D2158) involves evaporation of a liquefied petroleum gas sample, measuring the volume of residue and observing the residue for oil stain on a piece of filter paper.

Corrosive sulfur compounds can be detected by their effect on copper and the form in which the general copper strip corrosion test (ASTM D1838) for petroleum products is applied to liquefied petroleum gas. Hydrogen sulfide can be detected by its action on moist lead acetate paper, and a procedure is also used as a measure of sulfur compounds. The method follows the principle of the standard Doctor test.

4.6 Volatility and Vapor Pressure

The vaporization and combustion characteristics of liquefied petroleum gas are defined for normal applications by volatility, vapor pressure, and, to a lesser extent, specific gravity.

Volatility is expressed in terms of the temperature at which 95% of the sample is evaporated, and presents a measure of the least volatile component present (ASTM D1837). Vapor pressure (IP 410) is, therefore, a measure of the most extreme low temperature conditions under which initial vaporization can take place. By setting limits to vapor pressure and volatility jointly, the specification serves to ensure essentially single component products for the butane and propane grades (ASTM D1267, ASTM D2598, IP 410). By combining vapor pressure/volatility limits with specific gravity for propane-butane mixtures, essentially two-component systems are ensured.

The residue (ASTM D1025, ASTM D2158, IP 317), i.e., non-volatile matter, is a measure of the concentration of contaminants boiling above 37.8°C (100°F) that may be present in the gas.

For natural gasoline, the primary criteria are volatility (vapor pressure) and knock performance. Determination of the vapor pressure (ASTM D323, ASTM D4953, ASTM D5190, ASTM D5191) and distillation profile (ASTM D216, IP 191) are essential. Knock performance is determined by rating in knock test engines by both the motor method (ASTM D2700, IP 236) and the research method (ASTM D2699, IP 237). The knock characteristics of liquefied petroleum gases can also be determined (IP 238).

Other considerations for natural gasoline are copper corrosion (ASTM D130, IP 154, IP 411) and specific gravity (ASTM D1298, IP 160), the latter determination being necessary for measurement and transportation.

5 References

API. 2009. Refinery Gases Category Analysis and Hazard Characterization. Submitted to the EPA by the American Petroleum Institute, Petroleum HPV Testing Group. HPV Consortium Registration # 1100997 United States Environmental Protection Agency, Washington, DC. June 10.

ASTM. 2000. *Annual Book of ASTM Standards*. American Society for Testing and Materials, West Conshohocken, PA.

Austin, G.T. 1984. *Shreve's Chemical Process Industries*. 5th Edition. McGraw-Hill, New York. Chapter 6.

Cranmore, R.E., and Stanton, E. 2000. In *Modern Petroleum technology. Volume 1: Upstream*. R.A. Dawe (Editor). John Wiley & Sons Inc., New York. Chapter 9.

Crawford, D.B., Durr, C.A., Finneran, J.A., and Turner, W. 1993. "Chemicals from Natural Gas." In *Chemical Processing Handbook*. J.J. McKetta (Editor). Marcel Dekker Inc., New York. p. 2.

Drews, A.W. 1998. In *Manual on Hydrocarbon Analysis*. 6th Edition. A.W. Drews (Editor). American Society for Testing and Materials, West Conshohocken, PA. Introduction.

Fogg, P.G.T., and Sangster, J.M. (Editors). 2003. *Chemicals in the Atmosphere – Solubility, Sources, and Reactivity*. John Wiley & Sons Inc., Hoboken, New Jersey.

Francis, W., and Peters, M.C. 1980. *Fuels and Fuel Technology: A Summarized Manual*. Pergamon Press, New York. Section C.

Guthrie, V.B. 1967. In *Petroleum Processing Handbook*. W.F. Bland and R.L. Davidson (Editors). McGraw-Hill, New York. Section 11.

Hoffman, H.L. 1983. In *Riegel's Handbook of Industrial Chemistry*. 8th Edition. J.A. Kent (editor). Van Nostrand Reinhold Company, New York. Chapter 14.

Hydrocarbon Processing. 2002. *Gas Processing Handbook: A Special Report*. 81(5): 118.

Institute of Petroleum. 2001. IP Standard Methods 2001. The Institute of Petroleum, London, United Kingdom.

Johansen, N.G. 1998. In *Manual on Hydrocarbon Analysis*. 6th Edition. A.W. Drews (Editor). American Society for Testing and Materials, West Conshohocken, PA. Chapter 1.

Johnson, R.W., and Gordon, G.E. (Editors). 1987. *The Chemistry of Acid Rain: Sources and Atmospheric Processes*. Symposium Series No. 349. American Chemical Society, Washington, DC.

McKetta, J.J. 1993. "Gas Processing." In *Chemical Processing Handbook*. J.J. McKetta (Editor). Marcel Dekker Inc., New York. p. 59.

Mokhatab, S., Poe, W.A., and Speight, J.G. 2006. *Handbook of Natural Gas Transmission and Processing*. Elsevier, Amsterdam, Netherlands.

Moustafa, S.M.A. 1996. In *The Engineering Handbook*. R.C. Dorf (Editor). CRC Press, Boca Raton, Florida. Chapter 64.

Patnaik, P. (Editor). 2004. *Dean's Analytical Chemistry Handbook*. 2nd Edition. McGraw-Hill, New York.

Rawlinson, D., and Ward, E.R. 1973. In *Criteria for Quality of Petroleum Products*. J.P. Allinson (Editor). John Wiley & Sons Inc., New York. Chapter 3.

Ritter, T.J. 2000. In *Modern Petroleum technology. Volume 2: Downstream.* A.G. Lucas (Editor). John Wiley & Sons Inc., New York. Chapter 23.

Robinson, J.D., and Faulkner, R.P. 2000. In *Modern Petroleum technology. Volume 2: Downstream.* A.G. Lucas (Editor). John Wiley & Sons Inc., New York. Chapter 1.

Speight, J.G. 1993. *Gas Processing: Environmental Aspects and Methods.* Butterworth-Heinemann, Oxford, England.

Speight, J.G. 1996. *Environmental Technology Handbook.* Taylor & Francis, Washington, DC.

Speight, J.G., and Lee, S. 2000. *Environmental Technology Handbook.* 2nd Edition. Taylor & Francis, New York.

Speight, J.G. 2001. *Handbook of Petroleum Analysis.* John Wiley & Sons Inc., New York.

Speight, J.G. and Ozum, B. 2002. *Petroleum Refining Processes.* Marcel Dekker Inc., New York.

Speight, J.G. 2005. *Environmental Analysis and Technology for the Refining Industry.* John Wiley & Sons Inc., Hoboken, New Jersey.

Speight, J.G. 2007. *The Chemistry and Technology of Petroleum.* 4th Edition. CRC Press, Taylor & Francis Group, Boca Raton, Florida.

Stephenson, R.M., and Malanowski, S. 1987. *Handbook of the Thermodynamics of Organic Compounds.* Elsevier, New York.

5

Composition and Properties of Liquid Products

1 Introduction

A wide variety of liquid products are produced from petroleum that varies from the high-volatile naphtha to the low-volatile lubricating oil (see also Chapter 4, Chapter 6) (Guthrie, 1967; Speight, 2007). The liquid products are often characterized by a variety of techniques that include measurement of physical properties and fractionation into group types.

The purpose of this chapter section is to present examples of the various liquid effluents and the methods of analysis that can be applied to the analysis of various liquid effluents from refinery processes. The example chosen are those at the lower end of the petroleum product boiling range (naphtha) (Chapter 3), and at the higher end of the petroleum product boiling range (residual fuel oil) (Chapter 3). The environmental behavior of the products that are intermediate in boiling range between these two can be predicted, with some degree of caution, since behavior is very much dependent upon composition. Nevertheless, similar tests can be applied, from which reasonable deduction about behavior can be made.

The impact of the release of liquid products on the environment can, in part, be predicted from knowledge of the properties of the released liquid. Each part ocular liquid product from petroleum has its own set of unique analytical characteristics (Speight, 2002, 2007). Since these are well documented, there is no need for repetition here. The decision is to include the properties of the lowest boiling liquid product (naphtha) and a high boiling liquid product (fuel oil). From the properties of each product (as determined by analysis), a reasonable estimate can be made of other liquid products, but the relationship may not be linear, and is subject to the type of crude oil and the distillation range of the product.

Nevertheless, reference is made to the various test methods dedicated to these products, and which can be applied to the products boiling in the intermediate range. In the light of the various tests available for composition, such tests will be deemed necessary depending on the environmental situation and the requirements of the legislation, as well as at the discretion of the analyst.

The properties of the liquid products from petroleum have been described in detail elsewhere (Chapter 3) (Mushrush and Speight, 1995; Mushrush and Speight, 1998a; Speight, 2002, 2005, 2007), and there is no need to reproduce that material here.

Consequently, for the purposes of this chapter, naphtha and fuel oil have been selected as the liquids that are representative of the extremes of the boiling range of petroleum liquids. Naphtha is the fraction that commences boiling at room temperature, while fuel boils at the higher extreme of atmospheric distillation, and may even (especially residual fuel oil) contain non-volatile constituents that are not found in the equivalent-boiling range lubricating oil.

2 Naphtha

Naphtha is a liquid petroleum product that boils from about 30°C (86°F) to approximately 200°C (392°F), although there are different grades of naphtha within this extensive boiling range that have different boiling ranges (Guthrie, 1967; Goodfellow, 1973; Weissermel and Arpe, 1978; Francis and Peters, 1980; Hoffman, 1983; Austin, 1984; Chenier, 1992; Hori, 2000; Speight, 2007). The term *petroleum solvent* is often used synonymously with *naphtha*.

On a chemical basis, naphtha is difficult to define precisely, because it can contain varying amounts of the constituents (paraffins, naphthenes, aromatics, and olefins) in different proportions, in addition to the potential isomers of the paraffins that exist in the naphtha boiling range (Table 5.1 and Table 5.2). Naphtha is also represented as having a similar boiling range and carbon number to gasoline, being a precursor to gasoline.

The so-called *petroleum ether* solvents are specific boiling range naphtha, as is *ligroin*. Thus, the term *petroleum solvent* describes a special liquid hydrocarbon fractions obtained from naphtha and used in industrial processes and formulations (Weissermel and Arpe, 1978). These fractions

Table 5.1 General summary of product types and distillation range.

Product	Lower Carbon Limit	Upper Carbon Limit	Lower Boiling Point C	Upper Boiling Point C	Lower Boiling Point F	Upper Boiling Point F
Refinery gas	C1	C4	–161	–1	–259	31
Liquefied petroleum gas	C3	C4	–42	–1	–44	31
Naphtha	C5	C16	36	287	97	549
Gasoline	C4	C12	–1	216	31	421
Kerosene/diesel fuel	C8	C18	126	258	302	575
Aviation turbine fuel	C8	C16	126	287	302	548
Fuel oil	C12	>C20	216	421	>343	>649
Lubricating oil	>C20		>343		>649	
Wax	C17	>C20	302	>343	575	>649
Asphalt	>C20		>343		>649	
Coke	>C50*		>1000*		>1832*	

*Carbon number and boiling point difficult to assess; inserted for illustrative purposes only.

are also referred to as *industrial naphtha*. Other solvents include *white spirit*, that is sub-divided into *industrial spirit* (distilling between 30°C and 200°C, 86°F to 392°F) and *white spirit* (light oil with a distillation range of 135°C to 200°C, 275°F to 392°F). The special value of naphtha as a solvent lies in its stability and purity.

Naphtha is produced by any one of several methods that include (1) fractionation of straight-run, cracked, and reforming distillates, or even fractionation of crude petroleum; (2) solvent extraction; (3) hydrogenation of cracked distillates; (4) polymerization of unsaturated compounds

Table 5.2 Increase in the number of isomers with carbon number.

Number of Carbon Atoms	Number of Isomers
1	1
2	1
3	1
4	2
5	3
6	5
7	9
8	18
9	35
10	75
15	4,347
20	366,319
25	36,797,588
30	4,111,846,763
40	62,491,178,805,831

(olefins); and (5) alkylation processes. In fact, the naphtha may be a combination of product streams from more than one of these processes.

The more common method of naphtha preparation is distillation. Depending on the design of the distillation unit, either one or two naphtha steams may be produced: (1) a single naphtha, with an end point of about 205°C (400°F) and similar to straight-run gasoline, or (2) this same fraction, divided into a light naphtha and a heavy naphtha. The end point of the light naphtha is varied to suit the subsequent subdivision of the naphtha into narrower boiling fractions, and may be of the order of 120°C (250°F).

Sulfur compounds are most commonly removed or converted to a harmless form by chemical treatment with lye, doctor solution, copper chloride, or similar treating agents (Speight, 2007). Hydrorefining processes (Speight, 2007) are also often used in place of chemical treatment.

When used as a solvent, naphtha is selected for low sulfur content, and the usual treatment processes remove only sulfur compounds. Naphtha with a small aromatic content has a slight odor, but the aromatics increase the solvent power of the naphtha, and there is no need to remove aromatics unless odor-free naphtha is specified.

The variety of applications emphasizes the versatility of naphtha. For example, naphtha is used in paint, printing ink, and polish manufacturing, and in the rubber and adhesive industries as well, as in the preparation of edible oils, perfumes, glues, and fats. Further uses are found in the dry-cleaning, leather, and fur industries, and also in the pesticide field. The characteristics that determine the suitability of naphtha for a particular use are volatility, solvent properties (dissolving power), purity, and odor (generally lack thereof).

In order to meet the demands of a variety of uses, certain basic naphtha grades are produced, which are identified by boiling range. The complete range of naphtha solvents may be divided, for convenience, into four general categories:

1. Special boiling point spirits, having overall distillation range within the limits 30 to 165°C (86 to 329°F),
2. Pure aromatic compounds, such as benzene, toluene, ethylbenzene, xylenes, or mixtures (BTEX) thereof;
3. White spirit, also known as mineral spirit and naphtha, usually boiling within 150 to 210°C (302 to 410°F); and
4. High boiling petroleum fractions, boiling within limits 160 to 325°C (320 to 617°F).

Since the end use dictates the required composition of naphtha, most grades are available in both high and low solvency categories, and the various text methods can have major significance in some applications, and lesser significance in others. Hence, the application and significance of tests must be considered in the light of the proposed end use.

Odor is particularly important, since unlike most other petroleum liquids, many of the manufactured products containing naphtha are used in confined spaces, in factory workshops, and in the home.

On the other hand, and at the other end of the spectrum of petroleum liquids, *fuel oil* is applied not only to distillate products (*distillate fuel oil*), but also to residual material that is distinguished from distillate type fuel oil by boiling range, and hence, is referred to as *residual fuel oil* (ASTM D396).

Thus, *residual fuel oil* is fuel oil that is manufactured from the distillation residuum, and the term includes all residual fuel oils, including fuel oil obtained by visbreaking as well as by blending residual products from other operations (Gruse and Stevens, 1960; Guthrie, 1967; Kite and Pegg, 1973; Weissermel and Arpe, 1978; Francis and Peters, 1980; Hoffman, 1983;

Austin, 1984; Chenier, 1992; Hoffman and McKetta, 1993; Hemighaus, 1998; Warne, 1998; Charlot and Claus, 2000; Heinrich and Duée, 2000; Speight, 2007). The various grades of heavy fuel oils are produced to meet rigid specifications in order to assure suitability for their intended purpose.

Detailed analysis of residual products, such as residual fuel oil, is more complex than the analysis of lower molecular weight liquid products. As with other products, there are a variety of physical property measurements that are required to determine if the residual fuel oil meets specification. But the range of molecular types present in petroleum products increases significantly with an increase in the molecular weight (i.e., an increase in the number of carbon atoms per molecule). Therefore, characterization measurements or studies cannot, and do not, focus on the identification of specific molecular structures. The focus tends to be on molecular classes (paraffins, naphthenes, aromatics, polycyclic compounds, and polar compounds).

Several tests that are usually applied to the lower molecular weight colorless (or light-colored) products are not applied to residual fuel oil. For example, test methods such as those designed for the determination of the aniline point (or mixed aniline point) (ASTM D611, IP 2) and the cloud point (ASTM D2500, ASTM D5771, ASTM D5772, ASTM D5773) can suffer from visibility effects due to the color of the fuel oil.

Because of the high standards set for naphtha (McCann, 1998), it is essential to employ the correct techniques when taking samples for testing (ASTM D270, ASTM D4057, IP 51). Mishandling or the slightest trace of contaminant can give rise to misleading results. Special care is necessary to ensure that containers are scrupulously clean and free from odor. Samples should be taken with the minimum of disturbance so as to avoid loss of volatile components; in the case of low-boiling naphtha, it may be necessary to chill the sample. While awaiting examination, samples should be kept in a cool, dark place, so as to ensure that they do not lose volatile constituents or discolor and develop odors due to oxidation.

The physical properties of naphtha depend on the hydrocarbon types present; in general, the aromatic hydrocarbons have the highest solvent power, and the straight-chain aliphatic compounds, the lowest. The solvent properties can be assessed by estimating the amount of the various hydrocarbon types present. This method provides an indication of the solvent power of the naphtha, on the basis that aromatic constituents and naphthenic constituents provide dissolving ability that paraffinic constituents do not. Another method for assessing the solvent properties of naphtha measures the performance of the fraction when used as a solvent under specified conditions, such as, for example, by the Kauri Butanol test method (ASTM D1133). Another method involves measurement of the surface tension, from which the solubility parameter is calculated, and then provides an indication of dissolving power and compatibility. Such calculations have been used to determine the yield of asphaltenes from petroleum by use of various solvents (Mitchell and Speight, 1973; Speight, 2007). A similar principal is applied to determine the amount of insoluble material in lubricating oil using *n*-pentane (ASTM D893, ASTM D4055).

Insoluble constituents in lubricating oil can cause wear that can lead to equipment failure. Pentane insoluble materials can include oil-insoluble materials and some oil-insoluble resinous matter, originating from oil, additive degradation, or both. Toluene insoluble constituents arise from external contamination, fuel carbon, and highly carbonized materials from degradation of fuel, oil, and additives, or engine wear and corrosion materials. A significant change in pentane or toluene insoluble constituents indicates a change in oil properties that could lead to machinery failure. The insoluble constituents measured can also assist in evaluating the performance characteristics of used oil, or in determining the cause of equipment failure.

Thus, one test (ASTM D893) covers the determination of pentane and toluene insoluble constituents in used lubricating oils, using pentane dilution and centrifugation as the method of separation. The other test (ASTM D4055) uses pentane dilution, followed by membrane filtration to remove insoluble constituents that have a size greater than 0.8 micron.

2.1 Composition

The number of potential hydrocarbon isomers in the naphtha boiling range (Table 5.1 and Table 5.2) renders complete speciation of individual hydrocarbons impossible for the naphtha distillation range, and methods are used that identify the hydrocarbon types as chemical groups, rather than as individual constituents.

The data from the density (specific gravity) test method (ASTM D1298, IP 160) provides a means of identification of a grade of naphtha, but is not a guarantee of composition, and can only be used to indicate evaluate product composition or quality when used in conjunction with the data from other test methods. Density data are used primarily to convert naphtha volume to a weight basis, a requirement in many of the industries concerned. For the necessary temperature corrections, and also for volume corrections, the appropriate sections of the petroleum measurement tables (ASTM D1250, IP 200) are used.

The first level of compositional information is group-type totals, as deduced by adsorption chromatography (ASTM D1319, IP 156) to give volume percent saturates, olefins, and aromatics in materials that boil below 315°C (600°F). In this test method, a small amount of sample is introduced into a glass adsorption column packed with activated silica gel, of which a small layer contains a mixture of fluorescent dyes. When the sample has been adsorbed on the gel, alcohol is added to desorb the sample down the column, and the hydrocarbon constituents are separated according to their affinities into three types (aromatics, olefins, and saturates). The fluorescent dyes also react selectively with the hydrocarbon types, and make the boundary zones visible under ultraviolet light. The volume percentage of each hydrocarbon type is calculated from the length of each zone in the column.

There are other test methods available. Benzene content and other aromatics may be estimated by spectrophotometric analysis (ASTM D1017), and also by gas-liquid chromatography (ASTM D2267, ASTM D2600, IP 262). However, two test methods based on the adsorption concept (ASTM D2007, ASTM D2549) are used for classifying oil samples of initial boiling point of at least 200°C (392°F) into the hydrocarbon types of polar compounds, aromatics, and saturates, and recovery of representative fractions of these types. Such methods are unsuitable for the majority of naphtha samples because of volatility constraints.

An indication of naphtha composition may also be obtained from the determination of aniline point (ASTM D1012, IP 2), freezing point (ASTM D852, ASTM D1015, ASTM D1493), cloud point (ASTM D2500), and the solidification point (ASTM D1493). Although refinery treatment should ensure no alkalinity and acidity (ASTM D847, ASTM D1093, ASTM D1613, ASTM D2896, IP 1) and no olefins present, the relevant tests using bromine number (ASTM D875, ASTM D1159, IP 130), bromine index (ASTM D2710), and flame ionization absorption (ASTM D1319, IP 156) are necessary to ensure low levels (at the maximum) of hydrogen sulfide (ASTM D853), as well as the sulfur compounds in general (ASTM D130, ASTM D849, ASTM D1266, ASTM D2324, ASTM D3120, ASTM D4045, ASTM D6212, IP 107, IP 154) and especially corrosive sulfur compounds, such as are determined by the doctor test method (ASTM D4952, IP 30).

Aromatic content is a key property of low-boiling distillates, such as naphtha and gasoline, because the aromatic constituents influence a variety of properties, including boiling range (ASTM D86, IP 123), viscosity (ASTM D88, ASTM D445, ASTM D2161, IP 71), stability (ASTM D525, IP 40), and compatibility (ASTM D1133) with a variety of solutes. Existing methods use physical measurements and need suitable standards. Tests such as aniline point (ASTM D611) and Kauri-butanol number (ASTM D1133) are of a somewhat empirical nature, and can serve a useful function as control tests. Naphtha composition, however, is monitored mainly by gas chromatography, and although most of the methods may have been developed for gasoline (ASTM D2427, ASTM D6296), the applicability of the methods to naphtha is sound.

A multidimensional gas chromatographic method (ASTM D5443) provides for the determination of paraffins, naphthenes, and aromatics by carbon number in low olefinic hydrocarbon streams having final boiling points lower than 200°C (392°F). In the method, the sample is injected into a gas chromatographic system that contains a series of columns and switching values. First, a polar column retains polar aromatic compounds, bi-naphthenes, and high boiling paraffins and naphthenes. The eluent from this column goes through a platinum column that hydrogenates olefins, and then to a molecular sieve column that performs a carbon number separation based on the molecular structure, that is, naphthenes and paraffins. The fraction remaining on the polar column is further divided into three separate fractions that are then separated on a non-polar column by boiling point. A flame ionization detector detects eluting compounds.

In another method (ASTM D4420) for the determination of the amount of aromatic constituents, a two column chromatographic system connected to a dual filament thermal conductivity detector (or two single filament detectors) is used. The sample is injected into the column containing a polar liquid phase. The non-aromatics are directed to the reference side of the detector, and vented to the atmosphere as they elute. The column is backflushed immediately before the elution of benzene, and the aromatic portion is directed into the second column containing a non-polar liquid phase. The aromatic components elute in the order of their boiling points, and are detected on the analytical side of the detector. Quantitation is achieved by utilizing peak factors obtained from the analysis of a sample having a known aromatic content.

Other methods for the determination of aromatics in naphtha include a method (ASTM D5580) using a flame ionization detector, and methods in which a combination of gas chromatography and Fourier transform infrared spectroscopy (GC-FTIR) (ASTM D5986) and gas chromatography and mass spectrometry (GC-MS) (ASTM D5769) are used.

Hydrocarbon composition is also determined by mass spectrometry, a technique that has seen wide use for hydrocarbon-type analysis of naphtha and gasoline (ASTM D2789), as well as to the identification of hydrocarbon constituents in higher boiling naphtha fractions (ASTM D2425).

One method (ASTM D6379, IP 436) is used to determine the mono-aromatic and di-aromatic hydrocarbon contents in distillates boiling in the range from 50 to 300°C (122 to 572°F). In the method, the sample is diluted with an equal volume of hydrocarbon, such as heptane, and a fixed volume of this solution is injected into a high performance liquid chromatograph fitted with a polar column, where separation of the aromatic hydrocarbons from the non-aromatic hydrocarbons occurs. The separation of the aromatic constituents appears as distinct bands according to ring structure, and a refractive index detector is used to identify the components as they elute from the column. The peak areas of the aromatic constituents are compared with those obtained from previously run calibration standards to calculate the %w/w mono-aromatic hydrocarbon constituents and di-aromatic hydrocarbon constituents in the sample.

Compounds containing sulfur, nitrogen, and oxygen could possibly interfere with the performance of the test. Mono-alkenes do not interfere, but conjugated di- and poly-alkenes, if present, may interfere with the test performance.

Another method (ASTM D2425) provides more compositional detail (in terms of molecular species) than chromatographic analysis, and the hydrocarbon types are classified in terms of a Z-series, in which Z (in the empirical formula C_nH_{2n+z}) is a measure of the hydrogen deficiency of the compound. This method requires that the sample be separated into saturate and aromatic fractions before mass spectrometric analysis (ASTM D2549), and the separation is applicable to some fractions, but not others. For example, the method is applicable to high-boiling naphtha, but not to the low-boiling naphtha, since it is impossible to evaporate the solvent used in the separation without also losing the lower boiling constituents of the naphtha under investigation.

The percentage of aromatic hydrogen atoms and aromatic carbon atoms can be determined by high-resolution nuclear magnetic resonance spectroscopy (ASTM D5292) that gives the mol percent of aromatic hydrogen or carbon atoms. Proton (hydrogen) magnetic resonance spectra are obtained on sample solutions in either chloroform or carbon tetrachloride, using a continuous wave or pulse Fourier transform high-resolution magnetic resonance spectrometer. Carbon magnetic resonance spectra are obtained on the sample solution in chloroform-*d* using a pulse Fourier transform high-resolution magnetic resonance.

The data obtained by this method (ASTM D5292) can be used to evaluate changes in aromatic contents in naphtha, as well as kerosene, gas oil, mineral oil, and lubricating oil. However, results from this test are not equivalent to mass- or volume-percent aromatics determined by the chromatographic methods, since the chromatographic methods determine the percent by weight or percent by volume of molecules that have one or more aromatic rings, and alkyl substituents on the rings will contribute to the percentage of aromatics determined by chromatographic techniques.

Low-resolution nuclear magnetic resonance spectroscopy can also be used to determine percent by weight hydrogen in jet fuel (ASTM D3701), and in light distillate, middles distillate, and gas oil (ASTM D4808). As noted above, chromatographic methods are not applicable to naphtha, where losses can occur by evaporation.

The nature of the uses found for naphtha demands compatibility with the many other materials employed in formulation, with waxes, pigments, resins, etc.; thus, the solvent properties of a given fraction must be carefully measured and controlled. For most purposes, volatility is important, and, because of the wide use of naphtha in industrial and recovery plants, information on some other fundamental characteristics is required for plant design.

Although the focus of many tests is analysis of the hydrocarbon constituents of naphtha and other petroleum fractions, heteroatoms compounds that contain sulfur and nitrogen atoms cannot be ignored, and methods for their determination are available. The combination of gas chromatography with element selective detection gives information about the distribution of the element. In addition, many individual heteroatomic compounds can be determined.

Nitrogen compounds in middle distillates can be selectively detected by chemiluminescence. Individual nitrogen compounds can be detected down to 100 ppb nitrogen. Gas chromatography with either sulfur chemiluminescence detection or atomic emission detection has been used for sulfur selective detection.

Estimates of the purity of these products were determined in laboratories using a variety of procedures, such as freezing point, flame ionization absorbance, ultraviolet absorbance, gas chromatography, and capillary gas chromatography (ASTM D850, ASTM D852, ASTM D853, ASTM D848, ASTM D849, ASTM D1015, ASTM D1016, ASTM D1078, ASTM D1319, ASTM D2008, ASTM D22368, ASTM D2306, ASTM D2360, ASTM D5917, IP 156).

Gas chromatography (GC) has become a primary technique for determining hydrocarbon impurities in individual aromatic hydrocarbons and the composition of mixed aromatic hydrocarbons. Although a measure of purity by gas chromatography is often sufficient, gas chromatography is not capable of measuring absolute purity; not all possible impurities will pass through the gas chromatography column, and not all those that do will be measured by the detector. Despite some shortcomings, gas chromatography is a standard, widely used technique, and is the basis of many current test methods for aromatic hydrocarbons (ASTM D2306 ASTM D2360, ASTM D3054, ASTM D3750, ASTM D3797, ASTM D3798, ASTM D4492, ASTM D4534, ASTM D4735, ASTM D5060, ASTM D5135, ASTM D5713, ASTM D5917, ASTM D6144).

When classes of hydrocarbons, such as olefins, need to be measured, techniques, such as bromine index, are used (ASTM D1492, ASTM D5776).

Impurities other than hydrocarbons are of concern in the petroleum industry. For example, many catalytic processes are sensitive to sulfur contaminants. Consequently, there is also a series of methods to determine trace concentrations of sulfur-containing compounds (ASTM D1685, ASTM D3961, ASTM D4045, ASTM D4735).

Chloride-containing impurities are determined by various test methods (ASTM D5194, ASTM D5808, ASTM D6069) that have sensitivity to 1 mg/kg, reflecting the needs of industry to determine very low levels of these contaminants.

Water is a contaminant in naphtha, and should be measured using the Karl Fischer method (ASTM E-203, ASTM D1364, ASTM D1744, ASTM

D4377, ASTM D4928, ASTM D6304), by distillation (ASTM D4006), or by centrifuging (ASTM D96) and excluding by relevant drying methods.

Tests should also be carried out for sediment if the naphtha has been subjected to events (such as oxidation) that could lead to sediment formation and instability of the naphtha and the resulting products. Test methods are available for the determination of sediment by extraction (ASTM D473, IP 285) or by membrane filtration (ASTM D4807, IP 286) and the determination of simultaneously sediment with water by centrifugation (ASTM D96, ASTM D1796, ASTM D2709, ASTM D4007, IP 373, IP 374).

2.2 Density (Specific Gravity)

Density (the mass of liquid per unit volume at 15°C), and the related terms *specific gravity* (the ratio of the mass of a given volume of liquid at 15°C to the mass of an equal volume of pure water at the same temperature) and *relative density* (same as *specific gravity*), are important properties of petroleum products, as it is a part of product sales specifications, although it only plays a minor role in studies of product composition. Usually, a hydrometer, pycnometer, or digital density meter is used for the determination in all these standards.

The determination of density (specific gravity) (ASTM D287, ASTM D891, ASTM D941, ASTM D1217, ASTM D1298, ASTM D1555, ASTM D1657, ASTM D2935, ASTM D4052, ASTM D5002, IP 160, IP 235, IP 365) provides a check on the uniformity of the naphtha, and it permits calculation of the weight per gallon. The temperature at which the determination is carried out and for which the calculations are to be made should also be known (ASTM D1086). However, the methods are subject to vapor pressure constraints, and are used with appropriate precautions to prevent vapor loss during sample handling and density measurement. In addition, some test methods should not be applied if the samples are so dark in color that the absence of air bubbles in the sample cell cannot be established with certainty. The presence of such bubbles can have serious consequences for the reliability of the test data.

2.3 Evaporation Rate

Environmentally, the evaporation rate is an important property of naphtha, and although there is a significant relation between distillation range and evaporation rate, the relationship is not straightforward.

A simple procedure for determining the evaporation rate involves use of at least a pair of weighed shallow containers, each containing a weighed amount of naphtha. The cover-free containers are placed in a temperature-controlled and humidity-controlled draft-free area. The containers are reweighed at intervals, until the samples have completely evaporated or left a residue that does not evaporate further (ASTM D381, ASTM D1353, IP 131).

The evaporation rate can be derived either (1) by a plot of time versus weight, using a solvent having a known evaporation rate for comparison, or (2) from the distillation profile (ASTM D86, IP 123). Although the results obtained on the naphtha provide a useful guide, it is, wherever possible, better to carry out a performance test on the final product when assessing environmental effects.

2.4 Flash Point

The flash point is the lowest temperature at atmospheric pressure (760 mm Hg, 101.3 kPa) at which application of a test flame will cause the vapor of a sample to ignite under specified test conditions. The sample is deemed to have reached the flash point when a large flame appears and instantaneously propagates itself over the surface of the sample. The flash point data is used in shipping and safety regulations to define *flammable* and *combustible* materials. Flash point data can also indicate the possible presence of highly volatile and flammable constituents in a relatively non-volatile or non-flammable material.

Of the available test methods, the most common method of determining the flash point confines the vapor (closed cup) until the instant the flame is applied (ASTM D56, ASTM D93, ASTM D3828, 6450, IP 34, IP 94, IP 303). An alternate method that does not confine the vapor (open cup method (ASTM D92, ASTM D1310, IP 36) gives slightly higher values of the flash point.

Another test (ASTM E-659) is available that can be used as a complement to the flash point test, and involves determination of the auto-ignition temperature. However, the flash point should not be confused with auto-ignition temperature, which measures spontaneous combustion with no external source of ignition.

2.5 Odor and Color

In general, the paraffinic hydrocarbons possess the mildest odor, and the aromatics, the strongest, the odor level (ASTM D268, ASTM D1296, IP 89) being related to chemical character and volatility of the constituents. Odors due to the presence of sulfur compounds or unsaturated constituents are excluded by specification. Apart from certain high boiling aromatic fractions that are usually excluded by volatility from the majority of the naphtha fractions, which may be pale yellow in color, naphtha is usually colorless (water white).

Measurement of color (ASTM D156, ASTM D848, ASTM D1209, ASTM D1555, ASTM D5386, IP 17) provides a rapid method of checking the degree of freedom from contamination. Observation of the test for residue on evaporation (ASTM D381, ASTM D1353, IP 131) provides a further guard against adventitious contamination.

2.6 Volatility

One of the most important physical parameters that also relates to the evaporation rate is the boiling range distribution (ASTM D86, ASTM D1078, ASTM D2887, ASTM D2892, IP 123). The significance of the distillation test is the indication of volatility that dictates the evaporation rate, which is an important property for naphtha used in coatings and similar applications where the premise is that the naphtha evaporates over time, leaving the coating applied to the surface. In the basic test method (ASTM D86, IP 123), a 100-ml sample is distilled (manually or automatically) under prescribed conditions. Temperatures and volumes of condensate are recorded at regular intervals, from which the boiling profile is derived.

The determination of the boiling range distribution of distillates, such as naphtha and gasoline, by gas chromatography (ASTM D3710) not only helps identify the constituents, but also facilitates on-line controls at the refinery. This test method is designed to measure the entire boiling range of naphtha with either high or low Reid vapor pressure (ASTM D323, IP 69). In the method, the sample is injected into a gas chromatographic column that separates hydrocarbons in boiling point order. The column temperature is raised at a reproducible rate, and the area under the chromatogram is recorded throughout the run. Calibration is performed using a known mixture of hydrocarbons covering the expected boiling of the sample.

Another method is described as a method for determining the carbon number distribution (ASTM D2887, IP 321), and the data derived by this test method are essentially equivalent to those obtained by the test method for true boiling point distillation (ASTM D2892). The sample is introduced into a gas chromatographic column that separates hydrocarbons in boiling point order. The column temperature is raised at a reproducible rate, and the area under the chromatogram is recorded throughout the run. Boiling temperatures are assigned to the time axis from a calibration curve, obtained under the same conditions by running a known mixture of hydrocarbons covering the boiling range expected in the sample. From these data, the boiling range distribution may be obtained. However, this test method is limited to samples having a boiling range greater than 55°C (100°F) and having a vapor pressure (ASTM D323, ASTM D4953, ASTM D5190, ASTM D5191, ASTM D5482, ASTM D6377, ASTM D6378, IP 69, IP 394) sufficiently low to permit sampling at ambient temperature.

In studying the environmental effects of naphtha, it is necessary to relate volatility to the fire hazard associated with its use, storage, and transport, and also with the handling of the products arising from the process. This is normally based on the characterization of the solvent by flash point limits (ASTM D56, ASTM D93, IP 34, IP 170).

2.7 Environmental Impact

The main uses of petroleum naphtha fall into the general areas of (1) precursor to gasoline and other liquid fuels, (2) solvents (diluents) for paints, (3) dry-cleaning solvents, (4) solvents for cutback asphalts, (5) solvents in rubber industry, and (6) solvents for industrial extraction processes. Turpentine, the older and more conventional solvent for paints, has now been almost completely replaced by the cheaper and more abundant petroleum naphtha.

The term aliphatic naphtha refers to naphtha containing less than 0.1% benzene and with carbon numbers from C_5 through C_{16}. Aromatic naphtha has carbon numbers from C_6 through C_{16}, and contains significant quantities of aromatic hydrocarbons, such as benzene (>0.1%), toluene, and xylene. The final gasoline product as a transport fuel is a carefully blended mixture having a predetermined octane value. Thus, gasoline is a complex mixture of hydrocarbons that boils below 200°C (390°F). The hydrocarbon constituents in this boiling range are those that have four to twelve carbon atoms in their molecular structure.

Potential releases of naphtha from refineries and upgraders can be characterized as either controlled or unintentional releases. Controlled releases are planned releases from pressure relief valves, venting valves, and drain systems that occur for safety purposes or maintenance, are considered part of routine operations, and occur under controlled conditions. Unintentional releases are typically characterized as unplanned releases due to spills or leaks from various equipment, valves, piping, and flanges which result from equipment failure, poor maintenance, a lack of proper operating practices, adverse weather conditions, or other unforeseen factors. Refinery and upgrader operations are highly regulated, and regulatory requirements established under various jurisdictions, as well as voluntary non-regulatory measures implemented by the petroleum industry, are in place to manage these releases.

If spilled or discharged in the environment, naphtha represents a threat of the toxicity of the constituents to land and/or to aquatic organisms. A significant spill may cause long-term adverse effects in the aquatic environment. The constituents of naphtha predominantly fall in the C_5–C_{16} carbon range: alkanes, cycloalkanes, aromatics, and, if they are subject to a cracking process, alkenes as well. Naphtha may also contain a preponderance of aromatic constituents (up to 65%), with others containing up to 40% alkenes, while all of the others are aliphatic in composition, up to 100%.

Water solubility ranges from very low, for the longest-chain alkanes, to high, for the simplest mono-aromatic constituents. Generally, the aromatic compounds are more soluble than the same-sized alkanes, iso-alkanes, and cycloalkanes. This indicates that the components likely to remain in water are the one- and two-ring aromatics (C_6–C_{12}). The C_9–C_{16} alkanes, iso-alkanes, and one- and two-ring cycloalkanes are likely to be attracted to sediments, based on their low water solubilities and moderate to high octanol-water partition coefficient (log K_{ow}) and organic carbon-water partition coefficient (log K_{oc}) values.

Naphtha (especially the low-boiling naphtha) contains volatile organic compounds (VOCs) that are rapidly degraded in air, water, and soil. Considerable measures must be taken to prevent release of naphtha (constituents) to the atmosphere, and to minimize any exposure to the environment from activities in which naphtha is manufactured and used.

Constituents of naphtha can be carcinogenic, and frequently products sold as naphtha contain some impurities, which may also have harmful properties of their own. The method of manufacture means that there is a range of distinct chemicals in naphtha, which makes rigorous comparisons and identification of specific carcinogens difficult, and is further complicated by exposure to a significant range of other known and potential carcinogens.

3 Fuel Oil

Fuel oil is classified in several ways, but generally may be divided into two main types: (1) *distillate fuel oil* and (2) *residual fuel oil*. These classifications are still employed, but, of late, the terms distillate fuel oil and residual fuel oil have lost some of their significance, because fuel oils are now made for specific uses, and may be distillates, residuals, or mixtures of the two. The terms domestic fuel oils, diesel fuel oils, and heavy fuel oils are more indicative of the uses of fuel oils. More often than not, fuel oil is prepared by using a visbreaker unity (Figure 3–13) to perform mild thermal cracking on a residuum or on a high boiling distillate so that the product meets specifications.

Distillate fuel oil is vaporized and condensed during a distillation process; it has a definite boiling range and does not contain high-boiling oils or asphaltic components. A fuel oil that contains any amount of the residue from crude distillation or thermal cracking is a residual fuel oil. *Domestic fuel oil* is fuel oil that is used primarily in the home, and includes kerosene, stove oil, and furnace fuel oil. *Diesel fuel oil* is also a distillate fuel oil, but residual oil has been successfully used to power marine diesel engines, and mixtures of distillates and residuals have been used on locomotive diesels. *Furnace fuel oil* is similar to diesel fuel, but the proportion of cracked gas oil in diesel fuel is usually less, since the high aromatic content of the cracked gas oil reduces the cetane number of the diesel fuel.

Stove oil is a straight-run (distilled) fraction from crude oil, whereas other fuel oils are usually blends of two or more fractions. The straight-run fractions available for blending into fuel oils are heavy naphtha, light and heavy gas oils, and residua. Cracked fractions, such as light and heavy gas oils from catalytic cracking, cracking coal tar, and fractionator bottoms from catalytic cracking, may also be used as blends to meet the specifications of the different fuel oils.

Heavy fuel oil includes a variety of oils, ranging from distillates to residual oils, that must be heated to 260°C (500°F) or higher before they can be used. In general, heavy fuel oils consist of residual oils blended with distillates to suit specific needs. Included among heavy fuel oils are various industrial oils; when used to fuel ships, heavy fuel oil is called bunker oil. Historically, fuel oils were based on atmospheric residua (Chapter 3), and were known as straight run fuels. However, the increasing demand for transportation fuels such as gasoline, kerosine, and diesel has led to an increased value for the atmospheric residue as a feedstock for vacuum distillation and for cracking processes. As a consequence, most heavy fuel oils are currently based on vacuum residua (Chapter 3) as well as residues from thermal and catalytic cracking operations. These fuels differ in character from straight run fuels in that the density and mean molecular weight are higher, as is the carbon/hydrogen ratio. The density of some heavy fuel oils can be above 1,000 kg/m^3, which has environmental implications in the event of a spillage into fresh water.

To produce fuels that can be conveniently handled and stored in industrial and marine installations, and to meet marketing specification limits, the high viscosity residue components are normally blended with gas oils or similar lower viscosity fractions. In refineries with catalytic cracking units, catalytically cracked cycle oils are common fuel oil diluents. As a result, the composition of residual fuel oils can vary widely, and will depend on the refinery configuration, the crude oils being processed, and the overall refinery demand.

Fuel oil that is used for heating is graded from No. 1 Fuel Oil to No. 6 Fuel Oil, and covers light distillate oils, medium distillate, heavy distillate, a blend of distillate and residue, and residue oil. For instance, No. 2 and No. 3 Fuel oils refer to medium-to-light distillate grades used in domestic central heating (Speight, 2007).

The significance of the measured properties of residual fuel oil is dependent to a large extent on the ultimate uses of the fuel oil. Such uses include steam generation for various processes, as well as electrical power generation and propulsion. Corrosion, ash deposition, atmospheric pollution, and product contamination are side effects of the use of residual fuel oil, and in particular cases, properties such as vanadium, sodium, and sulfur contents may be significant.

The character of fuel oil generally renders the usual test methods for *total petroleum hydrocarbons* (Chapter 5) ineffective, since high proportions of the fuel oil (specifically residual fuel oil) are insoluble in the usual solvents employed for the test. In particular, the asphaltene constituents are insoluble in hydrocarbon solvents, and are only soluble in aromatic solvents and chlorinated hydrocarbons (chloroform, methylene dichloride, and the like). Residua and asphalt have high proportions of asphaltene constituents that render any test for *total petroleum hydrocarbons* meaningless unless a suitable solvent is employed in the test method (Speight, 2007).

Testing residual fuel oil does not suffer from the issues that are associated with sample volatility, but the test methods are often sensitive to the presence of gas bubbles in the fuel oil. An air release test is available for application to lubricating oil (ASTM D3427, IP 313), and may be applied, with modification, to residual fuel oil. However, with dark-colored samples, it may be difficult to determine whether all air bubbles have been eliminated. And, as with the analysis and testing of other petroleum products, the importance of correct sampling of fuel oil cannot be overemphasized, because no proper assessment of quality may be made unless the data are obtained on truly representative samples (ASTM D270, IP 51).

3.1 Asphaltene Content

The asphaltene fraction (ASTM D893, ASTM D2006, ASTM D2007, ASTM D3279, ASTM D4124, ASTM D6560, IP 143) is the highest molecular weight and most complex fraction in petroleum. The asphaltene content gives an indication of the amount of coke that can be expected during exposure to thermal conditions (Speight, 2001, Speight and Ozum, 2002; Speight, 2007).

In any of the methods for the determination of the asphaltene content (Speight et al., 1984), the residual fuel oil is mixed with a large excess (usually >30 volumes hydrocarbon per volume of sample) low-boiling hydrocarbon, such as *n*-pentane or *n*-heptane. For an extremely viscous sample, a solvent, such as toluene, may be used prior to the addition of the low-boiling hydrocarbon, but an additional amount of the hydrocarbon (usually >30 volumes hydrocarbon per volume of solvent) must be added to compensate for the presence of the solvent. After a specified time, the insoluble material (the asphaltene fraction) is separated (by filtration) and dried. The yield is reported as percentage (% w/w) of the original sample. Furthermore, different hydrocarbons (such as *n*-pentane or *n*-heptane) give different yields of the asphaltene fraction, and if the presence of the solvent is not compensated by use of additional hydrocarbon, the yield will be erroneous. In addition, if the hydrocarbon is not present in a large excess, the yields of the asphaltene fraction will vary, and will be erroneous (Speight, 2007).

Another method, not specifically described as an asphaltene separation method, is designed to remove pentane insoluble constituents by membrane filtration (ASTM D4055). In the method, a sample of oil is mixed with pentane in a volumetric flask, and the oil solution is filtered through a 0.8-micron membrane filter. The flask, funnel, and the filter are washed with pentane to completely transfer any particulates onto the filter, after which the filter (with particulates) is dried and weighed to give the pentane insoluble constituents as a percent by weight of the sample.

Particulates can also be determined by membrane filtration (ASTM D2276, ASTM D5452, ASTM D6217, IP 415).

The *precipitation number* is often equated to the asphaltene content, but there are several issues that remain obvious in its rejection for this purpose. For example, the method to determine the precipitation number (ASTM D91) advocates the use of naphtha for use with black oil or lubricating oil, and the amount of insoluble material (as a % v/v of the sample) is the precipitating number. In the test, 10 ml of sample is mixed with 90 ml of ASTM precipitation naphtha (that may or may nor have a constant chemical composition) in a graduated centrifuge cone, and centrifuged for 10 min at 600 to 700 rpm. The volume of material on the bottom of the centrifuge cone is noted until repeat centrifugation gives a value within 0.1 ml (the precipitation number). Obviously, this can be substantially different from the asphaltenes content.

If the residual fuel oil is produced by a thermal process, such as visbreaking, it may also be necessary to determine if toluene insoluble material is present by the methods, or modifications thereof, used to determine the toluene insoluble of tar and pitch (ASTM D4072, ASTM D4312). In the methods, a sample is digested at 95°C (203°F) for 25 minutes, and then extracted with hot toluene in an alundum thimble. The extraction time is eighteen hours (ASTM D4072) or three hours (ASTM D4312). The insoluble matter is dried and weighed.

3.2 Composition

The composition of residual fuel oils is varied, and is often reported in the form of four or five major fractions, as deduced by adsorption chromatography (Figure 5.1). In the case of cracked feedstocks, thermal decomposition products (carbenes and carboids) may also be present.

Column chromatography is used for several hydrocarbon type analyses that involve fractionation of viscous oils (ASTM D2007, ASTM D2549), including residual fuel oil. The former method (ASTM D2007) advocates the use of adsorption on clay and clay-silica gel, followed by elution of

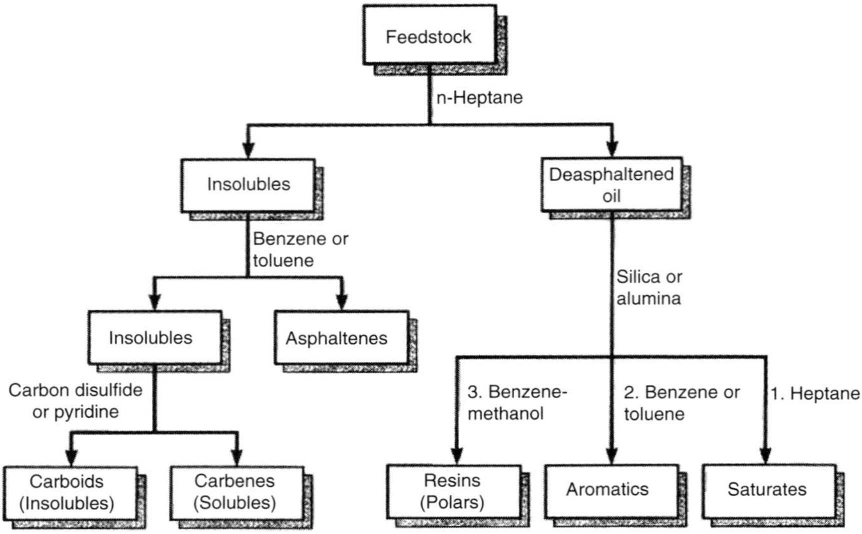

Figure 5.1 A SARA-type analysis (showing two additional fractions (carbenes and carboids) that are produced by thermal treatment of petroleum and petroleum products).

the clay with pentane to separate saturates, elution of clay with acetone-toluene to separate polar compounds, and elution of the silica gel fraction with toluene to separate aromatic compounds. The latter method (ASTM D2549) uses adsorption on a bauxite-silica gel column. Saturates are eluted with pentane; aromatics are eluted with ether, chloroform, and ethanol.

Correlative methods are derived relationships between fundamental chemical properties of a substance and measured physical or chemical properties. They provide information about oil from readily measured properties (ASTM D2140, ASTM D2501, ASTM D2502, ASTM D3238). One method (ASTM D2501) describes the calculation of the viscosity-gravity coefficient (VGC), a parameter derived from kinematic viscosity and density that has been found to relate to the saturate/aromatic composition. Correlations between the viscosity-gravity coefficient (or molecular weight and density) and refractive index to calculate carbon type composition in percent of aromatic, naphthenic, and paraffinic carbon atoms are employed to estimate of the number of aromatic and naphthenic rings present (ASTM D2140, ASTM D3238). Another method (ASTM D2502) permits estimation of molecular weight from kinematic viscosity measurements at 38 and 99°C (100 and 210°F) (ASTM D445). It is applicable to samples with molecular weights in the range from 250 to 700, but should not be applied indiscriminately for oils that represent extremes of composition for which different constants are derived (Moschopedis et al., 1976).

A major use for *gas chromatography* for hydrocarbon analysis has been simulated distillation, as discussed previously. Other gas chromatographic methods have been developed for contaminant analysis (ASTM D3524, ASTM D4291)

The aromatic content of fuel oil is a key property that can affect a variety of other properties, including viscosity, stability, and compatibility of with other fuel oil or blending stock. Existing methods for this work use physical measurements and need suitable standards. Thus, methods have been standardized using nuclear magnetic resonance (NMR) for hydrocarbon characterization (ASTM D4808, ASTM D5291, ASTM D5292). The nuclear magnetic resonance method is simpler and more precise. Procedures are described that cover light distillates with a 15 to 260°C boiling range, middle distillates and gas oils with boiling ranges of 200 to 370°C and 370 to 510°C, and residuum boiling above 510°C. One of the methods (ASTM D5292) is applicable to a wide range of hydrocarbon oils that are completely soluble in chloroform and carbon tetrachloride at ambient temperature. The data obtained by this method can be used to evaluate changes in aromatic contents of hydrocarbon oils due to process changes.

High ionizing voltage mass spectrometry (ASTM D2786, ASTM D3239) is also employed for compositional analysis of residual fuel oil. These methods require preliminary separation using elution chromatography (ASTM D2549). A third method (ASTM D2425) may be applicable to some residual fuel oil samples in the lower molecular weight range.

3.3 Density (Specific Gravity)

Density or specific gravity (relative density) is used whenever conversions must be made between mass (weight) and volume measurements. This property is often used in combination with other test results to predict oil quality, and several methods are available for measurement of density (or specific gravity). However, the density (specific gravity) (ASTM D1298, IP 160) is probably of least importance in determining fuel oil performance, but it is used in product control, in weight-volume relationships, and in the calculation of calorific value (heating value).

Two of the methods (ASTM D287, ASTM D1298) use an immersed hydrometer for measurement of density. The former method (ASTM D287) provides the results as API gravity. Two other methods (ASTM D1480, ASTM D1481) use a pycnometer to measure density or specific gravity, and have the advantage of requiring a smaller sample size, and can be used at higher temperatures than is normal, providing the vapor pressure of the liquid does not exceed specific limits at the temperature of the test. Two other test methods (ASTM D4052, ASTM D5002) measure density with a digital density analyzer. This device determines density by analysis of the change in oscillating frequency of a sample tube when filled with the test sample.

Another test method (ASTM D4052) covers the determination of the density or specific gravity of viscous oil, such as residual fuel oil, that are liquids at test temperatures between 15 and 35°C (59 and 95°F). However, application of the method is restricted to liquids with vapor pressures below 600 mm Hg and viscosity below 15,000 cSt at the temperature of test. In addition, and this is crucial for residual fuel oil, this test method should not be applied to samples so dark in color that the absence of air bubbles in the sample cell cannot be established with certainty.

3.4 Elemental Analysis

Elemental analysis of fuel oil often plays a more major role that it may appear to play in the lower boiling products. Aromaticity (through the atomic hydrogen/carbon ratio), sulfur content, nitrogen content, oxygen content, and metals content are all important features that can influence use of residual fuel oil.

Carbon content and *hydrogen content* can be determined simultaneously by the method designated for coal and coke (ASTM D3178), or by the method designated for municipal solid waste (ASTM E-777). However, as with any analytical method, the method chosen for the analysis may be subject to the peculiarities or character of the feedstock under investigation, and should be assessed in terms of accuracy and reproducibility. There methods that are designated for elemental analysis are:

1. *Carbon* and *hydrogen content* (ASTM D1018, ASTM D3178, ASTM D3343, ASTM D3701, ASTM D5291, ASTM E-777, IP 338);
2. *Nitrogen content* (ASTM D3179, ASTM D3228, ASTM D3431, ASTM E-148, ASTM E-258, ASTM D5291, ASTM E-778);
3. *Oxygen content* (ASTM E-385); and
4. *Sulfur content* (ASTM D124, ASTM D129, ASTM D139, ASTM D1266, ASTM D1552, ASTM D1757, ASTM D2622, ASTM D2785, ASTM D3120, ASTM D3177, ASTM D4045 and ASTM D4294, ASTM E-443, IP 30, IP 61, IP 103, IP 104, IP 107, IP 154, IP 155, IP 243).

The hydrogen content of fuel oil can also be measured by low-resolution magnetic resonance spectroscopy (ASTM D3701, ASTM D4808). The method is claimed to provide a simple and more precise alternative to existing test methods, specifically combustion techniques

(ASTM D5291) for determining the hydrogen content of a variety of petroleum-related materials.

Nitrogen occurs in residua, and, therefore, in residual fuel oil, and causes serious environmental problems as a result, especially when the levels exceed 0.5% by weight, as happens often in residua. In addition to the chemical character of the nitrogen, the amount of nitrogen in a feedstock determines the severity of the process, the hydrogen requirements, and to some extent, the sediment formation and deposition.

The determination of nitrogen in petroleum products is performed regularly by the Kjeldahl method (ASTM D3228), the Dumas method, and the microcoulometric (ASTM D3431) method. The chemiluminescence method is the most recent technique applied to nitrogen analysis for petroleum, and is used to determine the amount of chemically bound nitrogen in liquid samples.

In the method, the samples are introduced to the oxygen-rich atmosphere of a pyrolysis tube maintained at 975°C (1785°F). Nitrogen in the sample is converted to nitric oxide during combustion, and the combustion products are dried by passage through magnesium perchlorate [$Mg(ClO_4)_2$] before entering the reaction chamber of a chemiluminescence detector. In the detector, ozone reacts with the nitric oxide to form excited nitrogen dioxide:

$$NO + O_3 = NO_2{}^* + O2$$

Photoemission occurs as the excited nitrogen dioxide reverts to the ground state:

$$NO_2{}^* = NO_2 + h\nu$$

The emitted light is monitored by a photomultiplier tube to yield a measure of the nitrogen content of the sample. Quantitation is based on comparison with the response for carbazole in toluene standards.

Oxygen is one of the five (C, H, N, O, and S) major elements in fuel oil, but rarely exceeds 1.5 % by weight, unless oxidation has occurred during transportation and storage. Many petroleum products do not specify a particular oxygen content, but if the oxygen compounds are present as acidic compounds, such as phenols (Ar-OH) and naphthenic acids (cycloalkyl-COOH), they are controlled in different specifications by a variety of tests. The *total acidity* (ASTM D974, IP 139, IP 273) is determined for many products, especially fuels and fuel oil. Oxygen-containing impurities in the form of *gum* are determined by the *existent gum* test method (ASTM D381, IP 131) and *potential gum* test method (ASTM D873, IP 138). Elemental analysis of the gum can then provide its composition with some indication of the elements (other than carbon and hydrogen) that played a predominant role in its formation.

Being the third most common element (after carbon and hydrogen) in petroleum product, *sulfur* has been analyzed extensively. Analytical methods range from elemental analyses to functional group (sulfur-type) analyses to structural characterization to molecular speciation (Speight, 2001). Of the methods specified for the determination of sulfur (Speight, 2001), the method applied to the corrosion effect of sulfur is extremely important for liquid fuels. In this method (ASTM D1266, IP 154), fuel corrosivity is assessed by the action of the fuel on a copper strip (*the copper strip test*) that helps determine any discoloration of the copper due to the presence of corrosive compounds. The copper strip is immersed in the fuel and heated at 100°C (212°F) for two hours in a bomb. A test using silver as the test metal (IP 227) has also been published. Mercaptans are usually the corrosive reference sulfur compounds, and metal discoloration is due to formation of the metal sulfide. Thus, mercaptan sulfur is an important property of potential fuels. In addition to the copper strip test, the mercaptan sulfur (R-SH) content (ASTM D1219, IP 104) provides valuable information. As an alternative to determining the mercaptan content, a negative result in the *Doctor test* (ASTM D484, IP 30) may also be acceptable for the qualitative absence of mercaptans. The copper strip method (ASTM D130, ASTM D849, ASTM D4048, IP 154) may also be employed to determine the presence of corrosive sulfur compounds in residual fuel oil.

The Doctor test measures the amount of sulfur available to react with metallic surfaces at the temperature of the test. The rates of reaction are metal type, temperature, and time dependent. In the test, a sample is treated with copper powder at 149°C or 300°F. The copper powder is filtered from the mixture. Active sulfur is calculated from the difference between the sulfur contents of the sample (ASTM D129) before and after treatment with copper.

The determination of sulfur in liquid products by X-ray fluorescence (ASTM D2622, IP 336) has become an extremely well used method over the past two decades. This method can be used to determine the amount of sulfur in homogeneous liquid petroleum hydrocarbons over the range 0.1 to 6.0% by weight. Samples with a sulfur content above this range may be determined after dilution in toluene. The method utilizes the principle that when a sample is irradiated with a Fe^{55} source, fluorescent X-rays result. The sulfur $K\alpha$ fluorescence and a background correction at adjacent wavelengths are counted. A calibration of the instrument, wherein the integration time for counting is adjusted such that the displayed signal for the background corrected radiation equals the concentration of the calibration standard, gives a direct readout of the weight per cent sulfur in the sample. Interfering elements include aluminum, silicon, phosphorus, chlorine, argon, and potassium. Generally, the amounts of these elements are insufficient to affect sulfur x-ray counts in samples covered by this method. Atmospheric argon is eliminated by a helium purge.

It is also possible to determine nitrogen and sulfur simultaneously by chemiluminescence and fluorescence. An aliquot of the sample undergoes high temperature oxidation in a combustion tube maintained at 1050°C (1920°F). Oxidation of the sample converts the chemically bound nitrogen to nitric oxide (NO) and sulfur to sulfur dioxide (SO_2). In the nitrogen detector, ozone reacts with the nitric oxide to form excited nitrogen dioxide (NO_2). As the nitrogen dioxide reverts to its ground state, chemiluminescence occurs, and this light emission is monitored by a photomultiplier tube. The light emitted is proportional to the amount of nitrogen in the sample. In the sulfur detector, the sulfur dioxide is exposed to ultraviolet radiation, and produces a fluorescent emission. This light emission is proportional to the amount of sulfur, and is also measured by a photomultiplier tube. Quantitation is determined by a comparison to the responses given by standards containing carbazole and dimethyl sulfoxide in xylene.

Oxidative microcoulometry has become a widely accepted technique for the determination of low concentrations of sulfur in petroleum and petroleum products (ASTM D3120). The method involves combustion of the sample in an oxygen-rich atmosphere, followed by microcoulometric generation of tri-iodide ion to consume the resultant sulfur dioxide. It is intended to distinguish the technique from reductive microcoulometry, that converts sulfur in the sample to hydrogen sulfide that is titrated with coulometrically generated silver ion.

The bomb method for sulfur determination (ASTM D129) uses sample combustion in oxygen and conversion of the sulfur to barium sulfate, which is determined by mass. This method is suitable for samples containing 0.1 to 5.0% w/w sulfur, and can be used for most low-volatility petroleum products. Elements that produce residues insoluble in hydrochloric acid interfere with this method: this includes aluminum, calcium, iron, lead, and silicon, plus minerals such as asbestos, mica, and silica, and an alternate method (ASTM D1552) is preferred. This method describes three procedures: the sample is first pyrolyzed in either an induction furnace or a resistance furnace; the sulfur is then converted to sulfur dioxide, and either titrated with potassium iodate-starch reagent, or the sulfur dioxide is analyzed by infrared spectroscopy. This method is generally suitable for samples containing from 0.06 to 8.0% w/w sulfur that distill at temperatures above 177°C (351°F).

A variety of *miscellaneous elements* can also occur in residual fuel oil fraction. For example, *chlorine* is present as a chlorinated hydrocarbon, and can be determined (ASTM D808, ASTM D1317, ASTM D6160). A rapid test method suitable for analysis of samples by non-technical personnel is also available (ASTM D5384), and uses a commercial test kit where the oil sample is reacted with metallic sodium to convert organic halogens to halide, which is titrated with mercuric nitrate using diphenyl carbazone indicator. Iodides and bromides are reported as chloride.

Phosphorus is a common component of additives, and appears most commonly as a zinc dialkyl dithiophosphate or a tri-aryl phosphate ester, but other forms also occur. Two wet chemical methods are available, one of which (ASTM D1091) describes an oxidation procedure that converts phosphorus to aqueous ortho-phosphate anion. This is then determined by mass as magnesium pyrophosphate, or photochemically as molybdi-vanadophosphoric acid. In an alternate test (ASTM D4047), samples are oxidized to phosphate with zinc oxide, dissolved in acid, precipitated as quinoline phosphomolybdate, treated with excess standard alkali, and back-titrated with standard acid. Both of these methods are primarily used for referee samples. Phosphorus is most commonly determined using X-ray fluorescence (ASTM D4927) or ICP (ASTM D4951).

3.5 Flash Point

As for all petroleum products, considerations of safety in storage and transportation and, more particularly, contamination by more volatile products, are required. This is usually accommodated by the Pensky-Martens flash point test (ASTM D93, IP 34). For the fuel oil, a minimum flash point of 55°C (131°F) or 66°C (150°F) is included in most specifications.

3.6 Metals Content

The analysis for metal constituents in residual fuel oil can be accomplished by several instrumental techniques: Inductively Coupled Argon Plasma (ICAP) Spectrometry, Atomic Absorption (AA) Spectrometry, and X-ray Fluorescence (XRF) Spectrometry. Each technique has limitations, in terms of sample preparation, sensitivity, sampling, time for analysis, and overall ease of use. Thus, a variety of tests (ASTM D482, D1026, D1262, D1318, D1368, D1548, D1549, D2547 D2599, D2788, D3340, D3341, and D3605) either directly or as the constituents of combustion ash have been designated to determine metals in petroleum products based on a variety of techniques. At the time of writing, the specific test for the determination of metals in whole feeds has not been designated. However, this task can be accomplished by combustion of the sample so that only inorganic ash remains (ASTM D482). The ash can then be digested with an acid, and the solution examined for metal species by atomic absorption (AA) spectroscopy (IP 288, IP 285) or by inductively coupled argon plasma (ICP) spectrometry (ASTM C-1109, ASTM C-1111).

Atomic absorption provides very high sensitivity, but requires careful sub-sampling, extensive sample preparation, and detailed sample-matrix corrections. X-ray fluorescence requires little in terms of sample preparation, but suffers from low sensitivity and the application of major matrix corrections. Inductively coupled argon plasma spectrometry provides high sensitivity and few matrix corrections, but requires a considerable amount of sample preparation, depending on the process stream to be analyzed.

The inductively coupled argon plasma emission spectrometer (ICAP) method analyzes nickel, iron, and vanadium content of gas oil samples in the range from 0.1 mg/kg to 100 mg/kg. Thus, a 10 g sample of gas oil is charred with sulfuric acid, and subsequently combusted to leave the ash residue. The resulting sulfates are then converted to their corresponding chloride salts to ensure complete solubility. A barium internal standard is added to the sample before analysis. In addition, the use of the ICAP method for the analysis of nickel, vanadium, and iron present counteracts the two basic issues arising from metals analysis. The most serious issue is the fact that these metals are partly or totally in the form of volatile, chemically stable porphyrin complexes, and extreme conditions are needed to destroy the complexes without losing the metal that will be lost through volatilization of the complex. The second issue is that the alternate direct aspiration of the sample introduces large quantities of carbon into the plasma. This carbon causes marked and somewhat variable background changes in all direct measurement techniques.

Finally, the analytical method should be selected depending on the sensitivity required, the compatibility of the sample matrix with the specific analysis technique, and the availability of facilities. Sample preparation, if it is required, can present problems. Significant losses can occur, especially in the case of organo-metallic complexes, and contamination of environmental sample is of serious concern. The precision of the analysis depends on the metal itself, the method used, and the standard used for calibration of the instrument.

3.7 Pour Point and Viscosity

The pour point (ASTM D97, ASTM D5949, ASTM D5950, ASTM D5853, ASTM D5985, 1P 15) is the lowest temperature at which oil will flow, under prescribed conditions. The test method for determining the solidification point solidification point (ASTM D1493) might also be applied to residual fuel oil.

The pour point test is still included in many specifications, but not in some (ASTM D396, BS 2869) for residual fuel oil for assessing the mobility characteristics of residual fuel oil (ASTM D3245). Pour point procedures involving various preheat treatments prior to the pour point determination, and the use of viscosity at low temperatures has been proposed. The fluidity test (ASTM D1659) is one such procedure, as is the pumping temperature test (ASTM D3829), and another test, based upon viscosity measurements (IP 230), is also available.

Viscosity is an important property of residual fuel oils, since it provides information on the ease (or otherwise) with which a fuel can be transferred, under the prevailing temperature and pressure conditions, from storage tank to burner system. Viscosity data also indicates the degree to which a fuel oil needs to be preheated to obtain the correct atomizing temperature for efficient combustion. Most residual fuel oils function best when the burner input viscosity lies within a certain specified range.

The Saybolt Universal and Saybolt Furol viscometers are widely used in the USA, and the Engler in Europe. In the USA, viscosities on the lighter fuel grades are determined using the Saybolt Universal instrument at 38°C (100°F); for the heaviest fuels, the Saybolt Furol viscometer is used at 50°C (122°F). Similarly, in Europe, the Engler viscometer is used at temperatures of 20°C (68°F), 50°C (122°F), and in some instances at 100°C (212°F). The use of these empirical procedures for fuel oils is being superseded by the kinematic system (ASTM D396, BS 2869) specifications for fuel oils.

The determination of residual fuel oil viscosities is complicated by the fact that some fuel oils, containing significant quantities of wax, do not behave as simple Newtonian liquids, in which the rate of shear is directly proportional to the shearing stress applied. At temperatures in the region of 38°C (100°F), these fuels tend to deposit wax from solution, with a resulting adverse effect on the accuracy of the viscosity result, unless the test temperature is raised sufficiently high for all wax to remain in solution. While the present reference test temperature of 50°C (122°F) is adequate for use with the majority of residual fuel oils, there is a growing trend of opinion in favor of a higher temperature (82°C; 180°F), particularly in view of the availability of waxier fuel oils.

3.8 Stability

The problem of instability in residual fuel oil may manifest itself either as waxy sludge deposited on the soil or as fouling coastlines.

Asphaltene-type deposition may, however, result from the mixing of fuels of different origin and treatment, each of which may be perfectly satisfactory when used alone. For example, straight run fuel oils from the same crude oil are normally stable and mutually compatible, whereas fuel oils produced from thermal cracking and visbreaking operations may be stable but can be unstable or incompatible if blended with straight run fuels, and vice versa (ASTM (D1661).

Another procedure for predicting the stability of residual fuel oil involves the use of a spot test to show compatibility or cleanliness of the blended fuel oil (ASTM D2781, ASTM D4740). The former method (ASTM D2781) covers two spot test procedures for rating a residual fuel with respect to its compatibility with a specific distillate fuel. Procedure A indicates the degree of asphaltene deposition that may be expected in blending the components, and is used when wax deposition is not considered a fuel application problem. Procedure B indicates the degree of wax and asphalt deposition in the mixture at room temperature. The latter method (ASTM D4740) is applicable to fuel oils with viscosities up to 50 cSt at 100°C (212°F) to identify fuels or blends that could result in excessive centrifuge loading, strainer plugging, tank sludge formation, or similar operating problems. In the method, a drop of the preheated sample is put on a test paper and placed in an oven at 100°C. After one hour, the test paper is removed from the oven, and the resultant spot is examined for evidence of suspended solids, and rated for cleanliness using the procedure described in the method. In a parallel procedure for determining *compatibility*, a blend composed of equal volumes of the fuel oil sample and the blend stock is tested and rated in the same way as just described for the *cleanliness* procedure.

For oxidative stability, an important effect after a spill, a test method (ASTM D4636) is available to determine resistance to oxidation and corrosion degradation and their tendency to corrode various metals. The test method consists of one standard and two alternative procedures. In the method, a large glass tube containing an oil sample and metal specimens is placed in a constant temperature bath (usually from 100 to 360°C), and heated for the specific number of hours, while air is passed through the oil to provide agitation and a source of oxygen. Corrosiveness of the oil is determined by the loss in metal mass and microscopic examination of the sample metal surface(s). Oil samples are withdrawn from the test oil, and checked for changes in viscosity and acid number as a result of the oxidation reactions. At the end of the test, the amount of the sludge present in the oil remaining in the same tube is determined by centrifugation. Also, the quantity of oil lost during the test is determined gravimetrically. Metals used in the basic test and alternative test are aluminum, bronze, cadmium, copper, magnesium, silver, steel, and titanium. Other metals may also be specified, as determined by the history and storage of the fuel oil.

3.9 Environmental Impact

The most toxic components of fuel oils are the aromatics, such as benzene, toluene, xylene, naphthalene, and others. These aromatics are relatively highly soluble in water. After the aromatic fraction, toxicity decreases from olefins through naphthenes to paraffins. Within each of these groups, the lower molecular weight hydrocarbon tends to be more acutely toxic.

More pertinent to the present text, because of the publicity received over the years, heavy fuel oil is a blended product based on the residues from various refinery distillation and cracking processes. Heavy fuel oil is a viscous liquid with a characteristic odor, and requires heating for storage and combustion. Heavy fuel oil is used in medium to large industrial plants, marine applications, and power stations, in combustion equipment such as boilers, furnaces, and diesel engines.

Heavy fuel oil is a general term, and other names commonly used to describe this range of products include: residual fuel oil, bunker fuel, bunker C, fuel oil No. 6, industrial fuel oil, marine fuel oil, and black oil. In addition, terms such as heavy fuel oil, medium fuel oil, and light fuel oil are used to describe products for industrial applications to give a general indication of the viscosity and density of the product.

Short-term toxicity decreases as the type of fuel oil becomes less volatile (that is, No. 1 and No. 2 are moderately toxic, while toxicity decreases through No. 4, No. 5, and No. 6). Fuel oil No. 1 and No. 2 possesses moderate to high acute toxicity to biota with product-specific toxicity related to the type and concentration of aromatic compounds, while fuel oil No. 5 and No. 6 are considered to be less acutely toxic relative to other oil types. Fuel oil No. 4 has variable acute toxicity, depending on the amount of light fraction.

Like naphtha, potential releases of fuel oil from refineries and upgraders can be characterized as either controlled or unintentional releases. Controlled releases are planned releases from pressure relief valves, venting valves, and drain systems, that occur for safety purposes or maintenance, are considered part of routine operations, and occur under controlled conditions. Unintentional releases are typically characterized as unplanned releases due to spills or leaks from various equipment, valves, piping, and flanges, which result from equipment failure, poor maintenance, lack of proper operating practices, adverse weather conditions, or other unforeseen factors. Refinery and upgrader operations are highly regulated, and regulatory requirements established under various jurisdictions, as well as voluntary non-regulatory measures implemented by the petroleum industry, are in place to manage these releases.

Most fuel oil entering the environment comes from spills or leaking storage tanks. When spilled on soil, some components of fuel oil attach to soil. Furthermore, fuel spilled on water or soil evaporates into the air. Fuel oil can also contaminate soil sediments and private drinking water supplies. Other environmental impacts associated with oil production include blowouts, spills, brine disposal, and the production of hydrogen sulfide. Transportation of oil involves spill and leak hazards. Oil refining includes environmental effects, such as explosions, fires, air emissions, noise, odor, and water runoff.

Both solid particles and condensable liquid droplets are generated from most combustion sources, including fuel oil burners. Most of the particulate matter emitted by combustion sources is classified as fine particulate matter, with diameters less than 2.5 microns (PM2.5). Primary particulates include unburned carbonaceous materials (soot) that are directly emitted into the air. Secondary particulates, such as sulfates, are formed after sulfur dioxide is emitted into the air from combustion sources burning sulfur-containing fuels. Particulate matter less than 10 microns in size (PM10) is linked to a number of adverse health outcomes, including asthma, bronchitis, cardiac arrhythmia, and heart attacks (reference 9). Sulfates are also the primary cause of regional haze and acid deposition (acid rain). Direct PM emissions from residential and small commercial oil burners in the form of soot have decreased by approximately 95 percent over the past three decades (as will be discussed later in this section). Sulfates that condense in the outdoor air after being emitted by oil heating equipment are now the predominate form of particulate associated with emissions from fuel oil burners. Reducing the sulfur content of the fuel can lower sulfate emissions.

Lowering the sulfur content in fuel oil will significantly reduce the threats to public health and sensitive ecosystems posed by sulfur dioxide emissions. Emissions of nitrogen oxides (NOx), which contribute to a number of public health and environmental problems, will also decrease with lower sulfur heating oil. The use of cleaner (environmentally friendly) fuel oil has the potential to improve combustion efficiency by reducing fouling rates of boiler and furnace heat exchangers and other components. Further, the availability of low sulfur fuel oil will enable the introduction of highly efficient condensing furnace technology. Both outcomes will lower emissions of CO2 and other pollutants from this source sector by reducing fuel use.

Upon release into the environment, heavy fuel oil will break into small masses, and will not spread as rapidly as less viscous oil. The density of some heavy fuel oils means that they may sink on release to water, rather than float on the surface like other petroleum fuels. Loss of the lower molecular weight components due to volatility and dissolution will increase the density of the floating oil, causing it to sink. This heavy fraction will assume a tar-like consistency and stick to exposed substrates, or become adsorbed to particulates.

Weather conditions and temperature during the period after the spill will significantly influence the rate of dispersion; wind and wave action will tend to disperse oil into the water column, while higher temperatures will increase the rate of evaporation of lighter hydrocarbons. Water temperature is a major factor in determining the extent of the environmental impact following a heavy fuel oil spill, since higher temperatures will enhance loss of lower boiling constituents by evaporation as well as degradation processes.

4 Wastewaters

Wastewaters from petroleum refining consist of cooling water, process water, storm water, and sanitary sewage water (Beychok, 1967; Speight, 2005).

A large portion of water used in petroleum refining is used for cooling. Most cooling water is recycled over and over. Cooling water typically does not come into direct contact with process oil streams, and therefore contains fewer contaminants than process wastewater. However, it may contain some oil contamination, due to leaks in the process equipment.

Water used in processing operations accounts for a significant portion of the total wastewater. Process wastewater arises from desalting crude oil, steam stripping operations, pump gland cooling, product fractionator reflux drum drains, and boiler blowdown. Because process water often comes into direct contact with oil, it is usually highly contaminated. Storm water (i.e., surface water runoff) is intermittent, and will contain constituents from spills to the surface, leaks in equipment, and any materials that may have collected in drains. Runoff surface water also includes water coming from crude and product storage tank roof drains.

Refinery effluent water contains various hydrocarbon components, including gasoline blending stocks, kerosene, diesel fuel and heavier liquids (Speight, 2005). Also present may be suspended mineral solids, sand, salt, organic acids, and sulfur compounds. The nature of the components depends on the constituents of the inlet crude oil, as well as the processing scheme of the refinery. Most of these constituents would be undesirable in the effluent water, so it is necessary to treat the water to remove the contaminants.

Undesirable effects of hydrocarbons in water include taste and odor contamination, in addition to toxicity. Petroleum hydrocarbons can impart a perceptible unpleasant taste, and can adversely affect aquatic organisms. Naphthenic acid from refineries can have a toxic effect on plant and animal life, and aromatic hydrocarbons are toxic and/or carcinogenic to humans and animals, as well as to aquatic life.

Oil can also occur in a refinery wastewater stream, and may exist in one or more of three forms: (1) free oil, (2) emulsified oil, and (3) dissolved oil.

Free oil is oil in the form of separate oil globules of sufficient size that they can rise as a result of buoyancy force to the top of the water. Separators may readily be designed to remove this type of oil.

Emulsified oil is oil in the form of much smaller droplets or globules, with a diameter of 20 microns or less, which form a stable suspension in the water. A true emulsion will not separate by gravity, regardless of how long the emulsion stands under quiescent conditions. The term *emulsified oil* may also be applied to emulsions where the droplets are so small that they will not rise at a rate that allows a practical size separation device. It is possible to design enhanced gravity separators to treat waters containing this type of oil, but generally it is only practical for small flow rates.

Dissolved oil cannot be removed by gravity separation, and other methods must be adopted. Such means include (1) biological treatment (which is especially pertinent to this text), or (2) adsorption by activated carbon or other adsorbents or absorbents.

Waste waters are treated in onsite wastewater treatment facilities, and then discharged to publicly owned treatment works (POTWs), or discharged to surfaces waters under National Pollution Discharge Elimination System (NPDES) permits.

Many types of separation methods have been used to remove oil from refinery wastewater, with varying degrees of success (19). Some of the systems currently in use are: (1) API separators, (2) flocculation units, (3) dissolved and induced air flotation (DAF and IAF) units, (4) coalescing plate separators, and (5) multiple angle separators.

Petroleum refineries typically utilize primary and secondary wastewater treatment. Primary wastewater treatment consists of the separation of oil, water, and solids in two stages. During the first stage, an API separator, a corrugated plate interceptor, or other separator design is used. Wastewater moves very slowly through the separator, allowing free oil to float to the surface and be skimmed off, and solids to settle to the bottom and be scraped off to a sludge collecting hopper. The second stage utilizes physical or chemical methods to separate emulsified oils from the wastewater. Physical methods may include the use of a series of settling ponds with a long retention time, or the use of dissolved air flotation (DAF).

In the dissolved air flotation method, air is bubbled through the wastewater, and both oil and suspended solids are skimmed off the top. Chemicals, such as ferric hydroxide [$Fe(OH)_3$] or aluminum hydroxide [$Al(OH)_3$], can be used to coagulate impurities into a froth or sludge that can be more easily skimmed off the top. Some wastes associated with the primary treatment of wastewater at petroleum refineries may be considered hazardous and include: (1) API separator sludge, (2) primary treatment sludge, (3) sludge from other gravitational separation techniques, (4) float from dissolved air flotation units, and (5) wastes from settling ponds.

After primary treatment, the wastewater can be discharged to a publicly owned treatment works (POTW), or undergo secondary treatment before being discharged directly to surface waters under a National Pollution Discharge Elimination System (NPDES) permit. In secondary treatment, dissolved oil and other organic pollutants may be consumed biologically by microorganisms. Biological treatment may require the addition of oxygen through a number of different techniques, including activated sludge units, trickling filters, and rotating biological contactors. Secondary treatment generates bio-mass waste that is typically treated anaerobically and then dewatered.

Some refineries employ an additional stage of wastewater treatment called polishing to meet discharge limits. The polishing step can involve the use of activated carbon, anthracite coal, or sand to filter out any remaining impurities, such as biomass, silt, trace metals and other inorganic chemicals, as well as any remaining organic chemicals.

Certain refinery wastewater streams are treated separately, prior to the wastewater treatment plant, to remove contaminants that would not easily be treated after mixing with other wastewater. One such waste stream is the sour water drained from distillation reflux drums. Sour water contains dissolved hydrogen sulfide and other organic sulfur compounds and ammonia, which are stripped in a tower with gas or steam before being discharged to the wastewater treatment plant.

Wastewater treatment plants are a significant source of refinery air emissions and solid wastes. Air releases arise from fugitive emissions from the numerous tanks, ponds, and sewer system drains. Solid wastes are generated in the form of sludge from a number of the treatment units.

The refining industry will certainly continue to improve operations to minimize the amounts of hydrocarbons in wastewater streams. Some hydrocarbons will still enter wastewater streams because of small spills and leaks, and it will be necessary to recover these to further reduce the amount of hydrocarbons in refinery effluents.

5 References

ASTM. 2011. *Annual Book of ASTM Standards*. American Society for Testing and Materials, West Conshohocken, PA.

Austin, G.T. 1984. *Shreve's Chemical Process Industries*. 5th Edition. McGraw-Hill, New York. Chapter 37.

Beychok, M.R. 1967. *Aqueous Wastes from Petroleum and Petrochemical Plants*. John Wiley & Sons Inc., New York.

Chenier, 1992. *Survey of Industrial Chemistry*. 2nd Revised Edition. VCH Publishers, New York. Chapters 7 and 8.

Charlot, J-C., and Claus, G. 2000. In *Modern Petroleum Technology. Volume 2: Downstream*. A.G. Lucas (Editor). John Wiley & Sons Inc., New York. Chapter 21.

Francis, W., and Peters, M.C. 1980. *Fuels and Fuel Technology: A Summarized Manual*. Pergamon Press, New York. Section B.

Gruse, W.A., and Stevens, D.R. 1960. *Chemical Technology of Petroleum*. McGraw-Hill, New York. Chapter 11.

Goodfellow, A.J. 1973. In *Criteria for Quality of Petroleum Products*. J.P. Allinson (Editor). John Wiley & Sons Inc., New York. Chapter 4.

Guthrie, V.B. 1967. In *Petroleum Processing Handbook*. W.F. Bland and R.L. Davidson (Editors). McGraw-Hill, New York. Section 11.

Heinrich, H., and Duée, D. 2000. In *Modern Petroleum Technology. Volume 2: Downstream*. A.G. Lucas (Editor). John Wiley & Sons Inc., New York. Chapter 10.

Hemighaus, G. 1998. In *Manual on Hydrocarbon Analysis*. 6th Edition. A.W. Drews (Editor). American Society for Testing and Materials, West Conshohocken, PA. Chapter 3.

Hoffman, H.L. 1983. In *Riegel's Handbook of Industrial Chemistry*. 8th Edition. J.A. Kent (editor). Van Nostrand Reinhold Company, New York. Chapter 14.

Hoffman, H.L., and McKetta, J.J. 1993. "Petroleum Processing." In *Chemical Processing Handbook*. J.J. McKetta (Editor). Marcel Dekker Inc., New York. p. 851.

Hori, Y. 2000. In *Modern Petroleum technology. Volume 2: Downstream*. A.G. Lucas (Editor). John Wiley & Sons Inc., New York. Chapter 2.

Institute of Petroleum. 2011. *IP Standard Methods 2001*. The Institute of Petroleum, London, United Kingdom.

Kite, W.H. Jr., and Pegg, R.E. 1973. In *Criteria for Quality of Petroleum Products*. J.P. Allinson (Editor). John Wiley & Sons Inc., New York. Chapter 7.

McCann, J.M. 1998. In *Manual on Hydrocarbon Analysis*. 6th Edition. A.W. Drews (Editor). American Society for Testing and Materials, West Conshohocken, PA. Chapter 2.

Mitchell, D.L., and Speight, J.G. 1973. *Fuel*. 52: 149.

Moschopedis, S.E., Fryer, J.F., and Speight, J.G. 1976. *Fuel*. 55: 227.

Mushrush, G.W., and Speight, J.G. 1995. *Petroleum Products: Instability and Incompatibility*. Taylor & Francis. New York.

Mushrush, G.W., and Speight, J.G. 1998a. *Petroleum Chemistry and Refining*. J.G. Speight (Editor). Taylor & Francis, New York. Chapter 8.

Mushrush, G.W., and Speight, J.G. 1998b. *Reviews in Process Chemistry and Engineering* 1: 5.

Mushrush, G.W., Speight, J.G., Beal, E.J., and Hardy, D.R. 1999. *Preprints. Am. Chem. Soc. Div. Petrol. Chem*. 44(2): 175.

Speight, J.G. 2001. *Handbook of Petroleum Analysis*. John Wiley & Sons Inc., New York.

Speight, J.G. 2002. *Handbook of Petroleum Product Analysis*. John Wiley & Sons Inc., New York.

Speight, J.G. and Ozum, B. 2002. *Petroleum Refining Processes*. Marcel Dekker Inc., New York.

Speight, J.G. 2005. *Environmental Analysis and Technology for the Refining Industry*. John Wiley & Sons Inc., Hoboken, New Jersey.

Speight, J.G. 2007. *The Chemistry and Technology of Petroleum*. 4th Edition. CRC Press, Taylor & Francis Group, Boca Raton, Florida.

Speight, J.G., Long, R.B., and Trowbridge, T.D. 1984. *Fuel*. 63: 616.

Warne, T.M. 1998. In *Manual on Hydrocarbon Analysis*. 6th Edition. A.W. Drews (Editor). American Society for Testing and Materials, West Conshohocken, PA. Chapter 3.

Weissermel, K., and Arpe, H-J. 1978. *Industrial Organic Chemistry*. Verlag Chemie. New York. Chapter 13.

6

Composition and Properties of Solid Products

1 Introduction

The solid products from petroleum refining often take a back seat to the liquid products. Residua, asphalt, and coke do not receive the same amount of attention as the liquid fuels, and (incorrectly) are often considered as by-products, when their use makes them as important as many of the liquid fuels.

For example, asphalt has been used since the recorded beginnings of civilization (Speight, 2007). Approximately six thousand years ago (ca. 4000 B.C.), the Sumerians had a thriving shipbuilding industry that produced and used asphalt for caulking and waterproofing. As early as 2600 B.C., the Egyptians were using asphalt as a waterproofing material and also to impregnate the wrappings of mummies as a preservative. Ancient civilizations widely used asphalt as a mortar for building and paving blocks used in temples, irrigation systems, reservoirs, and highways.

The asphalt used by early civilizations occurred naturally, and was found in geologic strata as either soft, workable mortars or as solid black veins of rock formations. Natural asphalt (bitumen) was used extensively used until the early 1900s, when the discovery of asphalt production from petroleum refining and the increasing popularity of the automobile served to greatly expand the asphalt industry. Modern petroleum asphalt has the same durable qualities as bitumen, with the added advantage of being refined to a uniform condition, free from organic and mineral impurities.

On the other hand, petroleum coke is a black solid produced through the thermal decomposition of heavy petroleum process streams and residues. The feedstocks undergo cracking and carbonization to produce a product with a high carbon to hydrogen ratio, which may be granular or needle-like in appearance.

Petroleum cokes can be categorized generally as either (1) green coke or (2) calcined coke (Chapter 3). The initial product of the coking process, green coke, is used as a solid fuel. Further treatment of green coke at higher temperatures and pressures results in calcined coke, which is used in the manufacture of electrodes, in smelting applications, for graphite electrode production, and carbonization of steel.

2 Residua and Asphalt

Although asphalt is produced from the bottom-of-the-barrel residuum as a result of the fractional distillation process, its contribution to highway mobility makes it one of the most important products from petroleum refining.

The importance of residua and asphalt to the environmental analyst arises from spillage or leakage in the refinery or on the road. In either case, the properties of these materials are detrimental to the ecosystem in which the release occurred. As with other petroleum products, knowledge of the properties of residua and asphalt can help determine the potential cleanup methods, and may even allow regulators to trace the product to the refinery where it was produced. In addition, the character of residua and asphalt render the usual test methods for *total petroleum hydrocarbons* (Chapters 7 and 8) ineffective, since high proportions of asphalt and residua are insoluble in the usual solvents employed for the test. Application of the test methods for *total petroleum hydrocarbons* to fuel oil (Chapter 10) is also subject to similar limitations.

Residua are the dark-colored near-solid or solid products of petroleum refining that are produced by atmospheric and vacuum distillation (Figure 6.1) (Chapter 3). Asphalt is usually produced from a residuum, and is a dark brown to black cementitious material obtained from petroleum processing that contains very high molecular weight molecular polar species, called asphaltenes, that are soluble in carbon disulfide, pyridine, aromatic hydrocarbons, and chlorinated hydrocarbons (Chapter 3) (Gruse and Stevens, 1960; Guthrie, 1967; Broome and Wadelin, 1973; Weissermel and Arpe, 1978; Hoffman, 1983; Austin, 1984; Chenier, 1992; Hoffman and McKetta, 1993; Warne, 1998; Speight, 1992; Speight, 1999; Banaszewski and Blythe, 2000).

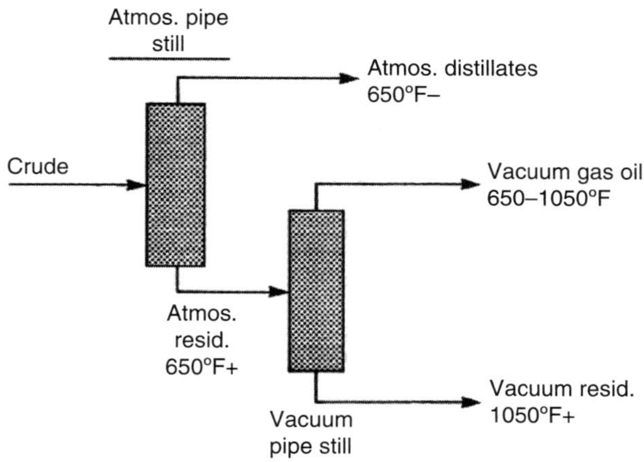

Figure 6.1 The distillation section of a refinery. (Source: Speight, J.G. 2007. *The Chemistry and Technology of Petroleum*. 4th Edition. CRC Press, Taylor & Francis Group, Boca Raton, Florida).

Residua and asphalt derive their characteristics from the nature of their crude oil precursor, the distillation process being a concentration process in which most of the heteroatoms and polynuclear aromatic constituents of the feedstock are concentrated in the residuum (Speight, 2001). Asphalt may be similar to its parent residuum, but with some variation possible by choice of manufacturing process. In general terms, residua and asphalt are a hydrocarbonaceous material that consist constituents (containing carbon, hydrogen, nitrogen, oxygen, and sulfur) that are completely soluble in carbon disulfide (ASTM D4). Trichloroethylene or 1,1,1-trichloroethane has been used in recent years as solvents for the determination of asphalt (and residua) solubility (ASTM D2042).

The residua from which asphalt are produced were once considered the garbage of a refinery, with little value and little use, other than as a road oil. In fact, the development of delayed coking (once the so-called the refinery garbage can) was with the purpose of converting residua to liquids (valuable products) and coke (fuel).

Asphalt manufacture involves distilling everything possible from crude petroleum until a residuum with the desired properties is obtained. This is usually done by stages, in which distillation at atmospheric pressure removes the lower boiling fractions, and yields an atmospheric residuum (*reduced crude*) that may contain higher boiling (lubricating) oils, wax, and asphalt. Distillation of the reduced crude under vacuum removes the oils (and wax) as overhead products, and the residuum remains as a bottom (or residual) product. The majority of the polar functionalities and high molecular weight species in the original crude oil, which tend to be non-volatile, concentrate in the vacuum residuum (Speight, 2000), thereby conferring desirable or undesirable properties on the residuum.

At this stage, the residuum is frequently, but incorrectly, referred to as pitch, and has a softening point (ASTM D36, ASTM D61, ASTM D2319, ASTM D3104, ASTM D3461) related to the amount of oil removed, which increases with increasing overhead removal. In character with the elevation of the softening point, the pour point is also elevated; the more oil distilled from the residue, the higher the softening point.

Propane deasphalting of a residuum also produces asphalt, and there are differences in the properties of asphalts prepared by propane deasphalting and those prepared by vacuum distillation from the same feedstock. Propane deasphalting also has the ability to reduce a residuum even further, and produce an asphalt product having a lower viscosity, higher ductility, and higher temperature susceptibility than other asphalts, although such properties might be anticipated to be very much crude oil dependent. Propane deasphalting is conventionally applied to low-asphalt-content crude oils, which are generally different in type and source from those processed by distillation of higher-yield crude oils. In addition, the properties of asphalt can be modified by air blowing in batch and continuous processes (Figure 6.2) (Speight, 1992 and references cited therein; Speight, 1999 and references cited therein). On the other hand, the

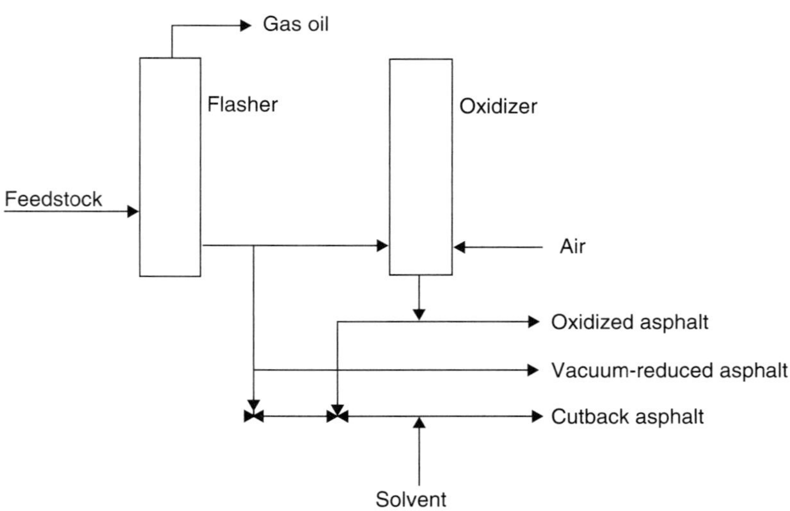

Figure 6.2 Asphalt manufacture including air blowing. (Source: Speight, J.G. 2007. The Chemistry and Technology of Petroleum. 4th Edition. CRC Press, Taylor & Francis Group, Boca Raton, Florida).

preparation of asphalts in liquid form by blending (cutting back) asphalt with a petroleum distillate fraction is customary, and is generally accomplished in tanks equipped with coils for air agitation, or with a mechanical stirrer or a vortex mixer.

An *asphalt emulsion* is a mixture of asphalt and an anionic agent, such as the sodium or potassium salt of a fatty acid. The fatty acid is usually a mixture, and may contain palmitic, stearic, linoleic, and abietic acids and/or high molecular weight phenols. Sodium lignate is often added to alkaline emulsions to affect better emulsion stability. Nonionic cellulose derivatives are also used to increase the viscosity of the emulsion if needed. The acid number is an indicator of its asphalt emulsification properties, and reflects the presence of high molecular weight asphaltic or naphthenic acids. Diamines, frequently used as cationic agents, are made from the reaction of tallow acid amines with acrylonitrile, followed by hydrogenation. The properties of asphalt emulsions (ASTM D977, ASTM D2397) allow a variety of uses. As with other petroleum products, sampling is an important precursor to asphalt analysis, and a standard method (ASTM D140) is available that provides guidance for the sampling of asphalts, liquid and semisolid, at point of manufacture, storage, or delivery.

The properties of residua and asphalt are defined by a variety of standard tests that can be used to define quality, while remembering that the properties of residua vary with cut-point (Speight, 2002), i.e., the volume % of the crude oil helps the refiner produce asphalt of a specific type or property (ASTM D496). Roofing and industrial asphalts are also generally specified in various grades of hardness, usually with a combination of softening point (ASTM D61, ASTM D2319, ASTM D3104, ASTM D3461) and penetration to distinguish grades (ASTM D312, ASTM D449).

The significance of a particular test is not always apparent by reading the procedure, and sometimes can only be gained through working familiarity with the test. The following tests are commonly used to characterize asphalts, but these are not the only tests used for determining the property and behavior of an asphaltic binder. As in the petroleum industry, a variety of tests are employed, having evolved through local, or company, use.

2.1 Acid Number

The acid number is a measure of the acidity of a product, and is used as a guide in the quality control of resid or asphalt properties. Since a variety of oxidation products contribute to the acid number, and the organic acids vary widely in service properties, the test is not sufficiently accurate to predict the precise behavior of asphalt in service.

Resids and asphalt contains a small amount of organic acids and saponifiable material that is largely determined by the percentage of naphthenic (cycloparaffinic) acids of higher molecular weight that are originally present in the crude oil. With increased hardness, asphalt from a particular crude oil normally decreases in acid number, as more of the naphthenic acids are removed during the distillation process. Acidic constituents may also be present as additives or as degradation products formed during service, such as oxidation products (ASTM D5770). The relative amount of these materials can be determined by titrating with bases. The acid number is used as a guide in the quality control of lubricating oil formulations. It is also sometimes used as a measure of lubricant degradation in service. Any condemning limits must be empirically established.

The *saponification number* expresses the amount of base that will react with 1 gm of the sample when heated in a specific manner. Since compounds of certain elements are sometimes added to asphalt and also consume alkali and acids, the results obtained indicate the effect of these extraneous materials in addition to the saponifiable material present. In the test method (ASTM D94, IP 136), a known weight of the sample is dissolved in methyl ethyl ketone or a mixture of suitable solvents, and the mixture is heated with a known amount of standard alcoholic potassium hydroxide for between 30 to 90 min at 80°C (176°F). The excess alkali is titrated with standard hydrochloric acid, and the saponification number is calculated.

2.2 Asphaltene Content

The asphaltene fraction (ASTM D2006, ASTM D2007, ASTM D3279, ASTM D4124, ASTM D6560, IP 143) is the highest molecular weight and most complex fraction in petroleum. The asphaltene content gives an indication of the amount of coke that can be expected during processing (Speight, 1999; Speight, 2001, Speight and Ozum, 2002).

In any of the methods for the determination of the asphaltene content, the crude oil or product (such as a residuum or asphalt) is mixed with a large excess (usually >30 volumes hydrocarbon per volume of sample) low-boiling hydrocarbon, such as *n*-pentane or *n*-heptane. For an extremely viscous sample, a solvent, such as toluene, may be used prior to the addition of the low-boiling hydrocarbon, but an additional amount of the hydrocarbon (usually >30 volumes hydrocarbon per volume of solvent) must be added to compensate for the presence of the solvent. After a specified time, the insoluble material (the asphaltene fraction) is separated (by filtration) and dried. The yield is reported as percentage (% w/w) of the original sample. In any of these tests, different hydrocarbons (such as *n*-pentane or *n*-heptane) will give different yields of the asphaltene fraction, and if the presence of the solvent is not compensated by use of additional hydrocarbon, the yield will be erroneous. In addition, if the hydrocarbon is not present in a large excess, the yields of the asphaltene fraction will vary, and will be erroneous (Speight, 1999).

The *precipitation number* is often equated to the asphaltene content, but there are several issues that remain obvious in its rejection for this purpose. For example, the method to determine the precipitation number (ASTM D91) advocates the use of naphtha for use with black oil or lubricating oil, and the amount of insoluble material (as a % v/v of the sample) is the precipitating number. In the test, 10 ml of sample is mixed with 90 ml of ASTM precipitation naphtha (that may or may not have a constant chemical composition) in a graduated centrifuge cone, and centrifuged for 10 min at 600 to 700 rpm. The volume of material on the bottom of the centrifuge cone is noted until repeat centrifugation gives a value within 0.1 ml (the precipitation number). Obviously, this can be substantially different from the asphaltene content.

In another test method (ASTM D4055), pentane insoluble materials above 0.8 micrometer in size can be determined. In the test method, a sample of oil is mixed with pentane in a volumetric flask, and the oil solution is filtered through a 0.8-micron membrane filter. The flask, funnel, and the filter are washed with pentane to completely transfer the particulates onto the filter, that is then dried and weighed to give the yield of pentane insoluble materials.

Another test method (ASTM D893) that was originally designed for the determination of pentane and toluene insoluble materials in used lubricating oils can also be applied to residua and asphalt. However, the method may need modification, by first adding a solvent (such as toluene) to the residuum or asphalt before adding pentane. The pentane insoluble constituents can include toluene insoluble materials. A significant change in the pentane insoluble constituents or toluene insoluble constituents indicates a change in properties of the resid or asphalt due to thermal or oxidative effects.

3.3 Carbon Disulfide Insoluble Constituents

The component of highest carbon content is the fraction termed *carboids*, and consists of species that are insoluble in carbon disulfide or in pyridine. The fraction that has been called *carbenes* contains molecular species that are soluble in carbon disulfide and soluble in pyridine, but are insoluble in toluene (Figure 6.3).

The carbene and carboids fractions are generated by thermal degradation or by oxidative degradation, and are not considered to be naturally occurring constituents of resids or asphalt. The test method for determining the toluene insoluble constituents of tar and pitch (ASTM D4072, ASTM D4312) can be used to determine the amount of carbenes and carboids in resids and asphalt.

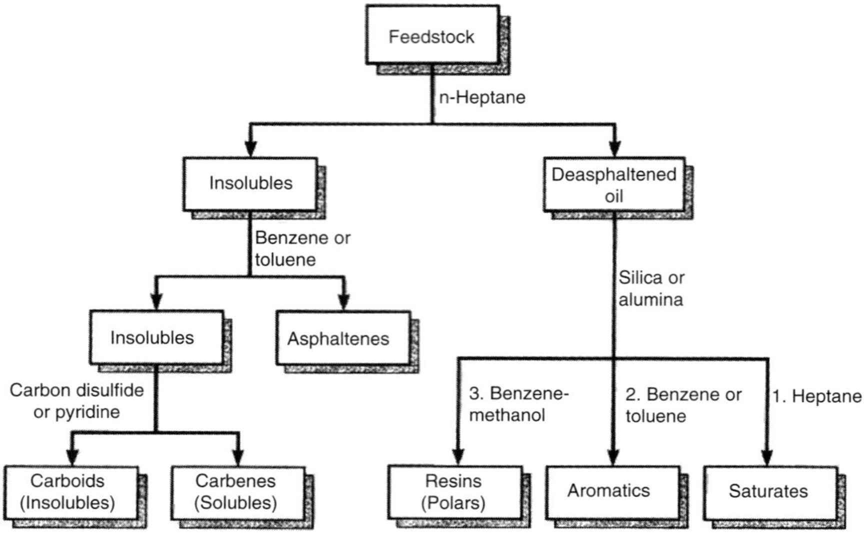

Figure 6.3 Asphalt fractionation. (Source: Speight, J.G. 2007. *The Chemistry and Technology of Petroleum*. 4th Edition. CRC Press, Taylor & Francis Group, Boca Raton, Florida).

2.4 Composition

Determination of the composition of resids and asphalt has always presented a challenge because of the complexity and high molecular weights of the molecular constituents. The principle behind composition studies is to evaluate resids and asphalt in terms of composition and performance.

The methods employed can be conveniently arranged into a number of categories: (a) fractionation by precipitation; (b) fractionation by distillation; (c) separation by chromatographic techniques; (d) chemical analysis using spectrophotometric techniques (infrared, ultraviolet, nuclear magnetic resource, X-ray fluorescence, emission, neutron activation), titrimetric and gravimetric techniques, elemental analysis; and (e) molecular weight analysis by mass spectrometry, vapor pressure osmometry, and size exclusion chromatography.

However, fractional separation has been the basis for most composition analysis, and the separation methods are used to produce operationally defined fractions (Figure 6.3). Three types of separation procedures are now in use: (a) chemical precipitation, in which n-pentane separation of an asphaltene fraction is followed by chemical precipitation of other fractions with sulfuric acid of increasing concentration (ASTM D2006); (b) adsorption chromatography using a clay-gel procedure, where, after removal of the asphaltene fraction, the remaining constituents are separated by selective adsorption/desorption on an adsorbent (ASTM D2007 and ASTM D4124); and (c) size exclusion chromatography, in which gel permeation chromatographic (GPC) separation of constituents occurs based on their associated sizes in dilute solutions (ASTM D3593).

The fractions obtained in these schemes are defined operationally or procedurally. The amount and type of the asphaltene constituents are, for instance, defined by the solvent used for precipitating them. Fractional separation does not provide well-defined chemical components, and the separated fractions should only be defined in terms of the particular test procedure (Speight, 1999, 2001). This is analogous to the definition of total petroleum hydrocarbons in which the composition is defined by the method of extraction (Chapter 7). However, these fractions are generated by thermal degradation or by oxidative degradation, and are not considered to be naturally occurring constituents. The test method for determining the toluene insoluble constituents of tar and pitch (ASTM D4072, ASTM D4312) can be used to determine the amount of carbenes and carboids in resids and asphalt.

In the methods, a sample is digested at 95°C (203°F) for 25 minutes, and then extracted with hot toluene in an alundum thimble. The extraction time is eighteen hours (ASTM D4072) or three hours (ASTM D4312). The insoluble matter is dried and weighed. Combustion will then show if the material is truly carbonaceous, or if it is inorganic ash from the metallic constituents (ASTM D482, ASTM D2415, ASTM D4628, ASTM D4927, ASTM D5185, ASTM D6443, IP 4).

Another method (ASTM D893) covers the determination of pentane and toluene insoluble constituents in used lubricating oils, and can also be applied. Pentane insoluble constituents include oil-insoluble materials, and toluene insoluble constituents can come from external contamination and highly carbonized materials from degradation. A significant change in pentane or toluene insoluble constituents indicates a change in properties that could lead to problems in further processing (for resids) or service (for asphalt).

There are two test methods used. Procedure A covers the determination of insoluble constituents without the use of coagulant in the pentane, and provides an indication of the materials that can be readily separated from the diluted material by centrifugation. Procedure B covers the determination of insoluble constituents that contains additives, and employs a coagulant. In addition to the materials separated by using Procedure A, this coagulation procedure separates some finely divided materials that may be suspended in the resid or asphalt. The results obtained by Procedures A and B should not be compared, since they usually give different values. The same procedure should be applied when comparing results obtained periodically or when comparing results determined in different laboratories.

In Procedure A, a sample is mixed with pentane and centrifuged, after which the resid or asphalt solution is decanted, and the precipitate is washed twice with pentane, dried, and weighed. For toluene insoluble constituents, a separate sample of the resid or asphalt is mixed with pentane and centrifuged. The precipitate is washed twice with pentane, once with toluene-alcohol solution, and once with toluene. The insoluble material is then dried and weighed. In Procedure B, Procedure A is followed, except that instead of pentane, a pentane-coagulant solution is used.

Many investigations of relationships between composition and properties take into account only the concentration of the asphaltene constituents, independently of any quality criterion. However, a distinction should be made between the asphaltene constituents that occur in straight run resids and those which occur in blown asphalts. Remembering that asphaltene constituents are a solubility class, rather than a distinct chemical class, means that vast differences occur in the makeup of this fraction when it is produced by different processes.

2.5 Density (Specific Gravity)

For clarification, it is necessary to understand the basic definitions that are used: (1) *density* is the mass of liquid per unit volume at 15.6°C (60°F), (2) *relative density* is the ratio of the mass of a given volume of liquid at 15.6°C (60°F) to the mass of an equal volume of pure water at the same temperature, and (3) *specific gravity* is the same as the relative density, and the terms are used interchangeably.

Density (ASTM D1298, IP 160) is an important property of petroleum products, since petroleum and especially petroleum products are usually bought and sold on that basis, or if sold on a volume basis, the volume is then converted to mass basis via density measurements. This property is almost synonymously termed as density, relative density, gravity, and specific gravity, all terms related to each other. Usually a hydrometer, pycnometer, or more modem digital density meter is used for the determination of density or specific gravity.

In the most commonly used method (ASTM D1298, IP 160), the sample is brought to the prescribed temperature, and transferred to a cylinder at approximately the same temperature. The appropriate hydrometer is lowered into the sample and allowed to settle, and after temperature equilibrium has been reached, the hydrometer scale is read, and the temperature of the sample is noted.

Although there are many methods for the determination of density, due to the different nature of petroleum itself and the different products, one test method (ASTM D5002) is used for the determination of the density or relative density of petroleum that can be handled in a normal fashion as liquids at test temperatures between 15 and 35°C (59 and 95°F). This test method applies to petroleum oils with high vapor pressures, provided appropriate precautions are taken to prevent vapor loss during transfer of the sample to the density analyzer. In the method, approximately 0.7 ml of crude oil sample is introduced into an oscillating sample tube, and the change in oscillating frequency caused by the change in mass of the tube is used in conjunction with calibration data to determine the density of the sample.

Another test determines density and specific gravity by means of a digital densimeter (ASTM D4052, IP 365). In the test, a small volume (approximately 0.7 ml) of liquid sample is introduced into an oscillating sample tube, and the change in oscillating frequency caused by the change in the mass of the tube is used in conjunction with calibration data to determine the density of the sample. The test is usually applied to of petroleum, petroleum distillates, and petroleum products that are liquids at temperatures between 15 and 35°C (59 and 95°F) that have vapor pressures below 600 mm Hg and viscosities below about 15,000 cSt at the temperature of test. However, the method should not be applied to samples so dark in color that the absence of air bubbles in the sample cell cannot be established with certainty.

Accurate determination of the density or specific gravity of crude oil is necessary for the conversion of measured volumes to volumes at the standard temperature of 15.56°C (60°F) (ASTM D1250, IP 200, Petroleum Measurement Tables). The specific gravity is also a factor reflecting the quality of crude oils.

The accurate determination of the API gravity of petroleum and its products (ASTM D287) is necessary for the conversion of measured volumes to volumes at the standard temperature of 60°F (15.56°C). Gravity is a factor governing the quality of crude oils. However, the gravity of a petroleum product is an uncertain indication of its quality. Correlated with other properties, gravity can be used to give approximate hydrocarbon composition and heat of combustion. This is usually accomplished though use of the API gravity that is derived from the specific gravity:

$$\text{API gravity, deg} = (141.5/\text{sp gr } 60/60°\text{F}) - 131.5$$

API gravity is a critical measure for reflecting the quality of petroleum, and can be related to many other physical properties (Speight, 2007).

API gravity, or density or relative density, can be determined using one of two hydrometer methods (ASTM D287, ASTM D1298). The use of a digital analyzer (ASTM D5002) is finding increasing popularity for the measurement of density and specific gravity.

In the method (ASTM D287), the API gravity is determined using a glass hydrometer for petroleum and petroleum products that are normally

Table 6.1 Effect of temperature and pressure on the density and API gravity of various residua.

Source of Residuum	Temperature									
	C	F	Pressure: psi	14.21	2,843	5,685	8,528	11,371	14,214	
			Pressure: atmos.	0.97	193	387	580	774	967	
			Pressure: MPa	0.098	19.6	39.2	58.8	78.4	98.0	
California	25	77		1.014	1.023	1.031	1.038	1.045	1.051	Density, gm/cc
				8.0	6.8	5.7	4.8	3.9	3.3	API gravity
	45	113		1.002	1.011	1.020	1.028	1.035	1.041	Density, gm/cc
				9.7	8.5	7.2	6.1	5.2	4.4	API gravity
	65	149		0.990	1.000	1.009	1.017	1.025	1.032	Density, gm/cc
				11.4	10.0	8.7	7.6	6.6	5.6	API gravity
Venezuela	25	77		1.024	1.032	1.040	1.048	1.054	1.061	Density, gm/cc
				6.7	5.6	4.6	3.5	2.7	1.9	API gravity
	45	113		1.012	1.020	1.029	1.037	1.044	1.051	Density, gm/cc
				8.3	7.2	6.0	5.0	4.0	3.1	API gravity
	65	149		1.000	1.009	1.018	1.027	1.034	1.041	Density, gm/cc
				10.0	8.7	7.5	6.3	5.3	4.4	API gravity

handled as liquids and that have a Reid vapor pressure of 26 psi (180 kPa) or less. The API gravity is determined at 15.6°C (60°F), or converted to values at 60°F, by means of standard tables. These tables are not applicable to non-hydrocarbons or essentially pure hydrocarbons, such as the aromatics.

For solid and semisolid materials, a pycnometer is generally used (ASTM D70), and a hydrometer is applicable to liquid materials (ASTM D3142). It is worthy of note at this point that the density (hence, the API gravity) of residua show pronounced changes due to the effects of temperature and pressure (Table 6.1). Therefore, isolation of the sample after leakage or spillage must also allow for equilibration to ambient conditions before measurements are made.

2.6 Elemental Analysis

Residua and asphalt are not composed of a single chemical species, but are rather a complex mixture of organic molecules that vary widely in composition, and are composed of carbon, hydrogen, nitrogen, oxygen, and sulfur, as well as trace amounts of metals, principally vanadium and nickel. The heteroatoms, although a minor component compared to the hydrocarbon moiety, can vary in concentration over a wide range, depending on the source.

Generally, most resids and asphalt have 79-88% w/w carbon, 7-13% w/w hydrogen, trace-8% w/w sulfur, 2-8% w/w oxygen, and trace-3% w/w nitrogen. Trace metals, such as iron, nickel, vanadium, calcium, titanium, magnesium, sodium, cobalt, copper, tin, and zinc, occur in crude oils. Vanadium and nickel are bound in organic complexes, and, by virtue of the concentration (distillation) process by which asphalt is manufactured, are also found in asphalt. The catalytic behavior of vanadium has prompted studies of the relation between vanadium content and sensitivity to oxidation (viscosity ratio).

Thus, elemental analysis is still of considerable value to determine the amounts of elements, and the method chosen for the analysis may be subject to the peculiarities or character of the material under investigation, and should be assessed in terms of accuracy and reproducibility. There methods that are designated for elemental analysis are:

1. *carbon* and *hydrogen content* (ASTM D1018, ASTM D3178, ASTM D3343, ASTM D3701, ASTM D5291, ASTM E-777, IP 338);
2. *nitrogen content* (ASTM D3179, ASTM D3228, ASTM D3431, ASTM E-148, ASTM E-258, ASTM D5291, and ASTM E-778);

3. *oxygen content* (ASTM E-385); and
4. *sulfur content* (ASTM D124, ASTM D129, ASTM D139, ASTM D1266, ASTM D1552, ASTM D1757, ASTM D2622, ASTM D2785, ASTM D3120, ASTM D3177, ASTM D4045 and ASTM D4294, ASTM E-443, IP 30, IP 61, IP 103, IP 104, IP 107, IP 154, IP 243).

The determination of *nitrogen* has been performed regularly by the Kjeldahl method (ASTM D3228), the Dumas method, and the microcoulometric (ASTM D3431) method. The chemiluminescence method is the most recent technique applied to nitrogen analysis for petroleum. The chemiluminescence method determines the amount of chemically bound nitrogen in liquid hydrocarbon samples. In the method, the samples are introduced to the oxygen-rich atmosphere of a pyrolysis tube maintained at 975°C (1785°F). Nitrogen in the sample is converted to nitric oxide during combustion, and the combustion products are dried by passage through magnesium perchlorate [$Mg(ClO_4)_2$] before entering the reaction chamber of a chemiluminescence detector.

Oxygen is one of the five (C, H, N, O, and S) major elements in resids and asphalt, although the level rarely exceeds 1.5 % by weight. Many petroleum products do not specify a particular oxygen content, but if the oxygen compounds are present as acidic compounds, such as phenols (Ar-OH) and naphthenic acids (cycloalkyl-COOH), they are controlled in different specifications by a variety of tests.

2.7 Float Test

The float test is used to determine the consistency of asphalt at a specified temperature. One test method (ASTM D139) is normally used for asphalt that is too soft for the penetration test (ASTM D5, ASTM D217, ASTM D937, ASTM D1403, IP 50 IP 179, IP 310).

2.8 Softening Point

The softening point of residua and asphalt is the temperature at which asphalt attains a particular degree of softness under specified conditions of test.

Resids and asphalt do not go through a solid-liquid phase change when heated, and therefore do not have true melting point. As the temperature is raised, the material gradually softens or becomes less viscous. For this reason, the determination of the softening point must be made by an arbitrary but closely defined method, if the test values are to be reproducible. Softening point determination is useful in determining the consistency of one element in establishing the uniformity of shipments or sources of supply.

There are several tests available to determine the softening point (ASTM D36, ASTM D61, ASTM D2319, ASTM D3104, ASTM D3461, IP 58). In the test method (ASTM D36, IP 58), a steel ball of specified weight is laid on a layer of sample contained in a ring of specified dimensions. The softening point is the temperature, during heating under specified conditions, at which the material surrounding the ball deforms and contacts a base plate.

2.9 Viscosity

Viscosity is a measure of flow characteristics, and is generally the most important controlling property for movement of resids from one unit to another, and for selection of asphalt to meet a particular application. A number of instruments are in common use for this purpose. The vacuum capillary (ASTM D2171) is commonly used to classify paving asphalt at 60°C (140°F). Kinematic capillary instruments (ASTM D2170, ASTM D4402) are commonly used in the 60°C to 135°C (140°F to 275°F) temperature range for both liquid and semisolid materials in the range of 30 cSt to 100,000 cSt. Saybolt tests (ASTM D88) are also used in this temperature range and at higher temperatures (ASTM E-102). At lower temperatures, the cone and plate instrument (ASTM D3205) has been used extensively in the viscosity range 1,000 poises to 1,000,000 poises. Other techniques include use of the sliding plate microviscometer and the rheogoniometer.

2.10 Weathering

This test (ASTM D529) evaluates the relative weather resistance of asphalt used for protective-coating applications, especially for roofing. No direct measure of outdoor life or service can be obtained from this test. Methods for preparing test panels (ASTM D1669) and failure end-point testing (ASTM D1670) are available.

2.11 Environmental Impact

Asphalt materials are categorized as asphalt cements or liquid asphalts. Asphalt cement is the heaviest fraction of asphalt. Consistency ranges from between solid to semiliquid at room temperature. Asphalt cements are typically used in hot mix technologies. Liquid asphalts are produced from the lighter fractions of the residual asphalt, or by dissolving asphalt cements in solvent or emulsifying asphalt cements in water.

Asphalt cement is designed to lower contaminant leaching rates. However, the amount the leaching is lowered depends on the physical and chemical characteristics of the particular environment. As a result, various formulations of asphalt and binders have been used to cover or encapsulate hazardous waste. Under certain conditions, solvents and road salts can accelerate breakdown of asphalt. In the aquatic environment, asphalt will sink to the bottom as a dark tarry substance.

Environmental protection laws have developed stringent codes limiting water flows and particulate and smoke emissions from oil refineries and asphalt processing plants. Not only dust, but sulfur dioxides, smoke, and many other emissions must be rigorously controlled. Electrostatic precipitators, primary dust collectors using single or multiple cone cyclones, and secondary collection units, consisting of fabric filter collectors commonly called "baghouses," are all required equipment to control emissions. Hydrocarbons formed in asphalt production, if unchecked, create odoriferous fumes and pollutants that will stain and darken the air. Pollutants emitted from asphalt production are controlled by enclosures that capture the exhaust and then recirculate it through the heating process. This not only eliminates the pollution, but also increases the heating efficiency of the process.

Higher costs of asphalt cement and mineral aggregate have forced the industry to increase efficiencies and recycle old asphalt pavements. In asphalt pavement recycling, materials reclaimed from old pavements are reprocessed along with new materials. The three major categories of asphalt recycling are (1) hot-mix recycling, where reclaimed materials are combined with new materials in a central plant to produce hot-mix paving mixtures, (2) cold-mix recycling, where reclaimed materials are combined with new materials, either onsite or at a central plant to produce cold-mix base materials, and (3) surface recycling, a process in which the old asphalt surface pavement is heated in place and re-laid. Organic asphalt recycling agents may also be added to help restore the aged asphalt to desired specifications.

Because of solvent evaporation and volatility, use of cutback asphalts, especially rapid-cure cutback asphalts which use gasoline or naphtha, is becoming more restricted or prohibited, while emulsified asphalts (in which only the water evaporates) are becoming more popular because of environmental regulations.

In addition, polymer-modified asphalt crack sealers are gaining in popularity, and many other asphalt modifiers are being developed. Modifiers are added to control pavement rutting, cracking, asphalt oxidation, and water damage. Some commercially available asphalt modifiers are polymers, including elastomers, metal complexes, elemental sulfur, fibers, hydrated lime, Portland cement, silicones, various fillers, and organic anti-strip agents.

Finally, bioasphalt is an asphalt alternative made from non-petroleum based renewable resources. These sources include sugar, molasses, rice, corn and potato starches, natural tree and gum resins, natural latex rubber and vegetable oils, lignin, cellulose, palm oil waste, coconut waste, peanut oil waste, canola oil waste, and dried sewage effluent. Asphalt can also be produced from waste vacuum tower bottoms produced in the process of cleaning used motor oils, which are normally burned or dumped into landfills.

3 Coke

Coke does not offer the same potential environmental issues as other petroleum products (Chapter 10 and above). It is used predominantly as a refinery fuel, unless other use for the production of a high-grade coke or carbon is desired. In the former case, the constituents of the coke that will release environmentally harmful gases, such as nitrogen oxides, sulfur oxides, and particulate matter, should be known. In addition, stockpiling coke on a site where it awaits use or transportation can lead to leachates, which are highly detrimental if rainfall (or acid rainfall) occurs. In such a case, application of the toxicity characteristic leaching procedure to the coke (TCLP, EPA SW-846 Method 1311), that is designed to determine the mobility of both organic and inorganic contaminants present in materials such as coke, is warranted before stockpiling the coke in the open.

Petroleum coke is the residue left by the destructive distillation of petroleum residua in processes such as the delayed coking process (Figure 6.4).

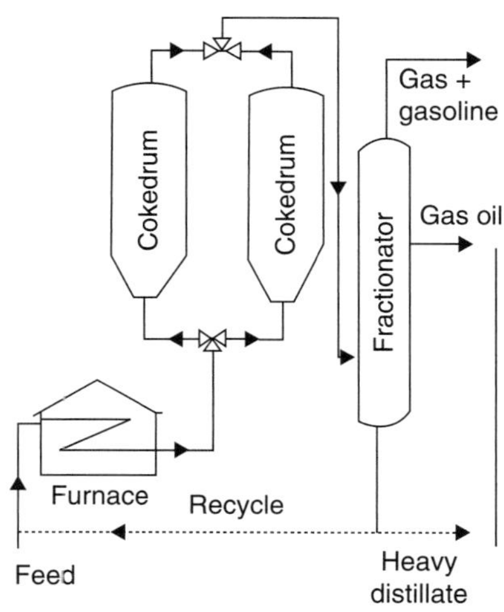

Figure 6.4 Coke production by the delayed coking process. (Source: http://www.osha.gov/dts/osta/otm/otm_iv/otm_iv_2.html)

That which is formed in catalytic cracking operations is usually non-recoverable, as it is often employed as fuel for the process.

Coke is a gray to black solid carbonaceous residue that is produced from petroleum during thermal processing; it is characterized by having a high carbon content (95%+ by weight) and a honeycomb type of appearance, and is insoluble in organic solvents (ASTM D121) (Chapter 2) (Gruse and Stevens, 1960; Guthrie, 1967; Weissermel and Arpe, 1978; Hoffman, 1983; Austin, 1984; Chenier, 1992; Hoffman and McKetta, 1993; Speight, 1992; Speight, 1999; Speight and Ozum, 2002).

Coke occurs in various forms, and the terminology reflects the type of coke (Table 6.2 and Table 6.3) that can influence behavior in the environment. But no matter what the form, coke usually consists mainly of carbon

Table 6.2 Description of delayed coke carbon forms.

Delayed Coke	
Needle coke	Ribbon like parallel ordered anisotropic domains that can also occur as folded structures
Lenticular/granular	Lenticular anisotropic domains of various sizes that are not aligned parallel to the particle surface
Mixed layer	Ribbon and lenticular anisotropic domains of various sizes in curved and irregular layered arrangements
Sponge	Porous microstructure with walls that are generally anisotropic but with pores and walls that vary in size
Shot	Ribbon and lenticular anisotropic domains arranged in concentric patterns to form shot-like coke
Amorphous	Isotropic carbon form closely associated with parent liquor. Higher in volatile matter than incipient mesophase
Incipient mesophase	Initial stage of mesophase formation. Transition stage between amorphous and mesophase
Mesophase	Nemitic liquid crystals. Lower in volatile matter than incipient mesophase

Table 6.3 Description of fluid coke carbon forms.

Fluid Coke	
Layered	Anisotropic carbon domains aligned in concentric layers parallel to the particle surface similar to an onion-like pattern
Non-layered	Anisotropic domains are not aligned parallel to the particle surface
Aggregates	Fragments of anisotropic domains
Amorphous	Isotropic carbon form closely associated with parent liquor. Higher in volatile matter than incipient mesophase
Incipient mesophase	Initial stage of mesophase formation. Transition stage between amorphous and mesophase
Mesophase	Nemitic liquid crystals. Lower in volatile matter than incipient mesophase

(greater than 90 percent, but usually greater than 95 per cent), and has a low mineral matter content (determined as ash residue). Coke is used as a feedstock in coke ovens for the steel industry, for heating purposes, for electrode manufacture, and for production of chemicals. The two most important qualities are *green coke* and *calcined coke*. This latter category also includes *catalyst coke* deposited on the catalyst during refining processes: this coke is not recoverable, and is usually burned as refinery fuel.

The test methods for coke are necessary for defining the coke as a fuel (for internal use in a refinery) or for other uses, particularly those test methods where prior sale of the coke is involved. Specifications are often dictated by environmental regulations, if not by the purchaser of the coke.

The test methods outlined below are the methods that are usually applied to petroleum coke, but should not be thought of as the only test methods. In fact, there are many test methods for coke (ASTM, 2000, Volume 05.06), and these test methods should be consulted either when more detail is required or a fuller review is required.

3.1 Ash

The ash content (that is, the ash yield, which related to the mineral matter content) (see also Section 2.6) is one of the properties used to evaluate coke, and indicates the amount of undesirable residue present. Some sample of coke may be declared to have an acceptable ash content, but this varies with the intended use of the coke.

For the test method, the preparation and sampling of the analytical sample must neither remove nor add mineral matter (ASTM D346). Improper dividing, sieving, crushing equipment, and some muffle furnace lining material can contaminate the coke, and lead to erroneous results. In addition, a high sulfur content of the furnace gases, regardless of the source of the sulfur, can react with an alkaline ash to produce erratic results. To counteract such an effect, the furnace should be swept with air.

In the test method (ASTM D4422), a sample of petroleum coke is dried, ground, and ashed in a muffle furnace at 700 to 775°C (1292 to 1427°F). The non-carbonaceous residue is weighed, and reported as the percent by weight ash. As already noted, the ash must not be understood to be the same as the mineral content of the petroleum coke.

In addition, ashing procedures can be used as a preliminary step for determination of the trace elements in coke, and by inference in the higher boiling fractions of the crude oil. Among the techniques used for trace element determinations are flameless and flame atomic absorption (AA) spectrophotometry (ASTM D2788, ASTM D5863) and inductively coupled argon plasma (ICP) spectrophotometry (ASTM D5708).

Inductively coupled argon plasma emission spectrophotometry (ASTM D5708) has an advantage over atomic absorption spectrophotometry (ASTM D4628, ASTM D5863), because it can provide more complete elemental composition data than the atomic absorption method. Flame emission spectroscopy is often used successfully in conjunction with atomic absorption spectrophotometry (ASTM D3605). X-ray fluorescence spectrophotometry (ASTM D4927, ASTM D6443) is also sometimes used, but matrix effects can be a problem. The method to be used for the determination of metallic constituents is often a matter of individual preference.

3.2 Composition

The composition of petroleum coke varies with the source of the crude oil, but in general, large amounts of high-molecular-weight complex hydrocarbons (rich in carbon, but correspondingly poor in hydrogen) make up a high proportion. The solubility of petroleum *coke* in carbon disulfide has been reported to be as high as 50-80%, but this is in fact a misnomer, since the coke is the insoluble, honeycomb material that is the end product of thermal processes.

Carbon and hydrogen in coke can be determined by the standard analytical procedures for coal and coke (ASTM D3178, ASTM D3179). However, in addition to carbon, hydrogen, and metallic constituents (*q.v.*), coke also contains considerable amounts of nitrogen and sulfur that must be determined prior to sale or use. These elements will appear as their respective oxides (NOx, SOx) when the coke is combusted, thereby causing serious environmental issues.

A test method (ASTM D5291) is available for simultaneous determination of carbon, hydrogen, and nitrogen in petroleum products and lubricants. There are at least three instrumental techniques available for this analysis, each based on different chemical principles. However, all involve sample combustion, components separation, and final detection.

In one of the variants of the method, a sample is combusted in an oxygen atmosphere, and the product gases are separated from each other by adsorption over chemical agents. The remaining elemental nitrogen gas is measured by a thermal conductivity cell. Carbon and hydrogen are separately measured by selective infrared cells as carbon dioxide and water. In another variant of the method, a sample is combusted in an oxygen atmosphere, the product gases are separated from each other, and the three gases of interest are measured by gas chromatography. In the third variant of the method, a sample is combusted in an oxygen atmosphere, the product gases are cleaned by passage over chemical agents, and the three gases of interest are chromatographically separated and measured with a thermal conductivity detector.

The nitrogen method is not applicable to samples containing <0.75% by weight nitrogen, or for the analysis of volatile materials such as gasoline, gasoline oxygenate blends, or aviation turbine fuels. The details of the method should be consulted along with those given in an alternate method for the determination of carbon, hydrogen, and nitrogen in coal and coke (ASTM D3179, ASTM D5373).

A test method (ASTM D1552) is available for sulfur analysis, and the method covers three procedures applicable to samples boiling above 177°C (350°F) and containing not less than 0.06 mass % sulfur. Thus, the method is applicable to most fuel oils, lubricating oils, residua, and coke, and coke containing up to 8% by weight sulfur can be analyzed. This is particularly important for cokes that originate from heavy oil and tar sand bitumen, where the sulfur content of the coke is usually at least 5% by weight.

In the iodate detection system (ASTM D1552), the sample is burned in a stream of oxygen at a sufficiently high temperature to convert about 97% by weight of the sulfur to sulfur dioxide. The combustion products are passed into an absorber containing an acidic solution of potassium iodide and starch indicator. A faint blue color is developed in the absorber solution by the addition of standard potassium iodate solution. As combustion proceeds, bleaching the blue color, more iodate is added. The sulfur content of the sample is calculated from the amount of standard iodate consumed during the combustion.

In the infrared detection system, the sample is weighed into a special ceramic boat that is then placed into a combustion furnace at 1371°C

(2500°F) in an oxygen atmosphere. Most of the sulfur present is converted to sulfur dioxide, which is then measured with an infrared detector after moisture and dust are removed by traps. The calibration factor is determined using standards approximating the material to be analyzed.

For the iodate method, chlorine in concentrations <1 mass % does not interfere. The isoprene rubber method can tolerate somewhat higher levels. Nitrogen when present >0.1 mass % may interfere with the iodate method; the extent of interference is dependent on the types of nitrogen compounds as well as the combustion conditions. It does not interfere in the infrared method. The alkali and alkaline earth metals, zinc, potassium, and lead, do not interfere with either method.

Determination of the physical composition can be achieved by any necessary test methods for determining the toluene insoluble constituents of tar and pitch (ASTM D4072, ASTM D4312). Furthermore, a variety of sample can be employed to give a gradation of soluble and insoluble fractions. The coke, of course, remains in the extraction thimble (Soxhlet apparatus), and the extracts are freed from the solvent and weight to give percent by weight yield(s).

Finally, one aspect that can play a role in compositional studies is the sieve (screening) analysis. Like all petroleum products, sampling is, or can be, a major issue. If not performed correctly, and poor sampling is the result, erroneous and very misleading data can be produced by the analytical method of choice. For this reason, reference is made to standard procedures, such as the *Standard Practice for Collection and Preparation of Coke Samples for Laboratory Analysis* (ASTM D346) and the *Standards Test Method for the Sieve Analysis of Coke* (ASTM D293).

3.3 Density

The *density* (*specific gravity*) of coke has a strong influence on the future use, and can affect the characteristics of the products, such as carbon and graphite.

The density (specific gravity) of coke can be conveniently measured by use of a pycnometer. In the test method (ASTM D5004), the mass of the sample is determined directly, and the volume is derived by determining the mass of liquid displaced when the sample is introduced into a pycnometer. Oil or other material sprayed on calcined petroleum coke to control dust will interfere. Such oil can be removed by flushing with a solvent, which also must be completely removed before the density determination.

The *real density* of coke is obtained when the particle size of the specimen is smaller than 75 mm. The real density (or the particle size) exerts a direct influence on the physical and chemical properties of the carbon and graphite products that are manufactured from the coke.

In the test method (ASTM D2638), a sample is dried and ground to pass a 75-mm screen. The mass of the volume is determined directly, and the volume is derived by the volume of helium displaced when the sample is introduced into a helium pycnometer. The ratio of the mass of the sample to the volume is reported as the real density.

The *vibrated bulk density* (VBD) is an indication of the porosity of calcined petroleum coke, which affects its suitability for use in pitch-bonded carbon applications. This property is strongly dependent upon average particle size and range, and tends to increase with decreasing coke size. In the test method (ASTM D4292), the coke is crushed and 100 g is measured after vibration and the bulk density is calculated. The procedure is limited to particles passing through a 6.68-mm opening sieve and retained on a 0.21-mm opening sieve.

3.4 Dust Control Material

Dust control material is applied to calcined coke to help maintain a dust free environment. It adds weight to the coke, and can have a negative effect on the quality of carbon and graphite artifacts made from the treated coke. Hence, a maximum amount may be specified.

In the test method (ASTM D4930), a weighed dry representative sample of 6.3 mm maximum sized coke is extracted using methylene chloride in a Soxhlet apparatus. The mass of the residue remaining after extraction and evaporation of the solvent is the mass of the dust control material. This test method is limited to those materials that are soluble in a solvent (for example, methylene chloride) that can be used in a Soxhlet extraction type of apparatus. Toluene and methyl chloroform have also been found to give equal results as methylene chloride.

3.5 Hardness

The Hardgrove Grindability Index (HGI) (ASTM D5003) is used to predict the ranking in industrial size mills used for crushing operations, and is commonly used to determine the hardness of coal samples (ASTM D409) (Speight, 1994). The rankings are based on energy required, feed rate, or both. With the introduction of petroleum coke in the coal market, this test method has been extended to the coke. In the current context, the Hardgrove Grindability Index is also used to select raw petroleum coke and coals that are compatible with each other when milled together in a blend, so that segregation of the blend does not occur during particle size reduction.

In the test method (ASTM D5003), the coke sample is crushed to produce a high yield of particles passing a No. 16 sieve and retained on a

No. 30 sieve. These particles are reduced in the Hardgrove grindability machine according to the test method for coal (ASTM D409). The quantity of particles retained on a No. 200 sieve is used to calculate the Hardgrove Grindability Index of the sample. Both this test method and the test method for coal (ASTM D409) produce the same results on petroleum coke samples.

3.6 Metals

The presence and concentration of various metallic elements in petroleum coke are major factors in the suitability of the coke for various uses.

In the test method (ASTM D5056), a sample of petroleum coke is ashed (thermally decomposed to leave only the ash of the inorganic constituents) at 525°C (977°F). The ash is fused with lithium tetra-borate or lithium meta-borate. The melt is then dissolved in dilute hydrochloric acid, and the resultant solution is analyzed by atomic absorption spectroscopy to determine the metals in the sample. However, spectral interferences may occur when using wavelengths other than those recommended for analysis, or when using multi-element hollow cathode lamps.

This test method can be used in the commercial transfer of petroleum coke to determine whether that lot of coke meets the specifications. This method can analyze raw and calcined coke for trace elements of aluminum, calcium, iron, nickel, silicon, sodium, and vanadium. The inductively coupled plasma atomic emission spectroscopy (ICPAES) method (ASTM DD 5600; ASTM D6357) is complementary to this method, and can also be used for determination of metals in petroleum coke.

In the inductively coupled plasma atomic emission spectroscopy (ICPAES) method (ASTM DD 5600), a sample of petroleum coke is ashed at 700°C 1292°F), and the ash is fused with lithium borate. The melt is dissolved in dilute hydrochloric acid, and the resultant solution is analyzed by inductively coupled plasma atomic emission spectroscopy using aqueous calibration standards. Because of the need to fuse the ash with lithium borate or other suitable salt, the fusibility of ash may need attention (ASTM D1857).

The wavelength dispersive X-ray spectroscopy method (ASTM D6376) provides a rapid means of measuring metallic elements in coke, and provides a guide for determining conformance to material specifications. A benefit of this method is that the sulfur content can also be used to evaluate potential formation of sulfur oxides, a source of atmospheric pollution. This test method specifically determines sodium, aluminum, silicon, sulfur, calcium, titanium, vanadium, manganese, iron, and nickel.

In the method, a weighed portion of a sample of coke is dried at 110°C (230°F) and crushed to pass a 200-mesh sieve, mixed with stearic acid, and then milled and compressed into a smooth pellet. The pellet is irradiated with an X-ray beam, and the characteristic X-rays of the elements analyzed are excited, separated, and detected by the spectrometer. The measured X-ray intensities are converted to elemental concentration by using a calibration equation derived from the analysis of the standard materials. The Kα spectral lines are used for all of the elements determined by this test method. This test method is also applicable to the determination of additional elements, provided appropriate standards are available for use and comparison.

3.7 Sulfur

In addition to metallic constituents (*q.v.*), coke also contains considerable amounts of sulfur (ASTM D1552, ASTM D3177, ASTM D4239) that must be determined prior to sale or use.

3.8 Environmental Impact

Because petroleum coke is a solid, fairly inert material, there has not been much concern about the related environmental effects. The typical battery of tests used to measure a chemical's impact on the environment, such as breakdown by sunlight, stability in water, and breakdown in the soil and volatility, cannot be measured for petroleum coke. In addition, the lack of significant adverse effects in animal studies suggests that the residual oil associated with petroleum coke either does not come off, or is present in amounts too low to cause harmful effects.

The specific chemical composition of any given batch of petroleum coke is determined by the composition of the feedstocks used in the coking process, which in turn are dependent upon the composition of the crude oil and refinery processing from which the feedstock is derived. Coke produced from feedstocks with high proportions of asphaltene constituents will contain higher concentrations of sulfur and metals than cokes produced from low-asphaltene feedstocks (Speight, 2007).

Most of the sulfur in coke exists as organic sulfur bound to the carbon matrix. However, the structure of organic sulfur compounds in petroleum coke is largely unknown. Metals, mainly vanadium and nickel, occur as metal chelates or porphyrins in the asphaltene fraction (Speight, 2007). Some metals are intercalated in the coke structure and are not chemically bonded, so they become part of the combustion ash and particulate matter. Metal concentrations in coke normally increase upon calcining, due to the weight loss from evolution of the volatile matter.

Typical parameters measured to define the chemical composition of petroleum coke are: (1) ash % w/w, (2) sulfur % w/w, (3) solvent-extractable material % w/w, (4) nickel % w/w (usually in parts per million – ppm), and (5) vanadium % w/w (usually in parts per million – ppm). Green coke (because of the lower temperature used in its production) contains higher levels of residual hydrocarbon than other grades of coke. Calcining (approximately 1200 to 1350°C, 2190 to 2460°F) removes essentially all of the residual solvent-extractable material.

If released to the environment, both forms of petroleum coke would not be expected to undergo many of the environmental fate pathways. Because petroleum coke is predominantly elemental carbon, it would not be subject to photolytic processes, nor would it be susceptible to biodegradation by microorganisms. Depending on the particle size and density of the material, terrestrial releases will become incorporated into the soil, or transported via wind or surface water flow. If released to the aquatic environment, petroleum coke will either incorporate into sediment or float on the surface, depending on the particle size and density in relation to water.

However, in order to reduce the environmental impact of petroleum coke (through combustion and/or storage in piles which are susceptible to run-off during period of rain and snow melt), gasification (which has a long history of commercialization) has been proposed as a means of converting petroleum coke (and other refinery waste streams) into power, steam, and hydrogen for use in the production of clean fuels. Gasification units are already in operation in a number of refineries, and it is expected that other refineries will add these units in the future (Marano, 2003; Speight, 2011a).

At the high temperatures that are employed in gasification applications, many gasification reactions are equilibrium controlled (Chadeesingh, 2011). Methanation is favored at lower temperatures; thus, little methane is produced. Since the gasifier is operated at adiabatic conditions, the heat liberated by exothermic reactions must balance with the heat required by the endothermic reactions and the heat required to heat the feed streams.

Gasification temperature is controlled by the addition of water or steam: for slurried feedstocks, the slurry water accomplishes this control, but for dry feedstocks, such as petroleum coke, steam must be injected with the feedstock to control temperature. Steam injection may also be used to adjust the composition of the product syngas (Chadeesingh, 2011).

4 Sludge

Solid products from refining operations often receive less attention than the liquid and gaseous products, for a variety of justifiable or unjustifiable reasons. In fact, the term *solid product is* a generic term that is often confusing and meant (incorrectly) to convey the idea of a *waste product*. However, in terms of waste definition, there are three basic approaches (as it pertains to petroleum, petroleum products, and non-petroleum chemicals) to defining petroleum or a petroleum product as hazardous: (1) a qualitative description of the waste by origin, type, and constituents; (2) classification by characteristics based upon testing procedures; and (3) classification as a result of the concentration of specific chemical substances (Chapter 1) (Speight, 2005).

Solid effluents are generated from many of the refining processes and petroleum handling operations, as well as wastewater treatment (Speight and Ozum, 2002; Speight, 2005, 2007). Both hazardous and non-hazardous wastes are generated, treated, and disposed. Refinery wastes are typically in the form of sludge (including sludge from wastewater treatment), spent process catalysts, filter clay, and incinerator ash. Treatment of these wastes includes incineration, land treating off-site, land filling onsite, land filling off-site, chemical fixation, neutralization, and other treatment methods (Speight, 2005).

As with any complex mixture, each component of the sludge can have its own environmental impact, or there is always the potential for associated reactions between the sludge components that change the potential for the environmental impact – usually to a higher potential.

Generally, sludge from petroleum refining operations has high pollution potentials, and an aerobic biological method is an efficient method of its treatment. The choice of aerobic method over anaerobic system lies in the fact that the petroleum industry produces a large volume of sludge coupled with a high concentration of biological oxygen demand (Asia et al., 2006).

A significant portion of the non-petroleum product outputs of refineries is transported off-site and sold as byproducts. These outputs include sulfur, acetic acid, phosphoric acid, and recovered metals. Metals from catalysts and from the crude oil that have deposited on the catalyst during the production often are recovered by third party recovery facilities.

Storage tanks are used throughout the refining process to store crude oil and intermediate process feeds for cooling and further processing. Finished petroleum products are also kept in storage tanks before transport off site. Storage tank bottoms are mixtures of sludge (typically emulsified oil and incompatible organic constituents, as well as iron rust from corrosion) which accumulates at the bottom of tanks. Liquid tank bottoms are periodically drawn off to prevent their continued build up; sludge is also removed during periodic cleaning of tanks for inspection. Tank bottoms may contain amounts of tetraethyl or tetramethyl lead (although this is increasingly rare due to the phase-out of leaded products), other metals, and phenols. Solids generated from leaded gasoline storage tank bottoms are listed as a RCRA hazardous waste.

Disposal of the solid waste can be achieved by installation of a gasifier (Marano, 2003). In fact, a wide variety of feedstocks can be considered for gasification, ranging from solids to liquids to gaseous streams, although when the feed is a gas or liquid, the operation is frequently referred to as partial oxidation (POX). From a process perspective, partial oxidation of gases and liquids is very similar to the gasification of solids.

The major requirement for a suitable feedstock is that it contains a significant content of carbon and hydrogen. Solid feedstocks include solid waste, residua, visbreaker bottoms, petroleum coke, and even biomass: the streams most commonly employed are generally low-value byproducts or waste streams generated by other processes (Speight, 2008, 2011a, 2011b).

5 References

Asia, I.O., Enweani, I. B., and Eguavoen, I.O. 2006. "Characterization and Treatment of Sludge from the Petroleum Industry." *African Journal of Biotechnology*. 5(5): 461–466.

Austin, G.T. 1984. *Shreve's Chemical Process Industries*. 5th Edition. McGraw-Hill, New York. Chapter 37.

Broome, D.C., and Wadelin, F.A. 1973. In *Criteria for Quality of Petroleum Products*. J.P. Allinson (Editor). John Wiley & Sons Inc., New York. Chapter 10.

Chadeesingh, R. 2011. "The Fischer-Tropsch Process." In *The Biofuels Handbook*. Royal Society of Chemistry, London, United Kingdom. Part 3, Chapter 5. pp. 476–517.

Chenier, P.J. 1992. *Survey of Industrial Chemistry*. 2nd Revised Edition. VCH Publishers, New York. Chapters 7.

Gruse, W.A., and Stevens, D.R. 1960. *Chemical Technology of Petroleum*. McGraw-Hill, New York. Chapter 15.

Guthrie, V.B. 1967. In *Petroleum Processing Handbook*. W.F. Bland and R.L. Davidson (Editors). McGraw-Hill, New York. Section 11.

Hoffman, H.L. 1983. In *Riegel's Handbook of Industrial Chemistry*. 8th Edition. J.A. Kent (Editor). Van Nostrand Reinhold Company, New York. Chapter 14.

Hoffman, H.L., and McKetta, J.J. 1993. "Petroleum Processing." In *Chemical Processing Handbook*. J.J. McKetta (Editor). Marcel Dekker Inc., New York. p. 851.

Institute of Petroleum. 2011. *IP Standard Methods 2001*. The Institute of Petroleum, London, United Kingdom.

Marano, J.J. 2003. *Refinery Technology Profiles: Gasification and Supporting Technologies*. National Energy Technology Laboratory and Energy Information Administration, United States Department of Energy, Washington, DC.

Speight, J.G. 1992. "Asphalt." *Kirk-Othmer Encyclopedia of Chemical Technology*. 4th Edition. 3: 689.

Speight, J.G. 1994. In *Asphaltenes and Asphalts, I. Developments in Petroleum Science,* 40. T.F. Yen and G.V. Chilingarian (editors), Elsevier, Amsterdam. Chapter 2.

Speight, J.G. 2000. *The Desulfurization of Heavy Oils and Residua*. 2nd Edition. Marcel Dekker Inc., New York.

Speight, J.G. 2001. *Handbook of Petroleum Analysis*. John Wiley & Sons Inc., New York.

Speight, J.G. and Ozum, B. 2002. *Petroleum Refining Processes*. Marcel Dekker Inc., New York.

Speight, J.G. 2005. *Environmental Analysis and Technology for the Refining Industry*. John Wiley & Sons Inc., Hoboken, New Jersey.

Speight, J.G. 2007. *The Chemistry and Technology of Petroleum*. 4th Edition. CRC Press, Taylor & Francis Group, Boca Raton, Florida.

Speight, J.G. 2008. *Synthetic Fuels Handbook: Properties, Processes, and Performance*. McGraw-Hill, New York.

Speight, J.G. 2011a. *The Refinery of the Future*. Gulf Professional Publishing, Elsevier, Oxford, United Kingdom.

Speight, J.G. (Editor). 2011b. *The Biofuels Handbook*. Royal Society of Chemistry, London, United Kingdom.

Weissermel, K., and Arpe, H-J. 1978. *Industrial Organic Chemistry*. Verlag Chemie. New York. Chapter 13.

7

Sample Collection and Preparation

1 Introduction

During any petroleum spill incident, the properties of the spilled oil must be known immediately. Unfortunately, the properties routinely measured by oil producers and refiners are not the ones that on-scene bioremediation scientists and engineers need to know most urgently. The data from analytical test methods carried out for the refining industry typically focus on the properties and characteristics of the crude oil as they pertain to refining and product slate. Similarly, the data from analytical test methods applied to petroleum products are focused on whether or not the product meets specifications and is suitable for sale.

Many of the specific chemicals in petroleum are hazardous because of their chemical reactivity, fire hazard, toxicity, and other properties. In fact, a simple definition of a hazardous chemical (or hazardous waste) is that it is a chemical substance (or chemical waste) that has been inadvertently released, discarded, abandoned, neglected, or designated as a waste material, and has the potential to be detrimental to the environment. Alternatively, a hazardous chemical may be a chemical that may interact with other (chemical) substances to give a product that is hazardous to the environment. Thus, it is essential to obtain analytical data relating to the properties and behavior of the spilled material.

Analytical data related to properties such as (1) evaporation rate, (2) the viscosity of the residual oil as lower boiling constituents evaporate, (3) to what extent dispersibility will occur naturally, or be enhanced with dispersants, (4) oxidation by aerial oxygen, (5) emulsion formation, (6) the density of the oil and any emulsions relative to the ecosystem – this is especially relevant for aqueous or marine spills – to indicate whether the oil is likely to sink or submerge, (7) the viscosity of the oil, at ecosystem temperatures, and (8) health hazards to on-site personnel from volatile organic compounds, and the toxicity to the indigenous flora and fauna.

While all of the above properties are important and relevant to bioremediation efforts, the fate of petroleum and petroleum products spilled into rivers, lacustrine water, and marine water is often difficult to predict, because of (among other issues) oxidation (Chapter 2). Petroleum and petroleum products oxidize on the surface, after which they form oil-in-water emulsions. The inclusion of polar functions, such as hydroxyl groups (-OH) or carbonyl groups (>C=O) (a result of the oxidation process), causes an increase in the density of the emulsion (relative to the original unoxidized petroleum) and an increased propensity to form emulsions. As a result, the emulsion sinks to various depths or even to the seabed, depending on the extent of the oxidation and the resulting density. This may give the appearance that the petroleum spill (as evidenced from the petroleum remaining on the surface of the water) is less than it actually was. The so-called *missing* oil will undergo further chemical changes, and eventually reappear on the water surface or on a distant beach.

Despite the nature of the environmental regulations and the precautions taken by the refining industry, the accidental release of crude petroleum and petroleum products into the environment has occurred, and, without being unduly pessimistic, will continue to occur. It is a situation that, to paraphrase *chaos theory*, no matter how well one prepares, the unexpected is inevitable.

Once the spill has occurred, the environmental analyst and bioremediation specialist have to identify the nature of the spilled material, and make deductions about the potential effects on the ecosystem(s) (Smith, 1999). Although petroleum itself and its various products are complex mixtures of many organic chemicals (Chapter 3), the predominance of one particular chemical or one particular class of chemicals may offer the environmental analyst or scientist an opportunity to monitor and study levels of pollutants in the ecosystem.

However, all analytical methods and the resulting data are severely affected by spatial heterogeneity, which requires using different analytical approaches and techniques so that topological and geographical properties are taken into account. Complex issues arise in spatial analysis, including the limitations of mathematical knowledge, the assumptions required by existing statistical techniques, and problems in computer based calculations. Hence, it is essential that a well-thought out sampling plan be designed according to valid statistical principles involving randomization and replication of treatments, in order to ensure that monitored results reflect reality in such a highly heterogeneous environment.

Furthermore, *intrinsic bioremediation*, the combined effect of natural destructive and non-destructive processes to reduce the mobility, mass, and associated risk of a contaminant (Chapter 1), is an important attenuation mechanism at contaminated field sites, because it limits pollutant migration and reduces contaminant mass in the subsurface. Quantifying the impact of intrinsic bioremediation at a field site involves conducting a comprehensive field sampling program in conjunction with extensive modeling of contaminant transport and fate.

2 Petroleum Chemicals

Briefly, for environmental purposes, petroleum constituents are subdivided into two classes: (1) *organic constituents* and (2) *inorganic constituents*. Furthermore subcategorization can be taken to the next level, insofar as organic chemicals are classified as (1) *volatile organic compounds*, (2) *semi-volatile organic compounds*, and (3) non-volatile compounds. The same subcategorization is also applied to the inorganic constituents of petroleum.

The first class of organic compounds, the *volatile organic compounds* (VOCs), is subdivided into *regulated compounds* and *unregulated compounds*. Regulated compounds have maximum contaminant levels, but unregulated compounds do not. Regulated compounds generally (but not always) have low boiling points, or low boiling ranges, and some are gases. Many of these chemicals can be detected at extremely low levels by a variety of instrumentation, including the human nose! In the case of the petroleum industry, sources for volatile organic compounds typically are petroleum refineries, fuel stations, naphtha (i.e., dry cleaning solvents, paint thinners, cleaning solvents for auto parts), and, in some cases, refrigerants that are manufactured from petrochemicals.

The second class of organic compounds, the *semi-volatile compounds*, typically have high boiling points, or high boiling ranges, and are not always easily detected by the instrumentation that may be used to detect the volatile organic compounds (including the human nose). Some of the common sources of contamination are high boiling petroleum products (e.g., lubricating oils), pesticides, herbicides, fungicides, wood preservatives, and a variety of other chemicals that can be linked to the refining industry.

The third group of constituents, the *non-volatile compounds*, offer a group of constituents that are difficult to define chemically. These are the extremely high-boiling constituents that continue to defy identification by a wide variety of techniques, leaving their chemical structure and behavior open to much speculation (Speight, 2007a).

Regulations are in place that set the maximum contamination concentration levels that are designed to ensure public safety. There are primary and secondary standards for inorganic chemicals. Primary standards are those chemicals that cause neurological damage, cancer, or blood disorders. Secondary standards are developed for other environmental reasons. In some instances, the primary standards are referred to as the *Inorganic Chemical Group*. The secondary standards are referred to as the *General Mineral Group* and *General Physical Testing Group*.

In the context of petroleum, the inorganic chemical group includes nickel, vanadium, iron, and copper, as well as sodium and potassium (from reservoir brines) and a variety of trace metals that vary depending upon the source of the crude petroleum (Speight, 2007a). The mineral group includes calcium, magnesium, sodium, potassium, bicarbonate, carbonate, chloride, sulfate, pH, alkalinity, hardness, electrical conductivity, total dissolved solids, surfactants, copper, iron, manganese, and zinc. The physical group includes turbidity, color, and odor. Many of these chemicals arise from desalting residues and from other processes where catalysts are used. A high level of any of these three chemicals in the soil or in the water is an indication that one or more specific processes (identified from the chemicals that have been released) or pollution prevention processes are not performing according to operational specifications.

Another source of toxic compounds is combustion (Speight, 2007a). In fact, some of the greater dangers of fires are from toxic products and by-products of combustion. The most obvious of these is carbon monoxide (CO), which can cause serious illness or death, because it forms carboxyhemoglobin with hemoglobin in the blood, so that the blood no longer carries oxygen to body tissues. Toxic sulfur dioxide and hydrogen chloride are formed by the combustion of sulfur compounds and organic chlorine compounds, respectively. A large number of noxious organic compounds, such as aldehydes, are generated as by-products of combustion, and, in addition to forming carbon monoxide, combustion under oxygen-deficient conditions produces polynuclear aromatic hydrocarbons consisting of fused-ring structures. Some of these compounds, such as benzo(a)pyrene, are pre-carcinogenic compounds, insofar as they are acted upon by enzymes in the body to yield cancer-producing metabolites.

While most investigations involving petroleum hydrocarbons are regulated by various agencies that may require methodologies, action levels, and cleanup criteria that are differ in some respects, the complex chemical composition of petroleum and petroleum products can make it extremely difficult to select the most appropriate analytical test methods for evaluating environmental samples, and to accurately interpret and use the data.

Accordingly, general methods of environmental analysis (Smith, 1999), i.e., analysis for the determination of petroleum or petroleum products that have been released, are available. The data determine whether or not a release of such chemicals will be detrimental to the environment, and may lead to regulations governing the use and handling of such chemicals. But first, sample collection, preservation, preparation, and handling protocols must be followed to the letter. This, of course, includes *chain of custody* or *sampling handling* protocols that will be defensible if and when legal issues arise. Thus, an accurate sample handling and storage log should be maintained, and should include the basic necessary information (Table 7.1). Attention to factors such as these enables standardized comparisons to be made when subsequent samples are taken (Speight, 2001, 2002; Dean, 2003; Speight, 2005).

Table 7.1 Suggested items for inclusion in a sampling log.

1.	The precise (geographic or other) location (or site or refinery or process) from which the sample was obtained.
2.	The identification of the location (or site or refinery or process) by name.
3.	The character of the bulk material (solid, liquid, or gas) at the time of sampling.
4.	The means by which the sample was obtained.
5.	The means and protocols that were used to obtain the sample.
6.	The date and the amount of sample that was originally placed into storage.
7.	Any chemical analyses (elemental analyses, fractionation by adsorbents or by liquids, functional type analyses) that have been determined to date.
8.	Any physical analyses (API gravity, viscosity, distillation profile) that have been determined to date.
9.	The date of any such analyses included in items 7 and 8.
10.	The methods used for analyses that were employed in 7 and 8.
11.	The analysts who carried out the work in 7 and 8.
12.	A log sheet showing the names of the persons (with the date and the reason for the removal of an aliquot) who removed the samples from storage and the amount of each sample (aliquot) that was removed for testing.

Whatever the case, methods of analysis must be available to determine the nature of the released chemical (waste), and from the data predict the potential hazard to the environment.

3 Sample Collection and Preparation

The ability to collect and preserve a sample that is representative of the site is a critically important step (Weisman, 1998; Dean, 2003; Patnaik, 2004). Obtaining representative environmental samples is always a challenge, due to the heterogeneity of different sample matrices. Additional difficulties are encountered with petroleum hydrocarbons, due to the wide range in volatility, solubility, biodegradation, and adsorption potential of individual constituents. Additionally, the procedures used for sample collection and preparation must be legally defensible.

There are many recommended sampling protocols (Table 7.2). The sampling methods used for petroleum hydrocarbons are generally thought of as methods for determination of the *total petroleum hydrocarbons*. In part due to the complexity of the components of the total petroleum hydrocarbons fractions, little is known about their potential for health or environmental impacts. As gross measures of petroleum contamination, the data related to the *total petroleum hydrocarbons* simply show that petroleum hydrocarbons are present in the sampled media. Measured total petroleum hydrocarbons values suggest the relative potential for human exposure and, therefore, the relative potential for human health effects.

Furthermore, in addition to the questionable aspect of the test method for the determination of total petroleum hydrocarbons (Speight, 2005), one aspect that is often ignored relates to the interaction of the various petroleum constituents and the petroleum product constituents with each other in the context of the ecosystem into which they are spilled: the potential for interactive behavior has not been studied to any great extent and should be addressed as soon as possible after the spill. While it is assumed that the constituents of petroleum as well as those of petroleum products are compatible, various effects in the ecosystem might induce incompatibility (Speight, 2001) thereby introducing errors in the analytical method.

Although most site investigations to determine the assessment of contamination of an ecosystem by petroleum hydrocarbons are regulated by local or by regional (state) governments, sample collection and preservation recommendations follow strict guidelines (Table 7.1), thereby reducing the potential for sample compromise. Therefore, before a sample is collected, the particular sample collection and preservation requirements must be investigated. And, because of holding (storage) time considerations, the laboratory must be selected and notified prior to the collection of the samples.

Table 7.2 Recommended sampling protocols (Environmental protection agency).

Analytical Parameter	Analytical Method(s)	Medium	Sample Container[a]		Preservatives[b]	Holding Time
			Volume	Type		
Trph[c]	EPA 418.1 (IR); gravimetric	Water	1 liter	Glass jar with Teflon-lined cap	Acid fix pH < 2; cool to 4°C	Extract in 7 days; analyze in 40 days
	GC/FID	Soil	125 mL	Widemouth glass with Teflon-lined cap	Cool to 4°C	Extract in 7 days; analyze in 40 days
Volatile petroleum hydrocarbons[d]	Various	Water	40 mL	Glass vial with Teflon-lined septum	Acid fix pH < 2; cool to 4°C	14 days
		Soil	40 mL	Glass vial with Teflon-lined septum	Cool to 4°C[e]	14 days
Extractable petroleum hydrocarbons[f]	Various	Water	1 liter	Glass jar with Teflon-lined cap	Acid fix pH < 2; cool to 4°C	Extract in 7 days; analyze in 40 days
		Soil	60 mL	Widemouth glass with Teflon-lined cap	Cool to 4°C	Extract in 7 days; analyze in 40 days

Source: Speight, 2001.

3.1 Soil Sampling

Soil testing has become an important method tool for assessing soil contamination and for arriving at proper cleanup recommendations (Weisman, 1998). Soil testing is also a valuable management aid for studying soil changes resulting from oil spills, and for diagnosing specific bioremediation methods. However, soil variability and dispersion of the oil spill are major concerns when deciding how to collect a representative soil sample. Soil samples submitted for analysis should be *representative* of the affected area. Therefore, sampling from a part of the affected area may not always produce a typical (or average) representation of the magnitude of the oil spill.

Soil samples can be collected using a variety of methods, depending on the depth of the desired sample, the type of sample required (disturbed vs. undisturbed), and the soil type. Soil sampling techniques range from subsampling or two-stage sampling, double sampling, composite sampling, random sampling, and stratified sampling. But the basis for most sampling plans in environmental sampling is the concept of random sampling. With random sampling, each sample point within the site has an equal probability of being selected (US EPA, 1992).

Therefore, in the study and selection of a site, the main consideration in physical soil sampling is the method of randomization of the soil samples, because the sample distribution usually depends on the degree of variability in a given area. Therefore, for those situations where there is inadequate information for developing a conceptual model for a site or for stratifying the site, it may be necessary to use a random sampling design.

Random sampling is the basis for all probability sampling techniques used in soil sampling, and serves as a reference point from which modifications to increase the efficiency of sampling are evaluated. Where there is a lack of information, as with oil spills, the simple random sampling design is the only design other than the systematic grid that can be used.

Soil samples used in various studies are usually collected from various locations and depths on the contaminated sites (Laleh et al., 2003). Random sampling has been employed in past investigations of oil spills, where the sampled soil was used for the analysis of total petroleum hydrocarbons (Reynolds et al., 1997), testing the effect of various concentrations of crude oil on fungal populations of soil (Obire and Anyanwu, 2009), hydrocarbon degradation of refined petroleum hydrocarbon in soils treated with 5% gasoline, kerosene, and diesel oil that was being investigated (Obire and Nwaubeta, 2001), or even crude oil that was isolated from oil-contaminated soil in Bangkok, Thailand (Palittapongarnpim et al., 1998).

Given the nature of the study, where there is a lack of information about the area to be studied and about the pollution distribution, the simple random sampling design may be the most suitable, but the disadvantages of such sampling must be recognized (Speight, 2005). Random sampling gives a suitable indication as to what is happening with time, and, from the results obtained from this sampling methodology, it is often possible to implement specific test methods for further analysis.

In summary, the random sampling method should be used with caution, especially when there is only limited data on the area being investigated. After which, deductions based on the results would assist in choosing a more appropriate sampling method for more accurate data collection.

Finally, the objective of sampling is to obtain a portion (spot sample) of material from a selected area within a container that is representative of the material in the area, or, in the case of running or all-level samples, a sample whose composition is representative of the total material in the container. A series of spot samples may be combined to create a representative sample (ASTM D4057). Hence, in all soil sampling procedures, the possibility of cross-contamination must be minimized by using disposable sampling equipment that is certified as clean for each sample collected. If disposable sampling tools are not available, specified cleaning procedures for reuse of the sample container should be used.

3.2 Sampling from an Aqueous Medium

Although in-situ bioremediation has been used for a number of years in the restoration of ground water contaminated by petroleum hydrocarbons, it has only been in recent years that this technology has been directed toward other classes of contaminants.

As with any sampling program, the goal of water sampling is to collect samples that are *representative* of the aqueous conditions, and to minimize changes in groundwater chemistry during sample collection and handling. Water sample collection and handling procedures can be a source of variability in water-quality concentrations, due to differences in sampling personnel, sampling procedures, and equipment.

Several different groundwater sampling procedures can be used, which vary primarily through the criteria used to determine when a sample is representative of water conditions. No single method or procedure is universally applicable to all types of groundwater-sampling programs; therefore, consideration should be given to a variety of factors when determining which method is best suited to site-specific conditions. These site-specific conditions include sampling objectives, equipment availability, site location, and physical constraints.

The design of a water-sampling program includes: (1) the horizontal flow direction, (2) the hydraulic conductivity contrasts of the geologic materials controlling contaminant transport from the site (thick units versus thin beds versus fractures), the types of contaminants to be sampled (such as volatile organic compounds, semi-volatile organic compounds, metals), and factors that could bias sampling results (such as turbidity for metals, co-solvation effects on organic compounds), and (3) lateral and vertical distribution of contamination (contaminants distributed throughout an entire unit being monitored versus localized distribution controlled by small scale features and other influences).

Being a dynamic system, the frequency of sampling from, for example, an aquifer, should be related to the time expected for significant changes to occur along the most contaminated flow path. Important considerations include the time interval required for water to move from injection wells to monitoring wells, seasonal variations in water table elevation or hydraulic gradient, changes in the concentration of dissolved oxygen or alternative electron acceptor, and costs of monitoring.

Continuous coring of subsurface solids and close interval (i.e., <1 m) sampling of water should be considered in many investigations for volatile organic compounds. To approach this level of depth detail in sampling, *push technologies* (which rely on hydraulic or hammer-driven, narrow diameter i.e., <2 inch probes for solid or water sampling) and/or multilevel sampling devices present very useful tools for site characterization.

Semi-annual or annual sampling of wells, particularly multi levels appropriately designed and completed, should be quite useful over the course of a long-term monitoring program. Sampling should track the down-gradient progress of risk-associated target compounds, and permit testing predictions of intrinsic bioremediation effects on risk reduction.

For both land-based (soil) and aqueous (river, lacustrine, and marine) media, the use of appropriate control samples, such as assays of untreated areas or areas outside the contaminant plume, is recommended to confirm the choice and effectiveness of the bioremediation technology (Suflita, 1989).

3.3 Sample Collection

The value of any analysis is judged by the characteristics of the sample as determined by laboratory tests. The sample used for the test(s) must be representative of the bulk material, or data will be produced that are not representative of the material, and will be, to be blunt, incorrect, no matter how accurate or precise the test method. In addition, the type and cleanliness of sample containers are important: if the container is contaminated, or is made of material that either reacts with the product or is a catalyst, the test results may be wrong.

Thus, the importance of the correct sampling of any sample destined for analysis should always be overemphasized. Incorrect sampling protocols can lead to erroneous analytical data, from which decisions about regulatory issues cannot be accurately made. In addition, adequate records of the circumstances and conditions during sampling have to be made: for example, in sampling from storage tanks, the temperatures and pressures of the separation plant and the atmospheric temperature would be noted.

At the other end of the volatility scale, samples that contain, or are composed of, high molecular weight paraffin hydrocarbons (wax) that are also in a solid state, may require judicious heating (to dissolve the wax) and agitation (homogenized, to ensure thorough mixing) before sampling. If room temperature sampling is the *modus operandi* and product cooling causes wax to precipitate, homogenization to ensure correct sampling is also necessary.

The first task in any analysis is to separate the analytes from the bulk of the sample. The traditional liquid extraction is the most common means employed, but alternate methods are also available (Table 7.3) (Dean, 1998 and references cited therein). A portion of the sample is mixed with an organic solvent, into which the analyte is preferentially partitioned. The idea of partitioning is the key to the success of this procedure. No organic analyte is completely removed from a sample by a single washing with an organic solvent.

Representative samples are prerequisite for the laboratory evaluation of any type of environmental sample, and many precautions are required in obtaining and handling representative samples (ASTM D270, ASTM D1265). The precautions depend upon the sampling procedure, the characteristics (low boiling or high boiling constituents) of the product being sampled, and the storage tank, container, or tank carrier from which the sample is obtained. In addition, the sample container must be clean, and the type to be used depends not only on the product, but also on the data to be produced.

However, obtaining a representative sample is not just a matter of relying on partition coefficients. Partitioning is thermodynamically controlled, and to achieve equilibrium requires a certain amount of intimate contact time between the sample matrix, the analyte molecules, and the organic solvent. Swirling an aqueous sample with the solvent is not effective for achieving a thermodynamic equilibrium. Mixing the two phases together as thoroughly as possible and then allowing the phases to separate achieves the desired equilibrium. The procedure should be repeated with additional portions of solvent. If equilibrium is not established, the efficiency of the extraction procedure is low, and the analytical data will be subject to error and indefensible (Dean, 1998 and references cited therein).

The basic objective of each procedure is to obtain a truly representative sample, or, more often, a composite of several samples that can be considered to be a representative sample. In some cases, because of the size of the storage tank and the lack of suitable methods of agitation, several samples are taken from large storage tanks in such a manner that the samples represent the properties of the bulk material from different locations

Table 7.3 Recommended extraction methods (Environmental protection agency).

Extraction Method	EPA SW-846 Method Number	Extraction Matrix	Compounds Extracted	Purpose
Separatory funnel	3510	Water	Semivolatile, nonvolatile	Laboratory
Continuous liquid-liquid	3520	Water	Semivolatile, nonvolatile	Laboratory
Solid-phase extraction	3535	Water	Semivolatile, nonvolatile	Laboratory / screening
Purge and trap	5030, 5035	Water, soil	Volatile	Laboratory / field preservation
Headspace	3810, 5021	Water, soil	Volatile	Screening / laboratory
Shake and vortex	*a*	Soil	Volatile, semivolatile, nonvolatile	Screening / laboratory
Soxhlet	3540, 3541	Soil	Semivolatile, nonvolatile	Laboratory
Sonication	3550	Soil	Semivolatile, nonvolatile	Laboratory
Supercritical fluid	3560, 3561	Soil	Semivolatile, nonvolatile	Laboratory
Subcritical fluid	3545	Soil	Semivolatile, nonvolatile	Laboratory

Source: Dean, 1998, and references cited therein.

[a]Not an EPA SW-846 approved method.

in the tank, and, thus, the composite sample will be representative of the entire lot being sampled. This procedure allows for differences in sample that might be due to the stratification of the bilk material due to tank size or temperature at the different levels of the storage tank. Solid samples require a different protocol that might involve melting (liquefying) the bulk material (assuming that thermal decomposition is not induced) followed by homogenization. On the other hand, the protocol used for coal sampling (ASTM D346, ASTM D2013) might also be applied to sampling petroleum products, such as coke, that are solid, and for which accurate analysis is required before sales.

Once the sampling procedure is accomplished, the sample container should be labeled immediately, to indicate the product, time of sampling, location of the sampling point, and any other information necessary for the sample identification. Also, if the samples were taken from different levels of the storage tank, the levels from which the samples were taken, and the amounts taken and mixed into the composite, should be indicated on the sample documentation.

Sampling records for any procedure must be complete, and should include, but are not restricted to, items relating to the origin of the sample, methods of storage, analytical tests preformed, the test methods used, and the analyst(s) who performed the test methods (Table 7.1).

In summary, there must be a means to identify the sample history as carefully as possible, so that each sample is tracked and defined in terms of source and activity. Thus, the accuracy of the data from any subsequent procedures and tests for which the sample is used will be placed beyond a *reasonable doubt*. If this is not the case, that data should be re-evaluated without personal bias, and (if unsuitability prevails) discarded as being inadequate for use.

3.3.1 Volatile Compounds

A volatile substance is one whose boiling point, or sublimation temperature, is such that it exists to a significant extent in the gaseous phase under ambient conditions. Some commonly encountered volatile hydrocarbons and semi-volatile hydrocarbons from petroleum and petroleum products organized by carbon number are:

Aliphatic	Aromatic
Pentane isomers	
Hexane isomers	Benzene
Heptane isomers	Toluene
Octane isomers	Ethylbenzene, xylenes
Nonane isomers	
Decane isomers	

There are several sampling procedures that are applicable to volatile compounds, but often, method application depends upon the compound(s) to be sampled (Dean, 2003). Part of the issue of sampling volatile compounds arises because some volatile substances sublime rather than boil, whereas other volatile substances emit significant quantities of vapor well below their boiling point.

For sampling volatile hydrocarbons in the field, two procedures are generally recommended: (1) the zero headspace method and (2) the solvent extraction method. However, these two procedures do not necessarily give equivalent results.

Zero headspace procedures involve the collection of a soil sample with immediate transfer to a container into which the sample fits exactly. The only space for gases is that within the soil pores. The volume of sample collected depends upon the concentration of volatiles in the soil. It is imperative that the container employed can be interfaced directly with the gas chromatograph. Several commercial versions of zero headspace sampling devices are available. The sample is transported to the laboratory at 4°C, where it is analyzed directly by purge and trap gas chromatography (EPA 5035) or other appropriate techniques, such as vacuum distillation (EPA 5032) or headspace (EPA 5021).

Solvent extraction procedures involve collection of sample by an appropriate device and subsequent immediate placement into a borosilicate glass vessel, which contains a known quantity of ultra pure methanol. The bottle is then transported to the laboratory at 4°C (39°F), and the methanol fraction analyzed by purge-and-trap gas chromatography (or similar procedure).

In general, the zero headspace procedures are employed when the concentrations of volatile compounds in the soil are relatively low, and the solvent extraction methods are used for more polluted soils. Irrespective of which procedure is used, quantitation of volatiles in soil is subject to serious errors if sufficient care is not taken with the sampling operation. Although direct purge-and-trap methods are frequently advocated for determination of volatiles in samples collected by zero headspace procedures, there are certain problems associated with this technique. Caution is advised, since the procedure collects that fraction of the volatile compounds that exists in a free form within the soil pore spaces, or is at least in a facile equilibrium with this fraction. Volatile compounds that are adsorbed on to the soil constituents will not be collected.

Gas chromatography detectors employed for the determination of volatile organics in soil are generally flame ionization detectors (FID), photo-ionization detectors (PID), or mass spectrometry. Flame ionization detectors will respond to all carbon compounds in the sample, whereas the photo-ionization detector is capable of some sensitivity, by virtue of the energy of the lamp employed. A stronger lamp will yield a more specific response to unsaturated (including aromatic) hydrocarbons, and may also be employed to give a complete BTEX (benzene, toluene, ethyl benzene and xylene) characterization at sites where this is likely to be an issue.

In regard to the columns used for the analysis of volatile hydrocarbons, a wide variety can be used. Large bore capillary columns of length typically about 105 meters are generally employed, although there may be some variation in choice of column according to the resolution required. There is some debate concerning appropriate limits for the gasoline range, and this is reflected in disparate test methods: for example, the upper range of the organic constituents of gasoline may be defined by naphthalene (boiling point: 218°C, 424°F) or dodecane (boiling point: 216°C, 420°F). While there is not a wide differential between the boiling points of these two compounds, a range of a few degrees can result in different conclusions of the data produced by the test method. When comparisons are made between the data from the two test methods, recognition of the differential is essential.

Typical gas chromatography conditions involve an oven temperature ramped between 40 and 240°C (104 and 464°F), with an injector at 200°C (390°F) and a detector maintained at 250°C (482°F). There are two methods of calibration for the gas chromatograph. One method consists of analyzing a mixture of individual hydrocarbons that bracket the gasoline range, and calculating an average response factor from the response for each individual component. The other method involves analyzing a standard that contains one or more gasoline.

3.3.2 Condensate Releases

Condensate releases might be equated to the release of volatile constituents, but are often named as such because of the specific constituents of the condensate, often with some reference to the gas condensate that is produced from certain petroleum wells and natural gas wells. However, the condensate is often restricted to the hydrocarbons up to and including n-octane, as well as traces of benzene, toluene, ethyl benzene, and xylenes (BTEX) family of compounds (Mokhatab et al., 2006; Speight, 2007a).

To determine the concentrations of benzene, toluene, ethyl benzene, and xylenes in condensate released, approved methods (such as EPA SW 846 8021B, EPA SW 846 8260) are not only recommended, but are insisted upon for regulatory issues. Polynuclear aromatic hydrocarbons (PAHs) may be present in condensate, and evaluation of condensate contamination should include the use of other test methods (EPA SW 846 8270, EPA SW 846 8310), provided that the detection limits are adequate to the task of soil and groundwater protection. Generally, at least one analysis may be required for the most contaminated sample location from each source area. Condensate releases in non-sensitive areas require analysis for naphthalene only. The analysts should ensure that the method has detection limits that are appropriate for risk determinations.

3.3.3 Semi-Volatile and Non-Volatile Compounds

In almost all cases of hydrocarbon contamination, some attention will have to be paid to the presence of semi- and non-volatile hydrocarbons. However, the collection, handling of samples and their ultimate preparation for analysis is entirely different from that used for volatile hydrocarbons. In general, it is not necessary to take such rigorous procedures to prevent loss of analyte following collection, although the procedures should still be verified using appropriate quality control measures (Dean, 2003).

Before analysis of semi- or non-volatile components can proceed, it is necessary that the hydrocarbon components be brought into solution. In a sample from a contaminated site, semi- and non-volatile molecules may exist in the soil pores in the free form within the pore spaces, but are far more likely to be adsorbed by organic matter attached to the soil. Indeed, the probability of such adsorption increases with increasing hydrophobicity of the molecules.

A number of procedures are available to help this dissolution, and include Soxhlet extraction (EPA 3540C), ultrasonic extraction (EPA 3550B), thermal extraction (EPA 8275A), and supercritical fluid extraction (EPA 3560, EPA 3561). Although these procedures are well documented, some of their important details are frequently overlooked, with the result that the extraction is unsatisfactory. In the case of ultrasonic extraction, the method (EPA 3550B) stipulates the use of an ultrasonic disrupter of the horn type, with a minimum power of 300 watts. Many laboratories, however, wrongly interpret this to mean an ultrasonic bath, used for cleaning glassware. Such baths are of far lower energy, and are not capable of separating the hydrocarbons from their association with humic material. As regards the use of supercritical fluid extraction, a methanol modifier is required to achieve complete extraction of polynuclear aromatic hydrocarbons, whereas supercritical carbon dioxide is sufficient to elute normal hydrocarbons (EPA 3560, EPA 3561).

For most analyses, it is necessary to separate the analytes of interest from the matrix (i.e., soil, sediment, and water). Extraction of analytes can be performed using one or more of the following methods: (1) extracting the analytes into a solvent, (2) heating the sample, as may be necessary to remove the solvent and for the analysis of volatile compounds, and (3) purging the sample with an inert gas, as is also used in the analyses of volatile compounds.

Soxhlet extraction, sonication, supercritical fluid extraction, sub-critical or accelerated solvent extraction, and purge and trap extraction have been introduced into a variety of methods for the extraction of contaminated soil. Headspace is recommended as a screening method. Shaking/vortexing is adequate for the extraction of petroleum hydrocarbons in most environmental samples. For these extraction methods, the ability to extract petroleum hydrocarbons from soil and water samples depends on the solvent and the sample matrix. Surrogates (compounds of known identity and quantity) are frequently added to monitor extraction efficiency. Environmental laboratories also generally perform matrix spikes (addition of target analytes) to determine if the soil or water matrix retains analytes.

Thus, solvents have different extraction efficiencies, and thus, extracting the same sample in the same manner by two different solvents may result in different concentrations. The choice of solvents is determined by many factors, such as cost, spectral qualities, method regulations, extraction efficiency, toxicity, and availability. Methylene chloride has been the solvent of choice for many semi-volatile analyses, due to its high extraction efficiency. Chlorofluorocarbon solvents, such as trichlorotrifluoroethane (Freon 113), have been used in the past for oil and grease analyses because of their spectral qualities (they do not absorb in the 2930 cm^{-1} infrared measurement wavelength) and low human toxicity.

The use of chlorofluorocarbons is to be questioned, because not all of the petroleum constituents are soluble in such solvents. Furthermore, the use of chlorofluorocarbons solvents is being reduced, even phased out of analytical methods, because of their detrimental effects on stratospheric ozone. Tetrachloroethylene and carbon tetrachloride are possible replacements, but caution is advised, since these solvents may be sensitive to light, and leave residual chlorine in the sample. Methanol is the most common solvent used to preserve and extract volatiles such as benzene, toluene, ethylbenzene, and xylene(s) in soil. But methanol is unsuitable for many constituents of petroleum and petroleum products. In short, there is no one solvent that will satisfy all of the criteria necessary of a complete and full extraction or solubility of petroleum and its products.

If the release has contaminated a water system, there are several methods that can be employed for sample separation (Table 7.4) (Patnaik, 2004,

Table 7.4 Methods applicable to the separation of volatile and semi-volatile organic compounds from water samples.

For Volatiles	For Semi-volatiles
Purge and trap	Separatory funnel extraction
Headspace	Continuous liquid-liquid extraction Solid phase extraction

Section 2). Volatile compounds (gasoline, solvents) in water are generally separated from the aqueous matrix by purging with an inert gas and trapping the compounds on a sorbent (EPA 5030, purge and trap analysis). The sorbent is later heated to release the volatile compounds, and a carrier gas sweeps the compounds into a gas chromatograph. Headspace analysis is recommended as a screening method (EPA 3810, EPA 5021), although it performs well in particular situations, especially field analysis. In this method, the water sample is placed in a closed vessel with a headspace and heated to drive volatiles into the gas phase, and instrument contamination is minimized, because only volatile compounds are introduced into the instrument. Addition of salts or acids may enhance this process.

Samples containing heavy oil, along with the volatile components, can severely contaminate purge and trap instrumentation, and caution is advised when interpreting the data. For such samples, it may be advisable to use a separatory funnel for the water extraction method for semi-volatiles involves extraction using a separatory funnel (EPA 3520). In this method, the sample is poured into a funnel-shaped piece of glassware, solvent is added, and the mixture is shaken vigorously. After layer separation, the extract (i.e., the solvent layer) is removed, filtered, dried with a desiccant, and concentrated. Multiple extractions on the same sample may increase overall recovery.

Another commonly used water extraction method for semi-volatiles involves continuous liquid-liquid extraction (EPA 3520). In this method, the sample (rather than being shaken with the solvent) is treated with a continuously heated solvent that is nebulized (broken into small droplets) and sprayed on top of the water. Liquid-liquid extraction is excellent for samples containing emulsion-forming solids, but it is more time-consuming than separatory funnel extractions. Nevertheless, time-consuming or not, the method must produced reliable data that can be used without question for monitoring or regulatory purposes.

Solid phase extraction (EPA 3535) also can be used for extraction and concentration of semi-volatile material. The technique involves passing the water sample through a cartridge or disk containing an adsorbent, such as silica or alumina. The adsorbent is often coated with compounds that impart selectivity for particular products or analytes, such as polynuclear aromatic hydrocarbons (PAHs or PNAs). After extraction, the analytes are separated from the solid phase by elution with a small amount of organic solvent. A variant of solid phase extraction involves dipping a sorbent-coated fiber into the water (solid phase micro-extraction). Adsorbed analytes are thermally desorbed directly into a heated chromatographic injection port. Generally, the solid phase extraction method requires much less solvent and glassware than separatory funnel and liquid-liquid extraction.

For the separation of samples from contaminated soil, there are also several possible methods, depending upon whether the contaminant is

Table 7.5 Methods applicable to the separation of volatile and semi-volatile organic compounds from soil samples.

For Volatiles	For Semi-volatiles
Purge and trap	Shaking or vortexing (can also be used for volatiles)
Headspace	Soxhlet Sonication Supercritical fluid Sub-critical fluid

volatile or semi-volatile (Table 7.5). Volatile compounds (such as BTEX, benzene, toluene, ethyl benzene, xylene, and gasoline) may be extracted from soil using, for example, methanol (EPA 5035, purge and trap analysis). In the method, the extraction is usually accomplished by mechanically shaking the soil with methanol. A portion of the methanol extract is added to a purge vessel, and diluted in reagent grade water. The extract is then purged similar to a water sample.

Headspace analysis (EPA 3810, EPA 5021), also works well for analyzing volatile petroleum constituents in soil. In the test method, the soil is placed in a headspace vial and heated to drive out the volatiles from the sample into the headspace of the sample container. Salts can be added for more efficient release of the volatile compounds into the headspace. Similar to water headspace analysis, the soil headspace technique is useful when heavy oils and high analyte concentrations are present, which can severely contaminate purge and trap instrumentation. Detection limits are generally higher for headspace analysis than for purge and trap analysis.

The simplest method to separate semi-volatile compounds from soil is to shake or vortex (vigorous mechanical stirring) the soil with a solvent. Adding a desiccant to the soil/solvent mixture can help to break up soil and increase the surface area. The word of caution here is to ensure that the drying agent does not adsorb the solute. Assuming that no such adsorption occurs, the extract can be analyzed directly. Simple shaking is quick and easy, making it an excellent field extraction technique. However, extraction efficiency will vary, depending on soil type and whether or not clay minerals (excellent adsorbents for many organic compounds) are present.

Soxhlet extraction (EPA SW-846 3540) is a very efficient extraction process that is commonly used for semi-volatile petroleum constituents. In the method, the solvent is heated and refluxed (recirculated) through the soil sample continuously for 16 hours, or overnight. This method generates a relatively large volume of extract that needs to be concentrated. Thus, it is more appropriate for semi-volatile constituents than for volatile constituents. Sonication extraction (EPA SW-846 3550) can also be used for semi-volatile compounds, and, as the name suggests, involves the use of sound waves to enhance analyte transfer from sample to solvent. Sonication is a faster technique than Soxhlet extraction, and it also can require less solvent.

Supercritical fluid extraction (EPA 3540, for total recoverable petroleum hydrocarbons; EPA 3561, for polynuclear aromatic hydrocarbons) is applicable to the extraction of semi-volatile constituents. Supercritical fluid extraction involves heating and pressuring a mobile phase to supercritical conditions (where the solvent has the properties of a gas and a liquid). The supercritical fluid is passed through the soil sample, and the analytes are concentrated on a sorbent, or trapped cryogenically. The analytes are eluted with a solvent, and analyzed with conventional techniques. Carbon dioxide is the most popular mobile phase.

Another method (EPA 3545, Accelerated Solvent Extraction) has been validated using a variety of soil matrices ranging from sand to clay. In the method, conventional solvents, such as methylene chloride (or a hexane-acetone mixture), are heated (100°C, 212°F) and pressurized (2000 psi), and then passed through the soil sample (this technique is also suitable for application to petroleum sludge and petroleum sediment). The method has the advantage of requiring smaller solvent volumes than traditional solvent extraction techniques.

In some cases, when petroleum and/or petroleum products are released to the environment, a free phase is formed, and sample(s) of the hydrocarbon material can be collected directly for characterization. The ability to analyze free product greatly aids the determination of product type and potential source. The samples may be diluted prior to analysis: EPA SW-846 3580, waste dilution, gives some guidelines for proper dilution techniques. However, caution is advised, since, as part of the initial sample collection procedure, water and sediment may be inadvertently included in the sample. There are several protocols involved in initial isolation and cleanup of the sample that must be recognized. In fact, considerable importance attaches to the presence of *water* or *sediment* in crude oil (ASTM D1796, ASTM D4007), for they lead to difficulties in other analyses.

Sediment usually consists of finely divided solids that may be dispersed in the oil or carried in water droplets. The solids may be drilling mud, sand, or scale picked up during the transport of the oil, or may consist of chlorides derived from evaporation of brine droplets in the oil. In any event, the sediment can lead to serious plugging of the equipment, corrosion due to chloride decomposition, and a lowering of residual fuel quality.

Water may be found in the crude, either in an emulsified form or in large droplets. The quantity is generally limited by pipeline companies and by refiners, and steps are normally taken at the wellhead to reduce the water content as low as possible. However, after a spill, water can be introduced by climatic conditions, and the relevant tests (ASTM D96, ASTM D954, ASTM D1796, IP 75) are regarded as important in crude oil analyses. Prior to analyses, it is often necessary to separate the water from a crude oil sample, and this is usually carried out by one of the procedures described in the preliminary distillation of crude petroleum (IP 24).

Overall, there are several methods that can be employed for organic semi-volatile sample preparation and clean-up procedures (Table 7.6).

3.3.4 Solids

For homogeneous materials, sampling protocols are relatively simple and straightforward, although caution is always advised, lest overconfidence cause errors in the method of sampling as well as introduction of extraneous material (EPA, 1998). On the other hand, the heterogeneous nature of soil and contaminated soil complicates the sampling procedures. If the soils and the sample are visibly heterogeneous, there is a very strong emphasis on need to obtain representative samples for testing and analysis.

Thus, the variable composition of contaminated soil as well as solid sample, such as petroleum coke, offers many challenges to environmental analysts, who must ensure that the sample under investigation is representative of the contaminated site or the coke. Furthermore, sample transportation can initiate (due to movement of the sample) processes that result in size and density segregation, in a manner analogous to variations in coal quality from sample to sample (ASTM D346; ASTM D2234; ASTM D4702; ASTM D4915; ASTM D4916; ASTM D6315; ASTM D6518; ASTM D6543; ISO 1988).

Therefore, the challenge in sampling solids for environmental analysis is to collect a relatively small portion of the sample that accurately represents the composition of the whole. This requires that sample increments must be collected in such a manner that no piece, regardless of position (or size) relative to the sampling position and implement, is selectively collected or rejected. Optimization of solids sampling is a function of the many variable constituents of coal, and is reflected in the methods by which an unbiased sample can be obtained, as is required by coal sampling (ASTM D197).

Thus, in order to test any particular environmental solids sample, there are two criteria that must be followed, and these are: (1) obtain a sample of the solid, (2) ensure that the sample is a true representative of the bulk material, and (last but by no means least) (3) ensure that the sample does not undergo any chemical or physical changes after completion of the sampling procedure and during storage prior to analysis.

Table 7.6 Organic semi-volatile sample preparation and clean-up procedures in SW-846.

EPA #	Methodology
3500B	Organic extraction and sample preparation
3510C	Separatory funnel liquid-liquid extraction
3520C	Continuous liquid-liquid extraction
3535	Solid phase extraction (SPE) (3535A in IVB)
3540C	Soxhlet extraction
3541	Automated Soxhlet extraction
3542	Extraction of semi-volatile analytes collected using modified Method 5 (Method 0010) sampling train
3545	Accelerated solvent extraction (ASE) (3545A in Update IVB)
3550B	Ultrasonic extraction
3560	Supercritical fluid extraction of total recoverable petroleum hydrocarbons (TRPH)
3561	Supercritical fluid extraction of polynuclear aromatic hydrocarbons
3562	Supercritical fluid extraction of PCB and organo-chlorine pesticides (Update IVB)
3580A	Waste dilution
3585	Waste dilution for volatile organics
3600C	Cleanup
3610B	Alumina cleanup
3611B	Alumina column cleanup and separation of petroleum wastes
3620B	Florisil cleanup
3630C	Silica gel cleanup
3640A	Gel-permeation cleanup
3650B	Acid-base partition cleanup
3660B	Sulfur cleanup
3665A	Sulfuric acid/permanganate cleanup

In short, the reliability of a sampling method is the degree of perfection with which the identical composition and properties of the entire whole coal are obtained in the sample. The reliability of the storage procedure is the degree to which the coal sample remains unchanged, thereby guaranteeing the accuracy and usefulness of the analytical data. At this point, a review of the sampling methods applied to coal, which under favorable conditions can be applied to environmental solids, is worthy of inclusion.

The sampling procedures (ASTM D346; ASTM D2234; ASTM D4702; ASTM D4915; ASTM D4916; ASTM D6315; ASTM D6518; ASTM D6543) are designed to give a precision such that if gross samples are taken repeatedly from a lot or consignment and prepared according to the standard test methods (ASTM D197; ASTM D2013), and one ash determination is made on the analysis sample from each gross sample, the majority (usually specified as 95 out of 100) of these determinations will fall within ±10% of the average of all of the determinations. When other precision limits are required, or when other constituents are used to specify precision, defined special-purpose sampling procedures may need to be employed.

Thus, when a property of the sample (which exists as a large volume of material) is to be measured, there usually will be differences between the analytical data derived from application of the test methods to a *gross lot* or *gross consignment* and the data from the *sample lot*. This difference (the *sampling error*) has a frequency distribution with a mean value and a variance. *Variance* is a statistical term, defined as the mean square of errors; the square root of the variance is more generally known as the *standard deviation* or the *standard error of sampling*.

Recognition of the issues involved in obtaining representative samples of coal and minimization of the *sampling error* has resulted in the designation of methods that dictate the correct manner for coal sampling (ASTM D346; ASTM D2234; ASTM D4702; ASTM D4915; ASTM D4916; ASTM D6315; ASTM D6518; ASTM D6543; ISO 1988; ISO 2309).

Finally, and this applies to all samples where separation of the solute or released material from the matrix is necessary, it must be recognized that the organic compounds, in, for example, soil or in a liquid matrix (water or an organic solvent), have an affinity for the sample matrix to one degree or another. It is this affinity that allows the sample to be retained by the matrix. The affinity may be due to adsorption on the surface of sample particles, or solvation in water or other solvent medium.

Therefore, it is necessary in an extraction process to recognize the potential solute-matrix interactions in order to overcome such interactions. A suitable choice of solvent is necessary, making sure that the solvent itself does not react with the solid or with the matrix. If the first extraction solvent (e.g., pyridine or carbon disulfide) of choice is too powerful, switch to a solvent of lesser ability, such as pentane, hexane, or petroleum ether, remembering that if it is a high-boiling petroleum sample that is being extracted, these solvents of lesser ability will leave some of the sample unextracted. In other cases, particularly when the sample is a hydrocarbon petroleum product, a polar solvent (e.g., methanol) can be used to extract polar organic materials like the phenols from the matrix.

On the other hand, organic acids (e.g., carboxylic acids, phenols) and organic bases (pyridines derivative) may be isolatable by adjustment of the pH to control the direction of partitioning. For example, acidification (pH <5) of the sample converts the organic bases into salts that move into the aqueous (water) phase, whereas adjusting the pH of the sample to basic (pH >9–11, depending on the nature of the solute) with a suitable base neutralizes the basic analytes and reverses the direction of partitioning to the organic phase. At the same time, organic acids are converted to water-soluble (hydrophilic) salts.

Thus, the simple expedient of separate analysis of the acid and basic extractions, rather than combining the extracts into a single sample extract, often serves to reduce matrix interference to a manageable level.

Although widely used, solvent extraction procedures have been demonstrated as sensitive to such variables as content of humic matter and moisture within samples. Supercritical fluid extraction appears to be a more robust procedure. Thermal extraction procedures are sensitive to the size of the soil sample, and in some cases the technique can result in cracking higher molecular weight constituents that do not volatilize out of the thermal zone. In the case of solvent extraction procedures, it is necessary to concentrate and also to clean up the samples. With complex mixtures of semi-volatile hydrocarbons, it is generally advisable to separate the aliphatic and aromatic fractions.

3.3.5 Water Samples

Samples of petroleum and petroleum products spilled into a water ecosystem must also be isolated for analysis. The most common methods for volatile materials are: (1) headspace analysis, and (2) *purge and trap* method. For semi volatile constituents, several methods are available, namely: (1) separatory funnel extraction, (2) continuous liquid-liquid extraction, and (3) solid phase extraction (Weisman, 1998).

Volatile compounds (gasoline, solvents) in water are generally separated from their matrix by purging with an inert gas and trapping the compounds on a sorbent (EPA 5030, purge and trap analysis). The sorbent is later heated to release the volatile compounds, and a carrier gas sweeps the compounds into a gas chromatograph.

Headspace analysis is recommended as a screening method by the US Environmental Protection Agency (Methods 3810, and 5021), although it performs well in particular situations, especially field analysis. In this method, the water sample is placed in a closed vessel with a headspace, and heated to drive volatiles into the gas phase. Addition of salts or acids may enhance this process. In headspace analysis, instrument contamination is minimized, because only volatile compounds are introduced into the instrument. Samples such as heavy oil and those with high analyte concentrations can severely contaminate purge and trap instrumentation.

The most commonly used water extraction method for semi-volatile contaminants is EPA Method 3510, separatory funnel extraction. The sample is poured into a funnel-shaped piece of glassware, solvent is added, and the mixture is shaken vigorously. After layer separation, the extract (i.e., the solvent layer) is removed, filtered, dried with a desiccant, and concentrated. Multiple extractions on the same sample may increase overall recovery.

Another commonly used water extraction method for semi-volatiles is EPA 3520, continuous liquid-liquid extraction. Rather than shaking solvent with the water sample, the solvent is continuously heated, nebulized (broken into small droplets), and sprayed on top of the water. Liquid-liquid extraction is excellent for samples containing emulsion-forming solids, but it is more time-consuming than separatory funnel extractions.

Solid phase extraction (SPE, EPA Method 3535) also can be used for extraction and concentration of semi-volatile material. The technique involves passing the water sample through a cartridge or disk containing an adsorbent, such as silica or alumina. The adsorbent is often coated with compounds that impart selectivity for particular products or analytes, such as PAHs. After extraction, the analytes are separated from the solid phase by elution with a small amount of organic solvent. A variant of SPE involves dipping a sorbent-coated fiber into the water (solid phase micro-extraction). Adsorbed analytes are thermally desorbed directly into a heated chromatographic injection port. SPE uses much less solvent and glassware than separatory funnel and liquid-liquid extraction.

3.4 Extract Concentration

Extract concentration is the one area in the isolation procedure that has the greatest potential for loss of analytes. The operative physical concept during concentration is vapor pressure, not boiling point. However, boiling is uncontrolled, insofar as all of the molecules in the sample are attempting to convert from the liquid state to the gaseous state. Therefore, equating sample concentration to boiling point leads to erroneous conclusions about choice of an appropriate method. On the other hand, vapor pressure is a continuous function that relates to the rate of evaporation, and evaporation is what is desired during sample concentration. Thus, the proper technique is to control the rate of solvent evaporation while minimizing analyte loss. A completely successful sample extraction can be performed that is completely negated by an inappropriate method for sample extract concentration.

In many cases, sample extracts are generally filtered, dried with desiccant, and concentrated before analysis. Concentration of the extract may allow for lower sample detection limits. Frequently, sample extracts must be concentrated to obtain detection limits low enough to meet regulatory action limits.

Concentration may be achieved by:

For Volatiles	*For Semi-volatiles*
Sorbent trapping	Snyder column
Cryogenic trapping	Kuderna Danish concentrator
Nitrogen evaporation	
Vacuum	

The trapping step in a purge and trap analysis is essentially a concentration step. Analytes are purged from the matrix into a gas stream, and captured on a sorbent trap. The analytes are released by heating the trap. Cryogenic trapping is also used in place of sorbent trapping. In cryogenic trapping, a very cold material (such as liquid nitrogen) surrounds a sample loop, and as analytes are purged and swept through the sample loop, they freeze in the sample loop. The analytes are released when the trap is heated.

Snyder columns are designed to allow highly volatile solvents to escape while retaining semi-volatile analytes of interest. Snyder columns are generally fitted onto the tops of flasks containing extracts, and column design permits solvent to escape as the flask is heated. The analytes of interest condense from a gas to a liquid phase, and fall back down into the solvent reservoir. The Kuderna-Danish concentrator is a Snyder column with a removable collection tube attached to the bottom. As solvent is evaporated, the extract is collected in the collection tube.

As an alternative to a Snyder column, the sample extracts may be concentrated with nitrogen evaporation by directing a slow stream of the gas over the extract surface at room temperature, resulting in minor loss of volatiles. Placing the extract container in warm water helps to speed the process, but then some loss of volatiles can occur. Concentration by evaporating excess solvent with a vacuum is not very common in environmental laboratories. Many semi-volatile analytes are lost in the procedure. Additionally, evaporating as a means of concentrating the sample cannot be used if the goal is to detect volatile analytes.

Cleanup steps are an important component of infrared (IR)-based and gravimetric methods, because these methods are very sensitive to non-petroleum hydrocarbon interferences. Cleanup steps are not always a part of the petroleum analytical process, but when they are necessary, the goals of extract cleanup steps typically include one or more of the following: (1) removal of non-petroleum compounds, (2) isolation of a particular petroleum fraction, and (3) concentration of analytes of interest.

The techniques employed to extract the analytes of interest can frequently extract interfering compounds. Polar compounds, such as animal and plant fats, proteins, and small biological molecules, may be improperly identified as petroleum constituents. Extract cleanup techniques can be used to remove them. In an ideal situation, only interfering compounds are removed. In reality, some polar petroleum constituents can also be removed.

Two techniques are used to clean petroleum extracts. In one technique, interfering compounds are removed by passing the extract through a glass column filled with sorbent. A second technique is to swirl the extract with loose sorbent, then remove the sorbent by filtration. Other methods involve trapping the interfering compounds on a sorbent column, such as alumina (EPA SW-846 3611), that is designed to remove interfering compounds and to fractionate petroleum wastes into aliphatic, aromatic, and polar fractions. The fractions can be analyzed separately, or combined for measurement of the total petroleum hydrocarbons. Alternatively, silica gel (EPA SW-846 3630) is commonly used for polynuclear aromatic hydrocarbons and phenols. Variations of this technique are used to clean (EPA 418.1) extracts before infrared analysis. In addition, the gel permeation technique has been used for cleanup (EPA SW-846 3640), and works on the principle of size exclusion. Large macromolecules, such as lipids, polymers, and proteins, are removed from the sample extract. Extracts obtained from soil that have (or have had) high biological activity may be cleaned by this method.

There are two non-column cleanup methods, one of which uses acid partition (EPA SW-846 3650) to separate the base/neutral and acid components by adjusting pH. This method is often used before alumina column cleanup to remove the acid components. The other method (EPA SW-846 3660) is used for sulfur removal, and uses copper, mercury, and tetrabutylammonium sulfite as desulfurization compounds. Sulfur is a common interfering compound for petroleum hydrocarbon analysis, particularly for sediments. Sulfur-containing compounds are very common in crude oil and heavy fuel oil. Elemental sulfur is often present in anaerobically biodegraded fuels. Thus, abnormally high levels of sulfur may be measured as part of the total petroleum hydrocarbons measurement if the cleanup technique is not used.

Even though cleanup procedures are advocated before sample analysis, there can be several limitations to various cleanup steps. The reasons for decreased effectiveness of cleanup procedures include (1) sample loading may exceed the capacity of cleanup columns, (2) non-petroleum compounds may have chemical structures similar to petroleum constituents, and may behave like a petroleum constituent, (3) analytes of interest may be removed during the cleanup, and (4) no single cleanup technique removes all of the chemical interferences.

3.5 Sample Cleanup

The use of any sample cleanup in environmental analysis is always accompanied by the possibility of analyte loss. Procedures that depend upon polarity interactions between the eluting solvent, the solid adsorbent, and the target analytes to achieve selective isolation are particularly prone to having the desired compounds ending up in the wrong fraction. Sources of these errors include mistakes in the preparation of the eluting solvent, use of the wrong or a deactivated absorbent, the presence of traces of polar solvents in the sample solution, and the structure of the analyte molecule that will determine the behavior of the analytes during separation (Speight, 1999). Attention to detail and procedure are required for successful use of sample cleanup techniques, and the aspects of the procedure that need to be examined include (1) suitability of the materials to achieve the cleanup that will be sample dependent, (2) introduction of laboratory contamination, and (3) success of the procedure on the individual sample.

The introduction of laboratory contamination is a significant, but often overlooked, concern in sample cleanup. Examples are the introduction of extractable materials from plastics used for joining tubes and contaminated solvents; these are only two of the potential points of laboratory contamination.

4 Measurement

The issues that face environment analysts include the need to provide higher-quality results. In addition, environmental regulations may influence the method of choice. Nevertheless, the method of choice still depends to a large extent on the boiling range (or carbon number) of the sample to be analyzed. For example, there is a large variation ion the carbon number range and boiling points (of normal paraffins) for some of the more common petroleum products, and thus, a variation in the methods that may be applied to these products (Speight, 2001, 2002).

The predominant methods of measuring the properties of petroleum products are covered by approximately seven test methods that are used in the determination of bulk quantities of liquid petroleum and its products (ASTM D96, ASTM D287, ASTM D1085, ASTM D1086, ASTM D1087, ASTM D1250, ASTM D1298).

Testing for suspended water and sediment (ASTM D96) is used primarily with fuel oils, where appreciable amounts of water and sediment may cause fouling of facilities for handling the oil and give trouble in burner mechanisms. Three standard methods are available for this determination. The centrifuge method gives the total water and sediment content of the sample by volume, the distillation method gives the water only volumetrically, and the extraction method gives the solid sediment in per cent by weight.

The determination of density of specific gravity (ASTM D287, ASTM D1298) in the measurement and calculation of volume of petroleum products is important, since gravity is an index of the weight of a measured volume of the product. Two scales are in use in the petroleum industry, specific gravity and API gravity, the determination being made in each case by means of a hydrometer of constant weight displacing a variable volume of oil. The reading obtained depends upon both the gravity and the temperature of the oil.

Gauging petroleum products (ASTM D1085, discontinued in 12996 but still in use) involves the use of procedures for determining the liquid contents of tanks, ships and barges, tank cars, and tank trucks. Depth of liquid is determined by gauging through specified hatches, or by reading gauge glasses or other devices. There are two basic types of gauges, innage and outage. The procedures used depend upon the type of tank to be gauged, its equipment, and the gauging apparatus.

An innage gauge is the depth of liquid in a tank measured from the surface of the liquid to the tank bottom, or to a datum plate attached to the shell or bottom. The innage gauge is used directly with the tank calibration table and temperature of the product to calculate the volume of product (ASTM D1250). On the other hand, an outage gauge is the distance between the surface of the product in the tank and the reference point above the surface, which is usually located in the gauging hatch. The outage gauge is used either directly or indirectly with the tank calibration table and the temperature of the product to calculate the volume of product. The amount of any free water and sediment in the bottom of the tank is also gauged, so that corrections can be made when calculating the net volume of the crude oil or petroleum product.

The liquid levels of products that have a Reid vapor pressure of 40 lb or more are generally determined by the use of gauge glasses, rotary or slip-tube gauges, tapes and bobs through pressure locks, or other types of gauging equipment. The type of gauging equipment depends upon the size and type of the pressure tank.

There are also procedures for determining the temperatures of petroleum and its products when in a liquid state. Temperatures are determined at specified locations in tanks, ships and barges, tank cars, and tank trucks. For a non-pressure tank, a temperature is obtained by lowering a tank thermometer of proper range through the gauging hatch to the specified liquid level. After the entire thermometer assembly has had time to attain the temperature of the product, the thermometer is withdrawn and read quickly. This procedure is also used for low-pressure tanks equipped with gauging hatches or standpipes, and for any pressure tank that has a pressure lock. For tanks equipped with thermometer wells, temperatures are obtained by reading thermometers placed in the wells with their bulbs at the desired tank levels. If more than one temperature is determined, the average temperature of the product is calculated from the observed temperatures. Electrical-resistance thermometers are sometimes used to determine both average and spot temperatures.

In general, the volume received or delivered is calculated from the observed gauge readings. Corrections are made for any *free* water and sediment as determined by the gauge of the water level in the tank. The resultant volume is then corrected to the equivalent volume at 15.6°C (60°F), by use of the observed average temperature and the appropriate volume correction table (ASTM D1250). When necessary, a further correction is made for any suspended water and sediment that may be present in materials such as crude petroleum and heavy fuel oils.

For the measurement of other petroleum products, a wide variety of tests is available. In fact, there are approximately three hundred and fifty tests (ASTM, 2000) that are used to determine the different properties of petroleum products. Each test has its own limits of accuracy and precision that must be adhered to if the data are to be accepted.

5 Accuracy

The *accuracy* of a test is a measure of how close the test result will be to the true value of the property being measured (ASTM, 2010; Patnaik, 2004). As such, the accuracy can be expressed as the *bias* between the test result and the true value. However, the *absolute accuracy* can only be established if the true value is known.

In the simplest sense, a convenient method to determine a relationship between two measured properties is to plot one against the other. Such an exercise will provide either a line fit of the points or a spread that may or may not be within the limits of experimental error. The data can then be used to determine the approximate accuracy of one or more points employed in the plot. For example, a point that lies outside the limits of experimental error (a *flyer*) will indicate an issue of accuracy with that test, and the need for a repeat determination.

However, the graphical approach is not appropriate for finding the absolute accuracy between more than two properties. The well-established statistical technique of regression analysis is more pertinent to determining the accuracy of points derived from one property and any number of other properties. There are many instances in which relationships of this sort enable properties to be predicted from other measured properties with as much precision as they can be measured by a single test. It would be possible to examine in this way the relationships between all the specified properties of a product, and to establish certain key properties from which the remainder could be predicted, but this would be a tedious task.

However, the application of statistical analysis to experiment data is not always straightforward, and may be fraught with inconsistencies, due to the assumptions that are involved in the statistical development an interpretation of the data. Statistical analysis of the data is only a part of the picture; the decision process has to be viewed as a coherent whole: a decision can only be passed by taking into account the complex interrelations among the chemical species being consumed and formed and the legal, economic, scientific, and environmental characteristics of the analytical process (Baker, 1966; Dixon and Massey, 1969; Alder and Roessler, 1972; Box et al., 1978; Caulcutt and Boddy, 1983; Jaffee and Spirer, 1987; Meier and Zünd, 2000; Patnaik, 2004, Section 3).

An alternative approach to that of picking out the essential tests in a specification using regression analysis is to take a look at the specification as a whole, and extract the essential features (termed *principal components analysis*).

Principal components analysis involves an examination of set of data as points in n-dimensional space (corresponding to n original tests) and determines (first) the direction that accounts for the biggest variability in the data (*first principal component*). The process is repeated until n principal components are evaluated, but not all of these are of practical importance, since some may be attributable purely to experimental error. The number of significant principal components shows the number of independent properties being measured by the tests considered.

Following from this, it is necessary to establish the number of independent properties that are necessary to predict product performance in service, with the goals of rendering any specification more meaningful and allowing a high degree of predictability of product behavior. On a long-term approach, it might be possible to obtain new tests of a fundamental nature to replace, or certainly to supplement, existing tests. In the short term, selecting the best of the existing tests to define product quality is the most beneficial route to predictability.

6 Precision

The *precision* of a test method is the variability between test results obtained on the same material, using a specific test method (ASTM, 2004; Patnaik, 2004). The precision of a test is usually unrelated to its accuracy. The results may be precise, but not necessarily accurate. In fact, the precision of an analytical method is the amount of scatter in the results obtained from multiple analyses of a homogeneous sample. To be meaningful, the precision study must be performed using the exact sample and standard preparation procedures that will be used in the final method. Precision is expressed as repeatability and reproducibility.

The precision of sampling, for example, solids, is a function of the size of increments collected and the number of increments included in a gross sample, improving as both are increased, and subject only to the constraint that increment size not be small enough to cause selective rejection of the largest particles present. The manner in which solids sampling is performed as it relates to the precision of the sample thus depends upon the number of increments collected from all parts of the lot and the size of the increments. In fact, the number and size of the increments are operating variables that can, within certain limits, be regulated by the sampler.

The *intra-laboratory precision* or the *within-laboratory precision* refers to the precision of a test method when the results are obtained by the same operator in the same laboratory using the same apparatus. In some cases, the precision is applied to data gathered by a different operator in the same laboratory using the same apparatus. Thus, intra-laboratory precision has an expanded meaning insofar as it can be applied to laboratory precision.

Repeatability or repeatability interval of a test (*r*) is the maximum permissible difference due to test error between two results obtained on the same material in the same laboratory.

$$r = 2.77 \times \text{standard deviation of test}$$

The repeatability interval (*r*) is, statistically, the 95% probability level or the differences between two test results are unlikely to exceed this repeatability interval more than five times in a hundred.

The *inter-laboratory precision* or the *between-laboratory precision* is defined in terms of the variability between test results obtained on the aliquots of the same homogeneous material in different laboratories using the same test method.

The term *reproducibility* or *reproducibility interval* (*R*) is analogous to the term repeatability, but it is the maximum permissible difference between two results obtained on the same material, but now in different laboratories. Therefore, differences between two or more laboratories should not exceed the reproducibility interval more than five times in a hundred.

$$\mathrm{R} = 2.77 \times \text{standard deviation of test}$$

The repeatability value and the reproducibility value have important implications for quality. As the demand for clear product specifications, and hence, control over product consistency, grows, it is meaningless to establish product specifications that are more restrictive than the reproducibility/repeatability values of the specification test methods.

7 Method Validation

Method validation is the process of proving that an analytical method is acceptable for its intended purpose. Many organizations provide a framework for performing such validations (ASTM, 2010). In general, methods for product specifications and regulatory submission must include studies on specificity, linearity, accuracy, precision, range, detection limit, and quantitation limit.

The process of method development and validation covers all aspects of the analytical procedure, and the best way to minimize method problems is to perform validation experiments during development.

In order to perform validation studies, the approach should be viewed with the understanding that validation requirements are continually changing and vary widely, depending on the type of product under testing and compliance with any necessary regulatory group.

In the early stages of new product development, it may not be necessary to perform all of the various validation studies. However, the process of validating a method cannot be separated from the actual development of the method conditions, because the developer will not know whether the method conditions are acceptable until validation studies are performed. The development and validation of a new analytical method may, therefore, be an iterative process. Results of validation studies may indicate that a change in the procedure is necessary, which may then require revalidation. During each validation study, key method parameters are determined and then used for all subsequent validation steps.

The first step in the method development and validation cycle should be to set minimum requirements, which are essentially acceptance specifications for the method. A complete list of criteria should be agreed on during method development and the end users before the method is developed so that expectations are clear. Once the validation studies are complete, the method developers should be confident in the ability of the method to provide good quantitation in their own laboratories. The remaining studies should provide greater assurance that the method will work well in other laboratories, where different operators, instruments, and reagents are involved, and where it will be used over much longer periods of time.

The remaining precision studies comprise much of what is often referred to as *ruggedness*. *Intermediate precision* is the precision obtained when an assay is performed by multiple analysts using several instruments on different days in one laboratory. Intermediate precision results are used to identify which of the above factors contribute significant variability to the final result.

The last type of precision study is *reproducibility* (*q.v.*), that is determined by testing homogeneous samples in multiple laboratories, often as part of inter-laboratory crossover studies. The evaluation of reproducibility results often focuses more on measuring bias in results than on determining differences in precision alone. Statistical equivalence is often used as a measure of acceptable inter-laboratory results. An alternative, more practical approach is the use of *analytical equivalence*, in which a range of acceptable results is chosen prior to the study and used to judge the acceptability of the results obtained from the different laboratories.

Performing a thorough method validation can be a tedious process, but the reliability of the data generated with the method is directly linked to the application of quality assurance and quality control protocols, which must be followed assiduously (Table 7.7) (Quevauviller, 2002).

Briefly, to assure quality assurance and quality control, samples are analyzed using standard analytical procedures. A continuing program of analytical laboratory quality control verifies data quality, and involves participation in inter-laboratory crosschecks, replicate sampling, and analysis. When applicable, it is advisable, even insisted upon by the Environmental Protection Agency, that analytical labs must be certified to complete the analysis requested.

Table 7.7 Comprehensive list of EPA quality control methods that apply to situations within and outside of the petroleum refining industry.

EPA #	Description of Method
814V96005	An Introduction to the ICR* Laboratory QC Database System
600987030	Availability Adequacy and Comparability of Testing Procedures for the Analysis of Pollutants Established under Section 304(h) of the Federal Water Pollution Control Act: Report to Congress
503990009	Citizen Volunteers in Environmental Monitoring: Summary Proceedings of The Second National Workshop New Orleans Louisiana
506690003	Contaminated Sediments: Relevant Statutes and EPA Program Activities
814P94001	DBP/ICR** Analytical Methods Guidance Manual: Public Comment

(Continued)

Table 7.7 (cont.) Comprehensive list of EPA quality control methods that apply to situations within and outside of the petroleum refining industry.

EPA #	Description of Method
625R02010	Developing and Implementing an Estuarine Water Quality Monitoring, Assessment and Outreach Program: The MYSound Project
625C02010	Developing and Implementing an Estuarine Water Quality Monitoring, Assessment and Outreach Program: The MYSound Project [CDROM]
600486011	Eastern Lake Survey Phase 1: Quality Assurance Report
600486008	Eastern Lake Survey Phase I: Quality Assurance Plan
600389013	Ecological Assessment of Hazardous Waste Sites: A Field and Laboratory Reference
600879006	EPA Manual for Organics Analysis Using Gas Chromatography-Mass Spectrometry
823B00007	Guidance for Assessing Chemical Contaminant Data for Use in Fish Advisories, Volume 1: Fish Sampling and Analysis, Third Edition
823R93002	Guidance for Assessing Chemical Contaminant Data for Use in Fish Advisories: Volume 1 - Fish Sampling and Analysis
600479019	Handbook for Analytical Quality Control in Water and Wastewater Laboratories
625689023	Handbook: Quality Assurance/quality Control (QA/QC) Procedures for Hazardous Waste Incineration
600R95178	ICR Microbial Laboratory Manual
814B96001	ICR Sampling Manual
600483004	Interim Guidelines and Specifications for Preparing Quality Assurance Project Plans
821R01027	Interlaboratory Validation Study Results for 'Cryptosporidium' Precision and Recovery for U.S. EPA Method 1622

(Continued)

Table 7.7 (cont.) Comprehensive list of EPA quality control methods that apply to situations within and outside of the petroleum refining industry.

EPA #	Description of Method
600R94134	Method 100.2. Determination of Asbestos Structures over 10 micrometers in Length in Drinking Water
821R02019	Method 1631, Revision E: Mercury in Water by Oxidation, Purge and Trap, and Cold Vapor Atomic Fluorescence Spectrometry
821R96013	Method 1632: Determination of Inorganic Arsenic in Water by Hydride Generation Flame Atomic Absorption
821R95028	Method 1632: Determination of Inorganic Arsenic in Water by Hydride Generation Flame Atomic Absorption [Draft]
821R95033	Method 1640: Determination of Trace Elements in Ambient Waters by On-Line Chelation Pre-concentration and Inductively Coupled Plasma-Mass Spectrometry
821B94004b	Method 1664: N-Hexane Extractable Material (HEM) and Silica Gel Treated N-Hexane Extractable Material (SGT-HEM) by Extraction and Gravimetry (Oil and Grease and Total Petroleum Hydrocarbons)
821R95026	Monitoring Trace Metals at Ambient Water Quality Criteria Levels: Briefing Book
503890006	Ocean Data Evaluation System (ODES) Data Submissions Manual
823B95001	QA/QC Guidance for Sampling and Analysis of Sediments Water and Tissue for Dredged Material Evaluations - Chemical Evaluations
810B92003	Quality Assurance Plan for the National Pesticide Survey of Drinking Water Wells: Analytical Method 2 - Chlorinated Pesticides
823R02006	Quality Assurance Project Plan for Analytical Control and Assessment Activities in the National Study of Chemical Residues in Lake Fish Tissue

(Continued)

Table 7.7 (cont.) Comprehensive list of EPA quality control methods that apply to situations within and outside of the petroleum refining industry.

EPA #	Description of Method
823R02005	Quality Assurance Project Plan for Sample Collection Activities for a National Study of Chemical Residues in Lake Fish Tissue
810B92001	Quality Assurance Project Plan for the National Pesticide Survey of Drinking Water Wells
810B92010	Quality Assurance Project Plan for the National Pesticide Survey of Drinking Water Wells: Analytical Method 1
810B92002	Quality Assurance Project Plan for the National Pesticide Survey of Drinking Water Wells: Analytical Method 1 - Nitrogen/Phosphorous Pesticides and Analytical Method 3 - Chlorinated Acid Herbicides
810B92011	Quality Assurance Project Plan for the National Pesticide Survey of Drinking Water Wells: Analytical Method 3
810B92005	Quality Assurance Project Plan for the National Pesticide Survey of Drinking Water Wells: Analytical Method 5 - Methyl Carbamates
810B92006	Quality Assurance Project Plan for the National Pesticide Survey of Drinking Water Wells: Analytical Method 6
810B92012	Quality Assurance Project Plan for the National Pesticide Survey of Drinking Water Wells: Analytical Method 6 - Ethylene Thiourea
810B92007	Quality Assurance Project Plan for the National Pesticide Survey of Drinking Water Wells: Analytical Method 7 - Fumigants
810B92008	Quality Assurance Project Plan for the National Pesticide Survey of Drinking Water Wells: Analytical Method 9 - Nitrate and Nitrite.
810B92015	Quality Assurance Project Plan for the National Pesticide Survey of Drinking Water Wells: Survey Statistics, Data Collection and Processing

(Continued)

Table 7.7 (cont.) Comprehensive list of EPA quality control methods that apply to situations within and outside of the petroleum refining industry.

EPA #	Description of Method
810B92014	Quality Assurance Project Plan for the National Survey of Pesticides in Drinking Water Wells: Well Sampling, Data Collection and Processing
430986004	Quality Assurance/Quality Control (QA/QC) for 301(h) Monitoring Programs: Guidance on Field and Laboratory Methods.
812B92004	Recalculation of Screening Level Concentrations for Nonpolar Organic Contaminants in Marine Sediments [Final Report]
600A94023	Regulation of Municipal Sewage Sludge under the Clean Water Act Section 503: A Model for Exposure and Risk Assessment for MSW-Compost
821R01028	Results of the Interlaboratory Method Validation Study Results for Determination of 'Cryptosporidium' and 'Giardia' Using U.S. EPA Method 1623
820R82103	Sampling Protocols for Collecting Surface Water, Bed Sediment, Bivalves, and Fish for Priority Pollutant Analysis- Final Report
823R92006	Sediment Classification Methods Compendium
832C86100	Sludge Compost Marketing and Distribution Regulatory Requirements in the United States
815R00002	Supplement A to the Unregulated Contaminant Monitoring Regulation Analytical Methods and Quality Control Manual
600491016	Test Methods for 'Escherichia coli' in Drinking Water: EC Medium with Mug Tube Procedure Nutrient Agar with Mug Membrane Filter Procedure
440183079c	Test Methods: Methods for Non-conventional Pesticides Chemicals Analysis of Industrial and Municipal Wastewater
815R01028	UCMR*** List 1 and List 2 Chemical Analytical Methods Quality Control Manual (1999)

(Continued)

Table 7.7 (cont.) Comprehensive list of EPA quality control methods that apply to situations within and outside of the petroleum refining industry.

EPA #	Description of Method
815R00006	Unregulated Contaminant Monitoring Regulation Analytical Methods and Quality Control Manual
815R99004	Unregulated Contaminant Monitoring Regulation Analytical Methods and Quality Control Manual
815R99004	Unregulated Contaminant Monitoring Regulation Analytical Methods and Quality Control Manual [Draft]
841B04005	Wadeable Streams Assessment: Quality Assurance Project Plan
600487037	Western Lake Survey: Phase 1 - Quality Assurance Report

*ICR: Information Collection Requests.

**DBP: Disinfectant By-Product.

***UCMR: Unregulated Contaminant Monitoring Rule.

However, in many cases, time constraints often do not allow for sufficient method validation. Many researchers have experienced the consequences of invalid methods, and realized that the amount of time and resources required to solve problems discovered later exceeds what would have been expended initially if the validation studies had been performed properly. *Putting in time and effort up front* will help any environmental analysts to find a way through the method validation maze, and eliminate many of the problems common to inadequately validated analytical methods.

One method that is often used for method validation is the addition of a surrogate to the sample matrix, and following the track of the surrogate through the separation procedure. It is appropriate that at this point, comments should be made on such procedures.

However, it is the nature of the sampling protocols, and the location of the site from which the sample are taken, that it should be assumed that no two samples are identical. Factors such as (1) depth from which the sample is taken, (2) distance of the sampling point from the initial location of the spill, and (3) ambient conditions play a role in determining sample character. In fact, each sample is a unique combination of matrix-analyte interactions. Adding surrogate compounds to each sample and then determining the recovery of the surrogates is used as the benchmark for gauging the success of the extraction procedure. The best surrogates are those that are most like the target analytes, and examples are the isotopically labeled versions of each of the target molecules. Whether or not use of such molecules is justified relates to the data that are required.

As an alternative, other similar chemical compounds can be used, but the choice of the surrogate compounds can limit, or it can maximize, data interpretation. For most analyses, it is possible to choose surrogates that will always generate excellent recoveries, regardless of the complexity of the sample. Choosing surrogate molecules to produce *acceptable quality control* is not realistic. The use of surrogates is to obtain reliable information about the overall strength and weaknesses of the analytical method. In addition, the surrogate(s) should reflect the chemical behavior and physical properties of the analytes. Important chemical behavior and properties includes (1) acidic and basic properties of the analytes, (2) the range of polarity of the analytes, (3) reactivity in chemical derivatization procedures, and (4) the sensitivity of the analytes to decomposition caused by extremes of chemical or physical environment.

Surrogates chosen to monitor any of these areas should ideally bracket the range of the property. However, it should be pointed out that very few individual methods specify surrogates that provide information on all of these areas, let alone the idea of bracketing the property. The analyst must determine the suitability of surrogates to reflect the properties and behavior of the analytes as well as the ability of the surrogate to be used for the collection of reliable and defensible data.

8 Quality Control and Quality Assurance

Quality control (QC) and quality assurance (QA) programs are key components of all analytical protocols in all areas of analysis, including environmental, pharmaceutical, and forensic testing, among others (Speight, 2001, 2002; Patnaik, 2004). These programs mandate that the laboratories follow a set of well-defined guidelines to achieve valid analytical results to a high degree of reliability and accuracy within an acceptable range. Although such programs may vary depending on the regulatory authority, certain key features of these programs are more or less the same.

However, there is often confusion between the terms *quality assurance* and *quality control*, perhaps because there is also considerable overlap between certain aspects of quality assurance and quality control programs.

8.1 Quality Control

Quality controls are single procedures that are performed in conjunction with the analysis to help assess in a quantitative manner the success of the individual analysis. Examples of quality controls are blanks, calibration, calibration verification, surrogate additions, matrix spikes, laboratory control samples, performance evaluation samples, and determination of detection limits. The success of the quality control is evaluated against an acceptance limit. The actual generation of the acceptance limit is a function of quality assurance; it would not be termed a quality control.

Unlike quality assurance plans, that mostly address regulatory requirements involving comprehensive documentation, quality control programs are science-based, the components of which may be defined statistically. The two most important components of quality control are (1) determination of precision of analysis and (2) determination of accuracy of measurement

Whereas *precision* (Section 5.0) measures the reproducibility of data from replicate analyses, the *accuracy* (Section 4.0) of a test estimates how accurate are the data, that is, how close the data would fall to probable true values, or how accurate is the analytical procedure to give results that may be close to true values. Both the precision and accuracy are measured on one or more samples selected at random from a given batch of samples for analysis. The precision of analysis is usually determined by running duplicate or replicate tests on one of the samples in a given batch of samples. It is expressed statistically as standard deviation, relative standard deviation (RSD), coefficient of variance (CV), standard error of the mean (M), and the relative percent difference (RPD).

The standard deviation in measurements, however, can vary with the concentrations of the analytes. On the other hand, RSD, which is expressed as the ratio of standard deviation to the arithmetic mean of replicate analyses and is given as a percent, does not have this problem, and is a more rational way of expressing precision:

$$\text{RSD} = -\ (\text{standard deviation/arithmetic mean of replicate analysis}) \times 100\%$$

The standard error of the mean, M, is the ratio of the standard deviation and the square root of the number of measurements (n):

$$\text{M} = \text{Standard deviation/n}$$

This scale, too, will vary in the same proportion as standard deviation with the size of the analyte in the sample.

In routine testing, many repeat analyses of a sample aliquot may not be possible. Alternatively, therefore, the precision of a test may be determined from duplicate analyses of sample aliquots, and expressed as RPD:

$$\text{RPD} = (a_1 - a_2)/[(a_1 + a_2)/2] \times 100\%$$

or

$$\text{RPD} = (a_2 - a_1)/[(a_1 + a_2)/2] \times 100\%$$

where a_1 and a_2 are the results of duplicate analyses of a sample. Since only two tests are performed on a selected sample, the RPD may not be as accurate a measure of precision as the relative standard deviation, and since the relative standard deviation does not vary with sample size, it should be used whenever possible to estimate precision of analysis from replicate tests.

The accuracy of an analysis can be determined by several procedures. One common method is to analyze a "known" sample, such as a standard solution or a quality control check standard solution that may be commercially available, or a laboratory-prepared standard solution made from a neat compound, and compare the test results with the true values (theoretically expected values). Such samples must be subjected to all analytical steps, including sample extraction, digestion, or concentration, similar to regular samples. Alternatively, accuracy may be estimated from the recovery of a known standard solution *spiked* or added into the sample, in which a known amount of the same substance that is to be tested is added to an aliquot of the sample, usually as a solution, prior to the analysis. The concentration of the analyte in the spiked solution of the sample is then measured. The percent spike recovery is then calculated. A correction for the bias in the analytical procedure can then be made, based on the percent spike recovery. However, in most routine analysis, such bias correction is not required. Percent spike-recovery then may be calculated as follows:

$$\text{Recovery, \%} = (\text{measured concentration})/(\text{theoretical concentration}) \times 100\%$$

The percent spike recovery to measure the accuracy of analysis may also be determined by the U.S. Environmental Protection Agency (EPA) method often used in environmental analysis:

$$\text{Recovery, \%} = [100(X_s - X_u)]/K$$

where X_s is the measured value for the spiked sample and X_u is the measured value for the unspiked sample, adjusted for the dilution of the spike, and K is the known value of spike in the sample.

8.2 Quality Assurance

Quality assurance is an umbrella term that is correctly applied to everything that the laboratory does to assure product reliability. As the product of a laboratory is information, anything that is done to improve the reliability of the generated information falls under quality assurance.

Quality assurance includes all of the quality controls, the generation of expectations (acceptance limits) from the quality controls, plus a great number of other activities such as (1) analyst training and certification, (2) data review and evaluation, (3) preparation of final reports of analysis, (4) information given to clients about tests that are needed to fulfill regulatory requirements, (5) use of the appropriate tests in the laboratory, (6) obtaining and maintaining laboratory certifications/accreditations, (7) conducting internal and external audits, (8) preparing responses to the audit results, (9) the receipt, storage, and tracking of samples, and (10) tracking the acquisition of standards and reagents.

Thus, the objective of quality assurance (usually in the form of a *quality assurance plan*) is to obtain reliable and accurate analytical results that may be stated with a high level of confidence (statistically), so that such results are legally defensible. The key features of any plan involve essentially documentation and record keeping. In short, a quality assurance program involves documentation of sample collection for testing, the receipt of samples in the laboratory, and their transfer to the individuals who perform the analyses. The information is recorded on chain-of-custody forms, stating the dates and times along with the names and signatures of individuals who carry out these tasks. Also, other pertinent information is recorded, such as any preservatives added to the sample to prevent degradation of test analytes, the temperature at which the sample is stored, the temperature to which the sample is brought prior to its analysis, the nature of the container (which may affect the stability of the sample), and its holding time prior to testing.

In fact, the performance of quality control is just one small aspect of the quality assurance program. The functions of quality assurance are embodied in the terms *analytically valid* and (in these days of perpetual litigation) *legally defensible*.

Analytically valid means that the target analyte has been (1) correctly identified and (2) quantified using fully calibrated tests. In addition, the sensitivity of the test (method detection limit) has been established. The analysts have also demonstrated that they are capable of performing the test. The accuracy and precision of the test on the particular sample must also have been determined, and the possibility of false positive and false negative results have been evaluated through performance of blanks and other test-specific interference procedures.

9 Method Detection Limit

The method detection limit (MDL) is the smallest quantity or concentration of a substance that the instrument can measure (Patnaik, 2004). It is related to the instrument detection limit (IDL) that depends on the type of instrument and its sensitivity, and on the physical and chemical properties of the test substance.

The method detection limit is, in reality, a statistical concept that is applicable only in trace analysis of certain type of substances, such as organic pollutants by gas chromatographic methods. The method detection limit measures the minimum detection limit of the method, and involves all analytical steps, including sample extraction, concentration, and determination by an analytical instrument. Unlike the instrument detection limit, the method detection limit is not confined only to the detection limit of the instrument.

In the environmental analysis of organic pollutants, the method detection limit is the minimum concentration of a substance that can be measured and reported with 99% confidence that the analyte concentration is greater than zero, and is determined from the analysis of a sample in a given matrix containing the analyte. For determination of the method detection limit, several replicate analyses are performed at the concentration level of the instrument detection limit, or at a level equivalent to two to five times the background noise level. The standard deviation of the replicate tests is found. The method detection limit is determined by multiplying the standard derivation by the *t-factor*.

In environmental analysis, however, periodic determination of the method detection limit (e.g., once per year or with any change in personnel, location, or instrument) is part of the quality control requirement.

10 References

Alder, H. L., and Roessler, E.B. 1972. *Introduction to Probability and Statistics*. W. H. Freeman, San Francisco, 1972.

ASTM, 2011. *Annual Book of ASTM Standards*. American Society for Testing and Materials, West Conshohocken, Pennsylvania.

Baker, C.C.T. 1966. *An Introduction to Mathematics*. Arco Publishing Company Inc., New York.

Box, G., Hunter, W., and Hunter, J. 1978. *Statistics for Experimenters*. John Wiley & Sons Inc., New York.

Caulcutt, R., and Boddy, R. 1983. *Statistics for Analytical Chemists*, 1983. Chapman and Hall, London, England.

Dean, J.R. 1998. *Extraction Methods for Environmental Analysis*. John Wiley & Sons, Inc., New York.

Dean, J.R. 2003. *Methods for Environmental Trace Analysis*. John Wiley & Sons Inc., Hoboken, New Jersey.

Dixon, W.J., and Massey, F.J. 1969. *Introduction to Statistical Analysis*, McGraw-Hill, New York.

EPA. 1998. *Test Methods for Evaluating Solid Waste - Physical/Chemical Methods*. EPA/SW-846, 3rd Edition, 1986, Update 1, 1992, Update II, 1994, Update III, 1996, Update IV, 1998. Environmental Protection Agency, Washington, DC.

EPA. 2004. Environmental Protection Agency, Washington, DC. Web site: http://www.epa.gov

IP. 2011. *IP Standard Methods 2011*. The Institute of Petroleum, London, United Kingdom.

Jaffe, A.J., and Spirer, H.F. 1987. *Misused Statistics-Straight Talk for Twisted Numbers*. Dekker, New York.

Laleh, S.R., Cimpoia, R., Sarrazin, M., Sunahara, G., Peisajovich, A., Leclair, G., and Serge, R.G. 2003. "Enhanced Biodegradation of Petroleum Hydrocarbons in Contaminated Soil." *Bioremediation Journal*, 7(1): 38.

Meier, P.C., and Zünd, R.E. 2000. *Statistical Methods in Analytical Chemistry*. John Wiley & Sons Inc., New York.

Mokhatab, S., Poe, W.A., and Speight, J.G. 2006. *Handbook of Natural Gas Transmission and Processing*. Elsevier, Amsterdam, Netherlands.

Palittapongarnpim, M., Pokethitiyook, P., Upatham, E., and Tangbanluekal, L. 1998. "Biodegradation of Crude Oil by Soil Microorganisms in the Tropics." *Biodegradation* 9: 83–90.

Patnaik, P. (Editor). 2004. *Dean's Analytical Chemistry Handbook*. 2nd Edition. McGraw-Hill, New York.

Quevauviller, P. 2002. *Quality Assurance for water analysis*. John Wiley & Sons Inc., Hoboken, New Jersey, USA.

Reynolds, C.M., Bhunia, P., and Koenen, B.A., 1997. *Soil Remediation Demonstration Project: Biodegradation of Heavy Fuel Oils*. Special Report 97–20. Cold Regions Research & Engineering Laboratory. US Army Corps of Engineers, Fort Belvoir, Virginia.

Smith, R.K. 1999. *Handbook of Environmental Analysis*. 4th Edition. Genium Publishing, Schenectady, NY.

Speight, J.G. 2001. *Handbook of Petroleum Analysis*. John Wiley & Sons Inc., New York.

Speight, J.G. 2002. *Handbook of Petroleum Product Analysis*. John Wiley & Sons Inc., New York.

Speight, J.G. 2005. *Environmental Analysis and Technology for the Refining Industry*. John Wiley & Sons Inc., Hoboken, New Jersey.

Speight, J.G. 2007a. *The Chemistry and Technology of Petroleum*. 4th Edition. CRC Press, Taylor & Francis Group, Boca Raton, Florida.

Speight, J.G. 2007b. *Natural Gas: A Basic Handbook*. GPC Books, Gulf Publishing Company, Houston, Texas, 2007.

US EPA. 1992. Preparation of Soil Sampling Protocols: Sampling Techniques and Strategies. EPA/600/R-92/128, Office of Research and Development, Environmental Protection Agency, Washington DC. July. pp. 3–7.

Weisman, W. 1998. *Analysis of Petroleum Hydrocarbons in Environmental Media*. Total Petroleum Hydrocarbons Criteria Working Group Series. Volume 1. Amherst Scientific Publishers, Amherst, MA. (See also: *Volume 2: Composition of Petroleum Mixtures*, Volume 2, 1998; *Volume 3: Selection of Representation Total Petroleum Hydrocarbons Fractions Based on Fate and Transport Considerations*, 1997; *Volume 4: Development of Fraction-Specific Reference Does and Reference concentrations for Total Petroleum Hydrocarbons*, 1997; and *Volume 5: Human Health Risk-Based Evaluation of Petroleum Contaminated Sites - Implementation of the Working Group Approach*, 1999).

8

Analytical Methods

1 Introduction

As already noted (Chapter 3), the chemical composition of petroleum and petroleum products is complex, and to make it even more complicated, the composition of a petroleum-related spilled material will change over time following release into the environment. These factors make it essential that the most appropriate analytical methods are selected from a comprehensive list of methods and techniques that are used for the analysis of environmental samples (Dean, 1998; Miller, 2000; Budde, 2001; Sunahara et al, 2002; Nelson, 2003; Smith and Cresser, 2003; Speight, 2005). However, once a method is selected, it may not be the ultimate answer to solving the problem of identification and, hence, behavior (Patnaik, 2004). There are significant numbers of petroleum-hydrocarbon impacted sites, and evaluation and remediation of these sites may be difficult, arising from the complexity of the issues (analytical, scientific, and regulatory, not to mention economic) regarding impacted water and soil.

Preliminary tests to assess feasibility of biotreatment of hydrocarbon contaminations are typically performed in the laboratory (for example, degradation experiments) or in the field (chiefly in situ respiration tests). In general, the quantity of soil used in laboratory experiments is relatively small compared to the quantity of soil that has to be treated on-site, and thus, spatial heterogeneity is rarely considered in laboratory experiments (Aichberger et al., 2005). This might in particular be true for small scale experiments and/or for inhomogeneous sites.

Furthermore, samples may be disturbed during sampling, storage and pre-treatment. Also, most laboratory experiments impose steady environmental conditions, while in an outdoor climate, these conditions show dynamic behavior (Frijer et al., 1996). In fact, data for biodegradation rates determined in the laboratory experiments might be higher than would be expected, thus favoring higher degradation rates compared to field data (Davis et al., 2003; Höhener et al., 2003). Laboratory tests to determine biodegradability potential indirectly via oxygen consumption or carbon dioxide may require corrections to the rates, that then have to be corrected by background respiration resulting from biodegradation of organic substances and other oxygen consuming processes (Balba et al., 1998; Baker et al., 2000). However, the manner in which the corrections are applied must be beyond reproach, or claims of falsification of the data will be the most likely result (Speight and Foote, 2011).

Field tests must take into consideration the various site conditions comprising local heterogeneity and changing environmental conditions. Permeability in the subsoil is rarely uniform, but typically reveals significant local differences that can only be assessed in situ (Davis et al., 2003). By the application of preliminary field tests, relevant information can be deducted on the definite system design, including actual flow rates, radii of influence, blower, and well layout (Leeson and Hinchee, 1997; Baker et al., 2000; Höhener et al., 2003).

Preliminary site assessments account for three critical prerequisites for bioremediation, namely for (1) hydrocarbon biodegradability of the predominating compounds, (2) contaminant bioavailability to indigenous microbial populations, and (3) environmental conditions present at a site.

First, the degree of *hydrocarbon biodegradation* under specific environmental conditions is mainly affected by the type of hydrocarbons in the contaminant matrix (Huesemann, 1995). Of the various petroleum fractions, n-alkanes and branched alkanes of intermediate length (C_{10} to C_{20}) are the preferred substrates to microorganisms, and tend to be most readily degradable. Longer chain alkanes ($>C_{20}$) are hydrophobic solids, and are difficult to degrade, due to their inherent recalcitrance and their poor water solubility (Chaîneau et al., 1995). Cycloalkanes are degraded more slowly than corresponding n-alkanes (unbranched) and branched alkanes.

Second, *contaminant bioavailability* may be determined by a series of preliminary tests. Strong interactions between soil matrix and hydrophobic pollutants can evolve, causing pollutant retention or even irreversible binding to sorbents. This phenomenon, known as aging, increases with time and significantly reduces bioavailability of hydrophobic contaminants in soil (Hatzinger and Alexander, 1995). Pollutant retention over time is governed by physical-chemical characteristics of the pollutant and by soil characteristics. Strong or even irreversible sorption onto soil is in general attributed to the soil organic matter (Luthy et al., 1997; Huang et al., 2003). However, the degree of hydrocarbon degradation is affected mainly by the type of hydrocarbon contaminants matrix, and only to a lesser extent by soil characteristics (Huesemann, 1995; Nocentini et al., 2000; Breedveld and Sparrevik, 2001). This might be true in particular for soils derived from further depths in the subsoil, where relatively low amounts of soil organic matter are present (Nierop and Verstraten 2003).

Third, *environmental conditions* have to be regarded, and include factors such as temperature, pH, moisture content, availability of mineral nutrients, and contaminant concentration. Most microorganisms can degrade specific types of hydrocarbons under extreme environmental conditions, even if rates might be lower and/or degradation might be incomplete (Margesin et al., 1997; Mohn and Stewart, 2000). Most petroleum-related hydrocarbons are readily degraded by means of aerobic microorganisms, although hydrocarbons can be degraded in the absence of oxygen (anaerobic conditions) if alternate electron acceptors, such as nitrate, manganese (IV), iron (III), sulphate, and carbon dioxide hydrocarbons, are present; however, the rate of anaerobic degradation may be substantially lower than the rate of aerobic degradation (Holliger and Zehnder, 1996; Heider et al., 1999; Grishchenkov et al., 2000; Boopathy, 2002; Massias et al., 2003).

The addition of nutrients has been reported to have a beneficiary effect on hydrocarbon degradation in soil (Dott et al., 1995; Breedveld and Sparrevik, 2001; Chaîneau et al., 2003), and typically a carbon/nitrogen/phosphorous (C/N/P) ratio of 100/10/1 is commonly proposed (Oudot and Dutrieux, 1989; Atagana et al., 2003), although water is often needed to promote optimal microbial activity (Margesin et al., 2000).

Above all, the long-term fate of petroleum hydrocarbons in areas where spills have occurred needs to be determined. This is only possible through knowledge of the constituents of petroleum and petroleum products, as well as application of the relevant text methods at the time of the spill. The data can then be used to determine whether or not hydrocarbon contamination will persist indefinitely.

The methods that are employed to measure the concentration of total petroleum hydrocarbons generate a single number that represents the combined concentration of all petroleum hydrocarbons in a sample that are measurable by the particular method (Speight, 2005). Therefore, the determination of the total petroleum hydrocarbons in a sample is method dependent. On the other hand, methods that measure a petroleum group type concentration separately are used to quantify different categories of hydrocarbons (e.g., saturates, aromatics, and polars/resins) (Speight, 2007). The results of petroleum group type analyses can be useful for product identification, because products such as gasoline, diesel fuel, and fuel oil have characteristic levels of various hydrocarbon structural moieties groups. Thus, the methods that measure identifiable petroleum fractions can be used to indicate and/or quantify the changes that have occurred through weathering of the sample.

Although these methods measure different petroleum hydrocarbon categories, there are several basic steps that are common to the analytical processes for all methods, no matter the method type or the environmental matrix. In general, these steps are: (1) sample collection and preservation: requirements specific to environmental matrix and analytes of interest, (2) extraction: separation of the analytes of interest from the sample matrix, (3) concentration: enhances the ability to detect analytes of interest, (4) cleanup, dependent upon the need to remove interfering compounds, and (5) measurement, or quantification, of the analytes (Dean, 1998). Each step affects the final result, and a basic understanding of the steps is vital to data interpretation.

2 Chemical And Physical Properties

During any oil spill incident, the properties of the spilled oil must be known immediately. Unfortunately, the properties routinely measured by oil producers and refiners are not the ones that on-site biodegradation teams need to know most urgently. Petroleum producers and refiners typically do not know to what extent the crude oil (or crude oil products) (1) will evaporate and how quickly, (2) the ability of the oil to disperse naturally, (3) the ability of the oil to disperse with the aid of dispersants, (3) the tendency of the oil to form emulsions, (4) the tendency of the oil (or its emulsions) to sink or submerge, (5) the viscosity of the oil at ambient temperatures, (6) the changing viscosity of the oil as the more volatile constituents evaporate, and (7) health hazard to on-site personnel, as well as the toxicity of the petroleum (petroleum product) constituents to land-based or aquatic flora and non-human fauna.

The importance of each of these properties is dependent upon the purity of the petroleum or petroleum product spilled, as well as the properties of the site (for example, soil composition, which amongst other properties is site specific) and the prevailing climatic conditions (arctic, sub-arctic, temperate, and tropical, which are also site specific). Because of this site specifity and the complex nature of petroleum and petroleum products (the latter is often refinery specific), generalizations of the interactions of petroleum and petroleum products on various sites are to be avoided.

In order to estimate the impact of a petroleum or petroleum product spill, several non-conventional (non-typical) properties must be assessed, and these properties are presented below in alphabetical order – and not necessarily in the order of importance.

2.1 Adhesion

Adhesion is the tendency of dissimilar particles and/or surfaces to cling to one another, while on the other hand, cohesion is the tendency of similar or identical particles/surfaces to cling to one another. The forces that cause adhesion and cohesion can be divided into several types: (1) chemical adhesion, (2) dispersive adhesion, and (3) diffusive adhesion.

It has long been recognized that petroleum and petroleum products tend to adhere to surfaces to a greater or lesser degree. Accordingly, a test has been developed, which uses a standard surface that gives a semi-quantitative measure of this adhesive property.

For the purposes of this test, petroleum adhesion is the mass of petroleum per unit area that will remain on a standard test surface after 'dunking and draining' for 30 minutes, under prescribed conditions (Jokuty et al., 1996). The standard procedure was developed using both fresh and evaporated oils with a wide range of viscosities. Test parameters that can be evaluated include temperature, oil viscosity, time, and test-surface area. A recent study has also determined that the relative adhesiveness of different oils is independent of the type of surface material used.

The usefulness of the test lies in the produced data, which allow the investigator to make reasonable deductions on the behavior of petroleum and its products under standard conditions, thereby removing much of the guesswork from the behavior of the spilled constituents. Such data, combined with a knowledge of the composition of the petroleum and/or the petroleum product, will assist the investigator to identify the principal constituents that are responsible for such behavior.

2.2 Biological Oxygen Demand

Biochemical oxygen demand (*biochemical oxygen demand*, BOD) is the amount of dissolved oxygen needed by aerobic biological organisms to break down (biodegrade) petroleum or a petroleum product present in a sample at certain temperature over a specific time period. The biological oxygen demand is commonly expressed in milligrams of oxygen consumed per liter of sample during 5 days of incubation at 20°C (68°F), and is often used as a robust surrogate of the degree of contamination. Alternatively, the unit lb/lb indicates the pounds of oxygen consumed by each pound of petroleum or petroleum product constituents per unit of time. When given in percent, the values indicate the pounds of oxygen consumed by each 100 pounds of chemical during the time stated (CHRIS, 1991).

There are two commonly recognized methods for the measurement of the biological oxygen demand: (1) the dilution method and (2) the manometric method.

In the *dilution method*, a small but measured amount of microorganism is added to each sample being tested, after which the sample is diluted with oxygen-saturated water, followed by inoculating with a fixed aliquot of the microorganism, measuring the dissolved oxygen, and then sealing the sample to prevent further oxygen dissolution. The petroleum or petroleum product sample is kept at 20°C (68°F) in the dark to prevent photosynthesis (and thereby the addition of oxygen) for five days, and the dissolved oxygen is measured again. The difference between the final dissolved oxygen and the initial dissolved oxygen is the biological oxygen demand. For measurement of the carbonaceous biological oxygen demand, a nitrification inhibitor is added after the dilution water has been added to the sample. The inhibitor hinders the oxidation of ammonia nitrogen.

In the *manometric method*, the sample is kept in a sealed container fitted with a pressure sensor. A substance that absorbs carbon dioxide (typically lithium hydroxide) is added in the container above the sample level. The petroleum or petroleum product sample is stored in conditions identical to the dilution method. Oxygen is consumed, and, as ammonia oxidation is inhibited, carbon dioxide is released. The total amount of gas, and thus, the pressure, decreases because carbon dioxide is absorbed. From the drop of pressure, the sensor electronics compute and display the consumed quantity of oxygen.

2.3 Boiling Point Distribution

In the petroleum refining industry, boiling range distribution data are used (1) to assess petroleum crude quality before purchase, (2) to monitor petroleum quality during transportation, (3) to evaluate petroleum for refining, and (4) to provide information for the optimization of refinery processes.

Traditionally, boiling range distributions of the various fractions (Table 8.1) have been determined by distillation. Yield-on-crude data are still widely reported in the petroleum assay literature, providing information on the yield of specific fractions obtained by distillation.

To some extent in the laboratory, atmospheric and vacuum distillation techniques have largely been replaced by *simulated* distillation methods, which use low resolution gas chromatography, and correlate retention times to hydrocarbon boiling points.

ASTM D2887 (Standard Test Method for Boiling Range Distribution of Petroleum Fractions by Gas Chromatography) and ASTM D3710 (Standard Test Method for Boiling Range Distribution of Gasoline and Gasoline Fractions by Gas Chromatography) use external standards composed of n-alkanes. ASTM D5307 (Standard Test Method for Determination of Boiling Range Distribution of Crude Petroleum by Gas Chromatography)

Table 8.1 Boiling ranges of petroleum fractions.

Product	TBP Cut Points, °C	TBP Cut Points, °F
Light straight-run gasoline	C5–70	C5–158
Light naphtha	70–100	158–212
Medium naphtha	100–150	212–302
Heavy naphtha	150–190	302–374
Light kerosene	190–235	374–455
Heavy kerosene	235–265	455–536
Atmospheric gas oil	265–343	536–650
Vacuum gas oil	343–565	650–1049
Atmospheric residue	>343	>650
Vacuum residue	>565	>1049

is very similar to ASTM D2887, but requires two determinations to be made with each sample, one of which uses an internal standard. The amount of material boiling above 538°C (1000°F) (reported as residue) is calculated from the differences between the two determinations.

From the point of view of petroleum and petroleum product analysis for environmental purposes, boiling range distributions provide an indication of volatility and component distribution. In addition, boiling range distribution data is also useful for the development of equations for predicting evaporative loss.

2.4 Chemical Dispersibility

The dispersibility of petroleum or petroleum products is a measure or indication of the petroleum's ability to spread on land or water. In some spill situations, and under appropriate conditions, dispersants may be an effective countermeasure for minimizing contamination of land-based sites and/or shorelines.

The *swirling flask test* (SFT) was developed for determining the effectiveness of various dispersants with different crude oils and products (Fingas et al., 1989a). Quantitation is accomplished through the use of gas chromatograph with flame ionization detection (GC/FID) (Fingas et al., 1995b).

In the test, a sample of the petroleum (or petroleum product) and dispersant are pre-mixed in a ratio of 25:1, and applied to salt water (3.3% w/w NaCl in water) in a ratio of 1:1200. Prior to 1994, quantitation was done by ultraviolet/visible spectrometry of solvent extracts of the dispersed oil. The test method can be employed to determine the effects of oil-to-water ratios and settling times (Fingas et al., 1989b) and energy (Fingas et al., 1991; Fingas et al., 1992; Fingas et al., 1993).

While this test method is useful for assessing the ability of petroleum or petroleum product to disperse on water, it must be remembered that unless specific experiment are carried out to determine the effects of oxidation on petroleum (or petroleum product) dispersibility, the deductions of petroleum behavior will be missing an important parameter.

2.5 Density, Specific Gravity, and API Gravity

Density is the mass per unit volume of a substance. It is most often reported for oils in units of g/mL or g/cm^3, and less often in units of kg/m^3. Density is temperature-dependent. It is an important property of petroleum and petroleum products, as it gives the investigator(s) indications of whether or not the contaminant(s) will float on water.

Two density-related properties of petroleum and petroleum products are often used: (1) specific gravity and (2) American Petroleum Institute (API) gravity. Specific gravity (*relative density*) is the ratio, at a specified temperature, of the oil density to the density of pure water. The API gravity scale (presented as °API) arbitrarily assigns an API gravity of 10° to pure water. Thus:

$$°\text{API} = [141.5/(\text{specific gravity at } 15.6\ °\text{C}) - 131.5]$$

The scale is commercially important for ranking petroleum quality: heavy oil typically has a gravity <20°API, medium oils are 20 to 35 °API, and light oils are 35 to 45°API, while liquid petroleum product can have an API gravity up to 65°. Tar sand bitumen typically has an API gravity <10°.

Petroleum with low density, and hence, low specific gravity, has a high API gravity. The price of a crude oil is usually based on its API gravity, with high gravity oils commanding higher prices (Speight 2011). API gravity, and density or specific gravity at 15°C (60°F), can be interconverted using Petroleum Measurements (American Petroleum Institute, 1982).

Petroleum (unless it is a specific heavy oil or tar sand bitumen) and petroleum products (unless it is a residual fuel oil or asphalt) will float on water if the density of the petroleum or petroleum product is less than that of the water. This behavior is typical of all crude oils and distillate products for both salt and fresh water. Some heavy oils, tar sand bitumens, and residual fuel oils may have densities greater than 1.0 g/mL, and their buoyancy behavior varies depending on the salinity and temperature of the water (Speight, 2009).

As stated above, density (which is temperature dependent) is an important property of petroleum and petroleum products, as it gives the investigator(s) indications of whether or not the contaminant(s) will float on water. However, such effects are not lasting.

The density of spilled oil will also increase with time, as the more volatile (and less dense) components are lost. After considerable evaporation, the density of some crude oils may increase enough for the oils to sink below the water surface. Density is also influenced by the oxidation of petroleum constituents. The inclusion of polar functions, such as hydroxyl groups (-OH) or carbonyl groups (>C=O) (a result of the oxidation process), causes an increase in the density of the petroleum or petroleum product (relative to the original material).

2.6 Emulsion Formation

A water-in-oil emulsion is a stable dispersion of small droplets of water in oil. When formed from crude oils spilled at sea, these emulsions can have very different characteristics from their parent crude oils. This has important implications for the fate and behavior of the oil and its subsequent cleanup. It is desirable, therefore, to determine if oil is likely to form an emulsion, and if so, whether that emulsion is stable, and the physical characteristics of the emulsion.

In an older test method, the tendency for a crude oil to form a water-in-oil emulsion was measured using a method based on the rotating flask apparatus (Mackay and Zagorski, 1982). All numerical values (mostly ones or zeroes) based on this method have subsequently been reduced to *yes* or *no*, respectively, and indicate the formation (or not) of an emulsion that remained stable 24 hours after settling. In a newer variation, the reproducibility is considerably improved, and several parameters, including (1) the water-to-oil ratio, (2) the fill volume, and (3) the orientation of the vessels, were found to be important parameters affecting emulsion formation.

However, such effects are not lasting. Emulsion formation and behavior is influenced by the oxidation of petroleum constituents (Chapter 2). The inclusion of polar functions, such as hydroxyl groups (–OH) or carbonyl groups (>C=O) (a result of the oxidation process), causes an increase in the density of the emulsion (relative to the original unoxidized petroleum) and an increased propensity to form emulsions. As a result, the emulsion forms and sinks to various depths, or even to the seabed, depending on the extent of the oxidation and the resulting density. This may give the erroneous appearance (leading to erroneous deductions with catastrophic consequences) that the petroleum spill (as evidenced from the petroleum remaining on the surface of the water) is less than it actually was. The so-called *missing* oil will undergo further chemical changes, and eventually reappear on the water surface or on a distant beach.

2.7 Evaporation

Evaporation is the removal of the lower-boiling constituents from petroleum or a petroleum product usually under ambient conditions, or, in the current context, under the conditions prevalent at the spill site.

Evaporation rate and loss are of importance for all volatile constituents of petroleum and petroleum products. While standard test methods, such as those designated for distillation and vapor pressure determination, are often used to determine evaporation properties (ASTM, 2011), test methods for determining evaporation loss are available for higher boiling petroleum products (ASTM D972: Standard Test Method for Evaporation Loss of Lubricating Greases and Oils; ASTM D2595: Standard Test Method for Evaporation Loss of Lubricating Greases Over Wide-Temperature Range). Although not necessarily applicable to petroleum and petroleum products in general, evaporation loss data can be obtained at any temperature in the range from 93 and 316°C (200 and 600°F).

Evaporation is a major process that contributes to the weathering of spilled oil. While pure compounds evaporate at constant rates, petroleum and petroleum products, which are composed of thousands of compounds, do not. Rapid initial loss of the more volatile fractions is followed by progressively slower loss of less volatile constituents. It is not uncommon for part of the total volume of an oil spill to evaporate within one day of the spill (Fingas et al., 1979).

Petroleum and petroleum products evaporate at a logarithmic rate with respect to time (Fingas, 1995, 1998). This is attributed to the overall logarithmic appearance of many components evaporating at different linear rates. Petroleum products with fewer constituents (such as diesel fuel) evaporate at a rate that is a square root with respect to time, which is a result of the number of components evaporating. The evaporation process, as evidenced by petroleum and petroleum products, is not strictly boundary-layer regulated, which is largely a result of the high saturation concentrations of oil components in air, and is associated with a high boundary-layer regulated rate. Some volatile crude oils and petroleum products show some effect of boundary-layer regulation at the start of the evaporation process, but after several minutes, evaporation slows because of the loss of the more volatile components, at which point evaporation ceases to be boundary-layer regulated.

It must also be recognized that as evaporation occurs, the density and viscosity of the residual petroleum or residual petroleum product increase, thereby causing behavioral changes in the contaminant. Such changes could be reflected in an increase in adhesion of the contaminant constituents to the soil or rock.

2.8 Fire Point and Flash Point

The *fire point* is the lowest temperature, corrected to one atmosphere pressure (14.7 psi), at which the application of a test flame to the petroleum or petroleum product sample surface causes the vapor of the oil to ignite and burn for at least five seconds.

At any time after a spill of petroleum or a petroleum product, fire should always be considered an imminent hazard. Related to fire point, the flash point is a measure of the tendency of the petroleum or a petroleum product to form a flammable mixture with air under controlled laboratory conditions. It is only one of a number of properties that should be considered in assessing the overall flammability hazard of a spilled material (ASTM D92).

The ignition temperature (sometimes called the *autoignition temperature*) is the minimum temperature at which the material will ignite without a spark or flame being present. The method of measurement is given by ASTM E659 (Standard Test Method for Autoignition Temperature of Liquid Chemicals).

Also related to fire point, the flammability limits of vapor in air are an expression of the percent concentration in air (by volume) given for the lower and upper limit. These values give an indication of relative flammability. The limits are sometimes referred to as *lower explosive limit* (LEL) and *upper explosive limit* (UEL).

The *flash point* of petroleum or a petroleum product is the temperature to which the sample must be heated to produce a vapor/air mixture above the liquid fuel that is ignitable when exposed to an open flame under specified test conditions. In North America, flash point is used as an index of fire hazard.

Flash point is an extremely important factor in relation to the safety of spill cleanup operations. Gasoline and other low-boiling liquid fuels can be ignited under most ambient conditions, and therefore pose a serious hazard when spilled. Many freshly spilled crude oils also have low flash points until the lighter components have evaporated or dispersed.

There are several ASTM methods for measuring flash points: ASTM D93/IP34 (Standard Test Methods for Flash Point by Pensky-Martens Closed Tester) and ASTM D56 (Standard Test Method for Flash Point by Tag Closed Tester) are among the most commonly used. The Pensky-Martens tester has an integral stirrer, but no cooling bath. The minimum flash point that can be determined by method D93/IP34 is 10°C (50°F). The Tag closed tester has an integral cooling bath, but no stirring mechanism.

Method ASTM D56 is intended for liquids with a viscosity less than 9.5 cSt at 25°C (77°F). The flash point and fire point (see below) of lubricating oil are determined by ASTM D92/IP 36 (Standard Test Method for Flash and Fire Points by Cleveland Open Cup).

2.9 Fractionation

Rather than quantifying a complex total petroleum hydrocarbon mixture as a single number, petroleum fractionation methods break the mixture into discrete fractions, thus providing data that can be used in a risk assessment, and in characterizing product type and compositional changes, such as those that may occur during weathering (oxidation). The fractionation methods can be used to measure both the volatile constituents and the extractable constituents.

In contrast to traditional methods for total petroleum hydrocarbons, that report a single concentration number for complex mixtures, the fractionation methods report separate concentrations for discrete aliphatic and aromatic fractions. The available petroleum fraction methods are GC-based, and are thus sensitive to a broad range of hydrocarbons. Identification and quantification of aliphatic and aromatic fractions allow one to identify petroleum products and evaluate the extent of product weathering. These fraction data also can be used in risk assessment.

One particular method is designed to characterize C_6 to C_{28+} petroleum hydrocarbons in soil as a series of aliphatic and aromatic carbon range fractions. The extraction methodology differs from other petroleum hydrocarbon methods, because it uses *n*-pentane, and not methylene chloride, as the extraction solvent. If methylene chloride is used as the extraction solvent, aliphatic and aromatic compounds cannot be separated.

The solvent n-Pentane extracts petroleum hydrocarbons in this range most efficiently. The whole extract is separated into aliphatic and aromatic petroleum-derived fractions (EPA SW-846 3611, EPA SW-846 3630). The aliphatic and aromatic fractions are analyzed separately by gas chromatography, and quantified by summing the signals within a series of specified carbon ranges that represent the fate and transport fractions. The gas chromatograph is equipped with a boiling point column (non-polar capillary column). Gas chromatographic parameters allow the measurement of a hydrocarbon range of n-hexane (C_6) to n-octacosane (C_{28+}), a boiling point range of approximately 65°C to 450°C (150 to 840°F).

2.10 Leachability and Toxicity

As a start and for regulatory and remediation purposes, a standard test is needed to measure the likelihood of toxic substances getting into the environment and causing harm to organisms. The test (required by the United States Environmental Protection Agency) is the *toxicity characteristic leaching procedure* (TCLP, EPA SW-846 Method 1311), designed to determine the mobility of both organic and inorganic contaminants present in liquid, solid, and multiphase wastes.

The method was developed to estimate the mobility of specific inorganic and organic contaminates that are destined for disposal in municipal landfills. The extraction is performed, using acetic as the extraction fluid. The pH of the acetic acid/sodium acetate buffer solution is maintained at 4.93. This sample/acetic acid mixture is subjected to rotary extraction, designed to accelerate years of material exposure in the shortest possible time. After extraction, the resulting liquid is subjected to analysis, utilizing a list of contaminants that includes metals, volatile organic compounds, semi-volatile organic compounds, pesticides, and herbicides.

An alternate procedure, the synthetic precipitation leaching procedure (SPLP, EPA SW-846 Method 1312), may be appropriate. This procedure is applicable for materials where the leaching potential due to normal rainfall is to be determined. Instead of the leachate simulating acetic acid mixture, nitric and sulfuric acids are utilized in an effort to simulate the acid rains resulting from airborne nitric and sulfuric oxides.

Toxicity values are often reported as the amount of chemical that is lethal to indigenous floral and faunal populations. Thus:

LC_{50}: the median lethal concentration is the estimated concentration of a compound that will cause death to 50 percent of the test population in a specified time after exposure. In most instances, LC_{50} is statistically derived by analysis of mortalities in various test concentrations after a fixed period of exposure.

EC_{50}: the median effective concentration is used when an effect other than death is the observed endpoint. EC_{50} is the estimated concentration of the compound in water that will have a specific effect on 50 percent of the test population in a specified time after exposure. As with LC_{50}, the EC_{50} is generally derived statistically.

TLm: the median tolerance limit for a floral and faunal species for the spilled chemical; this is a term sometimes used instead of EC_{50}.

2.11 Metals Content

Metals content in crude oils can provide valuable information about the origin of those oils, potentially aiding in identifying the source of oil spills.

Crude oil assays often include nickel and vanadium contents, due to the detrimental effects of these metals on catalysts used in cracking and desulphurization processes. In lubricating oils, metal contents can provide information on both the types of additives used in the oil and on the wear history of the equipment being lubricated. ASTM D5185, Standard

Test Method for Determination of Additive Elements, Wear Metals, and Contaminants in Used Lubricating Oils and Determination of Selected Elements in Base Oils by Inductively Coupled Plasma Atomic Emission Spectrometry (ICP-AES), can be used to determine over 20 different metals in a variety of petroleum products.

To overcome problems inherent in the direct analysis of xylene solutions by ICP/AES, a method was developed using microwave digestion of oils with nitric acid (Cao, 1992; Fingas et al., 1995).

2.12 Pour Point

The pour point of petroleum or a petroleum product is the lowest temperature at which the oil will just flow, under standard test conditions (ASTM D97). The failure to flow at the pour point is usually attributed to the separation of waxes from the oil, but can also be due to the effect of viscosity, in the case of very viscous oils. Also, particularly in the case of residual fuel oils, pour points may be influenced by the thermal history of the sample, that is, the degree and duration of heating and cooling to which the sample has been exposed.

From a spill response point of view, it must be emphasized that the tendency of the oil to flow will be influenced by the size and shape of the container, the head of the oil, and the physical structure of the solidified oil. The pour point of the oils are therefore an indication, and not an exact measure, of the temperature at which flow ceases (Dyroff, 1993).

2.13 Solubility in Aqueous Media

The solubility of petroleum or petroleum products in water (which should not be confused with the concept of emulsion formation) can be determined by bringing to equilibrium a volume of oil and water, and then analyzing the water phase. This analysis can be done by purging and trapping the dissolved hydrocarbons, or, alternatively, directly analyzing the headspace above the water.

Since oil is a complex mixture of components, and each component has a different solubility in water, the aqueous solubility is expressed as the cumulative concentration of the individually dissolved components. The composition and concentration of the solubilized mixture will depend upon conditions used during equilibration.

The term *solubility* as applied to petroleum and petroleum products is being replaced by the technically more precise term *water-soluble fraction*. The values reported in this catalogue were taken from those studies where an excess of oil was used (oil-to-water volume ratios of at least 1: 20), and where the processes of evaporation and oil-in-water emulsification were prevented.

2.14 Sulfur Content

The sulfur content of a crude oil is important for a number of reasons. Downstream processes, such as catalytic cracking and refining, will be adversely affected by high sulphur contents. During an oil spill, the sulfur content becomes a health and safety concern for cleanup personnel. In addition, if high sulphur oils are burning, they can produce dangerous levels of sulphur dioxide.

The total sulphur content of oil can be determined by numerous standard techniques. ASTM D129 (Standard Test Method for Sulfur in Petroleum Products) is applicable to petroleum products of low volatility and containing at least 0.1 mass percent sulfur. Sulfur contents are also determined in accordance with ASTM D4294 (Standard Test Method for Sulfur in Petroleum Products by Energy-Dispersive X-Ray Fluorescence Spectroscopy). This method is applicable to both volatile and non-volatile petroleum products with sulphur concentrations ranging from 0.05 to 5 mass percent.

2.15 Surface Tension and Interfacial Tension

Interfacial tension is the force of attraction between the molecules at the interface of two fluids. At the air/liquid interface, this force is often referred to as surface tension. The SI units for interfacial tension are millinewtons per meter (mN/m). These are equivalent to the former units of dynes per centimeter (dyne/cm).

The surface tension of petroleum (or a petroleum product), together with its viscosity, affects the rate at which an oil spill spreads. Air/oil and oil/water interfacial tensions can be used to calculate a spreading coefficient, which gives an indication of the tendency for the oil to spread. It is defined as:

$$\text{Spreading coefficient} = S_{WA} - S_{OA} - S_{WO}$$

S_{WA} is water/air interfacial tension, S_{OA} is oil/air interfacial tension, and S_{WO} is water/oil interfacial tension.

Spreading to a thin slick is likely if the spreading coefficient of an oil is greater than zero, and the higher the spreading coefficient, the faster the spreading will occur (Twardus, 1980).

Unlike density and viscosity, which show systematic variations with temperature and degree of evaporation, interfacial tensions of crude oils and oil products show no such correlations. Nor is there any correlation to viscosity (Jokuty et al., 1995).

A single test method (ASTM D971, Standard Test Method for Interfacial Tension of Oil against Water by the Ring Method), is applicable to the measurement of oil/water interfacial tensions. Unlike manually operated ring tensiometers, the maximum deformation of the lamella is detected electronically, and occurs before the ring pulls completely through the interface. This results in interfacial tensions that are slightly lower than those measured manually. Repeatability is +/–2% of the mean.

2.16 Total Petroleum Hydrocarbons

Total petroleum hydrocarbon (TPH) analyses (Table 8.2 and Table 8.3) are conducted to determine the total amount of hydrocarbon present in the environment (Pavlova and Ivanova, 2003; Speight, 2005). There are a wide variety of methods for measurement of the total petroleum hydrocarbon in a sample, but analytical inconsistencies must be recognized because of the definition of total petroleum hydrocarbons and the methods employed for analysis (Chapter 9) (Rhodes et al., 1994; Speight, 2001, 2002, 2007). Thus, in practice, the term total petroleum hydrocarbon is defined by the analytical method, since different methods often give different results because they are designed to extract and measure slightly different subsets of petroleum hydrocarbons.

The analysis for the total petroleum hydrocarbons (TPH) in a sample as a means of evaluating petroleum-contaminated sites is also an analytical method in common use. The data are used to establish target cleanup levels for soil or water; this is a common approach implemented by regulatory agencies in the United States, and in many other countries.

The data obtained by the analysis have become key remediation criteria, and it is essential that the environmental analyst (and others who may use the data) be knowledgeable about the various analytical methods. It is also important to know that minor method deviations may be found from region to region. For example, in terms of nomenclature, itself a complex and often ill-defined area of petroleum science (Speight, 2007), the analytical methods may refer to *total petroleum hydrocarbons* as *mineral oil, hydrocarbon oil, extractable hydrocarbon*, and *oil and grease*.

Thus, as often occurs in petroleum science (Speight, 2007), the definition of total petroleum hydrocarbons depends on the analytical method used, because the total petroleum hydrocarbons measurement is the total concentration of the hydrocarbons extracted and measured by a particular method. The same sample analyzed by different methods may produce different values. For this reason, it is important to know exactly how each determination is made, since interpretation of the results depends on understanding the capabilities and limitations of the method. If used indiscriminately, measurement of the total petroleum hydrocarbons in a sample can be misleading, leading to an inaccurate assessment of risk.

Thus, the choice of a specific method should be based on compatibility with the particular type of hydrocarbon contamination to be measured,

Table 8.2 Summary of analytical methods for determining the total petroleum hydrocarbons in a sample.

Method Name	Method Type/ Environmental Media	Typical Products Detected	Typical Carbon Ranges Detected	Methods
GC-based TPH methods	Primarily laboratory but also field applications—can be adapted for all media	Primarily gasolines, diesel fuel, and fuel oil No. 2; can be modified for heavier hydrocarbon mixtures (e.g., lubricating oils, heavy fuel oils)	Normally between C_6 and C_{25} or C_{36} (can be modified for higher carbon numbers)	EPA Method 8015B, state-modified 8015 methods
IR-based TPH methods	Laboratory and field screening—most appropriate for water and soil	Primarily diesel and fuel oils	Most hydrocarbons, with exception of volatile and very high hydrocarbons	EPA Method 418.1
Gravimetric TPH methods	Laboratory—most appropriate for wastewaters,sludges, and sediment	Most appropriate for heavier petroleum products (e.g., crude oils, lubricating oils)	Anything that is extractable, with exception of volatiles, which are lost	EPA Method 9071; EPA Method 1664
Immunoassay TPH methods	Field screening—most appropriate for soil and water	Various products (but yields only screening numbers)	Aromatic hydrocarbons (e.g., BTEX, PAHs)	EPA Method 4030

Source: EPA, 2004.

Table 8.3 Summary of methods for determining the total petroleum hydrocarbons in a sample.

Method Name	Matrix	Scope of Method
Gravimetric methods		
EPA method 413.1	Water and wastewater	Petroleum fuels from gasoline through No. 2 fuel oils are completely or partially lost in the solvent removal operation. Recoveries of some crude oils and heavy fuel oils will be low.
EPA method 9070	Solid waste	Applicable to determination of relatively nonvolatile hydrocarbons. Not applicable to measurement of light hydrocarbons; petroleum fuels, from gasoline through No. 2 fuel oils, are completely or partially lost. Recoveries of some crude oils and heavy fuel oils will be low.
EPA method 9071A	Sludge	Used to recover low levels of oil and grease from sludge. Used when relatively polar, heavy petroleum fractions are present, or when the levels of nonvolatile greases challenge the solubility limit of the solvent. Not recommended for measurement of low-boiling fractions.
Standard method 5520B	Water and wastewater	Suitable for biological lipids and mineral hydrocarbons. Not suitable for low-boiling fractions.
Standard method 5520D	Water and wastewater	Suitable for biological lipids and mineral hydrocarbons. Method D is the method of choice when relatively polar, heavy petroleum fractions are present, or when the levels of nonvolatile greases may challenge the solubility of the solvent.

(Continued)

Table 8.3 Summary of methods for determining the total petroleum hydrocarbons in a sample.

Method Name	Matrix	Scope of Method
Standard method 5520E	Sludge	Suitable for biological lipids and mineral hydrocarbons. Not suitable for low-boiling fractions. Method E is a modification of Method D.
Standard method 5520F	Water and wastewater	Suitable for biological lipids and mineral hydrocarbons. Not suitable for low-boiling fractions. May be used in conjunction with Method B, D, or E.
Infrared methods		
EPA method 413.2	Water and wastewater	Applicable to the measurement of light fuels, although loss of about half of any gasoline present during the extraction manipulation can be expected.
EPA method 418.1	Water and wastewater	Applicable to a wide range of hydrocarbons, although volatile components will be lost. Modifications exist for measurement of TPH in soil.

and furthermore, the choice may depend upon local or regional regulatory requirements for the type of hydrocarbon contamination that is known, or suspected, to be present. Additionally, the risk at a specific site will change with time as contaminants evaporate, dissolve, biodegrade, and/or become sequestered.

2.17 Viscosity

Viscosity is a measure of a fluid's resistance to flow; the lower the viscosity of a fluid, the more easily it flows. Like density, viscosity is affected by temperature. As temperature decreases, viscosity increases.

Viscosity is a very important property of oils, because it affects the rate at which spilled oil will spread, the degree to which it will penetrate shoreline substrates, and the selection of mechanical spill countermeasures equipment.

Viscosity measurements may be absolute or relative (sometimes called 'apparent'). Absolute viscosities are those measured by a standard method, with the results traceable to fundamental units. Absolute viscosities are distinguished from relative measurements made with instruments that measure viscous drag in a fluid, without known and/or uniform applied shear rates (Schramm, 1992). An important benefit of absolute viscometry is that the test results are independent of the particular type or make of viscometer used. Absolute viscosity data can be compared easily between laboratories worldwide.

Modern rotational viscometers are capable of making absolute viscosity measurements for both Newtonian and non-Newtonian fluids at a variety of well controlled, known, and/or uniform shear rates. Unfortunately, no ASTM standard method exists that makes use of these viscometers. Nonetheless, these instruments are in widespread use in many industries.

There are several ASTM Standard Methods for measuring the viscosity of oils. Of these, only methods ASTM D445 (Standard Test Method for Kinematic Viscosity of Transparent and Opaque Liquids) and ASTM D4486 (Standard Test Method for Kinematic Viscosity of Volatile and Reactive Liquids) will yield absolute viscosity measurements. Both of these methods make use of glass capillary kinematic viscometers, and will produce absolute measurements in units of centistokes (cSt) only for oils that exhibit Newtonian flow behavior (viscosity independent of the rate of shear).

Although now obsolete, at one time, the petroleum industry relied on measuring kinematic viscosity with the Saybolt viscometer and expressing kinematic viscosity in Saybolt Universal Seconds (SUS) or Saybolt Furol Seconds (SFS). Occasionally, Saybolt viscosities are still reported in the literature: ASTM D2161, Standard Practice for Conversion of Kinematic Viscosity to Saybolt Universal Viscosity or to Saybolt Furol Viscosity, establishes the official equations relating SUS and SFS to the SI kinematic viscosity units, mm^2/s.

2.18 Volatile Organic Compounds

Benzene, toluene, ethylbenzene, and xylenes (BTEX), and substituted benzenes are the most common aromatic compounds in petroleum, making up to a few percent of the total mass of some crude oils. They are the most soluble and mobile fraction of crude oil and many petroleum products, and, as such, frequently enter soil, sediments, and ground water because of accidental spills, leakage of petroleum fuels from storage tanks and pipelines, and improper oil waste disposal practices. BTEX are hazardous carcinogenic and neurotoxic compounds, and are classified as priority pollutants regulated by Environment Canada and the US Environmental Protection Agency.

A rapid, reliable, and effective method for direct determination of BTEX plus C_3-substituted benzenes has been developed, using gas chromatography with mass spectrometric detection (GC/MS) (Wang et al., 1995).

Vapor pressure, which is related to the amount of volatile organic compounds in the petroleum or petroleum product, is an important physical property of volatile liquids. It is the pressure that a vapor exerts on its surroundings. For volatile petroleum products, vapor pressure is used as an indirect measure of evaporation rate, and can be measured by a variety of methods, including Reid vapor pressure, dynamic vapor pressure, static vapor pressure, isoteniscopic vapor pressure, vapor pressure balance, and gas saturation.

The most commonly used method for petroleum and petroleum products is the Reid vapor pressure (ASTM D323 – Standard Test Method for Vapor Pressure of Petroleum Products). This test method determines vapor pressure at 37.8°C (100°F) of petroleum and petroleum products with an initial boiling point above 0°C (32°F). It is measured by saturating a known volume of oil in an air chamber of known volume and measuring the equilibrium pressure, which is then corrected to one atmosphere (14.7 psi).

2.19 Water Content

Some of the petroleum and petroleum products sampled contain substantial amounts of water. Because any process that would separate the oil and water would also change the composition of the oil, most properties are often determined on the oils as received. Therefore, for those oils with significant water contents (>5% w/w), many of the properties measured will not represent the properties of the *dry* oil.

Generally, water content is determined by the Karl Fischer titration (ASTM D6304). The procedure is adopted in many official standards as the standard test method for water determination in petroleum and petroleum products.

Water content has not been identified as a direct governing environmental factor in the biodegradation of petroleum and petroleum products (Frijer et al., 1996). However, an indirect effect of water content is that it influences the effective gas diffusion coefficient in soils. The rate of production of carbon dioxide is a means of expressing the mineralization rate of hydrocarbons.

3 Petroleum Group Analyses

Petroleum group analyses are conducted to determine amounts of the petroleum compound classes (e.g., saturates, aromatics, and polars/

resins) present in petroleum-contaminated samples. This type of measurement is sometimes used to identify fuel type, or to track plumes. It may be particularly useful for higher boiling products, such as asphalt. Group type test methods include multidimensional gas chromatography (not often used for environmental samples), high performance liquid chromatography (HPLC), and thin layer chromatography (TLC) (Miller, 2000; Patnaik, 2004).

Test methods that analyze individual compounds (e.g., benzene-toluene-ethylbenzene-xylene mixtures and polynuclear aromatic hydrocarbons) are generally applied to detect the presence of an additive, or to provide concentration data needed to estimate environmental and health risks that are associated with individual compounds. Common constituent measurement techniques include gas chromatography with second column confirmation, gas chromatography with multiple selective detectors, and gas chromatography with mass spectrometry detection (GC/MS) (EPA 8240).

Many common environmental methods measure individual petroleum constituents or *target compounds* rather than the whole signal from the total petroleum hydrocarbons. Each method measures a suite of compounds selected because of their toxicity and common use in industry.

For organic compounds, there are three series of target compound methods that must be used for regulatory purposes:

1. EPA 500 series: Organic Compounds in Drinking Water, as regulated under the Safe Drinking Water Act.
2. EPA 600 series: Methods for Organic Chemical Analysis of Municipal and Industrial Wastewater, as regulated under the Clean Water Act.
3. SW-846 series: Test Methods for Evaluating Solid Waste: Physical/Chemical Methods, as promulgated by the US EPA, Office of Solid Waste and Emergency Response.

The 500 and 600 series methods provide parameters and conditions for the analysis of drinking water and wastewater, respectively. One method (EPA SW-846) is focused on the analysis of nearly all matrices, including industrial waste, soil, sludge, sediment, and water miscible and non-water miscible wastes. It also provides for the analysis of groundwater and wastewater, but is not used to evaluate compliance of public drinking water systems.

Selection of one method over another is often dictated by the nature of the sample, and the particular compliance or cleanup program for which the sample is being analyzed. It is essential to recognize that capabilities and requirements vary between methods when requesting any analytical method or suite of methods. Most compound-specific methods use a gas

Table 8.4 Test methods for total petroleum hydrocarbons (Environmental protection agency).

SW-846 Method	Water/ Wastewater Method	Analytes	Primary Equipment	Sample Preparation[a]
4030[b]	n/a	TPHs	Immunoassay	Included in kit
4035[b]	n/a	PAHs	Immunoassay	Included in kit
8015[c]	n/a	Aliphatic and aromatic hydrocarbons; nonhalogenated VOCs	GC/FID	Extraction (SVOCs)[d]; purge-and-trap and headspace (VOCs)[d]; azeotropic distillation (nonhalogenated VOCs)[c, d]
8021[c, e]	502.2/602	Aromatic VOCs (not ethers or alcohols)	GC/PID	Purge-and-trap
8100	n/a	PAHs	GC/FID	Extraction[d]
8260[f]	524.2/624	VOCs	GC/MS	Purge-and-trap; static headspace; azeotropic distillation[d]
8270	525/625	SVOCs	GC/MS	Extraction[d]

(Continued)

Table 8.4 (cont.) Test methods for total petroleum hydrocarbons (Environmental protection agency).

SW-846 Method	Water/ Wastewater Method	Analytes	Primary Equipment	Sample Preparationa
8310	610	PAHs	High-performance liquid chromatography (HPLC)	Extraction[d]
8440[g]	418.1/[h]	TPH	IR spectro-photometer	Supercritical fluid extraction from soilsd[d]

[a]These are the standard methods of preparation for the corresponding method. They may vary depending on specific analytical needs.
[b]Screening method for soils.
[c]MTBE can be analyzed for with U.S. EPA SW-846 Method 8015 or 8021: however, 8021 has lower detection limits, is subject to less interference in highly contaminated samples; and tends to be more economical by providing BTEX data in the same analysis. Concerns about coelution with some alkanes requires at least one confirmatory analysis with SW-846 Method 8260 per site.
[d]See Chapter 4 of SW-846 for specific appropriate methods.
[e]8021 replaces 8010 and 8020.
[f]8260 replaces 8240.
[g]This method is similar to 418.1, however, perchlorethane (PCE) is used as an IR solvent instead of Freon-113.
[h]418.1 is used extensively, although it is not on the list of promulgated methods.

chromatographic selective detector, high performance liquid chromatography, or gas chromatography/mass spectrometry.

More correctly, group analytical methods are designed to separate hydrocarbons into categories, such as saturates, aromatics, resins, and asphaltenes (SARA), or paraffins, iso-paraffins, naphthenes, aromatics, and olefins (PIANO). These chromatographic, gas chromatographic, and high performance liquid chromatographic methods (HPLC) (Table 8.4) were developed for monitoring refinery processes or evaluating organic synthesis products. Column chromatographic methods that separate saturates from aromatics are often used as preparative steps for further analysis by gas chromatography/mass spectrometry. Thin layer chromatography is sometimes used as a screening technique for petroleum product identification.

3.1 Thin Layer Chromatography

In the environmental field, thin layer chromatography (TLC) is best used for screening analyses and characterization of semi-volatile and non-volatile petroleum products. The precision and accuracy of the technique is inferior to other methods (EPA 8015, EPA 418.1), but when speed and simplicity are desired, thin layer chromatography may be a suitable alternative. For characterizations of petroleum products such as asphalt, the method has the advantage of separating compounds that are too high boiling to pass through a gas chromatograph. While thin layer chromatography does not have the resolving power of a gas chromatograph, it is able to separate different classes of compounds. Thin layer chromatography analysis is fairly simple, and since the method does not give highly accurate or precise results, there is no need to perform the highest quality extractions.

In the method, soil samples are extracted by shaking or vortexing with the solvent. Water samples are extracted by shaking in a separatory funnel. If there is the potential for the presence of compounds that interfere with the method and make the data suspect, silica gel can be added to clean the extract. Sample extract aliquots are placed close to the bottom of a glass plate coated with a stationary phase. The most widely used stationary phases are made of an organic hydrocarbon moiety bonded to a silica backbone.

For the analysis of petroleum hydrocarbons, a moderately polar material stationary phase works well. The plate is placed in a sealed chamber with a solvent (mobile phase). The solvent travels up the plate, carrying compounds present in the sample. The distance a compound travels is a function of the affinity of the compound to the stationary phase relative to the mobile phase. Compounds with chemical structure and polarity similar to the solvent travel well in the mobile phase. For example, the saturated hydrocarbons seen in diesel fuel travel readily up a plate in a hexane mobile phase. Polar compounds, such as ketones or alcohols, travel a smaller distance in hexane than saturated hydrocarbons.

After a plate has been exposed to the mobile phase solvent for the required time, the compounds present can be viewed by several methods. Polynuclear aromatic hydrocarbons, other compounds with conjugated systems, and compounds containing heteroatoms (nitrogen, oxygen, or sulfur) can be viewed with long wave and short wave ultraviolet light. The unaided eye can see other material, or the plates can be developed in iodine. Iodine has an affinity for most petroleum compounds, including the saturated hydrocarbons, and stains the compounds a reddish/brown color.

The method is considered to be a qualitative and useful tool for rapid sample screening. Limitations of the method center on its moderate reproducibility, detection limits, and resolving capabilities. Variability between operators can be as high as 30%. Detection limits (without any concentration of the sample extract) are near 50 ppm (mg/kg) for most petroleum products in soils. When the aromatic content of a sample is high, as with bunker C fuel oil, the detection limit can be near 100 ppm. It is often not possible to distinguish between similar products, such as diesel and jet fuel. As with all chemical analyses, quality assurance tests should be run to verify the accuracy and precision of the method.

3.2 Immunoassay

A number of different testing kits based on immunoassay technology are available for rapid field determination of certain groups of compounds, such as benzene-toluene-ethylbenzene-xylene (EPA 4030) or polynuclear aromatic hydrocarbons (EPA 4035, Polycyclic Aromatic Hydrocarbons by Immunoassay). The immunoassay screening kits are self-contained, portable field kits that include components for sample preparation, instrumentation to read assay results, and immunoassay reagents.

Unless the immunoassay kit is benzene sensitive, the kit may display strong biases, such as the low affinity for benzene relative to toluene, ethylbenzene, xylenes, and other aromatic compounds. This will cause an underestimation of the actual benzene levels in a sample. And, since benzene is often the dominant compound in leachates, due to its high solubility, a low sensitivity for benzene is undesirable.

The quality of the analysis of polynuclear aromatic hydrocarbons is often dependent on the extraction efficiency. Clay and other cohesive soils lower the ability to extract polynuclear aromatic hydrocarbons. Another potential problem with polynuclear aromatic hydrocarbon analysis is that the test kits may have different responses for different compounds.

3.3 Gas Chromatography

Gas chromatography uses the principle of a stationary phase and a mobile phase. Much attention has been paid to the various stationary phases, and books have been written on the subject as it pertains to petroleum chemistry.

Briefly, *gas-liquid chromatography* (GLC) is a method for separating the volatile components of various mixtures (Altgelt and Gouw, 1975; Fowlis, 1995; Grob, 1995). It is, in fact, a highly efficient fractionating technique, and it is ideally suited to the quantitative analysis of mixtures when the possible components are known, and the interest lies only in determining the amounts of each present. In this type of application, gas chromatography has taken over much of the work previously done by the other techniques; it is now the preferred technique for the analysis of hydrocarbon gases, and gas chromatographic in-line monitors are having increasing application in refinery plant control. Gas-liquid chromatography is also used extensively for individual component identification, as well as percentage composition, in the gasoline boiling range.

The mobile phase is the carrier gas, and the gas selected has a bearing on the resolution. Nitrogen has very poor resolution ability, while helium or hydrogen is a better choice, with hydrogen being the best carrier gas for resolution. However, hydrogen is reactive, and may not be compatible with all sets of target analytes. There is an optimum flow rate for each carrier gas to achieve maximum resolution. As the temperature of the oven increases, the flow rate of the gas changes due to thermal expansion of the gas. Most modern gas chromatographs are equipped with constant flow devices that change the gas valve settings as the temperature in the oven changes, so changing flow rates are no longer a concern. Once the flow is optimized at one temperature, it is optimized for all temperatures.

For environmental analysis (Bruner, 1993), particularly the volatile samples found in total petroleum hydrocarbons (Table 8.5), the gas chromatograph is generally interfaced with a purge and trap system, as described in the section on gas chromatographic methods. The photoionization detector works by bombarding compounds with ultraviolet (UV) light, generating a current of ions. Compounds with double carbon bonds, conjugated systems (multiple carbon double bonds arranged in a specific manner), and aromatic rings are easily ionized with the ultraviolet light generated by the Photoionization detector lamp, while most saturated compounds require higher energy radiation.

One method (EPA 8020) that is suitable for volatile aromatic compounds is often referred to as benzene-toluene-ethylbenzene-xylene analysis, though the method includes other volatile aromatics. The method is similar to most volatile organic gas chromatographic methods. Sample preparation and introduction is typically by the purge and trap analysis

Table 8.5 Summary of gas chromatographic methods for determining the total petroleum hydrocarbons in a sample.

Method Name	Matrix	Scope of Method
Direct injection methods		
EPA method 8015B	Solid wastes	Used to determine the concentration of petroleum hydrocarbons, including gasoline range organics (GROs). Analysts should use the fuel contaminating the site for quantitation.
ASTM method D3328-90	Water	Petroleum oils such as distillate fuel, lubricating oil, and crude oil recovered from water or beaches. Identification of a recovered oil is determined by comparison with known oils, selected because of their possible relationship to the recovered oil.
Purge and trap and headspace methods		
EPA method 8015B	Solid wastes	Used to determine the concentration of petroleum hydrocarbons, including diesel range organics (DROs) and jet fuel. Analysts should use the fuel contaminating the site for quantitation.

(EPA 5030). Some oxygenates, such as methyl-t-butyl ether (MTBE), are also detected by the Photoionization detector, as well as olefins, branched alkanes, and cycloalkanes.

Certain false positives are common (EPA 8020). For example, trimethylbenzenes and gasoline constituents are frequently identified as chlorobenzenes (EPA 602, EPA 8020), because these compounds elute with nearly the same retention times from nonpolar columns. Cyclohexane is often mistaken for benzene (EPA 8015/8020), because both compounds are detected by a 10.2 eV photoionization detector and have nearly the same elution time from a nonpolar column (EPA 8015). The two compounds have very different retention times on a more polar column (EPA 8020), but a more polar column skews the carbon ranges (EPA 8015). False positives for oxygenates in gasoline are common, especially in highly contaminated samples.

For semi-volatile constituents of petroleum, the gas chromatograph is generally equipped with either a packed or capillary column. Either neat or diluted organic liquids can be analyzed via direct injection, and compounds are separated during movement down the column. The flame ionization detector uses a hydrogen-fueled flame to ionize compounds that reach the detector.

For polynuclear aromatic hydrocarbons, a method is available (EPA 8100), in which injection of sample extracts directly onto the column, and is the preferred method for sample introduction for this packed-column method.

A gas chromatography-flame ionization detector system can be used for the separation and detection of nonpolar organic compounds. Semi-volatile constituent are among the analytes that can be readily resolved and detected using the system. If a packed column is used, four pairs of compounds may not be adequately resolved, and are reported as a quantitative sum: anthracene and phenanthrene, chrysene and benzo(a)anthracene, benzo(b) fluoranthene and benzo(k)fluoranthene, and dibenzo(a,h) anthracene and indeno(1,2,3-cd)pyrene. This issue can be resolved through the use of a capillary column in place of a packed column.

3.4 High Performance Liquid Chromatography

A high performance liquid chromatography system can be used to measure concentrations of target semi-volatile and non-volatile petroleum constituents. The system only requires that the sample be dissolved in a solvent compatible with those used in the separation. The detector most often used in petroleum environmental analysis is the fluorescence detector. These detectors are particularly sensitive to aromatic molecules, especially polynuclear aromatic hydrocarbons. An ultraviolet detector may be used to measure compounds that do not fluoresce.

In the method, polynuclear aromatic hydrocarbons are extracted from the sample matrix with a suitable solvent, which is then injected into the chromatographic system. Usually, the extract must be filtered, because fine particulate matter can collect on the inlet frit of the column, resulting in high back-pressures and eventual plugging of the column. For most hydrocarbon analyses, reverse phase high performance liquid chromatography (i.e., using a nonpolar column packing with a more polar mobile phase) is used. The most common bonded phase is the octadecyl (C18) phase. The mobile phase is commonly aqueous mixtures of either acetonitrile or methanol.

After the chromatographic separation, the analytes flow through the cell of the detector. A fluorescence detector shines light of a particular wavelength (the excitation wavelength) into the cell. Fluorescent compounds absorb light and reemit light of other, higher wavelengths (emission wavelengths). The emission wavelengths of a molecule are mainly determined by its structure. For polynuclear aromatic hydrocarbons, the emission wavelengths are mainly determined by the arrangement of the rings, and vary greatly between isomers.

Some of the polynuclear aromatic hydrocarbons [such as phenanthrene, pyrene, and benzo(g,h,i)perylene] are commonly seen in products boiling in the middle to heavy distillate range. In a method for their detection and analysis (EPA 8310), an octadecyl column and an aqueous acetonitrile mobile phase are used. Analytes are excited at 280 nm and detected at emission wavelengths of >389 nm. Naphthalene, acenaphthene, and fluorene must be detected by a less-sensitive UV detector, because they emit light at wavelengths below 389 nm. Acenaphthylene is also detected by UV detector.

The methods using fluorescence detection will measure any compounds that elute in the appropriate retention time range and fluoresce at the targeted emission wavelength(s) (Falla Sotelo et al., 2008). In the case of one method (EPA 8310), the excitation wavelength excites most aromatic compounds. These include the target compounds and also many derivatized aromatics, such as alkylaromatics, phenols, anilines, and heterocyclic aromatic compounds containing the pyrrole (indole, carbazole, etc.), pyridine (quinoline, acridine, etc.), furan (benzofuran, naphthofuran, etc.), and thiophene (benzothiophene, naphthothiophene, etc.) structures. In petroleum samples, alkyl polynuclear aromatic hydrocarbons are strong interfering compounds. For example, there are five methylphenanthrenes and over 20 dimethylphenanthrenes. The alkyl substitution does not significantly affect either the wavelengths or intensity of the phenanthrene fluorescence. For a very long time after the retention time of phenanthrene, the alkylphenanthrenes will interfere, affecting the measurements of all later-eluting target polynuclear aromatic hydrocarbons.

Interfering compounds will vary considerably from source to source, and samples may require a variety of cleanup steps to reach required method detection limits. The emission wavelengths used (EPA 8310) are not optimal for sensitivity of the small ring compounds. With a modern electronically-controlled monochromator, wavelength programs can be used that tune excitation and emission wavelengths to maximize sensitivity and/or selectivity for a specific analyte in its retention time window.

3.5 Gas Chromatography-Mass Spectrometry

A gas chromatography-mass spectrometry system is used to measure concentrations of target volatile and semi-volatile petroleum constituents. It is not typically used to measure the amount of total petroleum hydrocarbons. The advantage the technique is the high selectivity, or ability to confirm compound identity through retention time and unique spectral pattern.

The current method (EPA SW-846 8260) for the analysis of volatile compounds reveals that most of the compounds listed in these methods are not typically found in petroleum products. However, a method that uses selected ion monitoring (SIM) involves a system set up to measure only selected target masses, rather than scanning the full mass range. This technique yields lower detection limits for specific compounds. At the same time, it gives the more complete information available from the total ion chromatogram and the full-mass-range spectrum of each compound. The technique is sometimes used to quantify compounds present at very low concentrations in a complex hydrocarbon matrix. It can be used if the target compound's spectrum has a prominent fragment ion at a mass that distinguishes it from the rest of the hydrocarbon compounds.

The most common method for GC/MS analysis of semi-volatile compounds (EPA SW-846 8270) includes 16 polycyclic aromatic compounds, some of which commonly occur in middle distillate to heavy petroleum products. The method also quantifies phenols and cresols, compounds that are not hydrocarbons, but may occur in petroleum products. Phenols and cresols are more likely found in crude oils and weathered petroleum products.

To reduce the possibility of false positives, the intensities of one to three selected ions are compared to the intensity of a unique target ion of the same spectrum. The sample ratios are compared to the ratios of a standard. If the sample ratios fall within a certain range of the standard, and the retention time matches the standard within specifications, the analyte is considered present. Quantification is performed by integrating the response of the target ion only.

Mass spectrometers are among the most selective detectors, but they are still susceptible to interferences. Isomers have identical spectra, while many other compounds have similar mass spectra. Heavy petroleum products can contain thousands of major components that are not resolved by the gas chromatograph. As a result, multiple compounds are simultaneously entering the mass spectrometer. Different compounds may share many of the same ions, confusing the identification process. The probability of misidentification is high in complex mixtures such as petroleum products.

4 Other Analytical Methods

4.1 Total Petroleum Hydrocarbons

Since the term *total petroleum hydrocarbons* (total petroleum hydrocarbons) includes any petroleum constituent that falls within the measurable amount of petroleum-based hydrocarbons in the environment, the information obtained for total petroleum hydrocarbons depends on the analytical method used. Therefore, the difficulty associated with measurement of the total petroleum hydrocarbons is that the scope of the methods varies greatly (Speight, 2005). Some methods are nonspecific, while others provide results for hydrocarbons in a boiling point range. Interpretation of analytical results requires an understanding of how the determination was made (Miller, 2000 and references cited therein; Speight, 2001, 2002; Dean, 2003; Speight, 2005).

Most of the analytical methods discussed here for total petroleum hydrocarbons have been developed within the framework of federal and state regulatory initiatives. The initial implementation of the Federal Water Pollution Control Act (FWPCA) focused on controlling conventional pollutants, such as oil and grease. Methods have been developed for monitoring wastewaters (EPA 413.1, EPA 413.2, EPA 418.1).

The method of analysis often used for total petroleum hydrocarbons (EPA 418.1) provides a *one-number* value of the total petroleum hydrocarbons in an environmental medium. It does not, by any stretch of the imagination, provide information on the composition (i.e., individual constituents or group analysis) of the hydrocarbon mixture.

There are indications that approved methods used for assessment of petroleum spills, including the method for total petroleum hydrocarbons (EPA 418.1), the methods for semi-volatile priority pollutant organics (EPA 625, EPA 8270), and the methods for volatile organic priority pollutant methods (EPA 602, EPA 1624, EPA 8240) are all inadequate for generating scientifically defensible information for natural resource damage assessment. These general organic chemical methods are deficient in chemical selectivity (types of constituents analyzed) and sensitivity (detection limits); the deficiencies in these two areas lead to an inability to interpret the environmental significance of the data in a scientifically defensible manner (Speight, 2005).

On the other hand, there are many analytical techniques available that measure total petroleum hydrocarbon concentrations in the environment, but no single method is satisfactory for the measurement of the entire range of petroleum-derived hydrocarbons. In addition, because the techniques vary in the manner in which hydrocarbons are extracted and detected, each method may be applicable to the measurement of different subsets of the petroleum-derived hydrocarbons present in a sample. The four most commonly used total petroleum hydrocarbon analytical methods include (1) gas chromatography (GC), (2) infrared spectrometry (IR), (3) gravimetric analysis, and (4) immunoassay (Table 8.2) (Miller, 2000 and references cited therein).

4.2 Gas Chromatography

Gas chromatographic methods are currently the preferred laboratory methods for measurement of total petroleum hydrocarbon measurement, because they detect a broad range of hydrocarbons, and provide both sensitivity and selectivity. In addition, identification and quantification of individual constituents of the total petroleum hydrocarbon mix is possible.

Methods based on gravimetric analysis (Table 8.3) are also simple and rapid, but they suffer from the same limitations as infrared spectrometric methods (Table 8.3). Gravimetric-based methods may be useful for oily sludge and wastewaters, which will present analytical difficulties for other more sensitive methods. Immunoassay methods for the measurement of total petroleum hydrocarbon are also popular for field testing, because they offer a simple, quick technique for in-situ quantification of the total petroleum hydrocarbons.

For methods based on gas chromatography (Table 8.4), the total petroleum hydrocarbons fraction is defined as any chemicals extractable by a solvent or purge gas and detectable by gas chromatography/flame ionization detection (GC/FID) within a specified carbon range. The primary advantage of such methods is that they provide information about the type of petroleum in the sample, in addition to measuring the amount. Identification of product type(s) is not always straightforward, however, and requires an experienced analyst of petroleum products (Sullivan and Johnson, 1993; Speight, 2001, 2002, 2007). Detection limits are method-dependent as well as matrix-dependent, and can be as low as 0.5 mg/L in water or 10 mg/kg in soil.

Chromatographic columns are commonly used to determine total petroleum hydrocarbon compounds approximately in the order of their boiling points. Compounds are detected by means of a flame ionization detector that responds to virtually all compounds that can burn. The sum of all responses within a specified range is equated to a hydrocarbon concentration by reference to standards of known concentration.

Two methods (EPA SW-846 8015 and 8015A) were, in the past, often quoted as the source of gas chromatography-based methods for the measurement of the total petroleum hydrocarbons in a sample. However, the original methods were developed for non-halogenated volatile organic compounds, and were designed to measure a short target list of chemical solvents, rather than petroleum hydrocarbons. Thus, because there was no universal method for total petroleum hydrocarbons, there were many variations of these methods. Recently, an updated method (EPA 8015B) provides guidance for the analysis of gasoline and diesel range organic compounds.

The current individual methods differ in procedure, compounds detected, extraction techniques, and extraction solvents used. Some methods may include a cleanup step to remove biogenic (bacterial or vegetation-derived) material, while others do not. The methods have in common a boiling point-type column and a flame ionization detector. Selection of a method depends on the type of hydrocarbon suspected to be in the sample.

For example, if gasoline is suspected to be the sole contaminant, the method will use purge/trap sample introduction. If higher boiling petroleum fractions (diesel, middle distillates, or motor oil) are the contaminants, the analysis will use direct injection and hotter oven temperatures. Mixtures or unknown contamination may require both volatile range and extractable range analyses. Alternately, a single injection can be used to analyze the whole sample, but the extraction method must not use a solvent evaporation step.

Gas chromatography-based methods can be broadly used for different kinds of petroleum contamination, but are most appropriate for detecting nonpolar hydrocarbons with carbon numbers between C6 and C25 or C36. Many lubricating oils contain molecules with more than 40 carbon atoms. In fact, crude oil itself contains molecules having 100 carbons or more. These high molecular weight hydrocarbons are outside the detection range of the more common gas chromatographic methods, but specialized gas chromatographs are capable of analyzing such high molecular weight constituents.

Accurate quantification depends on adjusting the chromatograph to reach as high a carbon number as possible, then running a calibration standard with the same carbon range as the sample. There should also be a check for mass discrimination, a tendency for higher molecular weight hydrocarbons to be retained in the injection port. If a sample is suspected to be heavy oil, or to contain a mixture of light oil and heavy oil, the most appropriate method must be used.

Gravimetric or infrared methods are often preferred for high molecular weight samples. These methods can even be used as a check on gas chromatographic data if it is suspected that high molecular weight hydrocarbons are present, but are not being detected.

Calibration standards vary. Most methods specify a gasoline calibration standard for volatile range total petroleum hydrocarbons, and a diesel fuel #2 standard for extractable range total petroleum hydrocarbons. Some methods use synthetic mixtures for calibration. Because most methods are written for gasoline or diesel fuel, total petroleum hydrocarbons methods may have to be adjusted to measure contamination by heavier hydrocarbons (e.g., heavy fuel oil, lubricating oil, or crude oil). Such adjustments may entail use of a more aggressive solvent, a wider gas chromatographic window that allows detection of molecules containing up to C36 or more, and a different calibration standard that more closely resembles the constituents of the sample under investigation.

Gas chromatographic methods can be modified and fine-tuned so that they are suitable for measurement of specific petroleum products or group types. These modified methods can be particularly useful when there is information on the source of contamination, but method results should be interpreted with the clear understanding that a modified method was used for detection of a specific carbon range.

Interpretation of gas chromatographic data is often complicated, and the analytical method should always be considered when interpreting concentration data. For example, a volatile range analysis may be very useful for quantifying total petroleum hydrocarbons at a gasoline release site, but a volatile range analysis will not detect the presence of lube oil constituents. In addition, a modified method that has been specifically selected for detection of gasoline-range organics at a gasoline-contaminated site may also detect hydrocarbons from other petroleum releases, because fuel carbon ranges frequently overlap. Gasoline is found primarily in the volatile range. Diesel fuel falls primarily in an extractable range. Jet fuel overlaps both the volatile and semi-volatile ranges. However, the detection of different kinds of petroleum constituents does not necessarily indicate that there have been multiple releases at a site. Analyses of spilled waste oil will frequently detect the presence of gasoline, and sometimes diesel. This does not necessarily indicate multiple spills, since all waste oils contain some fuel. As much as 10% of used motor oil can consist of gasoline.

If the type of contaminant is unknown, a *fingerprint* analysis can help in the identification procedure. A *fingerprint* or *pattern recognition* analysis is a direct injection analysis where the chromatogram is compared to chromatograms of reference materials. Certain fuels can be identified by characteristic, reproducible chromatographic patterns. For example, chromatograms of gasoline and diesel differ considerably, but many hydrocarbon streams may have similar fingerprints. Diesel No. 2 and No. 2 fuel oil both have the same boiling point range and chromatographic fingerprint. A fingerprint can be used to conclusively identify a mixture when a known sample of that mixture or samples of the mixture's source materials are available as references.

Furthermore, as a fuel evaporates or biodegrades, its pattern can change so radically that identification becomes difficult (Bartha, 1986). Consequently, a gas chromatographic fingerprint is not a conclusive diagnostic tool. The methods for total petroleum hydrocarbon analysis must stress calibration and quality control, while pattern recognition methods stress detail and comparability.

The gas chromatographic methods usually cannot quantitatively detect compounds below C6, because these compounds are highly volatile, and interference can occur from the solvent peak. As much as 25% of fresh gasoline can be below C6, but the problem is reduced for weathered gasoline and/or diesel range contamination, because most of the very volatile hydrocarbons (<C6) may no longer be present in the sample. Gas chromatographic methods may also be inefficient for quantification of polar constituents (nitrogen, oxygen, and sulfur containing molecules). Some of the polar constituents are too reactive to pass through a gas chromatograph, and thus will not reach the detector for measurement.

Oxygenated gasoline is sometimes analyzed by GC-based methods, but it should be noted that the efficiency of purge methods is lower for oxygenates, such as ethers and alcohols, because detector response to oxygenates is lower relative to hydrocarbons. Therefore, the data will be biased slightly low for ether-containing fuels, compared to equivalent amounts of traditional gasoline. Methanol and ethanol elute before hexane, and consequently, they are not quantified, and may not even be detected, due to co-elution with the solvent.

On the other hand, gas chromatographic methods may overestimate the concentration of total petroleum hydrocarbons in the sample, due to the detection of non-petroleum compounds. In addition, cleanup steps do not perfectly separate petroleum hydrocarbons from biogenic material, such as plant oils and waxes that are sometimes extracted from vegetation-rich soil. Silica gel cleanup may help to remove this interference, but may also remove some polar hydrocarbons.

Because petroleum is made up of so many isomers, many compounds, especially those with more than eight carbon atoms, co-elute with isomers of nearly the same boiling point. These unresolved compounds are referred to as the unresolved complex mixture. They are legitimately part of the petroleum signal, and unless otherwise specified, should be quantified. Quantifying such a mixture requires a baseline-to-baseline integration mode, rather than a peak-to-peak integration mode. The baseline-to-baseline integration quantifies all of the petroleum constituents in the sample, but in the peak-to-peak integration, only the individual resolved hydrocarbons (not including the unresolved complex mixture) are quantified.

4.3 Infrared Spectroscopy

Infrared methods measure the absorbance of the C-H bond, and most methods typically measure the absorbance at a single frequency (usually 2930 cm^{-1}) that corresponds to the stretching of aliphatic methylene (CH_2) groups. Some methods use multiple frequencies, including 2960 cm^{-1} (CH_3 groups) and 2900 to 3000 cm^{-1} (aromatic C-H bonds).

Therefore, for infrared spectroscopic methods, the total petroleum hydrocarbons are any chemicals extracted by a solvent that are not removed by silica gel and can be detected by infrared spectroscopy at a specified wavelength. The primary advantage of the infrared-based methods is that they are simple and rapid. Detection limits (e.g., for EPA 418.1) are approximately 1 mg/L in water and 10 mg/kg in soil. However, the infrared method(s) can often suffer from poor accuracy and precision, especially for heterogeneous soil samples. Also, the infrared methods give no information on the type of fuel present in the sample, and there is little, often no, information about the presence or absence of toxic molecules, and no specific information about potential risk associated with the contamination.

Samples are extracted with a suitable solvent (i.e., a solvent with no C-H bonds), and biogenic polar materials are removed with silica gel. Some polar petroleum constituents may be removed as part of the silica gel cleanup. The absorbance of the silica gel eluate is measured at the specified frequency, and compared to the absorbance of a standard or standards of known petroleum hydrocarbon concentration. The absorbance is a measurement of the sum of all the compounds contributing to the result. However, infrared methods cannot provide information on the type of hydrocarbon contamination.

For all IR-based TPH methods, the C-H absorbance is quantified by comparing it to the absorbance of standards of known concentration. An assumption is made that the standard has an aliphatic-to-aromatic ratio and an infrared response similar to that of the sample. Consequently, it is important to use a calibration standard as similar to the type of contamination as possible (EPA 418.1).

The infrared method that has been most frequently used (EPA 418.1) is appropriate only for water samples. A separatory funnel liquid/liquid extraction technique is used to extract the hydrocarbons from the water. A method (EPA 5520D) using a Soxhlet extraction technique is suitable for sludge. This extraction is frequently used to adapt the method (EPA 418.1) to soil samples. An infrared-based supercritical fluid extraction method for diesel range contamination (EPA 3560) is available.

Similar to gas chromatographic methods, the data from infrared methods must be interpreted after considering certain limitations and interferences that can affect data quality. For example, the C-H absorbance is not always measured in exactly the same way. Within the set of methods that specify a single infrared measurement, some methods call for the measurement at precisely 2930 cm-1, while others (including EPA 418.1) require measurement at the absorbance maximum nearest 2930 cm^{-1}. This variation can make a significant difference in the magnitude of the result, and can lead to confusion when comparing duplicate sample results. If only C-H absorbance is measured, infrared methods will potentially underestimate the concentration of total petroleum hydrocarbons in samples that contain petroleum constituents, such as benzene and naphthalene, which do not contain alkyl C-H groups.

Because an infrared result is calculated as if the aromatics in the sample were present in the same ratio as in the calibration standard, accuracy depends upon use of a calibration standard as similar to the type of contamination as possible. Use of a dissimilar standard will tend to create a positive bias in highly aliphatic samples and a negative bias in highly aromatic samples.

In summary, infrared methods are prone to interferences (positive bias) from non-petroleum sources, since many organic compounds have some type of alkyl group associated with them, whether petroleum-derived or not.

4.4 Gravimetry

Gravimetric methods measure all chemicals that are extractable by a solvent, not removed during solvent evaporation, and capable of being weighed. Some gravimetric methods include a cleanup step to remove biogenic material. The advantage of gravimetric methods is that they are simple and rapid. Detection limits are approximately 5-10 mg/L in water and 50 mg/kg in soil.

However, gravimetric methods are not suitable for measurement of low boiling hydrocarbons that volatilize at temperatures below 70-85°C. They are recommended for use with (1) oily sludge, (2) for samples containing heavy molecular weight hydrocarbons, or (3) for aqueous samples when hexane is preferred as the solvent.

Gravimetric methods give no information on the type of fuel present, no information about the presence or absence of toxic compounds, and no specific information about potential risk associated with the contamination.

In the method(s), petroleum constituents are extracted into a suitable solvent. Biogenic polar materials typically may be partially or completely removed with silica gel. The solvent is evaporated, and the residue is weighed. This quantity is reported as a percent of the total soil sample dry weight. These methods are better suited for heavy oil, because they include an evaporation step.

There are a variety of gravimetric oil and grease methods suitable for testing water and soil samples (e.g., EPA SW-846 9070, EPA 413.1, EPA 9071). Technically, the result is an oil and grease result because no cleanup step is used. One method (EPA 9071) is used to recover low levels of oil and grease by chemically drying a wet sludge sample and then extracting it using Soxhlet apparatus. Results are reported on a dry-weight basis. The method is also used when relatively polar high molecular weight petroleum fractions are present, or when the levels of non-volatile grease challenge the solubility limit of the solvent. Specifically, the method (EPA SW-846 9071) is suitable for biological lipids, mineral hydrocarbons, and some industrial wastewater.

Gravimetric methods for oil and grease (e.g., EPA SW-846 9071) measure anything that dissolves in the solvent and remains after solvent evaporation. These substances include hydrocarbons, vegetable oils, animal fats, waxes, soaps, greases, and related biogenic material. Gravimetric methods for total petroleum hydrocarbons (EPA 1664) measure anything that dissolves in the solvent and remains after silica gel treatment and solvent evaporation.

This method (EPA 1664) is a liquid/liquid extraction gravimetric procedure that employs n-hexane as the extraction solvent, in place of 1,1,2-trichloroethane (CFC-113) and/or 1,2,2-trifluoroethane (Freon-113), for determination of the conventional pollutant oil and grease. Because the nature and amount of material determined is defined by the solvent and by the details of the method used for extraction, oil and grease are *method-defined* analytes. The method may be modified to reduce interferences and take advantage of advances in instrumentation, provided that all method equivalency and performance criteria are met. However, n-hexane is a poor solvent for high molecular weight petroleum constituents (Speight, 2001, 2007). Thus, the method will produce erroneous data for samples contaminated with heavy oils.

All gravimetric methods measure any suspended solids that are not filtered from solution, including bacterial degradation products and clay fines. Method 9071 specifies using cotton or glass wool as a filter.

Because extracts are heated to remove solvent, these methods are not suitable for measurement of low boiling low molecular weight hydrocarbons (i.e., hydrocarbons having less than fifteen carbon atoms) that volatilize at temperatures below 70 to 85°C (158 to 185°F). Liquid fuels, from gasoline through No. 2 fuel oil, lose volatile constituents during solvent removal. In addition, soil results that are reported on a dry-weight basis suffer from potential losses of lower boiling hydrocarbon constituents during moisture determination, where the matrix is dried at approximately 103 to 105°C (217 to 221°F) for several hours in an oven.

4.5 Immunoassay

Immunoassay methods correlate total petroleum hydrocarbons with the response of antibodies to specific petroleum constituents. Many of the methods measure only aromatics that have an affinity for the antibody, benzene-toluene-ethylbenzene-xylene, and polynuclear aromatic hydrocarbon analysis (EPA 4030, Petroleum Hydrocarbons by Immunoassay).

The principle behind the test method(s) is that antibodies are made of proteins that recognize and bind with foreign substances (antigens) that invade host animals. Synthetic antibodies have been developed to complex with petroleum constituents. The antibodies are immobilized on the walls of a special cell or filter membrane. Water samples are added directly to the cell, while soils must be extracted before analysis. A known amount of labeled analyte (typically an enzyme with an affinity for the antibody) is added after the sample. The sample analytes compete with the enzyme-labeled analytes for sites on the antibodies. After equilibrium is established, the cell is washed to remove any unreacted sample or labeled enzyme. Color development reagents that react with the labeled enzyme are added. A solution that stops color development is added at a specified time, and the optical density (color intensity) is measured. Because the coloring agent reacts with the labeled enzyme, samples with high optical density contain low concentrations of analytes. Concentration is inversely proportional to optical density.

The antibodies used in immunoassay kits are generally designed to bond with selected compounds. A correction factor supplied by the manufacturer must be used to calculate the concentration of total petroleum hydrocarbons. The correction factor can vary depending on product type, because it attempts to correlate total petroleum hydrocarbons with the measured surrogates.

Immunoassay tests do not identify specific fuel types, and are best used as screening tools. The tests are dependent on soil type and homogeneity. In particular, for clay and other cohesive soils, the tests are limited by a low capacity to extract hydrocarbons from the sample.

5 References

Aichberger, H., Hasinger, M., Braun, R., and Loibner, A.P. 2005. "Potential of Preliminary Test Methods to Predict Biodegradation Performance of Petroleum Hydrocarbons in Soil." *Biodegradation*, 16: 115–125.

Altgelt, K.H., and Gouw, T.H. 1975. In *Advances in Chromatograph*. J.C. Giddings, E. Grushka, R.A. Keller, and J. Cazes (Editors). Marcel Dekker Inc., New York.

ASTM. 2011. *Annual Book of ASTM Standards*. American Society for Testing and Materials, West Conshohocken, Pennsylvania.

Atagana, H.I., Haynes, R.J., and Wallis, F.M. 2003. "Optimization of Soil Physical and Chemical Conditions ror the Bioremediation of Creosote-Contaminated Soil." *Biodegradation*, 14: 297–307.

Baker, R.J., Baehr, A.L., and Lahvis, M.A. 2000. "Estimation of Hydrocarbon Biodegradation Rates in Gasoline-Contaminated Sediment from Measured Respiration Rates." *J. Contam. Hydrol.*, 41: 175–192.

Balba, M.T., Al-Awadhi, N., and Al-Daher, R. 1998. "Bioremediation of Oil-Contaminated Soil: Microbiological Methods for Feasibility Assessment And Field Evaluation." *J. Microbiol. Methods*, 32: 155–164.

Bartha, R. 1986. "Biotechnology of Petroleum Pollutant Biodegradation." *Microb. Ecol.*, 12: 155–172.

Boopathy, R. 2002. "Use of Anaerobic Soil Slurry Reactors for the Removal of Petroleum Hydrocarbons in Soil." *Int. Biodeterioration Biodegradation*, 52: 161–166.

Breedveld, G.D., and Sparrevik, M. 2001. "Nutrient-Limited Biodegradation of PAH in Various Soil Strata at a Creosote Contaminated Site." *Biodegradation*, 11: 391–399.

Bruner, F. 1993. Gas Chromatographic Environmental Analysis: Principles, Techniques, and Instrumentation. John Wiley & Sons Inc., Hoboken, New Jersey.

Budde, W.L. 2001. *The Manual of Manuals*. Office of Research and Development, Environmental Protection Agency, Washington, DC.

Cao, J.R. 1992. *Microwave Digestion of Crude Oils and Oil Products for the Determination of Trace Metals and Sulphur by Inductively-Coupled Plasma Atomic Emission Spectroscopy*, Environment Canada Manuscript Report Number EE-140, Ottawa, Ontario, Canada.

Chaîneau, C.H., Yepremian, C., Vidalie, J.F., Ducreux, J., and Ballerini, D. 2003. "Bioremediation of a Crude Oil-Polluted Soil: Biodegradation, Leaching and Toxicity Assessments." *Water, Air, Soil Pollut.*, 144: 419–440.

Chaîneau, C.H., Morel, J.L., and Oudot, J. 1995. "Microbial Degradation in Soil Microcosms of Fuel Oil Hydrocarbons from Drilling Cuttings." *Environ. Sci. Technol.*, 29: 1615–1621.

CHRIS. 1991. Chemical Hazards Response Information System (CHRIS). United States Coast Guard, Department of Transportation, Washington, DC.

Davis, C., Cort, T., Dai, D., Illangasekare, T.H., and Munakata-Marr, J. 2003. "Effects of Heterogeneity and Experimental Scale on Biodegradation of Diesel." *Biodegradation*, 14: 373–384.

Dean, J.R. 1998. *Extraction Methods for Environmental Analysis*. John Wiley & Sons, Inc., New York.

Dott, W., Feidieker, D., Steiof, M., Becker, P.M., and Kämpfer, P. 1995. "Comparison of Ex Situ and In Situ Techniques for Bioremediation of Hydrocarbon-Polluted Soils." *Int. Biodeterioration & Biodegradation*, 301–316.

Dyroff, G.V. (Editor). 1993. *Manual on Significance of Tests for Petroleum Products: 6th Edition*, American Society for Testing and Materials, West Conshocken, Pennsylvania.

Falla Sotelo, F., Araujo Pantoja, P., López-Gejo, J., Le Roux, J.G.A.C., Quina, F.H., and Nascimento, C.A.O. 2008. "Application of Fluorescence Spectroscopy for Spectral Discrimination of Crude Oil Samples." *Brazilian Journal of Petroleum and Gas*, 2(2): 63–71.

Fingas, M.F., Duval, W.S., and Stevenson, G.B. 1979. *The Basics of Oil Spill Cleanup*, Environment Canada, Ottawa, Ontario, Canada.

Fingas, M.F., Dufort, V.M., Hughes, K.A., Bobra, M.A., and Duggan, L.V. 1989a. "Laboratory Studies on Oil Spill Dispersants." In *Chemical Dispersants - New Ecological Approaches*. M. Flaherty (Editor). Publication No. STP 1084, American Society for Testing and Materials, West Conshohocken, Pennsylvania. pp. 207–219.

Fingas, M.F., White, B., Stoodley, R.G., and Crerar, I.D. 1989b. "Laboratory Testing of Dispersant Effectiveness." *Proceedings*. 1989 Oil Spill Conference, American Petroleum Institute, Washington, D.C., pp. 365–373, 1989b.

Fingas, M.F., Kyle, D.A., Bier, I.E., Lukose, A., and Tennyson, E.J. 1991. "Physical and Chemical Studies on Oil Spill Dispersants: The Effect of Energy." *Proceedings*. 14th Arctic and Marine Oil Spill Program

Technical Seminar, Environment Canada, Ottawa, Ontario, Canada. pp. 87–106.

Fingas, M.F., Kyle, D.A., and Tennyson, E.J. 1992. "Physical and Chemical Studies on Oil Spill Dispersants: Effectiveness Variation with Energy." *Proceedings*. 15th Arctic and Marine Oil Spill Program Technical Seminar, Environment Canada. Ottawa, Ontario, Canada pp. 135–142.

Fingas, M.F., Kyle, D.A., and Tennyson, E.J. "Physical and Chemical Studies on Dispersants: The Effect of Dispersant Amount on Energy." *Proceedings*. 16th Arctic and Marine Oil Spill Program Technical Seminar, Environment Canada, Ottawa, Ontario, Canada. pp. 861–876.

Fingas, M.F., Fieldhouse, B., Gamble, L., and Mullin, J. 1995a. "Studies of Water-in-Oil Emulsions: Stability Classes and Measurement." *Proceedings*. 18th Arctic and Marine Oil Spill Program Technical Seminar, Environment Canada, Ottawa, Ontario, Canada. pp. 21–42.

Fingas, M.F., Kyle, D.A., Lambert, P., Wang, Z., and Mulling, J. 1995b. "Analytical Procedures for Measuring Oil Spill Dispersant Effectiveness in the Laboratory." *Proceedings*. 18th Arctic and Marine Oil Spill Program Technical Seminar, Environment Canada, Ottawa, Ontario, Canada. pp. 339–354.

Fingas, M.F., Ackerman, F., Lambert, P., Li, K., Wang, Z., Mullin, J., Hannon, L., Wang, D., Steenkammer, A., Hiltabrand, R. Turpin, R., and Campagna, P. 1995c. "The Newfoundland Offshore Burn Experiment: Further Results of Emissions Measurement." *Proceedings*. 18th Arctic and Marine Oil Spill Program Technical Seminar, Environment Canada, Ottawa, Ontario, pp. 915–995.

Fingas, M.F. 1995. "The Evaporation of Oil Spills." *Proceedings*. 18th Arctic and Marine Oil Spill Program Technical Seminar, Environment Canada, Ottawa, Ontario, Canada. pp. 43–60.

Fingas, M.F. 1998. "Studies on the Evaporation of Crude Oil and Petroleum Products. II. Boundary Layer Regulation." *J. Hazardous Materials*, 57(1–3): 41–58.

Fowlis, I.A. 1995. *Gas Chromatography*. 2nd Edition. John Wiley & Sons Inc., New York.

Frijer, J.I., De Jonge, H., Bounten, W., and Verstraten, J.M. 1996. "Assessing Mineralization Rates of Petroleum Hydrocarbons in Soils in Relation to Environmental Factors and Experimental Scale." *Biodegradation*, 7: 487–500.

Grishchenkov, V.G., Townsend, R.T., McDonald, T.J., Autenrieth, R.L., Bonner, J.S., and Boronin, A.M. 2000. "Degradation of Petroleum Hydrocarbons by Facultative Anaerobic Bacteria under Aerobic and Anaerobic Conditions." *Process Biochem.*, 35: 889–896.

Grob, R.L. 1995. *Modern Practice of Gas Chromatography*. 3rd Edition. John Wiley & Sons Inc., New York.

Hatzinger, P.B., and Alexander, M. 1995. "Effect of Ageing Chemicals in Soil upon Their Biodegradability and Extractability." *Environ. Sci. Technol.*, 29: 537–545.

Heider, J., Spormann, A.M., Beller, H.R., and Widdel, F. 1999. "Anaerobic Bacterial Metabolism of Hydrocarbons." *FEMS Microbiol. Rev.*, 22: 459–473.

Höhener, P., Duwig, C., Pasteris, G., Kaufmann, K., Dakhel, N., and Harms, H. 2003. "Biodegradation of Petroleum Hydrocarbon Vapors: Laboratory Studies on Rates and Kinetics in Unsaturated Alluvial Sand." *J. Contam. Hydrol.*, 1917: 1–23.

Holliger, C., and Zehnder, A.J.B. 1996. "Anaerobic Biodegradation of Hydrocarbons." *Curr. Opin. Biotechnol.*, 7: 326–330.

Huang, W., Pent, P., Yu, Z., and Fu, J. 2003. "Effects of Organic Matter Heterogeneity on Sorption and Desorption of Organic Contaminants by Soils and Sediments." *Appl. Geochem.*, 18: 955–972.

Huesemann, M.H. 1995. "Predictive Model for Estimating the Extent of Petroleum Hydrocarbon Biodegradation in Contaminated Soils." *Environ. Sci. Technol.*, 29: 7–18.

Jokuty, P., Fingas, M.F., Whiticar, S., and Fieldhouse, B. 1995. A Study of Viscosity and Interfacial Tension of Oils and Emulsions. Report No. EE-153. Environment Canada, Ottawa, Ontario, Canada.

Jokuty, P., Whiticar, S., McRoberts, K., and Mullin, J. 1996. "Oil Adhesion Testing - Recent Results." *Proceedings*. 19th Arctic and Marine Oil Spill Program Technical Seminar, Environment Canada, Ottawa, Ontario, Canada. pp. 9–27.

Leeson, A., and Hinchee, R.E. 1997. *Soil Bioventing, Principles and Practice*. CRC, Lewis Publishers, Boca Raton, Florida.

Luthy, R.G., Aiken, G.R., Brusseau, M.L., Cunningham, S.D., Gschwend, P.M., Pignatello, J.J., Reinhard, M., Traina, S.J., Weber, W.J., and Westall, J.C. 1997. "Sequestration of Hydrophobic Organic Contaminants by Geosorbents." *Environ. Sci. Technol.*, 31: 3341–3347.

Mackay, D. and Zagorski, W. 1982. Studies of Water-in-Oil Emulsions. Report No. EE-34. Environment Canada, Ottawa, Ontario, Canada.

Margesin, R., Zimmerbauer, A., and Schinner, F. 1997. "Efficiency of Indigenous and Inoculated Cold-Adapted Soil Microorganisms for Biodegradation of Diesel Oil in Alpine Soils." *Appl. and Environ. Microbiol.*, 63: 2660–2664.

Margesin, R., Zimmerbauer, A., and Schinner, F. 2000. "Monitoring of Bioremediation by Soil Biological Activities." *Chemosphere*, 40: 339–346.

Massias, D., Grossi, V., and Bertrand, J.C. 2003. "In Situ Anaerobic Degradation of Petroleum Alkanes in Marine Sediments: Preliminary Results." *Geoscience*, 335: 435–439.

Miller, M. (Editor). 2000. *Encyclopedia of Analytical Chemistry*. John Wiley & Sons Inc., Hoboken, New Jersey.

Mohn, W.M., and Stewart, G.R. 2000. "Limiting Factors for Hydrocarbon Biodegradation at Low Temperature in Arctic Soils." *Soil Biol. Biochem.*, 32: 1161–1172.

Nelson, P. 2003. Index to EPA Test Methods. US EPA New England Region, Boston MA.

Nierop, K.G.J., and Verstraten, J.M. 2003. "Organic Matter Formation in Sandy Subsurface Horizons of Dutch Coastal Dunes in Relation to Soil Acidification." *Org. Geochem.*, 34: 499–513.

Nocentini, M., Pinelli, D., and Fava, F. 2000. "Bioremediation of a Soil Contaminated by Hydrocarbon Mixtures: The Residual Concentration Problem." *Chemosphere*, 41: 1115–1123.

Oudot, J., and Dutrieux, E. 1989. "Hydrocarbon Weathering and Biodegradation in a Tropical Estuarine Ecosystem." *Mar. Environ. Res.*, 27: 195–213.

Patnaik, P. (Editor). 2004. *Dean's Analytical Chemistry Handbook*. 2nd Edition. McGraw-Hill, New York.

Pavlova, A., and Ivanova, R. 2003. "Determination of Petroleum Hydrocarbons and Polycyclic Aromatic Hydrocarbons in Sludge from Wastewater Treatment Basins." *J. Environ. Monit.*, 5: 319–323.

Rhodes, I.A., Hinojas, E.M., Barker, D.A., and Poole, R.A. 1994. "Pitfalls Using Conventional TPH Methods for Source Identification." *Proceedings*. Seventh Annual Conference: EPA Analysis of Pollutants in the Environment. Norfolk, VA. Environmental Protection agency, Washington, DC.

Schramm, L.L. (Editor). 1992. *Emulsions: Fundamentals and Applications in the Petroleum Industry*. American Chemical Society, Washington, DC.

Smith, K.A., and Cresser, M. 2003. *Soil & Environmental Analysis: Modern Instrumental Techniques*. Marcel Dekker Inc., New York.

Speight, J.G. 2001. *Handbook of Petroleum Analysis*. John Wiley & Sons Inc., Hoboken, New Jersey.

Speight, J.G. 2002. *Handbook of Petroleum Product Analysis*. John Wiley & Sons Inc., Hoboken, New Jersey.

Speight, J.G. 2005. *Environmental Analysis and Technology for the Refining Industry*. John Wiley & Sons Inc., Hoboken, New Jersey.

Speight, J.G. 2007. *The Chemistry and Technology of Petroleum*. 4th Edition. CRC Press, Taylor & Francis Group, Boca Raton, Florida.

Speight, J.G. 2009. *Enhanced Recovery Methods for Heavy Oil and Tar Sands*. Gulf Publishing Company, Houston, Texas.

Speight, J.G. 2011. *An Introduction to Petroleum Technology, Economics, and Politics*. Scrivener Publishing, Salem, Massachusetts.

Speight, J.G., and Foote, R. 2011. *Ethics in Science and Engineering*. Scrivener Publishing, Salem, Massachusetts, 2011.

Sullivan and Johnson, 1993. "'Oil' You Need to know about Crude – Implications of TPH Data for Common Petroleum Products." *Soil*, May, p. 8.

Sunahara, G.I., Renoux, A.Y., Thellen, C., Gaudet, C.L., and Pilon, A. (Editors). 2002. *Environmental Analysis of Contaminated Sites*. John Wiley & Sons Inc., New York.

Twardus, E.M. 1980. A Study to Evaluate the Combustibility and Other Physical and Chemical Properties of Aged Oils and Emulsions. Report Number EE-5. Environment Canada, Ottawa, Ontario, Canada.

US EPA. 2004. Environmental Protection Agency, Washington, DC. Web site: http://www.epa.gov

Wang, Z., Fingas, M.F., Landriault, M., Sigouin, L., and Xu, N. 1995. "Identification of Alkyl Benzenes and Direct Determination of BTEX and (BTEX + C3-Benzenes) in Oils by GC/MS." *Proceedings*. 18th Arctic and Marine Oil Spill Program Technical Seminar, Environment Canada, Ottawa, Ontario, Canada. pp. 141–164.

9

Biodegradation of Petroleum

1 Introduction

The contamination of soils and aquifers by spilled petroleum is a persistent and widespread pollution problem that causes ecological disturbances and the associated health implications (Bundy et al., 2002; Okoh, 2006; Salam et al., 2011). Once petroleum is released and comes into contact with water, air, the necessary salts, and microorganisms present in the environment, the natural process of petroleum biodegradation begins (Antai, 1990; Davies et al., 2001). However, some of the petroleum-related pollutants are carcinogenic and mutagenic (Miller and Miller, 1981; Obayori et al., 2009b).

The recognized mechanical and chemical methods for remediation of hydrocarbon-polluted environment are often expensive, technologically complex, and lack public acceptance (Speight, 1996; Speight and Lee, 2000; Vidali, 2001; Speight, 2005). Thus, bioremediation is often the method of choice for effective removal of hydrocarbon pollutants from a variety of ecosystems (Okoh and Trejo-Hernandez, 2006). In fact, petroleum and petroleum products are a rich source of carbon, and the reaction of the hydrocarbons contained therein with aerial oxygen (with the release of carbon-dioxide) is promoted by a variety of microorganisms (Odu, 1977; Atlas, 1981; Atlas and Bartha, 1992; Steffan et al., 1997).

However, the rate of microbial degradation of hydrocarbons in soils is affected by several physicochemical and biological parameters, including the number and species of microorganisms present, the conditions for microbial degradation activity (e.g., presence of nutrient, oxygen, PH, and temperature), the quality, quantity, and bioavailability or bioaccessibility of the contaminants, and the soil characteristics, such as particle size distribution (Chapter 1) (Atlas, 1991; Freijer et al., 1996; Margesin and Schinner, 1997a, 1997b; Dandie et al., 2010).

Hydrocarbon degrading bacteria and fungi are mainly responsible for the mineralization (conversion of hydrocarbons to carbon dioxide and water) of petroleum-related pollutants, and are distributed in diverse ecosystems (Leahy and Colwell, 1990; Song et al., 1990). Furthermore, the population of microorganisms found in a polluted environment will degrade petroleum-related constituents differently and at a different rate than microorganisms in a relatively clean environment (Obire, 1990; Obire, 1993; Obire and Okudo, 1997; Obire and Nwaubeta, 2001).

However, it is uncommon to find organisms that could effectively degrade both aliphatic constituents and aromatic constituents, possibly due to differences in metabolic routes and pathways for the degradation of the two classes of hydrocarbons. There are indications of the existence of bacterial species with propensities for simultaneous degradation of aliphatic hydrocarbons and aromatic hydrocarbons (Amund et al., 1987; Obayori et al., 2009b). This rare ability may be as a result of long exposure of the organisms to different hydrocarbon pollutants, resulting in genetic alteration and *acquisition* of the appropriate degradative genes.

Moreover, it is essential to recognize that the environmental impact of petroleum spills is dependent on previous hydrocarbon exposures and the adaptive status of the local microbiota (Greenwood et al., 2009). The different structural and functional response of microbial sub-groups to different hydrocarbons confirms that the overall response of biota is sensitive to petroleum composition. This suggests that the preferred response to anticipated contaminants may be engineered by pre-exposure to representative substrates. The controlled adaption of microbes to a threatening contaminant is the basis of proactive bioremediation technology, including the augmentation of newly contaminated sites with locally remediated soil in which the biota had already been adapted (bioaugmentation; Chapter 1).

Thus, establishing the chemical history of recently contaminated regions is an important aspect of environmental bioremediation. The premise being that microbial species adapted through a history of exposure to more bioavailable petroleum hydrocarbons are less severely impacted by a spill than microbes with no such pre-exposure or adaptation (Page et al., 1996; Peters et al., 2005). Indeed, the diversity of petroleum hydrocarbon degraders in most natural environments may be significant, but in the absence of a previous pollution history, the numbers of the microbes may be low due to lack of prior stimulus (Swanell et al., 1996).

2 Biodegradation of Specific Constituents

Petroleum and petroleum products are mixtures of differing molecular species hydrocarbons, and the constituents of these molecular categories are present in varied proportions, resulting in high variability in petroleum and petroleum products (Speight, 2007). In terms of bulk fractions (Chapter 2), the resin constituents and the asphaltene constituents are of particular interest (or notoriety) because these constituents generally resist degradation. After a spill, the constituents of petroleum and petroleum products are subjected to physical and chemical processes, such as evaporation or photochemical oxidation, which produce changes in the composition of the spilled material (Speight, 2007; Taghvaei Ganjali et al., 2007).

2.1 Alkanes

Alkanes are major constituents of conventional petroleum and petroleum products, and can be degraded by indigent or added bacteria.

Conventional (light) petroleum contains 10 to 40% w/w normal alkanes, but weathered and heavier oils may have only a fraction of a percent. Higher molecular weight alkanes constitute 5 to 20% w/w of light oils, and up to 60% w/w of the more viscous oils and tar sand bitumen. Aromatic hydrocarbons are those characterized by the presence of at least one benzene (or substituted benzene) ring. The low-molecular-weight aromatic hydrocarbons are subject to evaporation, and, although toxic to much marine life, are also relatively easily degraded. Conventional (light) petroleum typically contain between 2 and 20% w/w low-boiling aromatic compounds, whereas heavy oils contain less than 2% w/w aromatic compounds. As molecular weight and complexity increase, aromatics are less readily degraded. Thus, the degradation rate of polyaromatics is slower than that of monoaromatics.

Of these, the normal alkane series (straight-chain alkane series) is the most abundant and the most quickly degraded. Compounds with chains of up to 44 carbon atoms can be metabolized by microorganisms, but those having 10 to 24 carbon atoms (C_{10} to C_{24}) are usually the easiest to metabolize. Shorter chains (up to approximately C_8) also evaporate relatively easily. Only a few species can use C_1 to C_4 alkanes, and C_5 to C_9 alkanes are degradable by some microorganisms, but toxic to others.

Branched alkanes are usually more resistant to biodegradation than normal alkanes, but less resistant than cycloalkanes (naphthenes): those alkanes having carbon atoms in ring-like central structures. Branched alkanes are increasingly resistant to microbial attack as the number of branches increases. At low concentrations, cycloalkanes may be degraded at moderate rates, but some highly condensed cycloalkanes can persist for long periods after a spill.

Generally, with respect to the molecular composition of the aliphatic constituents of petroleum and petroleum-related products, microbial biodegradation attacks n-alkanes and isoprenoid alkanes. The polycyclic alkanes of sterane and triterpane type tend to be somewhat resistant to biodegradation. Since this is the case even for naphthenic type petroleum (which is originally depleted in *n*-alkanes), it has been concluded that the biodegradation of petroleum type pollutants, under natural conditions, will be restricted to n-alkanes and isoprenoids (Antić et al., 2006).

In aqueous systems, addition of acclimatized naturally occurring microorganisms (bioaugmentation) enhances the biodegradation of hydrocarbons. Since dissolved hydrocarbons are more available for microbiological degradation, application of dispersants and surfactants to increase the bioavailability significantly and enhance oil degradation (Mohn and Stewart, 2000; Zhang, 2008; Zahed et al., 2010). Other factors (such salinity and pH) have considerable effects on biodegradation of petroleum hydrocarbons in the marine environments as well.

For example, the different concentrations of sodium chloride (0 to 5% w/w) exert considerable influence on the biodegradation of petroleum and polynuclear aromatic hydrocarbons from the heavy crude oil-contaminated soil (Minai-Tehrani, 2009). Not surprisingly, increasing the concentration of sodium chloride in soil has a decreasing effect on petroleum biodegradation and the removal of polynuclear aromatic hydrocarbons. The biodegradation of total crude oil was higher in the absence of sodium chloride (41%), while the reduction in the biodegradation of polynuclear aromatic hydrocarbons was observed in the presence of 1% w/w sodium chloride (35%). A lower reduction of petroleum and polynuclear aromatic hydrocarbons was observed in the presence of 5% w/w sodium chloride (12% and 8%, respectively). The reduction of phenanthrene, anthracene, and pyrene reduction was higher in the presence of 1% w/w sodium chloride, while fluoranthene and chrysene reduction were higher in the absence of sodium chloride.

In a recent study (Lee et al., 2010), *Rhodococcus sp.* EH831, isolated from an oil-contaminated soil, has been shown to degrade a wide range of hydrocarbons and completely metabolize hexane. EH831 did not lose its activity at medium or low temperatures. Moreover, the biodegradation pathway of hexane by *Rhodococcus sp.* EH831 under aerobic conditions was revealed for the first time, which may be useful for the bioremediation of sites contaminated with various hydrocarbons or for the treatment of industrial discharge.

2.2 Aromatic hydrocarbons

Spills of aromatic products, such aromatic naphtha and leaks from underground fuel tanks, contribute significantly to the contamination of groundwater by aromatic compounds. Non-oxygenated monoaromatic hydrocarbons, such as benzene, toluene, ethylbenzene, and xylenes (BTEX), are of particular concern. The high water solubility of the BTEX species enables them to migrate in the subsurface and contaminate drinking water.

In a four-year study, intrinsic hydrocarbon bioremediation in a shallow aquifer underlying a natural gas reservoir was demonstrated (Gieg et al., 1999). All of the BTEX compounds were biodegraded under sulfate-reducing conditions, and toluene was also degraded under methanogenic conditions. Other authors (Weiner and Lovley, 1998) noted the persistence of benzene in the sulfate reduction zones of petroleum-contaminated aquifers. The persistence of the benzene was attributed to the lack of benzene-oxidizing sulfate reducers in the aquifer sediments, rather than to the inability of such microorganisms to grow and metabolize under freshwater conditions. In order to obtain accelerated benzene bioremediation in contaminated anaerobic aquifers, it was proposed (Weiner and Lovley, 1998) that aquifer sediments be supplemented with benzene-oxidizing sulfate reducers.

Two thermophilic aerobic bacteria (*Thermus aquaticus* and *Thermus* sp.) have been reported to degrade benzene, toluene, ethyl benzene, and xylene (BTEX) fractions co-metabolically (Chen and Taylor, 1995, 1997a). However, only small fractions of benzene and toluene were metabolized to carbon dioxide, and biodegradation was inhibited by higher BTEX concentrations, but was enhanced if strains were pre-grown on catechol and *o*-cresol, indicating that the pre-conditioning can enhance the performance of microbes (Chen and Taylor, 1995).

Two anaerobic consortia, consisting of unidentified bacterial cocci, could grow on all BTEX compounds as sole carbon and energy sources at 45 to 75°C (93 to 167°F), with 50°C (122°F) being the optimal temperature (Chen and Taylor, 1997b). Only a small fraction of toluene was mineralized to carbon dioxide. Biodegradation was coupled by both consortia to sulfate reduction as well as to generation of hydrogen sulfide. No growth or BTEX metabolism occurred when sulfate was omitted. Thus, sulfate-reducing bacteria are most likely the principal species that carry out the biodegradation, while other thermophilic species may use the early water-soluble BTEX metabolites.

The above-described aerobic and anaerobic thermophiles are used for a patented in situ thermally enhanced remediation strategy (Taylor et al., 1998). The authors propose the coupling of bioremediation to in situ dynamic underground stripping (DUS), a primary decontamination effort for the rapid treatment of contaminated soil at significant depths. A targeted subsurface volume is heated (by steam injection and electrical resistance) to vaporize the trapped contaminants; these are then removed by vacuum extraction. However, even VOCs, such as BTEX, are not completely removed during vaporization. Due to the use of heat, the entire underground environment remains at elevated temperatures of 50–70°C for at least 60 days after the heat treatment. This enables thermophilic bacteria to metabolize the residual contaminants.

Using the existing injection-withdrawal wells, the pre-grown biodegraders are inoculated not only to enhance the removal of low levels of residual BTEX from fuel leaks, but also to treat chlorinated hydrocarbons, such as trichloroethylene and chloroform, from past solvent cleaning practices.

The bioremediation process is terminated by lowering the temperature below 40°C (104°F). Such an in situ follow-up treatment could also be applied to fuel-contaminated plumes subjected to thermally enhanced vapor stripping as a primary treatment method, or as a stand-alone method, when the initial concentration of VOCs is low and the subsurface volume to be heated is small. In the latter case, thermophilic hydrocarbon degraders suspended in hot water are pumped into the subsurface.

The biodegradation of alkyltetralins has also been studied (Booth et al., 2007a). However, tetralin has been shown to be biodegraded by both mixed cultures of microbes (Strawinski and Stone, 1940; Soli and Bens, 1972), and by some strains able to utilize the compound as sole carbon and energy source (Schreiber and Winkler, 1983; Sikkema and Bont, 1991; Hernáez et al., 1999).

Recently, it has been demonstrated that *rhodococci* strains are able to degrade recalcitrant alkyltetralins (Frenzel et al., 2009). The identification of such bacteria capable of the biodegradation of alkyltetralins may be an important step toward the development of bioremediation strategies for sites contaminated by toxic aromatic hydrocarbons.

2.3 Polynuclear Aromatic Hydrocarbons

Polynuclear aromatic hydrocarbons, in the current context, are persistent organic compounds with two or more aromatic rings in various structural configurations. Polynuclear aromatic hydrocarbons constitute a large and diverse class of organic compounds. However, derivatives, such as tetralin (1,2,3,4-tetrahydronaphthalene) and decalin (decahydronaphthalene, bicyclo[4.4.0]decane), are not included in this group, but are include in the alkane group because of the saturated ring.

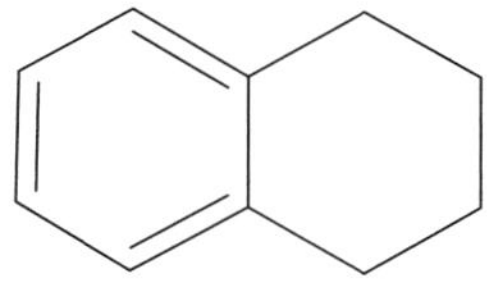

Tetralin

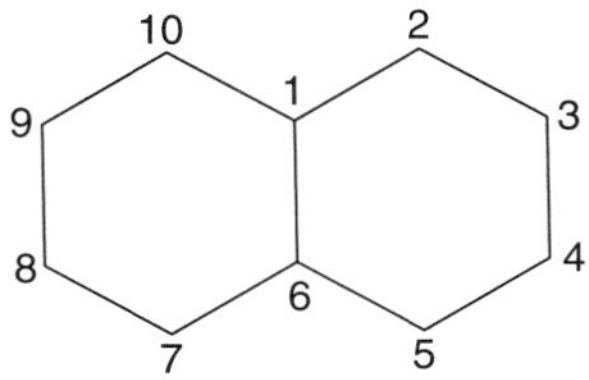

Decalin

The chemical properties, and hence the environmental fate, of polynuclear aromatic hydrocarbons are dependent in part upon molecular size (i.e., the number of aromatic rings and the pattern of ring linkage). Ring linkage patterns (also known as molecular topology) in polynuclear aromatic hydrocarbons may occur such that the tertiary carbon atoms are centers of two or three interlinked rings, as in the linear kata-condensed polynuclear aromatic hydrocarbon anthracene or the peri-condensed polynuclear aromatic hydrocarbon pyrene.

However, most polynuclear aromatic hydrocarbons occur as hybrids encompassing various structural components, such as in the polynuclear aromatic hydrocarbon benzo[*a*]pyrene.

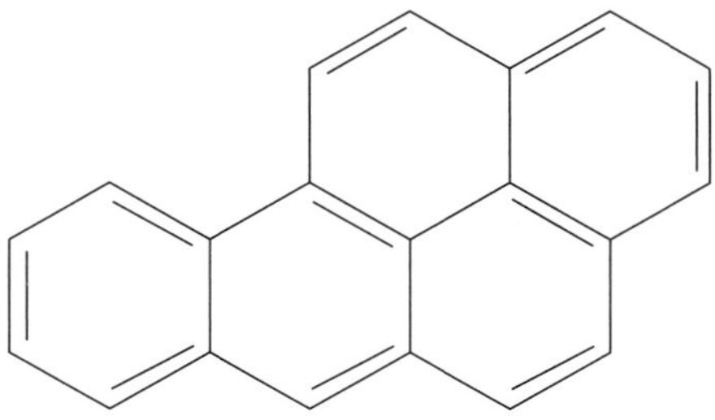

Benzo(a)pyrene

Generally, an increase in the size and angularity of a polynuclear aromatic hydrocarbon molecule results in a concomitant increase in hydrophobicity and electrochemical stability (Zander, 1983; Harvey, 1997).

Polynuclear aromatic hydrocarbon molecule stability and hydrophobicity are two primary factors which contribute to their persistence in the environment.

Polynuclear aromatic hydrocarbons are present as natural constituents in fossil fuels, are formed during the incomplete combustion of organic material, and are therefore present in relatively high concentrations in products of fossil fuel refining (Speight, 2007; Speight, 2012). Polynuclear aromatic hydrocarbons released into the environment may originate from petroleum products including gasoline, diesel fuel, and fuel oil (Pavlova and Ivanova, 2003). The concentration of polynuclear aromatic hydrocarbons in the environment varies widely, depending on the proximity of the contaminated site to the production source, the level of industrial development, and the mode(s) of polynuclear aromatic hydrocarbon transport.

The toxic, mutagenic and carcinogenic properties of polynuclear aromatic hydrocarbons have resulted in some of these compounds (including naphthalene, phenanthrene, and anthracene) to be designated as priority pollutants. In addition, the solubility of polynuclear aromatic hydrocarbons in aqueous media is very low (Luning Prak and Pritchard, 2002), which affects degradation of these compounds and can lead to biomagnification within an ecosystem.

Interest in the biodegradation mechanisms and environmental fate of polycyclic aromatic hydrocarbons (polynuclear aromatic hydrocarbons) is prompted by their ubiquitous distribution and their potentially deleterious effects on human health (Kanal and Harayama, 2000; Pavlova and Ivanova, 2003; Xia et al., 2006). The biodegradation of polynuclear aromatic hydrocarbons by microorganisms is the subject of many excellent reviews, and the biodegradation of polynuclear aromatic hydrocarbons composed of three rings is well documented (Cerniglia, 1984, Gibson and Subramanian, 1984; Cerniglia and Heitkamp, 1989; Cerniglia, 1992; Van der Meer et al., 1992; Shuttleworth and Cerniglia, 1995; Sutherland et al., 1995).

Active bioremediation strategies (such as biostimulation) for application to polynuclear aromatic-contaminated soils can be used to supply nutrients, oxygen, and other amendments to the subsurface to enhance indigenous microbial activity and contaminant biodegradation (Bamforth and Singleton, 2005; Borchert et al., 1995; Mohan et al., 2006). The benefits of adding oxygen and/or nutrients on the biodegradation of polynuclear aromatics hydrocarbons has been reported for contaminated soils from various sites (Breedveld and Sparrevik, 2000; Eriksson et al., 2000; Li et al., 2005; Liebeg and Cutright, 1999; Lundstedt et al., 2003; Talley et al., 2002). However, only a few studies have focused on the direct effects of biostimulation on the indigenous microbial community and polynuclear aromatic hydrocarbon-degrading bacteria (Ringelberg et al., 2001; Viñas et al., 2005).

Generally, microbial communities present in soils contaminated with polynuclear aromatic hydrocarbons are enriched by microorganisms able to use them as the only carbon source (Heitkamp and Cerniglia, 1988; Gallego et al., 2007). However, this process can be affected by a few key environmental factors (Roling-Wilfred et al., 2002; Simarro et al., 2011) that may be optimized to achieve a more efficient process.

Due to their lipophilic nature, polynuclear aromatic hydrocarbons have a high potential for bio-concentration (Clements et al., 1994; Twiss et al., 1999). In addition to increases in environmental persistence with increasing polynuclear aromatic hydrocarbon molecular size, evidence suggests that in some cases, polynuclear aromatic hydrocarbon toxicity also increases with size, up to at least four or five fused benzene rings (Cerniglia, 1992). The relationship between polynuclear aromatic hydrocarbon environmental persistence and increasing numbers of benzene rings is consistent with the results of various studies correlating environmental biodegradation rates and polynuclear aromatic hydrocarbon molecule size (Banerjee et al., 1995; Shuttleworth and Cerniglia, 1995).

The biodegradation of naphthalene (the simplest polynuclear aromatic hydrocarbon) process was optimized with preliminary experiments in slurry aerobic microcosms (Bestetti et al., 2003). From soil samples collected on a contaminated site, a *Pseudomonas putida* strain (designated as M8) capable of degrading naphthalene was selected. Microcosms were prepared with M8 strain by mixing non-contaminated soil and a mineral medium. Different experimental conditions were tested, varying naphthalene concentration, soil/water ratio, and inoculum density. The disappearance of hydrocarbon, the production of carbon dioxide, and the ratio of total heterotrophic and naphthalene-degrading bacteria were monitored at different incubation times. The kinetic equation that best fitted the disappearance of contaminant with time was determined. The results showed that the isolated strain enhanced the biodegradation rate with respect to the natural biodegradation.

Of the four-ring polynuclear aromatic hydrocarbons, fluoranthene, pyrene, chrysene, and benz[*a*]anthracene have been investigated to various degrees.

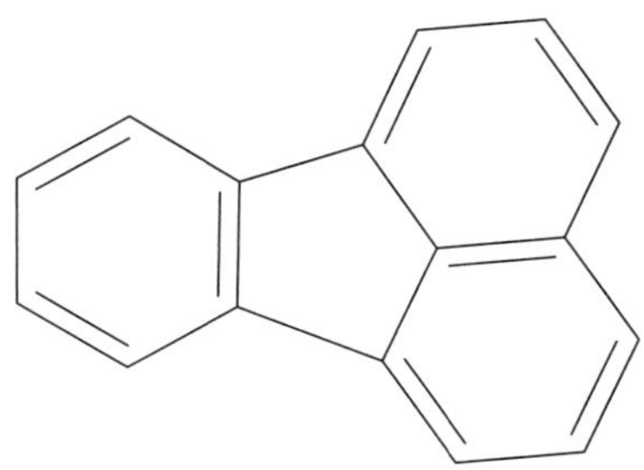

Fluoranthene

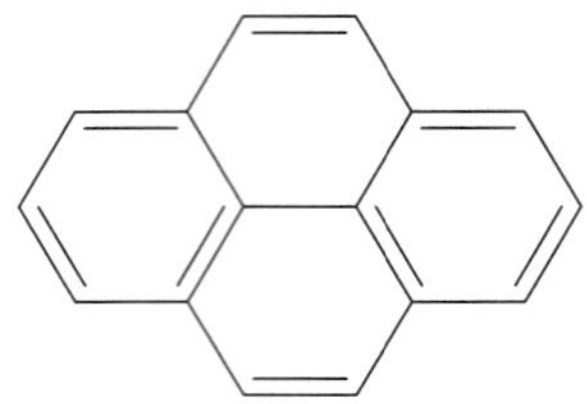

Pyrene

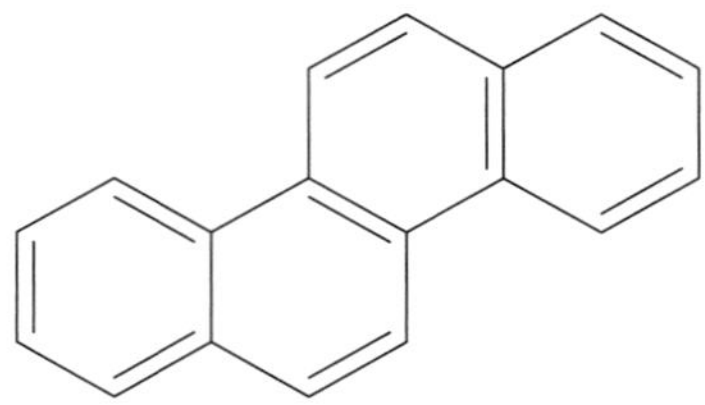

Chrysene

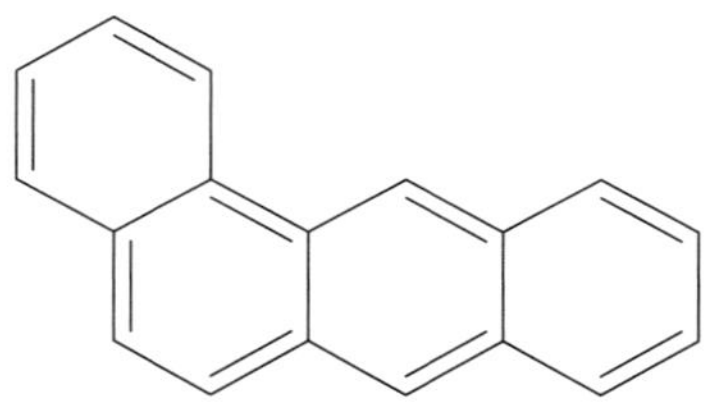

Benz(a)anthracene

Fluoranthene, a polynuclear aromatic hydrocarbon containing a five-membered ring, has been shown to be metabolized by a variety of bacteria, and pathways describing its biodegradation have been proposed (Mueller et al., 1990; Weissenfels, 1990; Weissenfels, 1991; Ye et al., 1996).

Fluoranthene has been used as a model compound in studies that have investigated the effects of surface-active compounds on polynuclear aromatic hydrocarbon biodegradation. Comparisons of the mineralization of fluoranthene by four fluoranthene-degrading strains in the presence of the nonionic surfactants triton x-100 and tween 80 showed that responses differed between strains (Willumsen et al., 1998).

The bacterial degradation of pyrene, a peri-condensed polynuclear aromatic hydrocarbon, has been reported by a number of groups, and some have identified metabolites and proposed pathways (Cerniglia and Heitkamp, 1990). Sediment microcosms inoculated with the mycobacterium showed enhanced mineralization of various polynuclear aromatic hydrocarbons, including pyrene and benzo(*a*)pyrene (Headlamp and Cerniglia, 1989).

Measuring the success of bioremediation of petroleum-related spills is based on several parameters, among them the degradation of polynuclear aromatic hydrocarbons in the crude oil. Though the lower *n*-alkanes are generally considered the most biodegradable compound class within crude oils (Leahy and Colwell, 1990; Atlas and Bartha, 1992; Prince, 1993), other studies point to exceptional conditions in which polynuclear aromatic hydrocarbons degrade preferentially to *n*-alkanes. Jones et al. (1983) showed that the biodegradation of alkylaromatic hydrocarbons was preferential to that of *n*-alkanes in crude oil when oil-contaminated sediments were aerobically incubated. Preferential biodegradation of the aromatic hydrocarbons was also observed in bitumen of the South Aquitaine Basin in which the *n*-alkane had not been completely biodegraded (Conan et al., 1980; Conan, 1981).

Generally, aromatic constituents with five or more rings are not easily attacked, and may persist in the environment for long periods. High-molecular-weight aromatics comprise 2 to 10% w/w conventional (light) petroleum, and up to 35% w/w of the more viscous petroleum.

Currently, there is only limited information regarding the bacterial biodegradation of polynuclear aromatic hydrocarbons with five or more rings in both environmental samples and pure or mixed cultures. Most studies have focused on the five-ring benzo(*a*)pyrene, due potential hazards to human health shown by this compound. Many studies have documented the environmental recalcitrance of benzo(*a*)pyrene to biodegradation (Cerniglia, 1992; Park et al., 1990; Wild and Jones, 1993; Goodin and Webber, 1995; Van Brummelen et al., 1996). Turnover times of in excess of three years in oil-contaminated freshwater sediments and possibly in excess of sixty years in uncontaminated sediments have been reported for the biotransformation of benzo(*a*)pyrene.

The efficiency of several chemical treatments as potential enhancers of the biodegradation of polynuclear aromatic hydrocarbons in contaminated soil has been evaluated by analyzing the mineralization of ^{14}C-labeled phenanthrene, pyrene, and benzo(a)pyrene (Piskonen and Itävaara, 2004). The effect of nonionic surfactants with Fenton oxidation and combinations of surfactants with the Fenton oxidation were evaluated in a micro-titer plate assay. The surfactants selected for the study were Tween 80, Brij 35, Tergitol NP-10, and Triton X-100. Phenanthrene mineralization was also positively induced by the Fenton treatments.

However, none of the treatments had a significant effect on benzo(a) pyrene mineralization. Surfactant additions at concentrations of 20% and 80% of the aqueous critical micelle concentration did not significantly affect the mineralization rates. When surfactant addition was combined with the Fenton oxidation, reduced mineralization rates were obtained when compared with mineralization after Fenton's treatment alone. The results indicated that the addition of Fenton's reagent may enhance the mineralization of PAHs in contaminated soil, whereas the addition of surfactants has no significant beneficial effect.

Increases in the understanding of the microbial ecology of polynuclear aromatic hydrocarbon-degrading communities and the mechanisms by which polynuclear aromatic hydrocarbon biodegradation occur will prove helpful for predicting the environmental fate of these compounds, and for developing practical polynuclear aromatic hydrocarbon bioremediation strategies in the future (Okerentugba and Ezeronye, 2003).

2.4 Phenolic Compounds

Phenolic compounds are common constituents of wastewaters from the oil industry. The bacteria involved in biodegradation may strongly influence the temperature-dependent performance of wastewater treatment processes. Onysko et al. (2000) determined the effect of temperature (10–25°C), appropriate for wastewater treatment, on growth and phenol degradation kinetics of psychrotrophic *Pseudomonas putida*. According to the substrate-inhibition model used, based on the Haldane equation and valid for substrate concentrations less than 400 mg l–1, the maximum specific growth rate (μmax) and the Monod constant (*K*s) increased with increasing temperature. The simultaneous increase of the inhibition constant (*K*i) indicated an increased degree of inhibition at low temperatures.

Catechol isomers are known to be toxic and carcinogenic for humans, and their contamination of soils and aquifers is of great environmental concern:

***Ortho* isomer**	***Meta* isomer**	***Para* isomer**
Catechol	Resorcinol	Hydroquinone
Pyrocatechol	1,3-benzenediol	1,4-benzenediol
1,2-benzenediol	*m*-benzenediol	*p*-benzenediol
o-benzenediol	1,3-dihydroxybenzene	1,4-dihydroxybenzene
1,2-dihydroxybenzene	*m*-dihydroxybenzene	*p*-dihydroxybenzene
o-dihydroxybenzene	resorcin	
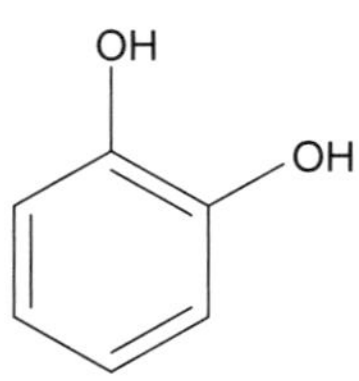	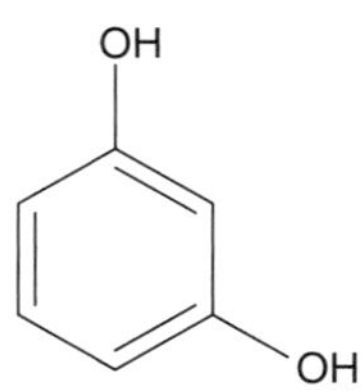	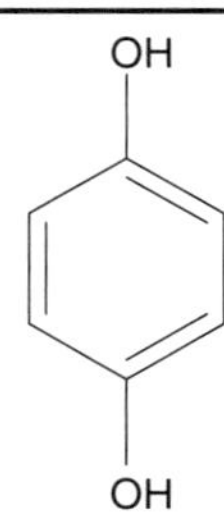

Soil microorganisms, like *Pseudomonas* spp. and *Mycobacterium*, were found to be capable of transforming and degrading toxic catechol isomers to easily absorbable metabolites. These abilities may be useful in removal of toxic organic compounds from the environment. The successful application of microorganisms to the bioremediation of contaminated sites requires a deeper understanding of how microbial degradation proceeds (Zeyaullah et al., 2009).

Another psychrotrophic *Pseudomonas putida* was reported to remove a wide variety of phenolic compounds from wastewater under aerobic and pH-neutral conditions at temperatures ranging from 1 to 35°C (Pillis and Davis, 1985). This strain can be used in trickling filter systems, activated sludge treatments, and outdoor lagoons, either alone or in combination with other microorganisms conventionally used in waste treatment. It can be cultured in wastewater using either a batch process, a semi-continuous, or a continuous process, for a sufficient amount of time (24 h to 4 weeks, depending on the temperature, the volume to be treated, and the concentration of the contamination) to achieve a significant reduction.

The resulting water is suitable for discharge into rivers and streams after conventional processing. Lagoon efficiency is often low in winter when microorganisms are less active. The use of this strain allowed a reduction in the amount of steam required in winter for lagoon heating in order to achieve normal lagoon operation, with a considerable saving in energy costs.

2.5 Spent Caustic

In the petroleum refining industry, caustic solutions (i.e., NaOH) are regularly used to remove H_2S and organic sulfur compounds from hydrocarbon streams. Once H_2S is reacted with NaOH, the solution becomes known as *spent caustic* or *spent sulfidic caustic*. Spent caustics typically have a pH value (>12) and high sulfide concentrations (2 to 3% w/w). Depending on the source, spent caustic may also contain phenols, mercaptans, amines, and other organic compounds that are soluble or emulsified in the caustic (Bechok, 1967).

Most spent caustics are sent off-site for commercial recovery or reuse, e.g., in pulp and paper mills, for treatment by wet air oxidation, or for disposal by deep-well injection (Kolhatkar and Sublette, 1996). The main methods dealing with caustic are chemical methods, such as neutralization and oxidation (Chernatskaya, 1974), but these need large investments, and have a high operating cost. And most importantly, they will lead to a serious environmental pollution and equipment corrosion (Shen, 1995).

Physico-chemical treatment of spent caustics, in general operated at high temperature and pressure, is characterized by high operating costs and large investment. Biological treatment could be an inexpensive alternative. Reports on use of biological treatment of spent caustic are rarely seen (Sipma et al., 2004). Biotechnology for wastewater treatment has a long history (Matveev, 1996): this method has been widely used due to soft degradation conditions (Eigenson et al., 1974), low cost, and no secondary pollution. If the spent caustic is placed directly into the biological treatment system without neutralization, it will affect the normal operation of subsequent biochemical systems because of its toxicity, high total dissolved solid (TDS), and complex composition (Wang and Yang, 2001), resulting in discharge above standard. Incomplete removal of thiols in spent caustic will lead to severe odor nuisance.

Thus, it is very important to find salt-tolerant microbial with excellent degradation for biotreatment of spent caustic.

2.6 Wastewater

In recent years, alkali/surfactant/polymer (ASP) flooding has been well developed for enhancement of oil recovery in some Chinese oilfields (Deng et al., 2005). This method has been proven to be a promising recovery approach to extract large volume of oil remaining in the ground after conventional oil recovery methods have been implemented.

There are various technologies that have been used to treat the conventional oily wastewater, including membrane filtration (Deriszadeh et al., 2010), adsorption (Fakhru'l-Razi et al., 2009), chemical precipitation and oxidation (Fakhru'l-Razi et al., 2009), and electrochemical treatment (Ramalho et al., 2010). A biological approach has been widely used to degrade oil and remediate oil-contaminated water (Lu et al., 2009; Ilori et al., 2011; Lu et al., 2011; Zekri et al., 2009; Zhou and Shen, 2010). However, it would be difficult to meet the requirement for flooding-produced water treatment in oilfield by using an approach alone, because of the highly complex characteristics of pollutants and large quantity of flooding-produced water. Hence, a combination of physicochemical and biological treatment is needed in order to enhance the treatment efficiency of flooding-produced water.

The application of the combination of hydrolysis acidification-dynamic membrane bioreactor-coagulation process to treating wastewater produced from polymer flooding has been reported to be effective (Zhang et al., 2010). In addition, use of a combination of the zero valent iron/EDTA/air system and activated sludge process, as well as coagulation and flocculation to remediate oilfield wastewater produced from polymer flooding, has also been claimed to be effective (Asia et al., 2006; Lu and Wei, 2011).

2.7 Chlorinated Compounds

For the contaminated sites containing volatile chlorinated hydrocarbons, the application of an anaerobic metabolism mode followed by an aerobic phase is more beneficial in bioremediation than using only either anaerobic or aerobic degradation. Polychlorinated biphenyls have widespread environmental occurrence, and have been suspected to be carcinogenic. For this reason, there is present research into developing and testing new treatments to break down polychloro-biphenyls (PCBs) into harmless compounds.

Polychloro-biphenyls can be biodegraded using sequential anaerobic-aerobic reactions. The anaerobic attack is a reductive dechlorination that results in the replacement of a chlorine atom with a hydrogen atom. The aerobic attack occurs on the ring, and the outcome is a breakage of the ring, and thus, destruction of the polychloro-biphenyl derivative. The anaerobic-aerobic attack is used in a sequence, because the aerobic attack only works on polychloro-biphenyl derivatives with a relatively low content of chlorine molecules. Thus, the anaerobic attack constitutes a pretreatment to remove chlorine atoms, and converts highly chlorinated polychloro-biphenyls to lower chlorinated polychloro-biphenyls, which are destroyed by aerobic treatment.

3 Petroleum Biodegradation

The biodegradability of any petroleum constituent is a measure of the ability of that constituent to be metabolized (or co-metabolized) by bacteria or other microorganisms through a series of biological process, which include ingestion by organisms as well as microbial degradation (Payne and McNabb, 1984). The chemical characteristics of the contaminants influence biodegradability; in addition, the location and distribution of petroleum contamination in the subsurface can significantly influence the likelihood of success for bioremediation.

The biodegradability of petroleum and petroleum product is inherently influenced by the composition of the substrate upon which the bacteria are acting (Chapter 1). For example, petroleum is quantitatively biodegradable, and kerosene, which consists almost exclusively of medium chain-length alkanes, is completely biodegradable under suitable conditions, but for heavy asphaltic crudes, approximately only 6 to 10% of the material oil may be biodegradable within a reasonable time period, even when the conditions are favorable for biodegradation (Bartha, 1986; Okoh et al., 2002; Okoh et al., 2003; Okoh, 2006). In addition, biodegradation of petroleum-related constituents can be enhanced by use of a consortium of different bacteria compared to the activity of single bacterium species (Ghazali et al., 2004; Milić et al., 2009).

In addition to the composition of the petroleum-related substrate, petroleum and petroleum products introduced to the environment are immediately subject to a variety of changes caused by physical, chemical, and biological effects, usually (incorrectly) referred to collectively as *weathering*. Physical and chemical processes include (1) evaporation, (2) dissolution of petroleum constituents in a water system (or aquifer), (3) dispersion, (4) photochemical oxidation, (5) formation of water-oil emulsions, and (6) adsorption onto suspended particulate material. These processes are not sequential, and typically occur simultaneously and cause important changes in the composition and properties of the original pollutant, which in turn may affect the rate or effectiveness of biodegradation.

Specifically, the biodegradation of petroleum typically: (1) raises the viscosity and decreases the API gravity, which adversely reduces the ability of the degraded product to flow, (2) decreases the hydrocarbon content, thereby increasing the residuum content, (3) increases the concentration of certain metals, (4) increases the sulfur content, and (5) increases oil acidity, and adds compounds, such as carboxylic acids and phenols. All of these changes are seen in the product of the product relative to the unchanged (non-biodegraded) petroleum (Miiller et al., 1987).

The commercial practice of bioremediation focuses primarily on the cleaning up of petroleum hydrocarbons (Del'Arco and de França, 1999). Thus, successful application of bioremediation technology to a contaminated ecosystem requires knowledge of the characteristics of the site and the parameters that affect the microbial biodegradation of pollutants (Sabate et al., 2004). However, a number of limiting factors have been recognized to affect the biodegradation of petroleum constituents (Table 9.1).

Despite the difficulty of degrading certain fractions, some hydrocarbons are among the most easily biodegradable naturally occurring compounds. Biodegradation gradually destroys petroleum-related spills by the sequential metabolism of various classes of compounds present in the oil (Bence et al., 1996). When biodegradation occurs in an oil reservoir, the process dramatically affects the fluid properties (Miiller et al., 1987), and hence, the value and producibility of an oil accumulation. Specifically, petroleum biodegradation typically raises viscosity of the residual material (which reduces oil producibility) and reduces the API gravity (which reduces the value of the produced oil). It increases the asphaltene content (relative to the saturated and aromatic hydrocarbon content and the starting material), the concentration of certain metals, the sulfur content, and oil acidity.

There are indications that petroleum biodegradation involves more biological components than just the microorganisms that directly attack petroleum constituents (the primary degraders), and shows that the primary degraders interact with these components (Head et al., 2006). In addition, primary degraders need to compete with other microorganisms

Table 9.1 Some common factors affecting petroleum hydrocarbon biodegradation (Bartha, 1986).

Factor	Comment
Petroleum hydrocarbon-composition (PHC).	Structure, amount, toxicity.
Physical state	Aggregation, spreading, dispersion, adsorption.
Weathering	Evaporation, photooxidation.
Water potential	Osmotic and matrix forces, exclusion of water from hydrophobic aggregates
Temperature	Influence on evaporation and degradation rates.
Oxidant	O_2 required to initiate oxidation, NO_3 or SO_4^{2} to sustain, PHC biodegradation.
Mineral nutrients	N, P, Fe may be limiting
Reaction	Low pH may be limiting
Microorganisms	PHC degraders may be absent or low in numbers

for limiting nutrients, and the non-petroleum-degrading microorganisms can be affected by metabolites and other compounds that are released by oil-degrading bacteria, and vice versa.

The environment, having been exposed to petroleum inputs for thousands of years, can assimilate the hydrocarbons under the proper conditions. However, areas of particular concern are low energy environments common to estuarine systems. These environments, such as marshes, mud flats, and subtidal areas, are vital to marine fisheries and estuarine productivity, and are especially sensitive to contaminant impacts. These systems are particularly vulnerable to impacts of petroleum where research has shown petroleum can persist in these systems for years.

The removal processes for petroleum in wetlands are (1) evaporation, (2) photo-oxidation, (3) dissolution of specific constituents, (4) microbial degradation, and (5) physical flushing. However, once incorporated into the sediment, biodegradation and dissolution are the primary removal mechanisms. Petroleum biodegradation in wetland environments can be limited by anoxia and nutrient availability. Consequently, estuarine wetlands are the most vulnerable of the low-energy intertidal areas to petroleum spills.

Following the 1994 San Jacinto River flood and spill in southeast Texas, a petroleum-contaminated wetland was reserved for a long-term research program to evaluate bioremediation as a viable spill response tool (Mills et al., 2003). Sediment samples from six test plots were collected eleven times over an eleven-month period to assess the concentration of petroleum remaining. The rapid degradation rates of the petroleum hydrocarbons are attributed to conditions favorable to biodegradation. It was suggested that elevated nutrient levels from the flood deposition and the unconsolidated nature of the freshly deposited sediment possibly provided a nutrient rich, oxic environment, and that an active and capable microbial community was present due to prior exposure to petroleum. These factors provided an environment conducive for the rapid bioremediation of the petroleum in the contaminated wetland, leading to bioremediation of 95% of the spilled material.

In 1996, crude oil was spilled from the Sea Empress at the entrance to the Milford Haven Waterway in Pembrokeshire, Wales: this was third largest oil spillage in or around the British Isles (and the twelfth largest in the world) at the time. The ship was en route to the Texaco oil refinery near Pembroke when she became grounded on mid-channel rocks at St. Ann's Head. Over the course of a week, approximately 432,000 barrels of petroleum were spilled into the sea within the Pembrokeshire Coast National Park, an important and sensitive wildlife and marine conservation areas.

Following the Sea Empress spill in the UK found enhanced biodegradation (statistical differences in GC-MS analyses of the control and the nutrient treatments) with no increased toxicity. Other more recent studies have included a biostimulation experiment involving a controlled spill on the St. Lawrence River and a petroleum-related spill in the United Kingdom (UK) (Mille et al., 2003). Using gas chromatography-mass spectrometer (GC–MS) analyses, there were no significant differences between the nutrient-enhanced and the control treatments, but evidence of elevated toxicity occurred in some of the nutrient-treated plots.

3.1 Effects of Biodegradation

These early stages of oil biodegradation (loss of n-paraffins followed by loss of acyclic iso-prenoids) can be readily detected by gas chromatography (GC) analysis of the petroleum. However, in heavily biodegraded crude oils, gas chromatographic analysis alone cannot distinguish differences in biodegradation, due to interference of the unresolved complex mixture that dominates the gas chromatographic traces of heavily degraded crude oils. Among such crude oils, differences in the extent of biodegradation can be assessed, using gas chromatography-mass spectrometry (GC-MS) to quantify the concentrations of biomarkers with differing resistances to biodegradation.

During biodegradation, the properties of the petroleum fluid changes because different classes of compounds in petroleum have different susceptibilities to biodegradation (Goodwin et al., 1983). The early stages of biodegradation (in addition to any evaporation effects) are characterized by the loss of n-paraffins (n-alkanes or branched alkanes) followed by loss of acyclic iso-prenoids (e.g., norpristane, pristane, and phytane). Compared with those compound groups, other compound classes (such as highly branched and cyclic saturated hydrocarbons, as well as aromatic compounds) are more resistant to biodegradation. However, even the more-resistant compound classes are eventually destroyed as biodegradation proceeds.

3.2 Conditions for Biodegradation

The composition and inherent biodegradability of the petroleum hydrocarbon pollutant are perhaps the first and most important considerations when the suitability of a cleanup approach is to be evaluated (Atlas, 1975). Heavier crude oil is generally much more difficult to biodegrade than lighter ones, just as heavier crude oils could be suitable for inducing increased selection pressure for the isolation of petroleum hydrocarbon degraders with enhanced efficiency. Also, the amount of heavy crude oil metabolized by some bacterial species increases with increasing concentration of the contaminant, while degradation rates may appear to be more pronounced within a specific concentration range (Okoh et al., 2002; Rahman et al., 2002).

Important aspects of the conditions for biodegradation at a spill site are the activity of microorganisms and the ability of the organisms to produce enzymes to catalyze metabolic reactions, which is governed by the genetic composition of the organism(s). Enzymes produced by microorganisms in the presence of carbon sources cause initial attack on the hydrocarbon constituents, while other enzymes are utilized to complete the breakdown of the hydrocarbon. Thus, lack of an appropriate enzyme either prevents attack or is a barrier to complete hydrocarbon degradation.

Biodegradation of petroleum-related constituents by bacteria can occur under both aerobic (oxic) and anaerobic (anoxic) conditions (Zengler et al., 1999), usually by the action of different consortia of microorganisms. In the subsurface, biodegradation occurs primarily under anaerobic conditions, mediated by sulfate reducing bacteria in cases where dissolved sulfate is present (Holba et al., 1996), or methanogenic bacteria in cases where dissolved sulfate is low (Bennett et al., 1993). Although subsurface oil biodegradation does not require oxygen, there is a requirement for the presence of essential nutrients (such as nitrogen, phosphorus, and potassium), which can be provided by dissolution/alteration of minerals in the water layer. In the absence of nutrients, the potential for hydrocarbon degradation in anoxic sediments is markedly reduced (Dibble and Bartha, 1976).

In-situ groundwater can be an effective medium for biodegradation of petroleum hydrocarbons. While there are some notable exceptions (such as MTBE, which is not a hydrocarbon; see Chapter 11), the short-chain, low molecular weight, more water-soluble constituents are degraded more rapidly, and to lower residual levels, than are long-chain, high molecular weight, less soluble constituents. However, as with all bioremediation efforts, petroleum and petroleum products (such as residual fuel oil and asphalt) typically have a residual high-boiling fraction composed of resin and asphaltene constituents, which is composed of complex, polynuclear aromatic systems (Speight, 2007).

Microbial utilization of hydrocarbons (being fully reduced substrates) requires an exogenous electron sink. In the initial attack, this electron sink has to be molecular oxygen. In the subsequent steps too, oxygen is the most common electron sink. In the absence of molecular oxygen, further biodegradation of partially oxygenated intermediates may be supported by nitrate or sulphate reduction.

Uptake and utilization of water insoluble substrates, such as petroleum alkanes, require specific physiological adaptations of the microorganisms. The synthesis of specific amphiphilic molecules (i.e., biosurfactants) is often surmised to be a prerequisite for either specific adhesion mechanisms to large oil drops or emulsification of oil, followed by uptake of submicron oil droplets. In fact, various species of bacteria have been observed to adopt the requisite strategy to deal with water insoluble substrates, such as hydrocarbons (Rosenberg, 1991). Hence, to facilitate hydrocarbon uptake through the hydrophilic outer membrane, many hydrocarbon-utilizing microorganisms produce cell wall-associated or extracellular surface-active agents (Haferburg et al., 1986). This includes low molecular weight compounds, such as fatty acids, triacyl-glycerol derivatives, and phospholipids, as well as the heavier glycolipids, examples of which include *emulsan* and *liposan* (Cirigliano and Carman, 1984).

Emulsan is the extracellular form of a polyanionic, cell-associated heteropolysaccharide produced by the oil-degrading bacterium *Acinetobacter calcoaceticus* RAG-1 (Rosenberg et al., 1979b; Zuckerberg et al., 1979). The biopolymer stabilizes emulsions of hydrocarbons in water, and has optimal activity when a mixture of aromatic and aliphatic components is present, such as in crude oil (Rosenberg et al., 1979a). The activity of the amphipathic emulsifier is due primarily to its high affinity for the oil-water interface (Zosim et al., 1982) and its ability to orient itself at the interface to form a hydrophilic film around the oil droplets (Zosim et al., 1982; Shabtai et al., 1986; Zosim et al., 1986).

Studies with bacteriophages, antibodies, and emulsan-deficient mutants have demonstrated that: (1) as the cells approach stationary phase, emulsan accumulates on the cell surface before release into the medium (Goldman et al., 1982); (2) cell-bound emulsan serves as a specific receptor and acts as stabilizer for the oil-water interface (Pines and Gutnick, 1981, 1984a); (3) this indicates that the cell-bound form of emulsan is required for growth on crude oil: species without cell-bound emulsan no longer grow well on petroleum-related (Rosenberg et al., 1983a; Rosenberg et al., 1983b; Pines and Gutnick, 1984b, 1986); and (4) the affinity of emulsan for the oil-water interface suggests that it might affect microbial degradation of emulsified oils (Gutnick and Minas, 1987).

In another work, crude oil was also treated with purified emulsan, the heteropolysaccharide bioemulsifier produced by *Acinetobacter calcoaceticus* RAG-1. A mixed bacterial population, as well as nine different pure cultures isolated from various sources, was tested for biodegradation of emulsan-treated and untreated crude oil. Biodegradation was measured both quantitatively and qualitatively. Biodegradation of linear alkanes and other saturated hydrocarbons, both by pure cultures and by the mixed population, was reduced after emulsan pretreatment. In addition, degradation of aromatic compounds by the mixed population was also reduced in emulsan-treated oil. In sharp contrast, aromatic biodegradation by pure cultures was either unaffected or slightly stimulated by emulsification of the oil (Foght et al., 1989).

3.3 Effect of Nutrients

Different types of nutrients (primarily nitrogen and phosphorus) have been applied to improve petroleum hydrocarbon degradation, including classic (water soluble) nutrients and oleophilic and slow-release fertilizers.

Bioavailability is one main factor that influences the extent of biodegradation of hydrocarbons. Generally, hydrocarbons have low-to-poor solubility in water, and, as a result, are adsorbed on to clay or humus fractions, so they pass very slowly to the aqueous phase, where they are metabolized by microorganisms. Cyclodextrins are natural compounds that form soluble inclusion complexes with hydrophobic molecules and increase degradation rate of hydrocarbons *in vitro*.

In the perspective of an *in situ* application, β-cyclodextrin does not increase eluviation (the lateral or downward movement of the suspended material in soil through the percolation of water) of hydrocarbons through the soil, and consequently does not increase the risk of groundwater pollution (Sivaraman et al., 2010). Furthermore, the combination of bioaugmentation and enhanced bioavailability due to β-cyclodextrin was effective for a full degradation (Bardi et al., 2003). Thus, in situ bioremediation of polynuclear aromatic hydrocarbon-polluted soil can be improved by the augmentation of degrading microbial populations and by the increase of hydrocarbon bioavailability (Bardi et al., 2007).

Inadequate mineral nutrients, especially nitrogen and phosphorus, often limit the growth of hydrocarbon utilizers in water and soils. Iron has been reported to be limiting only in clean, offshore seawater (Swannell et al., 1996). Sulfur, in form of sulfate ions, is plentiful in seawater, but could be limiting in some freshwater environments. The slight alkaline pH of seawater seems to be quite favorable for petroleum hydrocarbon degradation, but in acidic soils liming to pH 7.8 to 8.0 had a definite stimulatory effect.

Nutrients are very important ingredients for successful biodegradation of hydrocarbon pollutants, especially nitrogen, phosphorus, and in some cases, iron. Depending on the nature of the impacted environment, some of these nutrients could become limiting, thus affecting the biodegradation processes.

When a major oil spill occurs in freshwater and/or marine ecosystems, the supply of carbon is dramatically increased, and the availability of nitrogen and phosphorus generally becomes the limiting factor for oil degradation (Atlas, 1984). This is more pronounced in marine environments, due to the low background levels of nitrogen and phosphorus in seawater (Floodgate, 1984), unlike in freshwater systems that regularly fluctuate in nutrient status as result of perturbations and receipt of industrial and domestic effluents and agricultural runoff. Freshwater wetlands are typically considered to be nutrient limited, due to heavy demand for nutrients by the plants, which can be considered to be *nutrient traps*, since a substantial amount of nutrients is often found in the indigenous biomass (Mitsch and Gosselink, 1993).

Generally, the addition of nutrients is necessary to enhance the biodegradation of petroleum-related pollutants (Choi et al., 2002; Kim et al., 2005; Joshi and Pandey, 2011). In fact, even in harsh sub-Arctic climates, it has been observed that the effectiveness of fertilizers for petroleum increases the chemical, microbial and toxicological parameters compared to the used of various fertilizers in a pristine environment (Pelletier et al., 2004).

In another study using poultry manure as organic fertilizer in contaminated soil, biodegradation of petroleum-related hydrocarbons was reported to be enhanced, but the extent of biodegradation was influenced by the incorporation of alternate carbon substrates or surfactants (Okolo et al., 2005). However, excessive nutrient concentrations can inhibit the biodegradation activity (Challain et al., 2006), and there can be a negative effect on the biodegradation of hydrocarbons in the presence of high nitrogen-phosphorus-potassium levels (Oudot et al., 1998; Chaîneau et al., 2005): this effect is more pronounced on the bioremediation of aromatic hydrocarbons (Carmichael and Pfaender, 1997). The biodegradation of various aromatic hydrocarbons is also sensitive to acidity or alkalinity, and also to by-products of the biodegradation of the saturate fraction, which serves to explain the persistence of aromatic petroleum hydrocarbons in certain ecosystems.

On the other hand, in an investigation of the role of the nitrogen source in biodegradation of crude oil components by a defined bacterial consortium under cold, marine conditions (10°C/50°F), it was observed that nitrate did not affect the pH, whereas ammonium amendment led to progressive acidification, accompanied by an inhibition of the degradation of aromatic (particularly polynuclear aromatic) hydrocarbons (Foght et al., 1999). However, the aromatic systems were degraded or co-metabolized in the absence of nutrients, where the pH remained almost unchanged. The best overall biodegradation was observed in the presence of nitrate without ammonium, plus high phosphate buffering: a disadvantage of nitrate is that significant emulsification of the petroleum occurs. Generally, it is worth bearing in mind that acidity/alkalinity (pH) is an important factor that requires consideration, as it affects the solubility of both polynuclear aromatic hydrocarbons as well as the metabolism of the microorganisms, showing an optimal range for bacterial degradation between 5.5 and 7.8 (Bossert and Bartha, 1984; Wong et al., 2001).

The molar ratio of carbon, nitrogen, and phosphorus (C/N/P) is very important for the metabolism of the microorganisms, and, therefore, for degradation of polynuclear aromatic hydrocarbons (Bossert and Bartha, 1984; Alexander, 1994; Kwok and Loh, 2003). A molar ratio 100:10:1 is frequently considered optimal for contaminated soils (Bossert and Bartha, 1984; Alexander, 1994), while some authors have reported negative or no effects (Chaîneau et al., 2005). These contradictory results are due to the nutrient ratio required by bacteria that degrade polynuclear aromatic hydrocarbons, which depends on environmental conditions, type of bacteria, and type of hydrocarbon (Leys et al., 2005).

Furthermore, it is not surprising that the chemical form of those nutrients is also important, the soluble forms (i.e., iron or nitrogen in form of phosphate, nitrate, and ammonium) being the most frequent and efficient, due to their higher availability for microorganisms. Depending on the microbial community and their abundance, another factor that may improve polynuclear aromatic hydrocarbon degradation is the addition of readily assimilated carbon sources, such as glucose (Zaidi and Imam, 1999).

3.4 Effect of Temperature

Temperature plays an important role in the biodegradation of petroleum-related hydrocarbons, not only because of the direct effect on the chemistry of the pollutants, but also because of the effect on the physiology and diversity of the microbial surroundings (Atlas, 1975). In short, temperature can play the role of increasing a microbial reaction or inhibiting a microbial reaction in a similar manner to the general rules for the influence of temperature on chemical reactions.

Typically, biodegradation of petroleum and petroleum products occurs at temperatures less than 80°C (<176°F) (Conan, 1984; Barnard and Bastow, 1991): at higher temperatures (unless the microbes are of a specific thermophilic type), many of the microorganisms involved in subsurface oil biodegradation cannot exist.

Thus, the ambient temperature of an environment affects both the properties of spilled petroleum or petroleum product (Speight, 2007) and the activity or population of microorganisms (Venosa and Zhu, 2003). At low temperatures, the viscosity of the oil increases, while the volatility of toxic low-molecular weight hydrocarbons is reduced, delaying the onset of biodegradation (Atlas, 1981). Temperature also variously affects the solubility of hydrocarbons (Foght et al., 1996).

Although hydrocarbon biodegradation can occur over a wide range of temperatures, the rate of biodegradation generally decreases with decreasing temperature. Highest degradation rates generally occur in the range of 30 to 40°C (86 to 104°F) in soil environments, 20 to 30°C (68 to 86°F) in some freshwater environments, and 15 to 20°C (59 to 68°F) in marine environments (Bossert and Bartha, 1984). In fact, the biodegradability of petroleum is highly dependent not only on composition, but also on microbial incubation temperature (Atlas, 1975): at 20°C (68°F), conventional petroleum has higher abiotic losses and is more susceptible to biodegradation than heavy oil. As expected from petroleum chemistry and composition (Speight, 2007), the rate of mineralization for the heavy oil is significantly lower at 20°C (68°F) than for conventional oil.

During biodegradation, some preference is shown for removal of the paraffin constituents over the aromatic and asphaltic constituents, especially at low temperatures. Branched paraffins, such as pristane, are degraded at both 10 and 20°C (50 and 68°F). This was confirmed by showing that the residual material (after an incubation period of 42 days) had a lower relative percentage of paraffins and higher percentage of asphaltic constituents (usually resin and asphaltene constituents) than fresh or weathered oil (Atlas, 1975).

Finally, the relative resistance of conventional (light) petroleum and even petroleum distillate products to degradation at low temperatures should be considered in choosing shipping routes for these materials. Accidental spillages in the Arctic will most likely be of regional heavy type oil (such as the Prudhoe Bay crude oil), which will be subject to slow but constant microbial degradation (Atlas, 1975). As a quick note, the Prudhoe Bay oil field is a large oil field on North Slope of Alaska and contains an estimated 25 billion (25×10^9) barrels of petroleum (ca. 21° API) .

3.5 Effect of Dispersants

The effect of dispersants on the fate of dispersed oil has often been the subject of conflicting reports: some workers have proposing that dispersants had (1) little effect on oil biodegradation, (2) a positive effect, and (3) a negative effect (Robichauz and Myrick, 1972; Mulkins-Phillips and Stewart, 1974a; Traxler and Bhattacharya, 1979; Foght and Westlake, 1982; Lee et al., 1985; Litherathy et al., 1989).

On the other hand, and perhaps more specifically, it has been suggested that dispersants tend to increase oil biodegradation by increasing the surface area for microbial attack, and encouraging migration of the droplets through the water column, making oxygen and nutrients more readily available (Mulyono et al., 1994). However, dispersants can have a detrimental (toxic) effect on microbial processes, thereby retarding the rate of petroleum degradation (Mulyono et al., 1994; Varadaraj et al., 1995). It would appear that the dual capability of dispersants (increasing the surface area of dispersed oil and affecting the growth of hydrocarbon-degraders) is related to the chemistry of the dispersant, which influences the effectiveness of dispersants for bioremediation (Varadaraj et al., 1995; Davies et al., 2001).

It is clear that the introduction of external nonionic surfactants (the main components of oil spill dispersants) will influence the alkane degradation rate (Bruheim and Eimhjelle, 1998; Rahman et al., 2003). There are indications that the use of surfactants in situations of petroleum-related contamination may have a stimulatory, inhibitory, or neutral effect on the bacterial degradation of the petroleum constituents (Liu et al., 1995). Thus, there is the need to accurately characterize the roles of chemical and biological surfactants in order that performance in biological systems may be predicted (Rocha and Infante, 1997; Lindstrom and Braddock, 2002).

However, in contrast to chemical dispersants, which caused ecological damage after application for abatement of spilled petroleum-related constituents in marine ecosystems (Smith, 1968), biosurfactants from soil or freshwater microorganisms are less toxic and partially biodegradable (Poremba et al., 1991).

3.6 Rates of Oil Biodegradation

Different results for microbial activity measurements may be obtained in laboratory studies, depending on pretreatment and size of the sample, even when the environmental conditions are mimicked (Björklöf et al., 2008). These differences may be related to, among other factors, differences in the bioavailability of the contaminant in different analyses.

Thus, modeling chemical reactions to determine the rate of the reaction is a common practice in chemistry. However, blending of chemical reactants, which is a part of modeling studies, can have beneficial or adverse effects on chemical reactions. Blend time predictions are usually based on empirical correlations, and when a competitive side reaction is present, the final product distribution may be suspect. In biodegradation, the effect of the microbe in relation to the hydrocarbon stream on the reaction outcome is crucial. Also, the scale up of such reactions from the laboratory to the field may not be straightforward. Thus, there is a need for comprehensive review and planning, for optimistic caution must be exercised when models of biodegradation chemistry are used to predict information such as incubation period, product, byproducts, and whether or not the microbes will survive the duration of the treatment. It is also equally important to recognize the potential for interference of one chemical species with another, a situation that is not often determined in the laboratory when simple model substrates (such as single chemical entities or mixtures of two-to-twelve compounds that are not truly representative of petroleum and petroleum products) are used for the experiments.

The most important critical stage of petroleum degradation during the first 48 hours of a spill is usually evaporation, the process by which the lower molecular weight (lower-boiling constituents) constituents of petroleum volatilize into the atmosphere. Evaporation can be responsible for the loss of one- to two-thirds of the mass of spilled material (assuming that the spilled material is conventional crude oil or a distillate product) during this period, with the loss rate decreasing rapidly over time. The constituents of heavy oil, tar sand bitumen, and asphalt do not evaporate to the same extent, the lower boiling constituents being generally absent from these materials.

Evaporative loss is controlled by the composition and physical properties of the petroleum or petroleum derivative, the surface of the spill, wind velocity, and temperature (Payne and McNabb, 1984). Derivation of a universal model for such a process may insert inaccuracies because of the complex and changing nature of petroleum and petroleum products. In addition, the material left behind is richer in metals (mainly nickel and vanadium), waxes, resin constituents, and asphaltene constituents than the original oil. With evaporation, the specific gravity and viscosity of the original oil also increase: after several days, spilled conventional petroleum may begin to resemble Bunker C oil (heavy fuel oil) in composition and properties (Mielke, 1990).

Although, there has been a reported case of lack of correlation between degradation rates, specific growth rates, and concentration of the starter oil (Thouand et al., 1999), in such a case, it would appear that biomass was required only to a particular threshold, enough to produce the appropriate enzyme system that carry through the degradation process even when biomass production had ceased (Pitter and Chudoba, 1990), where production of variance with the theory of microbial growth in batch cells is totally dependent on the consumed carbon source.

Many reports on the effect of sunlight irradiation so far published have focused on the physicochemical changes on intact crude oil rather than to biodegraded crude oil (Jacquot et al., 1996; Nicodem et al., 1998). Recent studies have reported that photo-oxidation increases the biodegradability of petroleum hydrocarbon by increasing its bioavailability, and thus enhancing microbial activities (Maki et al., 2005).

In fact, bioremediation is a multi-variable process, and optimization through classical methods is subject to question. To overcome the disadvantages of the process, response surface methodology (RSM) has been advocated for analyzing the effects of several independent variables on the bioremediation process in order to assess the optimum conditions for the process (Nasrollahzadeh et al., 2007; Huang et al., 2008; Pathak et al., 2009; Vieira et al., 2009; Mohajeri et al., 2010; Zahed et al., 2010).

The outcome is the suggestion that the rates of biodegradation can be increased by modifying selected physical and chemical conditions that control biodegradation in multiphase systems, namely, (1) bioavailability and (2) terminal electron acceptor availability (Sandrin et al., 2006). With respect to bioavailability, model simulations suggest that: (1) increasing the interfacial area between the aqueous and solid phases, (2) increasing the rate of contaminant solubilization, and (3) minimizing the accumulation of the contaminant in non-aqueous phase liquids will result in significantly higher rates of biodegradation.

Possible approaches to increasing contaminant bioavailability will depend on the system in question. However, any approach that either increases the physical mixing of the non-aqueous phase liquids and aqueous phases or that increases the solubility of the contaminant in the aqueous phase will result in increased biodegradation. Such approaches include the addition of surfactants or co-solvents (to increase solubility), aeration, or hydraulic pulsing, to increase mixing of the two phases.

Due to the complex array of factors that influence biodegradation of petroleum-related products, it is not realistic to expect a simple rate model or kinetic model to provide precise and accurate descriptions of concentrations during different seasons and in different environments. Therefore, it is nearly impossible to predict the rates of biodegradation of such a process. To give a final answer on how much time remediation processes require and what the final mineral oil concentrations will be, experiments should be continued until the biodegradation processes have stopped completely. In future, it will be necessary to use complex models to yield a more exact assessment of soil remediation to the desired level (Maletić et al., 2009).

Finally, when predicting by modeling the behavior of an aged contaminant, it is relevant to adapt the models in use to correspond to conditions relevant at the contaminated sites. As with all efforts at modeling the outcome of complex processes, the variable parameters used in the models must be based on (1) the properties of the material (in this case, the petroleum-based contaminant), (2) data retrieved about the conditions of the actual site, and on (3) experiments performed using the original aged contaminant without any additions (model compounds or analytical *spikes*).

3.7 Effect of Weathering

Weathered petroleum (i.e., petroleum and petroleum related products) that has been exposed to air and oxidized and to other influences, such as evaporation, offers a differ challenge to bioremediation efforts.

In a study assessing the bioremediation efficiency of a weathered and recently contaminated soil in Brazil (Trindale et al., 2005), the authors reported low biodegradation efficiencies in the weathered soil contaminated with a high crude oil concentration compared to recently contaminated soil. Also, both soils (weathered and recently contaminated) submitted to bioaugmentation and biostimulation techniques presented biodegradation efficiencies approximately twice higher than the ones without natural attenuation.

Several areas at many sites, especially refinery sites with a long history of operation, have weathered oil floating on the surface of the groundwater and oil-saturated soil. However, site conditions can be manipulated to enhance bioremediation and speed up the degradation rates of the contaminants. There are several techniques that can be applied to enhance the biological degradation of contaminants:

1. Supplementation with suitable sources of nitrogen and phosphorus;
2. Manipulation of redox potential by the injection of air, oxygen, or nitrate to enhance aerobic biodegradation;
3. Addition of surfactants to make the contaminants bioavailable;
4. Site microbial inoculation; and
5. Injection of co-substrates, such as molasses or lactate, to enhance the biodegradation of chlorinated contaminants.

Oxygen is often the limiting factor in aerobic bioremediation at many sites. The degradation of petroleum hydrocarbons occurs much faster under aerobic conditions compared to anaerobic conditions. Therefore, the addition of oxygen can significantly increase the remediation rates.

Oxygen addition is most frequently used to address dissolved phase contamination, such as total petroleum hydrocarbons and BTEX, as well as contamination in the capillary fringe zone. Oxygen can only be effective if the hydrocarbons are bioavailable and there is no nutrient limitation.

The bioremediation of petroleum contaminated soil was investigated using a microscale landfarming (Yudono et al., 2011). The indigenous bacteria, *Pseudomonas pseudoalcaligenes*, *Bacillus megaterium*, and *Xanthobacter autotrophicus* were isolated from the contaminated sites Sungai Lilin Jambi Pertamina Ltd, and used further in the bioremediation experiments. The biodegradation rates of petroleum contaminated soil in the presence of the isolated bacteria were studied by using the chemical kinetics approach. The reaction orders were studied by using the differential method, and the reaction rate constants were studied by using the integral method. The results showed that the reaction orders were 1.0949, 1.3985, and 0.8823, and the reaction rate constants were 0.0189, 0.0204, and 0.0324 day-1, respectively. Considering the values of reaction orders and reaction rate constants, the biodegradation rate of contaminated soil by using each bacteria had significantly different value; *Xanthobacter Autotrophicus* bacteria could degrade the petroleum oil sludge faster than the others.

4 Application to Spills

The success of oil spill bioremediation depends on optimization of various physical, chemical, and biological conditions in the contaminated environment. There are two main approaches to bioremediation of petroleum-related spills: (1) bioaugmentation, in which oil-degrading microorganisms are added to supplement the existing microbial population, and (2) biostimulation, in which the growth of indigenous oil degraders is stimulated by the addition of nutrients or other growth-limiting co-substrates and/or habitat alteration (Chapter 1).

The first steps in assessing the potential of a petroleum-contaminated site (or petroleum product-contaminated site) as a potential bioremediation site is a detailed waste classification and a determination of the origin of the waste assist to assist in planning the removal activities.

Typically, an oil refinery waste consists of both Oil Pollution Act (OPA) and Comprehensive Environmental Response and Liability Act (CERCLA) wastes (oily pits from crude oil, refined products, tank bottoms, asbestos, corrosives, and small laboratory containers), using Oil Pollution Act or wastewater treatment wastes, as well as waste listed under the Resource Conservation and Recovery Act, and a careful waste classification is required to decide on the appropriate cleanup technology to remediate the site. Once this has been completed, representative samples of the land treatment unit (LTU) soil/waste should be collected and analyzed for, but

Table 9.2 Analytical methods for characterization of contaminated soil (US EPA, 2003).

Analyte	Method	Target Compound	Pro/Con/ Remark
TPH	EPA 418.1 (infrared), EPA 413.1 (gravimetric)	Mineral oil measurement. Gravimetric oil and grease.	Inexpensive and quick screening tool. Cannot be used to identify oil.
	Modified EPA SW846 8015B (GC/FID)	Total petroleum hydrocarbons & extractable hydrocarbons.	Hydro carbon quantification, basic product identification.
	Modified EPA SWS46 8015B (GC/FID)	C_8 to C_{40} normal and branched alkanes.	To determine weathering state and level of biodegradotion.
VOCs	Modified EPA SW846 8260B (GC/MS)	C_5 to C_{12}: analysis, gasoline additives.	Light product identification and degree of weathering.
Semi- volatiles	Modified EPA SW846 8270C (GC/MS) For PAH only, EPA SW846 8310 (HPLC)	3270C: semi-volatile compounds including parent and alkyl-substituted PAHs 8310: PAH	Quantification of all semi-volatile compounds, fingerprint inforniation. and long-term, weathering; expensive.

(Continued)

Table 9.2 Analytical methods for characterization of contaminated soil (US EPA, 2003).

Analyte	Method	Target Compound	Pro/Con/ Remark
Metals	Total EPA SW846 6010B Mercury SW846 7470A (liquids) and 7471A (solids)	6010B: antimony, arsenic, barium, beryllium, cadmium. chromium, cobalt, copper, lead, lithium. nickel, selenium, silver, thallium, vanadium, and zinc. 7470A & 7471A for mercury.	Quantification of all metals contained in soil; cost depends on the number of metals that are analyzed.
	TCLP using extraction method 1311	To test metals that may be a hazard to the environment.	Provides information on "stability" of metals in soil.
NORM	Direct-reading instrument to measure effective dose (Sievert)	Provides assessment of low levels of radiation.	Direct reading method: quick, inexpensive, does not identify the nature of the isotope.
	Laboratory analysis to measure concentration or activity (Bequerel)	Provides assessment of low levels of radiation	Laboratory analysis: expensive, provides accurate quantitative isotope characterization.

SW846 Methods are Available at http://www.epa.gov/epaoswer/hazwaste/test/main.htm

not necessarily limited to, volatile organic compounds (VOCs), polynuclear aromatic hydrocarbons (PAH), total petroleum hydrocarbons (TPH), metals, and naturally occurring radioactive material (NORM), in order to determine the most appropriate method for cleanup (Table 9.2).

4.1 General Application

Bioremediation products have been applied to clean up petroleum hydrocarbon contamination in various ecosystems and under a wide range of environmental conditions. Their applications include in-situ remediation of hydrocarbon contaminated marine shorelines, soil environments, surface water, groundwater, and water, and ex-situ treatment of hydrocarbon contaminated soil (e.g., use of land treatment units or other types of reactor systems such as compost piles, biopiles, slurry reactors, etc.) and water (e.g., in a bioreactor).

Bioremediation technology is typically used as a secondary polishing step after conventional mechanical cleanup options have been applied to remove free oil product. However, many case studies have demonstrated that bioremediation can also be used as a primary response strategy, especially for the cleanup of environmentally sensitive areas that are not amenable to conventional cleanup techniques and/or low-level petroleum hydrocarbon contamination.

Bioaugmentation appears to have little benefit for the treatment of spilled oil in an open environment. Microbial addition has not been shown to work better than nutrient addition alone in many field trials. However, case studies provided by vendors seem to suggest that application of bioaugmentation products could still have some potential in the treatment of specific oil components, isolated spills in confined areas, or certain environments where oil-degrading microorganisms are deficient. Unfortunately, the evidence for such a conclusion is not strong, and in most cases, is scientifically deficient.

Biostimulation has been proven to be a promising tool to treat certain aerobic oil-contaminated marine shorelines. One of the key factors for the success of oil biostimulation is to maintain an optimal nutrient level in the interstitial pore water. In other words, background nutrient concentrations at the contaminated site should be one of the primary determining factors in the decision to apply nutrients, and biostimulation might not always be necessary if sufficient nutrients are naturally present at a spill site to supply non-limiting concentrations to the degrading populations. However, effects of nutrients are also highly site-specific. For example, the availability of oxygen rather than nutrients is often the limiting factor in wetland environments, where addition of nutrient products has not been successful in enhancing oil biodegradation (although it has been successful in accelerating the restoration of the affected plant biomass to an abundant and rich recovery).

Different nutrient products have shown variable effectiveness, depending on oil properties, the nature of the nutrient products, and the characteristics of the contaminated environments. Based on limited field trials, it appears that slow-release fertilizers may be an excellent choice if the nutrient release rates are balanced against physical loss rates, water-soluble fertilizers may be more cost-effective in low-energy shorelines and fine-grained sediments where water transport is limited, and oleophilic fertilizers may be more suitable for use on hard, rocky shorelines, although further research is still required to confirm this suggestion. In general, commercial oleophilic nutrient products have not shown clear advantages over common agricultural fertilizers in stimulating oil biodegradation.

Bioremedial approaches may have a role in treating hydrocarbon contamination for non-point sources. Limited available information appears to suggest that application of bioremediation agents could show promise for the treatment of hydrocarbon contamination in stormwater, especially used in conjunction with other stormwater countermeasures, such as wet detention ponds. Bioremediation agents may also be effective for the treatment of bilge water, although, due to the lack of any systematic investigation into its effectiveness, it is still uncertain whether this approach could compete with other existing technologies. Further field tests are needed to provide stronger evidence on the potential of this strategy.

The extreme uncertainty associated with the efficacy of bioremediation agents is due in large part to the poorly designed field tests that have been conducted to demonstrate efficacy. Much of the reported literature lacked proper controls and treatment randomization and replication, or the data were incorrectly analyzed. If there is any hope for advancement of commercial bioremediation for the environments described in this report, especially estuaries, experiments based on sound scientific principles are needed. Unfortunately, resources for field-testing commercial bioremediation agents are scarce, and field studies are extremely expensive to carry out. That is why it is best to rely on laboratory microcosm or mesocosm studies to provide needed data to support this technology. When spills occur and the on-scene coordinator, in conjunction with the Regional Response Team, decides to implement commercial bioremediation for cleanup, they should try to set aside control areas if at all possible to allow a more effective evaluation of treatment success (Zhu et al., 2004).

Generally, indigenous hydrocarbon-degrading marine micro-organisms promote oil dispersion and biodegradation of crude oil, provided that sufficient nutrients are available for microbial growth. However, this may be a relatively slow process, depending on the nature of the oil tested. The application of dispersants to an oil slick at sea can have a dual environmental benefit, stimulating both oil dispersion and oil biodegradation (Mulyono et al., 1994; Varadaraj et al., 1994).

The success of a dispersant at stimulating oil biodegradation is dependent on the ability of the dispersant to promote the growth of indigenous hydrocarbon-degrading microorganisms. The presence of suspended clay particles substantially increases the rate of oil dispersion and subsequent biodegradation, presumably by providing surfaces on which the microorganisms can grow and maintain higher populations than that observed in seawater in the absence of clay (Lunel et al., 1995; Lee et al., 1997a; Bragg and Yang, 1995). These mixtures are neutrally buoyant, and may promote the decomposition of oil residues.

Since the contaminants of concern in crude oil are readily biodegradable under appropriate conditions, the success of oil-spill bioremediation depends on our ability to establish those conditions in the contaminated environment. The most important requirement is that bacteria with appropriate metabolic capabilities must be present. If they are, their rates of growth and hydrocarbon biodegradation can be maximized by ensuring that adequate concentrations of nutrients and oxygen are present, and that the pH is between about 6 and 9. The physical and chemical characteristics of the oil are also important determinants of bioremediation success. Heavy crude oils that contain large amounts of resin and asphaltene compounds are less amenable to bioremediation than are light-or medium-weight crude oils that are rich in aliphatic components. Finally, the oil surface area is extremely important, because growth of oil degraders occurs almost exclusively at the oil-water interface (Atlas and Bartha, 1972).

Obviously, some of these factors can be manipulated more easily than others. For example, nothing can be done about the chemical composition of the oil, and no adequate engineering approaches are currently available for providing oxygen to oil-contaminated surficial sediments in the intertidal zone. Therefore, the two main approaches to oil-spill bioremediation are: (1) *bioaugmentation*, in which oil-degrading bacteria are added to supplement the existing microbial population, and (2) *biostimulation*, in which nutrients or other growth-limiting co-substrates are added to stimulate the growth of indigenous oil degraders. Since oil-degrading bacteria usually grow at the expense of one or more components of crude oil, and these organisms are ubiquitous (Mulkins-Phillips and Stewart, 1974b; Roubal and Atlas, 1978; Atlas, 1981), there is usually no reason to add hydrocarbon degraders unless the indigenous bacteria are incapable of degrading one or more important contaminants. The size of the hydrocarbon-degrading bacterial population usually increases rapidly in response to oil contamination, and it is very difficult, if not impossible, to increase the microbial population over that which can be achieved by biostimulation alone (Westlake et al., 1978; Lee et al., 1997b).

The carrying capacity of most environments is probably determined by factors such as predation by protozoans, the oil surface area, or scouring of attached biomass by wave activity that are not affected by bioaugmentation, and added bacteria seem to compete poorly with the indigenous population (Tagger et al., 1983; Lee and Levy, 1989). Therefore, it is unlikely that they will persist in a contaminated beach, even when they are added in high numbers. As a result, bioaugmentation has never been shown to have any long-term beneficial effects in shoreline cleanup operations.

Biostimulation involves the addition of rate-limiting nutrients to accelerate biodegradation by indigenous microorganisms. When an oil spill occurs, it results in a huge influx of carbon into the impacted environment. Carbon is the basic structural component of living matter, and in order for the indigenous microorganisms to be able to convert this carbon into more biomass, they need significantly more nitrogen and phosphorus than is normally present in the environment. Both of these elements are essential ingredients of protein and nucleic acids of living organisms. The main challenge associated with biostimulation in oil-contaminated coastal areas or tidally influenced freshwater rivers and streams is maintaining optimal nutrient concentrations in contact with the oil.

5 References

Aislabie, J.M., McLeod, M., and Fraser, R. 1998. "Potential for Biodegradation of Hydrocarbons in Soil from the Ross Dependency, Antarctica." *Appl. Microbiol. Biotechnol.*, 49: 210–214.

Aislabie, J.M., Foght, J., and Saul, D. 2000. "Aromatic Hydrocarbon-degrading Bacteria from Soil near Scott base, Antarctica." *Polar Biol.*, 23: 183–188.

Aislabie, J.M., Balks, M.R., Foght, J.M., and Waterhouse, E.J. 2004. "Hydrocarbon Spills on Antarctic Soils: Effects and Management." *Environ. Sci. Technol.*, 38(5): 1265–74.

Al-Darbi, M.M., Saeed, N.O., Islam, M.R., and Lee, K. 2005. "Biodegradation of Natural Oils in Seawater." *Energy Sources*, 27:19–34.

Alexander, M. 1994. *Biodegradation and Bioremediation*. Academic Press Inc., New York.

Amund, O.O., Adewale, A.A., and Ugoji, E.O. 1987. "Occurrence and Characterization Of Hydrocarbon Utilizing Bacteria In Nigerian Soils Contaminated with spent motor oil." *Indian J. Microbiol.*, 27: 63–87.

Antai, S.P. 1990. "Biodegradation of Bonny light crude oil by Bacillus sp and Pseudomonas sp." *Waste Management*, 10: 61–64.

Antić, M.P., S.Jovančićević, B.S., Ilić, M., Vrvić, M.M., and Schwarzbauer, J. 2006. "Petroleum Pollutant Degradation by Surface Water Microorganisms." *Environ. Sci. Pollut. Res.*, 13(5): 320–327.

Asia, I.O., Enweani, I.B., and Eguavoen, I.O. 2006. "Characterization and Treatment of Sludge from the Petroleum Industry." *African Journal of Biotechnology*, 5(5): 461–466.

Atlas, R.M., and Bartha, R. 1972. "Degradation and Mineralization of Petroleum in Sea Water: Limitation by Nitrogen And Phosphorus." *Biotech. Bioeng.*, 14: 309–318.

Atlas, R.W. 1975. "Effects of Temperature and Crude Oil Composition on Petroleum Biodegradation." *Applied Microbiology*, 30(3): 396–403.

Atlas, R.M. 1978. "An Assessment of the Biodegradation of Petroleum in the Arctic." In *Microbial Ecology*. M.W. Loutit and J.A.R. Miles (Editors). Springer-Verlag, Berlin. pp. 86–90.

Atlas, R.M. 1981. "Microbial Degradation of Petroleum Hydrocarbons: An Environmental Perspective." *Microbiological Rev.*, 45: 180–209.

Atlas, R.M. 1991. "Microbial Hydrocarbon Degradation – Bioremediation of oil spills." *J. Chem. Technol.-Biotechenol.*, 52: 149–156.

Atlas, R.M., and Bartha, R. 1992. "Hydrocarbon Biodegradation and Oil Spill Bioremediation." *Adv. Microbial Ecol.*, 12: 287–338.

Atlas, R.M. 1995. "Petroleum Biodegradation and Oil Spill Bioremediation." *Mar. Pollut. Bull.*, 31: 178–182.

Bagherzadeh-Namazi, A., Shojaosadati, S.A., and Hashemi-Najafabadi, S. 2008. "Biodegradation of used engine oil using Mixed and Isolated Cultures." *Int. J. Environ. Res.*, **2**, 431–440.

Bamforth, S.M., and Singleton, I. 2005. "Bioremediation of Polycyclic Aromatic Hydrocarbons: Current Knowledge and Future Directions." *J. Chem. Technol. Biotechnol.*, 80: 723–736.

Bardi, L., Ricci, R., and Marzona, M. 2003. "In Situ Bioremediation of a Hydrocarbon Polluted Site with Cyclodextrin as a Coadjuvant to Increase Bioavailability." *Water, Air, And Soil Pollution: Focus*, 3: 15–23.

Bardi, L., Martini, C., Opsi, F., Bertolone, E., Belviso, S., Masoero, G., Marzona, M., and Ajmone Marsan, F. 2007. "Cyclodextrin-Enhanced in Situ Bioremediation of Polyaromatic Hydrocarbons-Contaminated Soils and Plant Uptake." *J. Incl. Phenom. Macrocycl. Chem.*, 57: 439–444.

Barnard, P.C., and Bastow, M.A. 1991. "Hydrocarbon generation, migration, alteration, entrapment and mixing in the Central and Northern North Sea." In *Petroleum Migration*. W.A. England, and A.J. Fleet (Editors). Special Publication. Geological Society, London, united Kingdom. pp. 167–190.

Bartha, R. 1986. *Microbial Ecology: Fundamentals and Applications*. Addisson-Wesley Publishers, Reading, Massachusetts.

Bechok, M.R. 1967. *Aqueous Wastes from Petroleum and Petrochemical Plants*. John Wiley & Sons Inc., New York.

Bej, A.K., Saul, D., and Aislabie, J. 2000. "Cold-tolerant Alkane-degrading *Rhodococcus* Species from Antarctica." *Polar Biol.*, 23: 100–105.

Bence, A.E., Kvenvolden, K.A., and Kennicutt, M.C. 1996. "Organic Geochemistry Applied to Environmental Assessments of Prince William Sound, Alaska, after the Exxon Valdez Oil Spill – A Review." *Organic Geochemistry*, 24: 7–42.

Bennett, P.C., Siegel, D.E., Baedecker, M.J., and Hult, M.F. 1993. "Crude Oil in a Shallow Sand and Gravel Aquifer 1. Hydrogeology and Inorganic Geochemistry." *Applied Geochemistry*, 8: 529–549.

Bertrand, J.C., Almallah, M., Acquaviva, M., and Mille, G. 1990. "Biodegradation of Hydrocarbons by an Extremely Halophilic Archaebacterium." *Lett. Appl. Microbiol.*, 11: 260–263.

Bestetti, G., Collina, E., Di Gennaro, P., Lasagni, M., and Pitea, D. 2003. "Kinetic Study of Naphthalene Biodegradation in Aerobic Slurry Phase Microcosms for the Optimization of the Process." *Water, Air, And Soil Pollution: Focus*, 3: 223–231.

Björklöf, K., Salminen, J., Sainio, P., and Jørgensen, K. 2008. "Degradation rates of aged petroleum hydrocarbons are likely to be mass transfer dependent in the Field." *Environ Geochem Health*, 30: 101–107.

Borchert, S., Mueller, J., Alesi E., Leins, C., and Haninger, V. 1995. "In Situ Bioremediation Application Strategies for Soil and Groundwater Impacted by PAHs." *Land Contam. Reclam.*, 3: 6–1–6–4.

Bossert, I., and Bartha, R. 1984. "The Fate of Petroleum in Soil Ecosystems." In *Petroleum Microbiology*. R.M. Atlas (Editor). Macmillan, New York. pp. 453–473.

Bragg, J.R., and Yang, S.H. 1995. "Clay-Oil Flocculation and its Role in Natural Cleansing in Prince William Sound Following the Exxon Valdez Oil Spill." In *Exxon Valdez Oil Spill: Fate and Effects in Alaskan Waters*. ASTM STP No. 1219. P.G. Wells, J.N. Butler, and J.S. Hughes (Editors). American Society for Testing and Materials, West Conshohocken, Pennsylvania.

Braddock, J.F., and McCarthy, K.A. 1996. "Hydrologic and Microbiological Factors Affecting Persistence and Migration of Petroleum Hydrocarbons Spilled in a Continuous-Permafrost Region." *Environ Sci. Technol.*, 30: 2626–2633.

Braddock, J.F., Lindstrom, J.E., Yeager, T.R., Rasley, B.T., and Brown, E.J. 1996. "Patterns of Microbial Activity in Oiled and Unoiled Sediments in Prince Williams Sound." *Proceedings*. Am. Fish Soc. Symp., 18: 94–108.

Braddock, J.F., Ruth, M.L., Walworth, J.L., and McCarthy, K.A. 1997. "Enhancement and Inhibition of Microbial Activity in Hydrocarbon-Contaminated Arctic Soils: Implications for Utrientamended Bioremediation." *Environ Sci. Technol.*, 31: 2078–2084.

Bradley, P.M., and Chapelle, F.H. 1995. "Rapid Toluene Mineralization by Aquifer Microorganisms at Adak, Alaska: Implications for Intrinsic

Bioremediation in Cold Environments." *Environ Sci. Technol.*, 29: 2778–2781.

Breedveld, G.D., and Sparrevik, M. 2000. "Nutrient-Limited Biodegradation of PAH in Various Soil Strata at a Creosote Contaminated Site." *Biodegradation*, 11: 391–399.

Bruheim, P., and Eimhjelle, K. 1998. "Chemically Emulsified Crude Oil as Substrate for Bacterial Oxidation: Differences In Species Response." *Can. J. Microb.*, 44(2): 195–199.

Carmichael, L.M., and Pfaender, F.K. 1997. "The Effect of Inorganic and Organic Supplements on The Microbial Degradation Of Phenanthrene And Pyrene In Soils." *Biodegradation*, 8: 1–13.

Cerniglia, C.E. 1984. "Microbial Metabolism of Polycyclic Aromatic Hydrocarbons." *Adv. Appl. Microbiol.*, 30: 31–71.

Cerniglia, C.E. 1992. "Biodegradation of Polycyclic Aromatic Hydrocarbons." *Biodegradation*, 3: 351–368.

Cerniglia, C.E., and Heitkamp, M.A. 1989. "Microbial Degradation of Polycyclic Aromatic Hydrocarbons in the Aquatic Environment." In *Metabolism of Polycyclic Aromatic Hydrocarbons in the Aquatic Environment*. U. Varanasi (Editor). CRC Press, Inc., Boca Raton, Florida. pp. 41–68.

Cerniglia, C.E., and Heitkamp, M.A. 1990. "Polycyclic Aromatic Hydrocarbon Degradation by *Mycobacterium*." *Methods Enzymol.*, 188: 148–153.

Chaîneau, C.H., Rougeux, G., Yepremian, C., and Oudot, J. 2005. "Effects of Nutrient Concentration on the Biodegradation of Crude Oil and Associated Microbial Populations in the Soil." *Soil Biol. Biochem.*, 37: 1490–1497.

Chen, C.I., and Taylor, R.T. 1995. "Thermophilic Biodegradation of BTEX by Two *Thermus* Species." *Biotechnol. Bioeng.*, 48: 614–624.

Chen, C.I., and Taylor, R.T. 1997a. "Batch and Fed-Batch Bioreactor Cultivations of a *Thermus* Species with Thermophilic BTEX-degrading Activity." *Appl. Microbiol. Biotechnol.*, 47: 726–733.

Chen, C.I., and Taylor, R.T. 1997b. "Thermophilic Biodegradation of BTEX by Two Consortia of Anaerobic Bacteria." *Appl. Microbiol. Biotechnol.*, 48: 121–128.

Chernatskaya, N. 1974. "Modern Methods for Determination of Contaminants in Refinery Waste Waters." *Chemistry and Technology of Fuels and Oils*, 10(9): 692–695.

Choi, S-C., Kwon, K.K., Sohn, J.H., and Kim, S-J. 2002. "Evaluation of Fertilizer Additions to Stimulate Oil Biodegradation in Sand Seashore Mescocosms." *J. Microbiol. Biotechnol.*, 12: 431–436.

Chugunov, V.A., Ermolenko, Z.M., Martovetskaya, I.I., Mironava, R.I., Zhirkova, N.A., Kholodenko, V.P., and Urakov, N.N. 2000.

"Development and Application Of A Liquid Preparation With Oil-Oxidizing Bacteria." *Appl. Biochem. Microbiol.*, 36: 577–581.

Cirigliano, M.C., and Carman, G.M. 1984. "Isolation of a Bioemulsifier from *Candida lipolytica*." *Appl. Environ. Microb.*, 48: 747–750.

Clements, W.H., Oris, J.T., and Wissing, T.E. 1994. "Accumulation and Food Chain Transfer of Fluoranthene and Benzo[*a*]pyrene in *Chironomus riparius* and *Lepomis macrochirus*." *Arch. Environ. Contam. Toxicol.*, 26: 261–266.

Connan, J., Restle, A., and Albrecht, P. 1980. "Biodegradation of Crude Oils in the Aquitaine Basin." In *Advances in Organic Geochemistry*, 1979. A.G. Douglas and J.R. Maxwell (Editors). Pergamon Press, Oxford, United Kingdom. pp. 1–17.

Connan, J. 1981. "Un Exemple De Biodegradation Preferentielle des Hydrocarbures Aromatique dans des Asphaltes du Bassin Sud-Aquitain (France)." *Bull Cent Rech Explor-Prod Elf Aquitaine*, 5: 151–171.

Connan, J. 1984. "Biodegradation of crude oils in reservoirs." In *Advances in Petroleum Geochemistry*. J. Brooks, and D.H. Welte (Editors). Academic Press, London, United Kingdom. Volume 1, pp. 299–335.

Dandie, C.E., Weber, J., Aleer, S., Adetutu, E.M., Ball, A.S., and Juhasz, A.L. 2010. "Assessment of Five Bioaccessibility Assays for Predicting the Efficacy of Petroleum Hydrocarbon Biodegradation in Aged Contaminated Soils." *Chemosphere*, 81: 1061–1068.

Das, K., and Mukherjee, A.K. 2007. "Crude Petroleum-Oil Biodegradation Efficiency of Bacillus Subtilis and *Pseudomonas eruginosa* Strains Isolated from a Petroleum-Oil Contaminated Soil from North-East India." *Bioresource Technol.*, 98: 1339–1345.

Davies, L., Daniel, F., Swannell, R., and Braddock, J. 2001. Biodegradability of Chemically-Dispersed Oil. Report No. AEAT/ENV/R0421. AEA Technology Environment, Abingdon, Oxfordshire, United Kingdom.

Del'Arco, J.P., and de França, F.P. 1999. "Biodegradation of Crude Oil in a Sandy Sediment." *International Biodeterioration and Biodegradation*, 44: 87–92.

Delille, D., Bassères, A., and Dessommess, A. 1998. "Effectiveness of Bioremediation for Oil-Polluted Antarctic Seawater." *Polar Biol.*, 19: 237–241.

Delille, D., Pelletier, E., Rodriguez-Blanco, A. and Ghiglione, J. 2009. "Effects of Nutrient and Temperature on Degradation of Petroleum Hydrocarbons in Sub-Antarctic Coastal Seawater." *Polar Biol.*, 32: 1521–1528.

Delille, D., and Delille, B. 2000. "Field Observations on the Variability of Crude Oil Impact in Indigenous Hydrocarbon-Degrading Bacteria from Sub-Antarctic Intertidal Sediments." *Mar. Environ. Res.*, 49: 403–417.

Delille, D., Coulon, F., and Pelletier, E. 2004. "Effects of Temperature Warming During a Bioremediation Study of Natural and Nutrient-Amended Hydrocarbon-Contaminated Sub-Antarctic Soils." *Cold Reg. Sci. Tech.*, 40: 61–70.

Deng, S., Yu, G., Jiang, Z., Zhang, R., and Ting, Y. 2005. "Destabilization of Oil Droplets in Produced Water from ASP Flooding." *Colloids Surf. A.*, 252: 113–119.

Deriszadeh, A., Husein, M.M., and Harding, T.G. 2010. "Produced Water Treatment by Micellar-Enhanced Ultrafiltration." *Environ. Sci. Technol.*, 44: 1767–1772.

Dibble, J.T., and Bartha, R. 1976. "The Effect of Iron on the Biodegradation of Petroleum in Seawater." *Appl. Environ. Microb.*, 31: 544–550.

Eigenson, S., Loakimis, E.G., and Lukinskayal, N.T. 1974. "Prospects for Development in Water Usage and Water Discharge In Petroleum Refineries And Petrochemical Plants." *Chemistry and Technology of Fuels and Oils*, 10(9): 665–667.

Eja, M.E., Arikpo, G.E., and Udo, S.M. 2006. "The Bioremediation Potentials of Fungal Species Isolated from Soils Polluted by Petroleum Products in Cross River University of Technology, Calabar, Nigeria." *International Journal of Natural and Applied Sciences (IJNAS)*, 1(1): 15–20.

Eriksson, M., Dalhammar, G., and Borg-Karlson, A.K. 2000. "Biological Degradation of Selected Hydrocarbons in an Old PAH/Creosote Contaminated Soil from a Gas Work Site." *Appl. Microbiol. Biotechnol.*, 53: 619–626.

Fakhru'l-Razi, A., Pendashteh, A., Abdullah, L.C., Biak, D.R.A., Madaeni, S.S., and Abidin, Z.Z. 2009. "Review of Technologies for Oil and Gas Produced Water Treatment." *J. Hazard. Mater.*, 170: 530–551.

Floodgate, G.D. 1984. "The Fate of Petroleum in Marine Ecosystems." In R.M. Atlas, *Petroleum Microbiology*. MacMillan, New York. pp. 355–397.

Foght, J.M., and Westlake, D.W.S. 1982. "Effect of the Dispersant Corexit 9527 on the Microbial Degradation of Prudhoe Bay Oil." *Canadian Journal of Microbiology*, 28: 117–122.

Foght, J.M., Gutnick, D.L., and Westlake, D.W.S. 1989. "Effect of Emulsan on Biodegradation of Crude Oil by Pure and Mixed Bacterial Cultures." *Applied and Environmental Microbiology*, January: 36–42.

Foght, J.M., Westlake, D., Johnson, W.M., and Ridgway, H.F. 1996. "Environmental Gasoline-Utilizing Isolates and Clinical Isolates of *Pseudomonas aeruginosa* are Taxonomically Indistinguishable by Chemotaxonomic and Molecular Techniques." *Microbiol.*, 142: 1333–1338.

Foght, J., Semple, K., Gauthier, C., Westlake, D.S., Blenkinsopp, S., Sergy, G., Wang, Z., and Fingas, M. 1999. "Effect of nitrogen Source on Biodegradation of Crude Oil by a Defined Bacterial Consortium

Incubated under Cold, Marine Conditions." *Environ Technol.*, 20: 839–849.

Freijer, J.I., de Jonge, H., Bouten, W., and Verstraten, J.M. 1996. "Assessing Mineralization Rates of Petroleum Hydrocarbons in Soils in Relation to Environmental Factors and Experimental Scale." *Biodegradation* 7: 487–500.

Frenzel, M., James, P., Burton, S.K., Rowland, S.J., and Lappin-Scott, H.M. "Towards Bioremediation of Toxic Unresolved Complex Mixtures of Hydrocarbons: Identification of Bacteria Capable of Rapid Degradation Of Alkyltetralins." *J. Soils Sediments*, 9:129–136.

Gallego, J.L., García, M.J., Llamas, J.F., Belloch, C., Pelaez, A.I., and Sanchez, J. 2007. "Biodegradation of Oil Tank Botton Sludge using Microbial Consortia." *Biodegradation*, 18: 269–281.

Ghazali, F.M., Rahman, R.N.Z.A., Salleh, A.B., and Basri, M. (2004). "Biodegradation of Hydrocarbons in Soil By Microbial Consortium." *Intl. Biodeter. Biodegrad.*, 54: 61–67.

Gibson, D.T., and Subramanian, V. 1984. "Microbial Degradation of Aromatic Hydrocarbons." In *Microbial Degradation of Organic Compounds*. D.T. Gibson (Editor). Marcel Dekker, Inc., New York, N.Y. pp. 181–252.

Gieg, L.M., Kolhatkar, R.V., McInerney, M.J., Tanner, R.S., Harris, S.H. Jr., Sublette, K.L., and Suflita, J.M. 1999. "Intrinsic Bioremediation of Petroleum Hydrocarbons in a Gas Condensate-Contaminated Aquifer." *Environ Sci. Technol.*, 33: 2550–2560.

Goldman, S., Shabtai, Y., Rubinovitz, C., Rosenberg, E., and Gutnick, D.L. 1982. "Emulsan in *Acinetobacter calc oaceticus* RAG-1: Distribution of Cell-Free and Cell-Associated Cross-Reacting Material." *Appl. Environ. Microbiol.*, 44: 165–170.

Goodin, J.D., and Webber, M.D. 1995. "Persistence and Fate of Anthracene and Benzo[*a*]pyrene In Municipal Sludge Treated Soil." *J. Environ. Qual.*, 24: 271–278.

Goodwin, N.S., Park, P.J.D., and Rawlinson, A.P. 1983. "Crude Oil Biodegradation under Simulated and Natural Conditions." In *Advances in Organic Geochemistry*. 1981. M. Bjorøy (Editor). John Wiley & Sons, Inc. New York. pp. 650–658.

Greenwood, P.F., Wibrow, S., George, S.J., and Tibbett, M. 2009. "Hydrocarbon Biodegradation and Soil Microbial Community Response to Repeated Oil Exposure." *Organic Geochemistry*, 40: 293–300.

Gutnick, D.L., and Minas, W. 1987. "Perspectives on Microbial Surfactants." *Biochem. Soc. Trans.*, 15: 22S-35S.

Haferburg, D., Hommel, R., Claus, R., Kleber, H.P. 1986. "Extracellular Microbial Lipids as Biosurfactants." *Adv. Biochem. Eng. Biotech.*, 33: 53–93.

Harvey, R.G. 1997. *Polycyclic Aromatic Hydrocarbons*. John Wiley & Sons, Inc., VCH, New York.

Head, I.M., Jones, D.M., and Röling, W.F.M. 2006. "Marine Microorganisms make a Meal of Oil." *Nature Reviews Microbiology*, 4: 173–182.

Heitkamp, M.A., and Cerniglia, C.E. 1988. "Mineralization of Polycyclic Aromatic Hydrocarbons by a Bacterium Isolated from Sediment below an Oil Field." *Applied and Environmental Microbiology*, 54: 1612–1614.

Heitkamp, M.A., and Cerniglia, C.E. 1989. "Polycyclic Aromatic Hydrocarbon Degradation by a *Mycobacterium* Sp. In Microcosms Containing Sediment and Water from a Pristine Ecosystem." *Appl. Environ. Microbiol.*, 55: 1968–1973.

Hernáez, M.J., Reineke, W., and Santero, E. 1999. "Genetic Analysis of Biodegradation of Tetralin by a *Sphingomonas* Strain." *Appl. Environ. Microbiol.*, 65(4): 1806–1810.

Holba, A.G., Dzou, I.L., Hickey, J.J., Franks, S.G., May, S.J., and Lenney, T. 1996. "Reservoir Geochemistry of South Pass 61 Field, Gulf of Mexico: Compositional Heterogeneities Reflecting Filling History and Biodegradation." *Organic Geochemistry*, 24: 1179–1198.

Huang, L., Ma, T., Li, D., Liang, F., Liu, R.-., and Li, G. 2008. "Optimization of Nutrient Component for Diesel Oil Degradation by *Rhodococcus erythropolis*." *Mar. Pollut. Bull*, 56: 1714–1718.

Huesemann, M.H., Hausmann, T.S., and Fortman, T.J. 2004. "Does Bioavailability Limit Biodegradation? A Comparison of Hydrocarbon Biodegradation and Desorption Rates in Aged Soils." *Biodegradation*, 15: 261–274.

Ilori, M. O., Adebusoye, S.A., Obayori, O.S., Oyetibo, G.O., Ajidahun, O., James, C., and Amund, O.O. 2011. "Extensive Biodegradation of Nigerian Crude Oil (Escravos Light) by Newly Characterized Yeast Strains." *Pet. Sci. Technol.*, 29: 2191–2208.

Jacquot, F., Guiliano, M., Doumenq, P., Munoz, D., and Mille, G. 1996. "In Vitro Photo-oxidation Of Crude Oil Maltenic Fractions: Evolution of Fossil Biomarkers and Polycyclic Aromatic Hydrocarbons." *Chemosph.*, 33: 671–681.

Jean, J.M., Lee, Sh., Wang, Chattopadhyay, P. and Maity, J. 2008. "Effects of Inorganic Nutrient Levels on the Biodegradation of Benzene, Toluene, and Xylene (BTX) by Pseudomonas Spp in a Laboratory Porous Media Sand Aquifer Model." *Bioresource Technol.*, 99: 7807–7815.

Jones, D.M., Douglas, A.G., Parkes, R.J., Taylor, J., Giger, W., and Schaffner, G. 1983. "The Recognition Of Biodegraded Petroleum-Derived Aromatic Hydrocarbons In Recent Marine Sediments." *Mar Poll Bull*, 14: 103–108.

Joshi, P.A., and Pandey, G.B. 2011. "Screening of Petroleum Degrading Bacteria from Cow Dung." *Research Journal of Agricultural Sciences*, 2(1): 69–71.

Kanal, R.A., and Harayama, S. 2000. "Biodegradation of High-Molecular-Weight Polycyclic Aromatic Hydrocarbons by Bacteria." *Journal of Bacteriology*, 182(8): 2059–2067.

Kim, S., Choi, D.H., Sim, D.S., and Oh, Y. 2005. "Evaluation of Bioremediation Effectiveness on Crude Oil-Contaminated Sand." *Chemosph.*, 59: 845–852.

Kim, M., Bae, S.S., Seol, M., Lee, J., and Oh, Y. 2008. "Monitoring Nutrient Impact on Bacterial Community Composition during Bioremediation of Anoxic PAH-Contaminated Sediment." *J. Microbiol*, 46: 615–623.

Knezevich, V.O., Koren, E., Ron, Z. and Rosenberg, E. 2006. "Petroleum Bioremediation in Seawater Using Guano as the Fertilizer." *Bioremediation J.*, 10: 83–91.

Kolhatkar, A., Sublette, K. L. 1996. "Biotreatment of Refinery Spent Sulfidic Caustic by Specialized Cultures and Acclimated Activated Sludge." *Appl. Biochem. Biotechnol.*, 58(57): 945–957.

Kwok, C-K., and Loh, K-C. 2003. "Effects of Singapore Soil Type on the Bioavailability of Nutrients in Soil Bioremediation." *Advance of Environmental Research*, 7: 889–900.

Leahy, J.G., and Colwell, R.R. 1990. "Microbial Degradation of Hydrocarbons in the Environment." *Microbiol. Rev.*, 54: 305–315.

Lee, K., Wong, C.S., Cretney, W.J., Whitney, F.A., Parsons, T.R., Lalli, C.M., and Wu, J. 1985. "Microbial Response to Crude Oil and Corexit 9527: SEAFLUXES enclosure study." *Microbial Ecology*, 11: 337–351.

Lee, K., and Levy, E.M. 1989. "Enhancement of the Natural Biodegradation of Condensate and Crude Oil on Beaches of Atlantic Canada." *Proceedings*. 1989 Oil Spill Conference. American Petroleum Institute, Washington, DC. pp. 479–486.

Lee, K., Lunel, T., Wood, P., Swannell, R., and Stoffyn-Egli, P. 1997a. "Shoreline Cleanup by Acceleration of Clay-Oil Flocculation Processes." *Proceedings*. 1997 International Oil Spill Conference, American Petroleum Institute, Washiongton, DC. pp. 235–240.

Lee, K., Tremblay, G.H., Gauthier, J., Cobanli, S.E., and Griffin, M. 1997b. "Bioaugmentation and Biostimulation: A Paradox between Laboratory and Field Results." *Proceedings*. 1997 International Oil Spill Conference. American Petroleum Institute, Washington, DC. pp. 697–705.

Lee, E-H, and Kim, J., Cho, K-S., Ahn, Y.G., and Hwang, G-S. 2010. "Degradation of Hexane and Other Recalcitrant Hydrocarbons by a Novel Isolate, *Rhodococcus Sp.* EH831." *Environ Sci. Pollut. Res.*, 17: 64–77.

Lepo, J.E., and Cripe, C.R. 1999. "Microbial Biosystems: New Frontiers." *Proceedings*. 8th International Symposium on Microbial Ecology. C.R. Bell, M. Brylinsky, and P. Johnson-Green (Editors). Atlantic Canada Society for Microbial Ecology, Halifax, Nova Scotia, Canada.

Leys, M.N., Bastiaens, L., Verstraete, W., and Springael, D. 2005. "Influence of the carbon/nitrogen/phosphorus ratio on polycyclic aromatic hydrocarbons degradatiojn by *Mycobacterium* and *Sphingomonas* in soil." *Applied and Microbiology and Biotechnology*, 66: 726–736.

Li, J., Pignatello, J.J., Smets, B.F., Grasso, D., and Monserrate, E. 2005. "Bench-Scale Evaluation of In Situ Bioremediation Strategies for Soil at a Former Manufactured Gas Plant Site." *Environ. Toxicol. Chem.*, 24: 741–749.

Liebeg, E.W., and Cutright, T.J. 1999. "The Investigation of Enhanced Bioremediation through the Addition of Macro and Micro Nutrients in a PAH Contaminated Soil. Int. Biodeterior." *Biodegrad.*, 44: 55–64.

Lindstrom, J.E., and Braddock, J.F. 2002. "Biodegradation of Petroleum Hydrocarbons at Low Temperature in the Presence of the Dispersant Corexit 9500." *Mar. Poll. Bull.*, 44: 739–747.

Liu, Z., Jacobson, A.M., and Luthy, R.G. 1995. "Biodegradation of Naphthalene in Aqueous Nonionic Surfactant Systems." *Appl. Environ. Microb.*, 61: 45–151.

Litherathy, P., Haider, S., Samhan, O., and Morel, G. 1989. "Experimental Studies on Biological and Chemical Oxidation of Dispersed Oil in Seawater." *Water Science and Technology*, 21: 845–856.

Lu, M., Zhang, Z., Yu, W., and Zhu, W. (2009). "Biological Treatment of Oilfield-Produced Water: A Field Pilot Study." *Int. Biodeter. Biodegr.*, 63: 316–321.

Lu, M., and Wei, X. 2011. "Treatment of Oilfield Wastewater 211 Containing Polymer by the Batch Activated Sludge Reactor Combined with a Zerovalent Iron/EDTA/Air System." *Bioresour. Technol.*, 102: 2555–2562.

Lu, M., Wei, X., and Su, Y. 2011. "Aerobia Treatment of Oilfield Wastewater with a Bio-Contact Oxidation Reactor." *Desalination and Water Treatment*, 27: 334–340.

Lundstedt, S., Haglund, P., and Oberg, L. 2003. "Degradation and Formation of Polycyclic Aromatic Compounds during Bioslurry Treatment of an Aged Gasworks Soil." *Environ. Toxicol. Chem.*, 22: 1413–1420.

Lunel, T. "Understanding the Mechanism of Dispersion through Droplet Size Measurements at Sea." In *The Use of Chemicals in Oil Spill Response*. ASTM STP No. 1252. P. Lane (Editor). American Society for Testing and Materials, West Conshohocken, Pennsylvania.

Luning Prak, D.J., and Pritchard, P.H. 2002. "Solubilization of Polycyclic Aromatic Hydrocarbon Mixtures in Micellar Nonionic Surfactant Solution." Water *Research*, 36: 3463–3472.

Maila, M.P., and Cloete, T.E. 2004. "Bioremediation of Petroleum Hydrocarbons through Landfarming: Are Simplicity and Cost-Effectiveness the Only Advantages?" *Reviews in Environmental Science & Bio/Technology*, 3: 349–360.

Maki, H., Sasaki, T., and Haramaya, S. 2005. "Photo-oxidation of Biodegradable Crude Oil and Toxicity of the Photo-oxidized Products." *Chemosph.*, 44: 1145–1151.

Maletić, S., Dalmacija, B., Rončević, S., Agbaba, J., and Petrović, O. 2009. "Degradation Kinetics of an Aged Hydrocarbon-Contaminated Soil." *Water Air Soil Pollut.*, 202: 149–159.

Margesin, R., and Schinner, F. 1997a. "Efficiency of Indigenous and Inoculated Cold-Adapted Soil Microorganisms for Biodegradation of Diesel Oil in Alpine Soils." *Appl. Environ. Microbiol.*, 63: 2660–2664.

Margesin, R., and Schinner, F. 1997b. "Bioremediation of Diesel-Oil-Contaminated Alpine Soils at Low Temperatures." *Appl. Microbiol. Biotechnol.*, 47: 462–468.

Margesin, R., and Schinner, F. (Editors). 1999a. *Cold-adapted Organisms.* Springer, New York.

Margesin, R., and Schinner, F. 1999b. "Biological Decontamination of Oil Spills in Cold Environments." *J. Chem. Technol. Biotechnol.*, 74: 381–389.

Margesin, R. 2000. "Potential of Cold-Adapted Microorganisms for Bioremediation of Oil-Polluted Alpine Soils." *Int. Biodeterior. Biodegrad.*, 46: 3–10.

Margesin, R., and Schinner, F. 2001. "Potential of Halotolerant and Halophilic Microorganisms for Biotechnology." *Extremophiles*, 5: 73–83.

Marín, J.A., Moreno, J.L., Hernández, T., and García, C. 2006. "Bioremediation by Composting of Heavy Oil Refinery Sludge in Semiarid Conditions." *Biodegradation*, 17: 251–261.

Matveev, M.S. 1996. "Biological Purification of Waste Water from Oil Refineries." Chemistry and Technology of Fuels and Oils, 2(5): 329–332.

Mielke, J.E. 1990. *Oil in the Ocean: The Short- and Long-Term Impacts of a Spill.* Congressional Research Service, 90–356 SPR, July 24, p. 11.

Mihara, C., Yano, T., Kozaki, S., and Imamura, T. 1999. Microbial Strain, Method for Biodegrading Organic Compounds and Method for Environmental Remediation. United States Patent 5,962,305. October 5.

Miiller, D.E., Holba, A.G., and Huges, W.B. 1987. "Effects of biodegradation on crude oils." In *Exploration for Heavy Crude Oil and Natural Bitumen*. R.F. Meyer (Editor). AAPG Studies in Geology #25. American Association of Petroleum Geologists, Tulsa, Oklahoma. pp. 233–241.

Milić, J.S., Beškoski, V.P., Ilić, M.V. Ali, S.A.M., Gojgić-Cvijović, G.Đ., and Vrvić, M.M. 2009. "Bioremediation of Soil Heavily Contaminated with Crude Oil and its Products: Composition of the Microbial Consortium." *J. Serb. Chem. Soc.*, 74(4): 455–460.

Miller, E.C., and Miller, J.A. 1981. "Search for the Ultimate Chemical Carcinogens and Their Reaction with Cellular Macromolecules." *Cancer*, 47: 2327–2345.

Mills, M.A., Bonner, J.S., McDonald, T.J., Page, C.A., and Autenrieth, R.L. 2003. "Intrinsic Bioremediation of a Petroleum-Impacted Wetland." *Marine Pollution Bulletin*, 46: 887–899.

Minai-Tehrani, D., Minoui, S., and Herfatmanesh, A. 2009. "Effect of Salinity on Biodegradation of Polycyclic Aromatic Hydrocarbons

(PAHs) of Heavy Crude Oil in Soil." *Bull. Environ. Contam. Toxicol.*, 82: 179–184.

Mitsch, W.J., and Gosselink, J.G. 1993. *Wetlands*. 2nd Edition. John Wiley & Sons Inc., New York.

Mohajeri, L., Aziz, H.A., Isa, M.H., and Zahed, M.A. (2010). "A Statistical Experiment Design Approach for Optimizing Biodegradation of Weathered Crude Oil in Coastal Sediments." *Bioresource Technol.*, 101: 893–900.

Mohammed, D., Ramsubhag, A.S., and Beckles, D.M. 2007. "An Assessment of the Biodegradation of Petroleum Hydrocarbons in Contaminated Soil Using Non-indigenous, Commercial Microbes." *Water Air Soil Pollut.*, 182: 349–356.

Mohan, S.V., Kisa, T., Ohkuma, T., Kanaly, R.A., and Shimizu, Y. 2006. "Bioremediation Technologies for Treatment of PAH-Contaminated Soil and Strategies to Enhance Process Efficiency." *Rev. Environ. Sci. Biotechnol.*, 5: 347–374.

Mohn, W.W., and Stewart, G.R. 2000. "Limiting Factors for Hydrocarbon Biodegradation at Low Temperature in Arctic Soils." *Soil Biol. Biochem.*, 32: 1161–1172.

Mueller, J.G., Chapman, P.J., Blattmann, B.O., and Pritchard, P.H. 1990. "Isolation and Characterization of a Fluoranthene-Utilizing Strain Of *Pseudomonas paucimobilis*." *Appl. Environ. Microbiol.*, 56: 1079–1086.

Mulkins-Phillips, G.J., and Stewart, J.E. 1974a. "Effect of Environmental Parameters of Bacterial Degradation on Bunker C Oil, Crude Oil and Hydrocarbons." *Applied and Environmental Microbiology*, 28: 547–552.

Mulkins-Phillips, G.J., and Stewart, J.E. 1974b. "Distribution of Hydrocarbon-Utilizing Bacteria in Northwestern Atlantic Waters and Coastal Sediments." *Can. J. Microbiol.*, 20: 955–962.

Mulyono, M., Jasjfi, E., and Maloringan, M. 1994. "Oil Dispersants: Do They Do Any Good?" *Proceedings*. 2nd International Conference on Health, Safety, and Environment in Oil and Gas Exploration and Production, Jakarta. Volume 1, pp. 539–549.

Nasrollahzadeh, H.S., Najafpour, G.D., and Aghamohammadi, N. 2007. "Biodegradation of Phenanthrene by Mixed Culture Consortia in a Batch Bioreactor using Central Composite Face-Entered Design." *Int. J. Environ. Res.*, 1: 80–87.

Nicodem, D.E., Guedes, C.L.B., and Correa, R.J. 1998. "Photochemistry of Petroleum I. Systematic Study of a Brazilian Intermediate Crude Oil." *Mar. Chem.*, 63: 93–104.

Nikolopouloua, M., and Kalogerakis, N. 2008. "Enhanced Bioremediation of Crude Oil Utilizing Lipophilic Fertilizers Combined with Biosurfactants and Molasses." *Mar. Pollut. Bull.*, 56: 1855–1861.

Obayori, O.S., Ilori, M.O., Adebusoye, S.A., Oyetibo, G.O., Omotayo, A.E., and Amund, O.O. 2009. "Degradation of hydrocarbons and biosurfactant production by *Pseudomonas sp.* strain LP1." *World J. Microbiol. Biotechnol.*, 25: 1615–1623.

Obire, O. 1990. "Bacterial Degradation of three Different crude oils in Nigeria." *Nig. J. Bot.*, 1: 81–90.

Obire, O. 1993. "The Suitability of Various Nigerian Petroleum Fractions as Substrate for Bacterial Growth." *Discov. Innov.*, 9: 25–32.

Obire, O., and Okudo, I.V. 1997. "Effects of Crude Oil on a Freshwater Stream in Nigeria." *Discov. Innov.*, 9: 25–32.

Obire, O., and Nwaubeta, O. 2001. "Biodegradation of Refined Petroleum Hydrocarbons in Soil." *J. Appl. Sci. Environ. Mgt.*, 5(1): 43–46.

Odu, C.T.I. 1977. "Pollution and the Environment." *Bulletin of the Science Association of Nigeria*, 3(2): 284–285.

Okerentugba, P.O., and Ezeronye, O.U. 2003. "Petroleum Degrading Potentials of Single and Mixed Microbial Cultures Isolated from Rivers and Refinery Effluent in Nigeria." *African Journal of Biotechnology*, 2(9): 288–292.

Okoh, A.I. 2003. "Biodegradation of Bonny light Crude Oil in Soil Microcosm by Some Bacterial Strains Isolated from Crude Oil Flow Stations Saver Pits in Nigeria." *Afr. J. Biotech.*, 2(5): 104–108.

Okoh, A.I., Ajisebutu, S., Babalola, G.O., and Trejo-Hernandez, M.R. 2001. "Potentials of *Burkholderia cepacia* Strain RQ1 in The Biodegradation of Heavy Crude Oil." *Intl. Microb.*, 4: 83–87.

Okoh, A.I., Ajisebutu, S., Babalola, G.O., and Trejo-Hernandez, M.R. 2002. "Biodegradation of Mexican Heavy Crude Oil (Maya) by *Pseudomonas aeruginosa*." *J. Trop. Biosci.*, 2(1): 12–24.

Okoh, A.I. 2006. "Biodegradation Alternative in the Cleanup of Petroleum Hydrocarbon Pollutants." *Biotechnology and Molecular Biology Review*, 1(2): 38–50.

Okoh, A.I., and Trejo-Hernandez, M.R. 2006. "Remediation of Petroleum Hydrocarbon Polluted Systems: Exploiting the Bioremediation Strategies." *African J. Biotechnol.*, 5: 2520–2525.

Okolo, J.C., Amadi, E.N., and Odu, C.T.I. 2005. "Effects of Soil Treatments Containing Poultry Manure on Crude Oil Degradation in a Sandy Loam Soil." *Appl. Ecol. Environ. Res.*, 3(1): 47–53.

Onwurah, I.N E., Ogugua, V.N., Onyike, N.B., Ochonogor, A.E., and Otitoju, O.F. 2007. "Crude Oils Spills in the Environment, Effects and Some Innovative Clean-Up Biotechnologies." *Int. J. Environ. Res.*, 1: 307–320.

Onysko, K.A., Budman, H.M., and Robinson, C.W. 2000. "Effect of Temperature on the Inhibition Kinetics of Phenol Biodegradation by *Pseudomonas putida* Q5." *Biotechnol. Bioeng.*, 70: 291–299.

Oudot, J., Merlin, F.X., and Pinvidic, P. 1998. "Weathering Rates of Oil Components in a Bioremediation Experiment in Estuarine Sediments." *Mar. Environ. Res.*, 45: 113–125.

Page, D.S., Boehm, P.D., Douglas, G.S., Bence, A.E., Burns, W.A., and Mankiewic, P.J. 1996. "The Natural Petroleum Hydrocarbon Background in Subtidal Sediments of Prince William Sound Alaska, USA." *Environment Toxicology and Chemistry*, 15: 1266–1281.

Pavlova, A., and Ivanova, R. 2003. "Determination of Petroleum Hydrocarbons and Polycyclic Aromatic Hydrocarbons in Sludge from Wastewater Treatment Basins." *J. Environ. Monit.*, 5: 319–323.

Peters, K.E., Walters, C.C., and Moldowan, J.M. 2005. *The Biomarker Guide – Edition II*. Cambridge University Press, Cambridge, United Kingdom.

Park, K.S., Sims, R.C., DuPont, R.R., Doucette, W.J., and Matthews, J.E. 1990. "Fate of Polynuclear Aromatic Hydrocarbon Compounds in Two Soil Types: Influence of Volatilization, Abiotic Loss, and Biological Activity." *Environ. Toxicol. Chem.*, 9: 187–195.

Payne, J.R., and McNabb, G.D. Jr. 1984. "Weathering of Petroleum in the Marine Environment." *Marine Technology Society Journal*, 18(3): 24.

Pathak, H., Kantharia, D., Malpani, A., and Madamwar, D. 2009. "Naphthalene Biodegradation using Pseudomonas Sp. HOB1: In Vitro Studies and Assessment of Naphthalene Degradation Efficiency in Simulated Microcosms." *J. Hazard. Mater.*, 166: 1466–1473.

Pelletier, E., Delille, D., and Delille, B. 2004. "Crude Oil Bioremediation in Sub-Antarctic Intertidal Sediments: Chemistry and Toxicity of Oil Residues." *Mar. Environ. Res.*, 57: 311–327.

Pillis, L.J., and Davis, L.T. 1985. Microorganism Capable of Degrading Phenolics. United States Patent 4,556,638. December 3.

Pines, O., and Gutnick, D.L. 1981. "Relationship between Phage Resistance and Emulsan Production. Interaction Of Phages With The Cell-Surface of *Acinctobacter alcoaceticus* RAG-1." *Arch. Microbiol.*, 130: 129–133.

Pines, O., and Gutnick, D.L. 1984a. "Specific Binding of a Bacteriophage at a Hydrocarbon-Water Interface." *Appl. Environ. Microbiol.*, 157: 179–183.

Pines, O., and Gutnick, D.L. 1984b. "Alternate Hydrophobic Sites on the Cell Surface of *Acinetobacter calcoaceticuls* RAG-1." *FEMS Microbiol. Lett.*, 22: 307–311.

Pines, O., and Gutnick, D.L. 1986. "Role for Emulsan in Growth of *Acinetobacter calcoaceticus* RAG-1 on Crude Oil." *Appl. Environ. Microbiol.*, 51: 661–663.

Piskonen, R., and Itävaara, M. 2004. "Evaluation of Chemical Pretreatment of Contaminated Soil for Improved PAH Bioremediation." *Appl. Microbiol. Biotechnol.*, 65: 627634.

Pitter, P., and Chudoba, J. 1990. *Biodegradability of Organic Substances in the Aquatic Environment*. CRC Press, Boca Raton, Florida. pp. 7–83.

Poremba, K., Gunkel, W., Lang, S., and Wagner, F. 1991. "Marine biosurfactants. III. Toxicity Testing With Marine Microorganisms and Comparison With Synthetic Detergents." *Zeitschrift. Natureforsch.*, 46c: 210–216.

Prince, R. 1993. "Petroleum Spill Bioremediation in Marine Environments." *Crit. Rev. Microbiol.*, 19: 217–242.

Rahman, K.S.M., Thahira-Rahman, J., Lakshmanaperumalsamy, P., and Banat, I.M. 2002. "Towards Efficient Crude Oil Degradation by a Mixed Bacterial Consortium." *Biores. Tech.*, 85: 257–261.

Rahman, K.S.M., Thahira-Rahman, T., Kourkoutas, Y., Petsas, I., Marchant, R., and Banat, I.M. 2003. "Enhanced Bioremediation of n-Alkanes in Petroleum Sludge using Bacterial Consortium Amended with Rhamnolipid and Micronutrients." *Biores Tech.*, 90(2): 159–168.

Ramalho, A.M.Z., Martinez-Huitle, C.A., and Da Silva, D.R. 2010. "Application of Electrochemical Technology for Removing Petroleum Hydrocarbons from Produced Water Using a DSA-Type Anode at Different Flow Rates." *Fuel*, 89: 531–534.

Ramirez, L.C., Iturbe, R., and Torres, L.G. 2008. "Effect of Nutrient and Surfactant Addition on Polyaromatic Hydrocarbon (PAH) Biodegradation in Contaminated Soils." *Land Contamination and Reclamation*, 16, 1–11.

Reynolds, C.M., Bhunia, P., and Koenen, B.A. 1997. Soil Remediation Demonstration Project: Biodegradation of Heavy Fuel Oils. Special Report 97–20. Cold Regions Research & Engineering Laboratory. US Army Corps of Engineers, Fort Belvoir, Virginia.

Ringelberg, D.B., Talley, J.W., Perkins, E.J., Tucker, S.G., Luthy, R.G., Bouwer, E.J., and Fredrickson, H.L. 2001. "Succession of Phenotypic, Genotypic, and Metabolic Community Characteristics during In Vitro Bioslurry Treatment of Polycyclic Aromatic Hydrocarbon-Contaminated Sediments." *Appl. Environ. Microbiol.*, 67: 1542–1550.

Robichauz, T.J., and Myrick, H.N. 1972. "Chemical Enhancement of the Biodegradation of Crude Oil Pollutants." *Journal of Petroleum Technology*, 24: 16–20.

Rocha, C., and Infante, C. 1997. "Enhanced oily sludge biodegradation by a tensio-active agent isolated from *Pseudomonas aeruginosa* USBCS1." *Appl. Microb. Biotech.*, 47: 615–619.

Roling-Wilfred, F.M., Milner, M., Jones, D.M., Lee, K., Daniel, F., and Swanell-Richard, J.P. 2002. "Robust Hydrocarbon Degradation and Dynamic of Bacterial Communities during Nutrients – Enhanced Oil Spill Bioremediation." *Applied and Environmental Microbiology*, 68: 5537–5548.

Rosenberg, E., Perry, A., Gibson, D.T., and Gutnick, D.L. 1979a. "Emulsifier of Arthrobacter RAG-1: Specificity of Hydrocarbon Substrate." *Appl. Environ. Microbiol.*, 37: 409–413.

Rosenberg, E., Zuckerberg, A., Rubinovitz, C., and Gutnick, D.L. 1979b. "Emulsifier of Arthrobacter RAG-1: Isolation and Emulsifying Properties." *Appl. Environ. Microbiol.*, 37: 402–408.

Rosenberg, E., Kaplan, N., Pines, O., Rosenberg, M., and Gutnick, D.L. 1983a. "Capsular Polysaccharides Interfere with Adherence of

Acinetobacter calcoaceticus to Hydrocarbon." *FEMS Microbiol. Lett.*, 17: 157–160.

Rosenberg, E., Gottlieb, A., and Rosenberg, M. 1983b. "Inhibition of Bacterial Adherence to Hydrocarbons and Epithelial Cells by Emulsan." *Infect. Immun.*, 39: 1024–1028.

Rosenberg, E. 1991. "The Hydrocarbon-Oxidizing Bacteria." In *The Prokaryotes*. 2nd Edition. A. Bolows, H.G. Truger, M. Dworkin, K.H. Schleifer, and W. Harder (Editors). Springer-Verlag, New York. Volume 1, pp. 446–458.

Roubal, G. and Atlas, R.M. 1978. "Distribution of Hydrocarbon-Utilizing Microorganisms and Hydrocarbon Biodegradation Potentials in Alaskan Continental Shelf Areas." *Appl. Environ. Microbiol.*, 35: 897–905.

Ruberto, L., Dias, R., Lo Balbo, A., Vazquez, S.C., Hernandez, E.A., and Mac Cormack, W.P. 2009. "Influence of Nutrients Addition and Bioaugmentation on the Hydrocarbon Biodegradation of A Chronically Contaminated Antarctic Soil." *J. Appl. Microbiol.*, 106: 1101–1110.

Sabate, J., Vinas, M., and Solanas, A.M. 2004. "Laboratory-Scale Bioremediation Experiments on Hydrocarbon-Contaminated Soils." *Internl. Biodeter. Biodegrad.*, 54: 19–25.

Salam, L.B., Obayori, O.S., Akashoro, O.S., and Okogie, G.O. 2011. "Biodegradation of Bonny Light Crude Oil by Bacteria Isolated from Contaminated Soil." *Int. J. Agric. Biol.*, 13(2): 245–250.

Salas, N., Ortiz, L., Gilcoto, M., Varela, M., Bayona, J.M., Groom, S., Álvarez-Salgado, X.A., and Albaigés, J. 2006. "Fingerprinting Petroleum Hydrocarbons in Plankton and Surface Sediments during the Spring and Early Summer Blooms in the Galician Coast (NW Spain) after the Prestige Oil Spill." *Mar. Environ. Res.*, 62: 388–413.

Sandrin, T.R., Kight, W.B., Maier, W.J., and Maier, R.M. 2006. "Influence of a Non-aqueous Phase Liquid (NAPL) on Biodegradation of Phenanthrene." *Biodegradation*, 17: 423–435.

Schreiber, A.F., and Winkler, U.K. 1983. "Transformation of Tetralin by Whole Cells of *Pseudomonas stutzeri* As39." *Appl Microbiol. Biotechnol.*, 18(1): 6–10

Shabtai, Y., Pines, O., and Gutnick, D.L. 1986. "Emulsan: A Case Study of Microbial Capsules as Industrial Products." *Dev. Ind. Microbiol.*, 26: 291–307.

Shen, Z. 1995. "Alkaline Residue of Gasoline from FCC used to Purify Straight Run Gas oil." *Environmental Protection in Petrochemical Industry*, 2: 38–41.

Shuttleworth, K.L., and Cerniglia, C.E. 1995. "Environmental Aspects of Polynuclear Aromatic Hydrocarbon Biodegradation." *Appl. Biochem. Biotechnol.*, 54: 291–302.

Sikkema, J., and Bont, J.A.M. 1991. "Isolation and Initial Characterization of Bacteria Growing On Tetralin." *Biodegradation*, 2(1): 15–23.

Simarro, R., González, N., Bautista, L.F., Sanz, R., and Molina, M.C. 2011. "Optimisation of Key Abiotic Factors of PAH (Naphthalene, Phenanthrene and Anthracene) Biodegradation Process by a Bacterial Consortium." *Water Air Soil Pollut.*, 217: 365–374.

Sipma, J., Svitelskaya, A., van der Mark, B., Pol, L.W.H., Lettinga, G., Buisman, C.J.N., and Janssen, A.J.H. 2004. "Potentials of Biological Oxidation Processes for The Treatment Of Spent Sulfidic Caustics Containing Thiols." *Water Res.*, 38: 4331–4340.

Siron, R., Pelletier, E., and Brochu, C. 1995 "Environmental Factors Influencing the Biodegradation of Petroleum Hydrocarbons in Cold Seawater." *Arch. Environ. Contam. Toxicol.*, 28: 406–416.

Sivaraman, C., Ganguly, A., and Mutnuri, S. 2010. "Biodegradation of Hydrocarbons in the Presence of Cyclodextrins." *World J. Microbiol. Biotechnol.*, 26: 227–232.

Smith, J. 1968. *Torrey Canyon – Pollution and Marine Life*. Report by the Plymouth Laboratory of the Marine Biological Association of the United Kingdom, London. Cambridge University Press, London, United Kingdom. pp. 1906.

Soli, G., and Bens, E.M. 1972. "Bacteria which Attack Petroleum Hydrocarbons in a Saline Medium." *Biotechnol. Bioeng.*, 14:319–330.

Song, H., Wang, X., and Bartha, R. 1990. *Appl. Environ Microbiol.*, 56: 652.

Sotsky, J.B., and Atlas, R.M. 1994. "Frequency of Genes in Aromatic and Aliphatic Hydrocarbon Biodegradation Pathways within Bacterial Populations from Alaskan Sediments." *Can J Microbiol*, 40: 981–985.

Speight, J.G. 1996. *Environmental Technology Handbook*. Taylor & Francis, Washington, DC.

Speight, J.G., and Lee, S. 2000. *Environmental Technology Handbook*. 2nd Edition. Taylor & Francis, New York.

Speight, J.G. 2005. *Environmental Analysis and Technology for the Refining Industry*. John Wiley & Sons Inc., Hoboken, New Jersey.

Speight, J.G. 2007. *The Chemistry and Technology of Petroleum*. 5th Edition. CRC Press, Taylor & Francis Group, Boca Raton, Florida.

Speight, J.G. 2012. *The Chemistry and Technology of Coal*. 3rd Edition. CRC Press, Taylor & Francis Group, Boca Raton, Florida.

Steffan, R.J, Mccloy, K., Vainberg, S., Condee, C.W., and Zhang, D. 1997. "Biodegradation of the Gasoline Oxygenates Methyle tert-Butyl Ether, Ethyl tert-Butyl Ether and tert – Amyl Methyl Ether by Propane-Oxidizing Bacteria." *Appl. Environ. Microbiol.*, 63(11): 4216–4222.

Strawinski, R.J., and Stone, R.W. 1940. "The Utilization of Hydrocarbons by Bacteria." *J. Bacteriol.*, 40(3): 461.

Sutherland, J.B., Rafii, F., Khan, A.A., and Cerniglia, C.E. 1995. "Mechanisms of polycyclic aromatic hydrocarbon degradation." In *Microbial transformation and degradation of toxic organic chemicals*. L.Y. Young, and C.E. Cerniglia (Editors), John Wiley & Sons Inc.-Liss, New York. pp. 269–306.

Swannell, R.P.J., Lepo, J.E., Lee, K., Pritchard, P.H., and Jones, D.M. 1995. "Bioremediation of Oil-Contaminated Fine-Grained Sediments in Laboratory Microcosms." *Proceedings*. 2nd International Oil Spill Research and Development Forum, 23 – 26 May 1995. International Maritime Organisation, London, United Kingdom. pp. 45–55.

Swanell, R.P.J., Lee, K., and McDonagh, M. 1996. "Field Evaluation of Marine Oil Spill Bioremediation." *Microb. Rev.*, June: 342–365.

Tagger, S., Bianchi, A., Juillard, M., LePetit, J., and Roux, B. 1983. "Effect of Microbial Seeding of Crude Oil in Seawater in a Model System." *Mar. Biol.*, 78: 13–20.

Taghvaei Ganjali, S., Nahri Niknafs, B., and Khosravi, M. et al., 2007. "Photo-oxidation of Crude Petroleum Maltenic Fraction I Natural Simulated Conditions and Structural Elucidation of Photoproducts." *Iran J. Environ. Health Sci. Eng.*, 4(1): 37–42.

Talley, J.W., Ghosh, U., Tucker, S.G., Furey, J.S., and Luthy, R.G. 2002. "Particle-Scale Understanding of the Bioavailability Of PAHs in Sediment." *Environ. Sci. Technol.*, 36: 477–483.

Taylor, T.R., Jackson, K.J., Duba, A.G., and Chen, C.I. 1998. In Situ Thermally Enhanced Biodegradation of Petroleum Fuel Hydrocarbons and Halogenated Organic Solvents. United States Patent 5,753,122. May 19.

Thouand, G., Bauda, P., Oudot, J., Kirsch, G., Sutton, C., and Vidalie, J.F. 1999. "Laboratory Evaluation of Crude Oil Biodegradation with Commercial or Natural Microbial Inocula." *Can. J. Microb.*, 45(2): 106–115.

Traxler, R.W., and Bhattacharya, L.S. 1979. "Effect of a Chemical Dispersant on Microbial Utilization of Petroleum Hydrocarbons." In *Chemical Dispersants for the Control of Oil Spills*. L.T. McCarthy, G.P. Lindblom, and H.F. Walter, American Society for Testing and Materials, West Conshohocken, Pennsylvania.

Trindade, P.V.O., Sobral, L.G., Rizzo, A.C.L., Leite, S.G.F., and Soriano, A.U. 2005. "Bioremediation of a Weathered And a Recently Oil-Contaminated Soils from Brazil: A Comparison Study." *Chemosph.*, 58: 515–522.

US EPA. 2003. *Aerobic Degradation of Oily Wastes: A Field Guidance Book for Federal On-Scene Coordinators*. Version 1.0. US Environmental Protection Agency Response and Prevention Branch, Region 6 South Central, Dallas, Texas.

Van Brummelen, T.C., Verweij, R.A., Wedzinga, S.A., and Van Gestel, C.A.M. 1996. "Enrichment of Polycyclic Aromatic Hydrocarbons in Forest Soils near a Blast Furnace Plant." *Chemosphere*, 32: 293–314.

Van der Meer, J.R., de Vos, W.M., Harayama, S., and Zehnder, A.J.B. 1992. "Molecular Mechanisms of Genetic Adaptation to Xenobiotic Compounds." *Microbiol. Rev.*, 56: 677–694.

Varadaraj, R., Robbins, M.L., Bock, J., Pace, S., and MacDonald, D. 1995. "Dispersion and Biodegradation of Oil Spills on Water." *Proceedings*.

1995 Oil Spill Conference, American Petroleum Institute, Washington, DC. pp. 101–106.
Venosa, A.D., Suidan, M.T., Wrenn, B.A., Strohmeier, K.L., Haines, J.R., Eberhart, B.L., King, D., and Holder, E. 1996. "Bioremediation of an Experimental Oil Spill on the Shoreline of Delaware Bay." *Environmental. Sci. and Technol.*, 30(5): 1764–1775.
Venosa, A.D., and Zhu, X. 2003. "Biodegradation of Crude Oil Contaminating Marine Shorelines and Freshwater Wetlands." *Spill Sci. Tech. Bull.*, 8(2):163–178.
Vidali, M. 2001. "Bioremediation: An Overview." *Pure Appl. Chem.*, 73: 1163–1172.
Vieira, P.A., Faria, S.R., Vieira, B., De Franca, F.P., and Cardoso, V.L. 2009. "Statistical Analysis and Optimization of Nitrogen, Phosphorus, and Inoculum Concentrations for the Biodegradation of Petroleum Hydrocarbons by Response Surface Methodology." *J. Microbiol. Biotechnol.*, 25: 427–438.
Viñas, M., Sabate, J., Espuny, M.J., and Solanas, A.M. 2005. "Bacterial Community Dynamics and Polycyclic Aromatic Hydrocarbon Degradation during Bioremediation of Heavily Creosote-Contaminated Soil." *Appl. Environ, Microbiol.*, 71: 7008–7018.
Wang, H., and Yang, M. 2001. "Screening, Identification and Domestication of a Bacterium able to Degrade Organic Dross." *Shandong Science*, 14(2): 42–45.
Weiner, J.M., and Lovley, D.R. 1998. "Anaerobic Benzene Degradation in Petroleum-Contaminated Aquifer Sediments after Inoculation with a Benzene-Degrading Enrichment." *Appl. Environ. Microbiol.*, 64: 775–778.
Weissenfels, W.D., Beyer, M., and Klein, J. 1990. "Degradation of Phenanthrene, Fluorene and Fluoranthene by Pure Bacterial Cultures." *Appl. Microbiol. Biotechnol.*, 32: 479–484.
Weissenfels, W.D., Beyer, M., Klein, J., and Rehm, H.J. 1991. "Microbial Metabolism of Fluoranthene: Isolation and Identification of Ring Fission Products." *Appl. Microbiol. Biotechnol.*, 34: 528–535.
Westlake, D.W.S., Jobson, A.M., and Cook, F.D. 1978. "In Situ Degradation of Oil in a Soil of the Boreal Region of the Northwest Territories." *Can. J. Microbiol.*, 24: 245–260.
Whyte, L.G., Bourbonnière, L., and Greer, C.W. 1997. "Biodegradation of Petroleum Hydrocarbons by Psychrotrophic *Pseudomonas* Strains Possessing both Alkane (*alk*) and Naphthalene (*nah*) Catabolic Pathways." *Appl. Environ. Microbiol.*, 63: 3719–3723.
Whyte, L.G., Hawari, J., Zhou, E., Bourbonnière, L., Inniss, W.E., and Greer, C.W. 1998. "Biodegradation of Variable-Chain-Length Alkanes at Low Temperatures by a Psychrotrophic *Rhodococcus* sp." *Appl. Environ. Microbiol.*, 64: 2578–2584.

Whyte, L.G., Bourbonnière, L., Bellerose, C., and Greer, C.W. 1999. "Bioremediation Assessment of Hydrocarbon-Contaminated Soils from the High Arctic." *Bioremediation J.*, 3: 69–79.

Wild, S.R., and Jones, K.C. 1993. "Biological and Abiotic Losses of Polynuclear Aromatic Hydrocarbons from Soils Freshly Amended with Sewage Sludge." *Environ. Toxicol. Chem.*, 12: 5–12.

Williams, W.A., and May, R.J. 1997. "Low-temperature Microbial Aerobic Degradation of Polychlorinated Biphenyls in Sediments." *Environ. Sci. Technol.*, 31: 3491–3496.

Willumsen, P.A., Karlson, U., and Pritchard, P.H. 1998. "Response of Fluoranthene-Degrading Bacteria to Surfactants." *Appl. Microbiol. Biotechnol.*, 50: 475–483.

Wolfe, D.A., Hameedi, M.H., Galt, J.A., Watabayashi, G., Shrot, J., O'Claire, C., Rice, S., Michel, J., Payne, J.R., Braddock, J., Hanna, S., and Sale, D. 1994. "The Fate of the Oil Spilled from the Exxon Valdez." *Environ Sci. Technol.*, 28: 561A-568A.

Wong, J.W.C., Lai, K.M., Wan, C.K., Ma, K.K., and Fang, M. 2001. "Isolation and Optimization of PAHs-Degradative Bacteria from Contaminated Soil for PAHs Bioremediation." *Water, Air, and Soil Pollution*, 13: 1–13.

Wrenn, B.A., Suidan, M.T., Strohmeir, K.L., Eberhart, B.L., Wilson, G.J., and Venosa, A.D. 1997a. "Nutrient Transport during Bioremediation of Contaminated Beaches: Evaluation with Lithium as a Conservative Tracer." *Water Res.*, 31(3): 515–524.

Wrenn, B.A., Boufadel, M.C., Suidan, M.T., and Venosa, A.D. 1997b. "Nutrient Transport during Bioremediation of Crude Oil Contaminated Beaches." In *In Situ and On-Site Bioremediation*. Volume 4, pp. 267–272. Battelle Memorial Institute, Columbus, Ohio.

Xia, X.H., Yu, H., Yang, Z.F., and Huang, G.H. 2006. "Biodegradation of Polycyclic Aromatic Hydrocarbons in the Natural Waters of the Yellow River: Effects of High Sediment Content on Biodegradation." *Chemosphere*, 65: 457–466.

Yakimov, M.M., Giuliano, L., Bruni, V., Scarfi, S., and Golyshin, P.N. 1999. "Characterization of Antarctic Hydrocarbon-Degrading Bacteria Capable of Producing Bioemulsifiers." *Microbiologica.*, 22: 249–256.

Ye, D., Siddiqi, M.A., Maccubbin, A.E., Kumar, S., and Sikka, H.C. 1996. "Degradation of Polynuclear Aromatic Hydrocarbons by *Sphingomonas paucimobilis*." *Environ. Sci. Technol.*, 30: 136–142.

Yudono, B., Said, M., Sabaruddin, Napoleon, A., and Fanani, Z. 2011. "Kinetics Approach of Biodegradation of Petroleum Contaminated Soil by using Indigenous Isolated Bacteria." *J. Trop. Soils*, 16(1): 33–38.

Yumoto, I., Nakamura, A., Iwata, H., Kojima, K., Kusumuto, K., Nodasaka, Y., and Matsuyama, H. 2002. "*Dietzia psychralcaliphila sp. nov.*, A Novel Facultatively Psychrophilic Alkaliphile that Grows On Hydrocarbons." *Int. J. Syst. Evol. Microbiol.*, 52: 85–90.

Zahed, M.A., Aziz, H.A., Isa, M.H., and Mohajeri, L. 2010. "Enhancement of Biodegradation of n-alkanes from Crude Oil Contaminated Seawater." *Int. J. Environ. Res.*, 4(4): 655–664.

Zaidi, B.R., and Imam, S.H. 1999. "Factors Affecting Microbial Degradation of Polycyclic Aromatic Hydrocarbon Phenanthrene in Caribbean Coastal Water." *Marine Pollution Bulletin*, 38: 738–749.

Zander, M. 1983. "Physical and Chemical Properties of Polycyclic Aromatic Hydrocarbons." In *Handbook of polycyclic aromatic hydrocarbons*. A. Bjørseth (Editor). Marcel Dekker, Inc., New York. pp. 1–26.

Zekri, A.Y., Abou-Kassem, J.H., and Shedid, S.A. 2009. "A New Technique for Measurement of Oil Biodegradation." *Pet. Sci. Technol.*, 27:666–677.

Zengler, K., Richnow, H.H., Rossello-Mora, R., Michaelis, W., and Widdel, F. 1999. "Methane Formation from Long-Chain Alkanes by Anaerobic Microorganisms." *Nature.*, 401: 266–269.

Zeyaullah, M.D., Abdelkafel, A.S., Ben Zabya, W., and Ali, A. 2009. "Biodegradation of Catechols by Micro-Organisms – A Short Review." *African Journal of Biotechnology*, 8(13): 2916–2922.

Zhang, G-L., Wu, Y-T., Qian, X-P., and Meng, Q. 2005. "Biodegradation of Crude Oil by *Pseudomonas aeruginosa* in the Presence Of Rhamnolipids." *J Zhejiang Univ. Sci.*, 6B(8): 725–730.

Zhang, Y., Gao, B., Lu, L., Yue, Q., Wang, Q., and Jia, Y. 2010. "Treatment of Produced Water from Polymer Flooding in Oil Production by the Combined Method of Hydrolysis Acidification-Dynamic Membrane Bioreactor–Coagulation Process." *J. Petrol., Sci. Eng.*, 74: 14–19.

Zhao, L., Cao, X.L., Wang, H.Y., Liu, X., and Jiang, S.X. 2008. "Determination of Petroleum Sulfonates in Crude Oil by Column-Switching Anion-Exchange Chromatography." *Chinese Chem. Lett.*, 19: 219–222.

Zhou, Q., and Shen, B. 2010. "Biodegradation potential and influencing factors of a special microorganism to treat petrochemical wastewater." *Pet. Sci. Technol.*, 28: 135–145.

Zhu, X., Venosa, A.D., and Suidan, M.T. 2004. Literature Review on the Use of Commercial Bioremediation Agents for Cleanup of Oil-Contaminated Environments. Report No. EPA/600/R-04/075. National Risk Management Research Laboratory, United States Environmental Protection Agency, Cincinnati, Ohio. July.

Zosim, Z., Gutnick, D.L., and Rosenberg, E. 1982. "Properties of Hydrocarbon-in-Water Emulsions Stabilized by *Acinetobacter* RAG-1 Emulsan." *Biotechnol. Bioeng.*, 24: 281–292.

Zosim, Z., Rosenberg, E., and Gutnick, D.L. 1986. "Changes in Hydrocarbon Emulsification Specificity of the Polymeric Bioemulsifier Emulsan: Effects of Alkanols." *Colloid Polym. Sci.*, 264: 218–223.

Zuckerberg, A., Diver, A., Peeri, Z., Gutnick, D.L., and Rosenberg, E. 1979. "Emulsifier of Arthrobacter RAG-1: Chemical and Physical Properties." *Appl. Environ. Microbiol.*, 37: 414–420.

10

Biodegradation of Naphtha and Gasoline

1 Introduction

Naphtha and gasoline spills are common, and can cause water contamination issues as hydrocarbons dissolve in ground water and travel offsite in the aquifer. Hydrocarbons naturally degrade in the subsurface due to microbial-mediated reactions. However, the reaction rates are slow, because electron acceptors, like oxygen, are quickly depleted in contaminated ground water and are slowly recharged. Contaminated ground water has significant hydrocarbon concentrations, but depleted electron acceptors, whereas the overlying unsaturated zone contains oxygen, but lower hydrocarbon concentrations. Early response and intervention are the keys to minimizing extent and costs of remedial action for gasoline and its components, and are essential to protecting public health and the environment.

2 Identity and Origin

The term naphtha refers to any one of several low-boiling flammable liquid mixtures of hydrocarbons that is (in the current context) a distillation product from petroleum; naphtha is also produced by the destructive distillation of coal and/or oil shale.

Overall, the terminology for naphtha, like much of the petroleum industry terminology, is not standardized, is refinery dependent, and can be confusing. For example, naphtha is a low-boiling distillate fraction, usually boiling below 250°C (482°F), but often with a fairly wide boiling range, and there is a whole range of special-purpose hydrocarbon fractions that can be described as naphtha. The petroleum refining industry describes the 0 to 100°C (32 to 212°F) fraction from the distillation of crude oil as *light virgin naphtha* and the 100 to 200°C (212 to 390°F) fraction as *heavy virgin naphtha*. The product stream from the fluid catalytic cracker (FCC) is often subdivided into three fractions: (1) light FCC naphtha, the fraction boiling below 105°C (221°F), (2) intermediate FCC naphtha (boiling range: 105 to 160°C (221 to 320°F)), and (3) heavy FCC naphtha (boiling range: 160 to 200°C (320 to 390°F)).

On the other hand, other terminology includes *full range naphtha*, which is defined as the fraction distilled from petroleum with hydrocarbons in petroleum boiling between 30 and 200°C (68 and 390°F) or 30 and 220°C (68 and 428°F). Naphtha is a complex mixture of hydrocarbons generally having between 5 and 12 carbon atoms per molecule. *Light naphtha* is the fraction boiling between 30 and 90°C (68 and 194°F), and consists of molecules with 5 to 6 carbon atoms. *Heavy naphtha* boils between 90 and 220°C, and consists of molecules with 6 to 12 carbons.

Naphtha may be further defined by the boiling range, and may be further characterized by a high aliphatic content (aliphatic naphtha) or by a high aromatic content (aromatic naphtha), using the following parameters: (1) density (g/ml or specific gravity), (2) PONA analysis, which is paraffins, olefins, naphthenes, and aromatics, (3) PIONA analysis, which is paraffins, iso-paraffins, olefins, naphthenes, and aromatics, or (4) PIANO, which is paraffins, iso-paraffins, aromatics, naphthenes, and olefins. The analysis typically involves use of capillary gas chromatography, and the data are given in % v/v/ or % w/w.

Generally, the lower density naphtha will have higher paraffin content (paraffinic naphtha), and the main application for this type of naphtha is as a feedstock in the petrochemical production of olefins. This type of naphtha is also called (incorrectly) *straight run gasoline*, but is inadequate for use in an automobile engine. Blending with other refinery streams and additives is required before gasoline suitable for sales is produced (Speight, 2007).

When used as feedstock in petrochemical steam crackers, naphtha is heated in the presence of water vapor and the absence of oxygen or air until the hydrocarbon molecules break apart. The primary products of the cracking process are olefins (ethylene, propylene, butene, and butadiene). When naphtha is used as a feedstock in catalytic reforming, the primary products are aromatics, including benzene, toluene, and the isomeric xylenes.

Higher density naphtha is typically richer in naphthenes and/or aromatics, and may be referred to as naphthenic and aromatic naphtha. These naphtha types can also be used in the petrochemical industry, but more often are used as a feedstock for refinery catalytic reformers, where they convert the lower octane naphtha to a higher octane product (reformate) (Speight, 2007).

When there has been a naphtha spill at any site, the investigator would be wise to discover the source of the naphtha and request a listing of the properties. The MSDS sheet may not be sufficient to define the properties, so a full list giving the origin of the naphtha and the properties will be in order.

Gasoline is not a direct product of petroleum refining, but is a composite of several product streams and additives (Speight, 2007) (Chapter 3). Thus, the gasoline produced by a refinery is a blend of all the appropriate available streams, including, but not limited to: (1) straight-run naphtha, (2) naphtha, from the fluid catalytic cracking unit, (3) hydrocracked naphtha, and (4) reformate. The straight-run gasoline may be acceptable as is for blending, or may need minor upgrading for use in the gasoline blend.

Typically, gasoline is composed of 41 to 62% aliphatic hydrocarbons, with the remainder being aromatic hydrocarbons, including a benzene, toluene, ethylbenzene, and xylene (BTEX) fraction of 10 to 59%.

The aromatic hydrocarbons in naphtha and gasoline are generally more toxic than aliphatic compounds with a similar number of carbon atoms, and have more mobility in water, due to their solubility being 3 to 5 times higher. By virtue of its properties, gasoline contains a high proportion of volatile organic compounds (VOCs).

3 Remediation

When attempting to clean up naphtha or gasoline spills, timely and comprehensive source control and associated hydrogeological investigations are needed once a release is detected. These include: (1) immediate control and cessation of the release, (2) repair or removal of the release source, such as a tank or pipe, (3) removal/recovery of free product in both the saturated/unsaturated zones, and (4) removal of residual free product from the subsurface soils. Any remedial action initiated before the source is controlled is ineffective, and has the potential of expanding the scope of the remedial action as uncontrolled sources continue to migrate in the subsurface.

Because of the low-boiling nature of naphtha and gasoline, evaporation can play a significant role in site cleanup, but there is always the criticism that allowing considerable amounts of hydrocarbon vapor to enter the atmosphere will have serious environmental consequences.

Alternatively, physical action, such as excavation, may be an acceptable approach to source control of small releases. Contaminated soils removed by excavation can be treated by disposal (asphalt batching, daily landfill cover), or physical (thermal desorption) or biological (biopiles) treatment (Speight, 1996; Speight and Lee, 2000; Speight, 2005). For larger releases, however, alternatives include free-phase recovery, barrier installation, and hydraulic control of the ground-water plume. A variety of single and multi-phase extraction techniques moves both liquid and gas phases to the surface for treatment. At the surface, dewatering and subsequent recycling treats the higher concentrations of recovered material. Direct on-site thermal catalytic processes destroy lower concentrations, and absorbents such as granular activated charcoal to clean the air or water before discharge (Speight, 1996; Speight and Lee, 2000; Speight, 2005).

On the other hand, and for a variety of reasons, bioremediation might be the cleanup method of choice.

3.1 Site Remediation

Two major objectives of site remediation include destruction of residual or dissolved naphtha or gasoline constituents or their removal from the impacted area. Destruction can range from total mineralization/oxidation or reduction to inorganic components or transformation to some unlisted form. Chemical reaction, biological means, or thermal processes accomplish destruction.

This is a brief description of the myriad of innovative techniques developed to refine and enhance the implementation of four basic technologies that have evolved for the active remediation of gasoline released to the subsurface: (1) subsurface ventilation, (2) pump-and-treat technology, (3) in-situ chemical oxidation, and (4) in-situ bioremediation.

Specific techniques refining these technologies were developed to protect the environment, reduce treatment time, and achieve lower decontamination objectives (Table 10.1). In many cases, the techniques expand the matrix (i.e., soil, water, and air) capability of the technology. For example, under subsurface ventilation, the combination of air sparging with soil-vapor extraction expands the capability of subsurface ventilation (SVE) to remove gasoline vapors from both the saturated and the vadose zones. In contrast to basic technology, specific techniques exploit unique characteristics of the site, as well as those of the contaminant, to expedite remediation.

Several technologies, implemented in a sequence based on field progress results, schedule, and cost, usually make up the typical remediation project. The mix of unique surface and subsurface site conditions and the physical and chemical characteristics of the contaminants will determine selection of the appropriate remedial technologies and refinement

Table 10.1 Technologies and techniques for remediation of soil and groundwater contaminated with gasoline.

Tech-nology	Vadose or Unsaturated Zone	Groundwater & Saturated Soils	Techniques for Refinement and Enhancement
Subsurface Ventilation (vacuum or pressure)			
	X		Soil vapor extraction (SVE)
	X		Bioventing
	X	X	SVE with thermal enhancement
		X	Air sparging
	X	X	SVE & AS/ air injection
	X	X	In-Situ air stripping (ISAS)
	X	X	SVE & air sparging w/ Ozone
	X	X	Biosparging
	X	X	Hydraulic/pneumatic fracturing
	X	X	SVE & dewatering
Pump and Treatment			
		X	Pump and treat
		X	Pulse pumping
		X	Reverse flow pumping
	X	X	In situ flushing, surfactants
	X		Single phase vacuum extraction
	X	X	Vacuum extraction (multi-phase) (MPE)

(Contiued)

Table 10.1 (cont.) Technologies and techniques for remediation of soil and groundwater contaminated with gasoline.

Tech-nology	Vadose or Unsaturated Zone	Groundwater & Saturated Soils	Techniques for Refinement and Enhancement
		X	MPE Hydraulic/ pneumatic fracturing
		X	MPE thermal enhancements
In situ chemical oxidation			
	X	X	Hydrogen Peroxide
	X	X	Fenton's Reagent
	X	X	Controlled Fenton's using stabilized peroxide/chelated iron
	X	X	Permanganates of sodium or potassium
In situ biodegradation			
	X	X	Aeration, biosparging
	X	X	Dissolved oxygen injection
		X	Oxygen diffusion
		X	Oxygen release compound (ORC)
		X	Denitrifying
	X	X	Nutrient gas injection
	X	X	Monitored natural attenuation
	X	X	Enhanced natural attenuation
	X	X	Monitored natural attenuation

techniques. For example, the physical characteristics of methyl tert-butyl ether (MTBE) are well suited to traditional, physical remedial approaches that have proven to be effective with the other components of gasoline.

For in-situ treatment, recent demonstrations have confirmed the effectiveness of both chemical and biological oxidation processes for the destruction of gasoline. Refinements in the formulation of chemical oxidants (e.g., stabilized hydrogen peroxide, chelated catalysts) for in-situ treatments provide greater control, extended longevity, and more thorough treatment than earlier, uncontrolled formulations. This, together with uniform delivery of the oxidant using an approach called *deep remediation injection systems* (DRIS), has avoided many of the bounce-back problems observed with well-point injection programs. For in-situ bioremediation, providing higher concentrations of oxygen than available from simple aeration accelerates biological processes. Oxygen concentrators, pure oxygen sources, and oxygen release compounds now deliver abundant electron acceptors to the biologically active zone.

Chemically or biologically mediated processes can destroy gasoline in the subsurface, but bioremediation is a potentially advantageous serious option (Speight, 1996; Speight and Lee, 2000; Speight, 2005). While the simplest approach is generally the most efficient and cost effective, contaminant characteristics and site conditions of geology and hydrogeology often dictate more complicated place-transfer or phase-transfer systems.

For ex-situ destruction, ventilation blowers or pumping systems transfer gasoline constituents directly from the subsurface to the surface. Indirect phase-transfer processes may require several steps to move the gasoline constituents to the surface, where they can subsequently be removed from the air or water matrix before discharge. Consider the differences in the techniques discussed below with respect to direct and indirect processes of place-transfer and phase-transfer to affect ex situ destruction.

In the *subsurface ventilation evacuation method* (vacuum or pressurization), gas exchange is increased in the subsurface, and is an effective method of moving vapor-phase gasoline hydrocarbons to the surface for destruction. Another form of the subsurface ventilation evacuation method is performed in situ, in the unsaturated (or vadose) zone. This type of soil-remediation technology, in which a vacuum is applied to the soil, pulls gas-phase volatiles and some semi-volatile contaminants from the soil through extraction wells to the surface. The subsurface ventilation evacuation method is also known as *in-situ soil venting*, in-situ volatilization, enhanced volatilization, or soil vacuum extraction. The gas leaving the soil may be treated to recover or destroy the contaminants, depending on local and state air discharge regulations.

Vertical extraction vents are typically used at depths of 1.5 meters (5 feet) or greater, and have been successfully applied as deep as 91 meters (300 feet). Horizontal extraction vents (installed in trenches or horizontal borings) can be used as warranted by contaminant-zone geometry, drill-rig access, or other site-specific factors.

Bioventing (BV) is a process of stimulating the natural, in-situ biodegradation of contaminants in soil by providing air or oxygen to existing soil microorganisms. Bioventing uses low air-flow rates to provide only enough oxygen to sustain microbial activity in the vadose zone. Oxygen is most commonly supplied through direct air injection into residual contamination in soil.

In addition to degradation of adsorbed fuel residuals, volatile compounds move slowly through biologically active soil or shallow sediments where the vapors are biodegraded. The primary mechanism of volatile removal by the subsurface ventilation evacuation method is mass transfer. However, circulating air in the vadose zone stimulates microbial activity to degrade volatile and some semi-volatile compounds directly in the soil or sediment matrix. The major difference between the subsurface ventilation evacuation method and the bioventing method is the volume of air moved through the subsurface. Low-flow ventilation and gas exchange favor biological activity, while higher flows and extraction favor volatile removal.

Bioventing can reduce vapor-treatment costs, and can be also result in the remediation of semi-volatiles that cannot be removed by direct volatilization alone. Bioventing uses the same blowers as systems used in the subsurface ventilation evacuation method, to provide specific distribution and flux of air through the contaminated vadose zone to stimulate the indigenous microorganism to degrade hydrocarbons. The subsurface ventilation evacuation method thermal enhancement direct heating of subsurface soils and sediments by radio frequency (RF) heating or by resistance heating with electrode pairs or indirect heating through the injection of steam or hot air through the injection wells enhances the subsurface ventilation evacuation method performance dramatically, if the system is designed and operated properly.

Heated air or steam helps to "loosen" some less volatile compounds from the soil or sediments, in a process similar to steam distillation. When sufficient power is available, direct heating of soils with electrode pairs has been very effective for the focused remediation of small areas at several sites.

Air sparging (AS) injection of compressed air at controlled pressures and volumes into the water-saturated soils or sediments remediates the sediments and ground water by three processes:

1. In-situ air stripping of dissolved volatile organic compounds;
2. Volatilization of trapped- and adsorbed-phase contamination present below the water table and in the capillary fringe; and
3. Aerobic biodegradation of both dissolved- and adsorbed-phase contaminants.

Stripping and volatilization are the dominant removal mechanisms, with biodegradation becoming more significant over the long term. Air sparging is a low-cost technique applicable to a wide range of contaminant concentrations, and is flexible enough to accommodate diverse geological and hydrogeological limitations. For air sparging to be successful, the soil or sediment in the saturated zone must have sufficient permeability to allow the injected air to readily escape up into the unsaturated zone. Air sparging, therefore, will work fastest at sites with coarse-grained soil or sediment, like sand and gravel. The Henry's law constant for a given contaminant defines, in part, the effectiveness of air sparging. Empirical approaches are typically used to evaluate and optimize the effectiveness of air sparging in the field, because air channeling and residence time, water mixing, and equilibrium considerations complicate interpretation of the inter-phase mass-transfer efficiency.

Variations in the sparging cycle, from continuous sparging to pulse sparging, to avoid channeling and permeability changes, have met with various degrees of success, depending largely on the geology and hydrogeology. The combination of the subsurface ventilation evacuation method and the air sparging method expedites removal of dissolved volatile organic compounds from the subsurface. The subsurface ventilation evacuation method sweeps from the vadose zone the organic compounds that are stripped from the aqueous phase and volatilized from the saturated zone. Like air sparging alone, air stripping and volatilization are the dominant removal factors; however, overall ventilation of the subsurface stimulates the indigenous microflora to degrade residual and semi-volatile compounds on the soils.

3.2 In-Situ Air Stripping

In situ air stripping (ISAS) combines three technologies, air sparging, horizontal wells, and soil-vapor extraction. In situ air stripping uses horizontal wells to inject (sparge) air into the ground water. The horizontal wells provide more effective access to a horizontal ground-water plume. As the air comes into contact with contaminants, they volatize and rise through the soil sediment. The volatile organic compounds are then extracted from overlying soils by standard soil-vapor extraction. The air sparging process eliminates the need for surface ground-water treatment systems, such as air strippers.

The subsurface ventilation method and air sparging (replacing the air with the ozone) in the subsurface leads to the in situ destruction of many gasoline constituents. Ozone, the third strongest direct oxidant after fluorine and hydroxyl radical, can oxidize organic contaminants by direct oxidation, in addition to the oxidation by free hydroxyl radical. In-situ destruction of volatile organic compounds reduces surface treatment costs, and often pretreats adsorbed semi-volatile compounds to improve their susceptibility to biodegradation. Subsequent oxygen release also stimulates in-situ biodegradation.

Biosparging, like its sister technique bioventing, uses lower airflow (0.5 to 3 cubic ft^3/minute per injection point) than needed for air stripping and volatilization. The objective is to provide enough oxygen and gas exchange to drive in situ aerobic biodegradation in the saturated zone without stripping the hydrocarbons. This approach is particularly effective and appropriate for biodegradable, but poorly stripped compounds, i.e., those with a low Henry's Law constant, like acetone and MTBE. Microflora in the vadose zone generally treat volatile organic compounds stripped during bio-sparging, so capturing and treating them at the surface is usually not an issue.

3.3 Pump-and-Treat Systems

Pump-and-treat systems operate by pumping ground water containing dissolved or non-aqueous phase liquid hydrocarbon to the surface. Free-phase hydrocarbons are separated out, dissolved constituents are removed, and the water is either reinjected or discharged to a surface water body or municipal sewage plant. Contaminant recovery is limited by the behavior of the contaminant in the subsurface (primarily solubility and the adsorption /partition coefficients), site geology and hydrogeology, and the extraction system design. It is further complicated by residual contamination in the saturated zone and adsorbed contamination in the capillary fringe, as well as insoluble non-aqueous phase liquids (NAPLS) floating on the water table.

Pump-and-treat systems are also used to contain contaminated ground water, provide hydraulic control, and recover contaminant mass in either gas or liquid phase by creating a capture zone around the pumping well. The natural hydrogeological property of the site and the rate at which ground water is extracted limit the capture zone of the recovery well. As refinements to the pump-and-treat system, *pulse pumping/reverse flow pumping* address low-permeability formations, channeling, capillary fringe, cost, and the delivery of nutrients to stimulate in situ bioremediation.

3.4 Vacuum Enhanced Recovery

The application of a vacuum to the extraction point (*vacuum enhanced recovery*, VER) provides a method to further enhance the capture zone. A high-vacuum or negative pressure applied to a recovery well and to the formation enhances liquid recovery of the well by increasing the net effective drawdown. VER increases the mass removal of the volatile and semi-volatile contaminants, by maximizing dewatering and facilitating volatilization from previously saturated sediments via increased air movement. Physical removal of significant hydrocarbon mass increases subsurface oxygen levels for aerobic biodegradation of residual contaminant. Vacuum enhanced recovery is cost-effective to enhance the overall recovery of contaminants, especially under low-permeability conditions.

3.5 Single-Phase Vacuum Extraction

In the *single-phase vacuum extraction* (SPVE) method, a single pump removes fluid, via a drop tube, and applies a vacuum to the well and formation. The well produces both liquid- and vapor-phase material. Compared to pumping alone, single-phase vacuum extraction increases the capture zone, and therefore, reduces the number of recovery wells needed, and accelerates the recovery of both liquid and residual contaminants. Although limited to depths of less than 25 ft, this technique is one of few enhancements for mass removal from low permeability sites.

3.6 Dual-Phase Vacuum Extraction

Similar in its advantages to the single-phase system, the *dual-phase vacuum extraction* (DPVE) (also called the *multiphase-phase vacuum extraction*) uses a submersible pump for liquid recovery at greater depths (>25 ft), and a diaphragm or liquid-ring pump for evacuation of the formation. A single-vacuum pump can be used for multiple wells under a variety of design strategies. When coupled with surfactants, a pump-and-treat system can be used as an in-situ soil washing system.

3.7 Surfactant Enhanced Aquifer Redemption

Surfactant enhanced aquifer redemption (SEAR) techniques have been particularly effective at removing non-aqueous phase liquids and the residual material (from non-aqueous phase liquids) from highly permeable subsurface systems. Surfactant selection must be paired with the target constituents, and must exhibit some short-term resistance to biodegradation by the indigenous microflora.

3.8 In Situ Chemical Oxidation

Chemical oxidants for the oxidation of organic compounds in the subsurface are selected in part based on their oxidizing power (Speight, 2005).

3.8.1 Hydrogen Peroxide/Fenton's Reagent

Hydrogen peroxide reacts catalytically with naturally occurring or injected ferrous iron to form the highly reactive hydroxyl radical. Fenton's reagent is the iron-catalyzed hydrogen peroxide. This strong oxidizer, exceeded only by fluorine in oxidizing power, oxidizes organic constituents in gasoline to carbon dioxide and water. Recent advances in chemical stabilization of the peroxide and controlled release of the iron by chelation has extended the reaction time and provided more control of this chemistry.

3.8.2 Permanganates

Potassium or sodium permanganate is an important oxidizer for odor control in sewage treatment plants. The oxidation potential of permanganate increases with decreasing pH. The temperature-sensitive solubility of the dark-purple crystals of potassium permanganate presents handling challenges circumvented by the soluble sodium permanganate. Like Fenton's reagent, permanganates are capable of oxidizing gasoline constituents ranging from alkenes to PAHs, but are less sensitive to pH than Fenton's reagents, exhibit slower reaction kinetics, and produce manganese oxide as a product.

3.9 In Situ-Enhanced Bioremediation

Bioremediation uses indigenous or introduced (augmented) microorganisms (primarily bacteria) to degrade organic contaminants into harmless substances, biomass, or carbon dioxide and water. The kinetics of biodegradation are limited by substrate availability, electron donors, nutrients, pH, moisture content, other carbon and energy sources, and other factors beyond the scope of this summary of techniques employed in remediation, often involving some form of bioremediation.

The end-products of the biodegradation of petroleum hydrocarbon degradation are carbon-dioxide and water, which are also measures of microbial respiration and activity in soils. Higher rates of biodegradation by soil microbial activity as shown from the carbon-dioxide production occurred during the first four weeks; the increase in carbon-dioxide production was progressive initially, but started decreasing after week four. This is also collaborated by the rates of biodegradation, which was highest in the first four weeks (Obire and Nwaubeta, 2001).

The amount of carbon dioxide produced in the control and in the hydrocarbon contaminated soil is in the order:

$$\text{Gasoline} > \text{kerosene} > \text{diesel oil.}$$

Respiration of microbes occurred very rapidly during the initial period of incubation, when the lighter and more readily degraded fractions were degraded, but slowed down as the residue became more difficult to degrade on account of the increase of the heavier fractions.

Carbon-dioxide production, which is a measure of microbial activity and respiration, occurred in all soils. This confirmed the fact that the microorganisms metabolized the petroleum hydrocarbons, which resulted in a progressive increase in carbon dioxide evolved, which decreased later, following exhaustion of metabolizable fractions.

The indigenous microbial flora of the soil showed hydrocarbon degradation potentials, as was revealed by total hydrocarbon data obtained and carbon-dioxide production in the soils. There was significant difference between control soil and gasoline–contaminated soil at both 5% and 1% level of probability (Obire and Nwaubeta, 2001).

While it is known that n-alkanes become more readily depleted by biodegradation than branched and cyclic alkanes, details are limited about the relative susceptibilities of branched alkanes, alkylcyclohexane derivatives, alkyl cyclopentane derivatives, and aromatic hydrocarbons to biodegradation, and about how the accompanying carbon-isotope distribution is affected.

In order to address these specific issues, the analysis of a suite of 18 oils from the Barrow Island oilfield, Australia, and a non-biodegraded reference oil were analyzed compositionally, in order to detail the effect of minor to moderate biodegradation on the naphtha range C_5 to C_9 hydrocarbons (George et al., 2002). Carbon isotopic data for individual low molecular weight hydrocarbons were also obtained for six of the oils.

Biodegradation resulted in strong depletion of n-alkanes (>95%) from most of the oils. Benzene and toluene were partially or completely removed from the Barrow Island oils by water washing. However, hydrocarbons with lower water solubility were either not affected by water washing, or water washing had only a minor effect. There are three main controls on the susceptibility to biodegradation of cyclic, branched, and aromatic low molecular weight hydrocarbons: (1) carbon skeleton, (2) degree of alkylation, and (3) position of alkylation.

Ring preference ratios at C_6 and C_7 showed that iso-alkanes are retained preferentially relative to alkyl cyclohexane derivatives and, to some extent, alkyl cyclopentane derivatives. Dimethylpentane derivatives were substantially more resistant to biodegradation than most dimethylcyclopentane derivatives, but methylhexane derivatives were depleted faster than methylpentane derivatives and dimethylcyclopentane derivatives. For the C_8 and C_9 hydrocarbons, alkylcyclohexane derivatives were more resistant to biodegradation than linear alkanes. There was also a trend of lower susceptibility to biodegradation with greater alkyl substitution for iso-alkanes, alkyl cyclohexane derivative derivatives, alkyl cyclopentane derivative derivatives, and alkylbenzene derivatives.

It was also noted that the position of alkylation had a strong control, with adjacent methyl groups reducing the susceptibility of an isomer to biodegradation. 1,2,3-Trimethylbenzene was the most resistant of the C_3

alkylbenzene isomers during moderate biodegradation. 2-Methylalkane derivatives were the most susceptible branched alkanes to biodegradation, 3-methylalkane derivatives were the most resistant, and 4-methylalkane derivatives had intermediate resistance. Therefore, terminal methyl groups are more prone to bacterial attack compared to mid-chain isomers, and C_3 carbon chains are more readily utilized than C_2 carbon chains. 1,1-Dimethylcyclopentane and 1,1-dimethylcyclohexane were the most resistant of the alkyl cyclohexane derivatives and alkyl cyclopentane derivatives to biodegradation.

It was of interest to note that biodegradation led to enrichment in ^{13}C for each remaining hydrocarbon, due to preferential removal of ^{12}C. Differences in the rates of biodegradation of low molecular weight hydrocarbons shown by compositional data were also reflected in the level of enrichment in ^{13}C. The carbon isotopic effects of biodegradation show a decreasing level of isotopic enrichments in ^{13}C with increasing molecular weight. This suggested that the kinetic isotope effect associated with biodegradation may be site specific, and often related to a terminal carbon, where its impact on the isotopic composition becomes progressively diluted with increasing carbon number.

Another investigation focused on determination of the kinetic rate laws for the aerobic biodegradation of a mixture of twelve volatile petroleum hydrocarbons (typical of constituents of the naphtha fraction and gasoline product) and methyl tert-butyl ether (MTBE) in unsaturated alluvial sand. Laboratory column and batch experiments were performed at room temperature under aerobic conditions, and a reactive transport model for vapors of volatile organic compounds in soil gas coupled to degradation kinetics was used for data interpretation (Höhener et al., 2003).

In the column experiment, an acclimatization period of twenty-three days was used before steady-state diffusive vapor transport through the horizontal column was achieved. Kinetic parameters were derived from the concentration profiles of toluene, m-xylene, n-octane, and n-hexane, because substrate saturation was approached with these compounds under the experimental conditions. The removal of cyclic alkanes, isooctane, and 1,2,4-trimethylbenzene followed first-order kinetics over the whole concentration range applied. Batch experiments suggested first-order disappearance rate laws for all volatile organic compounds except n-octane, which decreased following zero-order kinetics in live batch experiments.

Biodegradation of a broad range of linear and branched alkanes, parent and alkyl alicyclic hydrocarbons, and benzene and alkyl-substituted benzenes has been observed over the past decades in various types of environments (Yadav and Hassanizadeh, 2011). In addition to naphtha and gasoline, this also applies to natural gas condensate (Prince and Suflita, 2007).

The biodegradability under aerobic conditions of volatile hydrocarbons (4 to 6 carbon atoms per molecule) contained in gasoline and consisting of n-alkanes, iso-alkanes, cycloalkanes, and alkenes, has been investigated (Solano-Serena et al., 2000; Plaza et al., 2008).

In one case (Solano-Serena et al., 2000), activated sludge was used as the reference microflora. The biodegradation test involved the degradation of the volatile fraction of gasoline in closed tasks under optimal conditions. The kinetics of biodegradation was monitored by carbon dioxide production. Final degradation was determined by gas chromatographic analysis of all measurable hydrocarbons (12 compounds) in the mixture after sampling the headspace of the tasks. The degradation of individual hydrocarbons was also studied with the same methodology. When incubated individually, all hydrocarbons used as carbon sources, except 2,2-dimethylbutane and 2,3-dimethylbutane, were completely consumed in 30 days or less with different velocities and initial lag periods. When incubated together as constituents of the light gasoline fraction, all hydrocarbons were metabolized, often with higher velocity than for individual compounds. Co-metabolism was involved in the degradation of dimethyl iso-alkanes.

In another case (Plaza et al., 2008), crude oil and its distillation products (generally these were naphtha and gasoline constituents) were efficiently degraded by selected microbial strains. The mixture of the isolates enhanced slightly the degradation efficiency of crude oil and its distillation products. The short alkanes, as well as long alkanes and BTEX, were well degraded during crude oil degradation. The cultures of the bacteria and their mixture were not found to be toxic to *Vibrio fischeri* during crude oil degradation. *Ralstonia picketti* SRS and *Alcaligenes piechaudii* SRS produce biosurfactants, but there was no evidence that biosurfactants produced by the strains were indeed involved in the degradation of crude oil and the distillate fractions.

When sediment and groundwater samples collected from a gas condensate-contaminated aquifer were incubated under methanogenic, and especially under sulfate-reducing, conditions, even though no exogenous nitrogen or phosphorus was added, biodegradation was observed. Furthermore, the addition of gasoline to the gas condensate samples had minimal impact on the biodegradation of saturated compounds, but substantially increased the diversity and extent of aromatic compounds undergoing transformation. This is attributed to the promotion or induction of biodegradation pathways in the indigenous microflora following the addition of the gasoline components. The chemical species that promoted the biodegradation are not precisely known, but may have been present in the initial condensate, and reduced in concentration by various mechanisms (such as dissolution, biodegradation), such that their concentration in the aquifer fell below necessary levels. A variety of aromatic hydrocarbons would appear to be likely candidates (Prince and Suflita, 2007).

Techniques to stimulate in-situ bioremediation have also focused on providing the factors limiting the biological activity of the system.

The oxygen demand for aerobic biodegradation of gasoline constituents has been satisfied by a technique called *direct oxygen injection*, or *oxygen diffusion using oxygen release compounds* (ORC).

Activated sludge represents a natural consortium of broadly diverse active bacteria, which are capable of degrading organic waste compounds. Encapsulated-activated sludge beads have been applied to the biodegradation of a number of environmental pollutants, including 2,4,6-trichlorophenol in groundwater (Razavi-Shiravi and Veenstra, 2000), phenol in wastewater (Shishido et al., 1995), and mixtures of acetaldehyde and propionaldehyde in waste gas (Adly Ibrahim et al., 2001). Non-encapsulated activated sludge cultures have been previously used for the biodegradation of gasoline hydrocarbons (Solano-Serena et al., 1999).

Gasoline is a major groundwater contaminant as the result of release from leaking underground storage tanks, accidental spills, leaking process equipment, and other industrial activities (Dowd, 1984; Cherry, 1987; Day et al., 2001). The use of encapsulated sludge microbeads for the biodegradation of gasoline hydrocarbons may be feasible as suspension in a pump-and-treat surface bioreactor system (ex situ bioremediation) or as slurry, filling a subsurface permeable reactive barrier (in situ bioremediation). Alternatively, the encapsulated sludge microbeads may have the potential to be used for the start-up of bioreactors or the maintenance of high biodegradation activities in bioreactors (bioaugmentation).

The emulsification–internal gelation procedure (Moslemy et al., 2004) illustrates a novel technique for producing sludge-encapsulating gellan gum microbeads. The biodegradation of gasoline by gellan gum-encapsulated and non-encapsulated activated sludge from a wastewater treatment plant was proven. A high degree of biodegradation of gasoline at initial concentrations of the encapsulated activated sludge is required to degrade petroleum hydrocarbons under aerobic conditions. Encapsulated sludge cultures demonstrated comparable biodegradation activity with non-encapsulated sludge, but at approximately four times lower biomass levels, and seemed to offer certain advantages, such as protection from toxicity of contaminants.

3.10 Monitored Natural Attenuation

Natural attenuation, also known as passive bioremediation, intrinsic bioremediation, or intrinsic remediation, is a passive remedial approach that depends upon natural processes to degrade and dissipate petroleum constituents in soil and ground water. Some of the processes involved in natural attenuation of petroleum products include aerobic and anaerobic biodegradation, dispersion, volatilization, and adsorption. In general, for petroleum hydrocarbons, biodegradation is the most important natural attenuation mechanism; it is the only natural process that results in an actual reduction of petroleum.

4 BTEX and MTBE

Modern gasoline formulations are commonly supplemented with the additive methyl tert-butyl ether (MTBE), which allows a reduction on the amount of BTEX (Mays, 1989). Although less toxic than BTEX, MTBE appears to be more recalcitrant under natural conditions (Table 10.2) (Deeb et al., 2001). BTEX and MTBE are the most water-soluble components of gasoline, and therefore, these compounds predominate in groundwater contaminant plumes from recent accidental gasoline spills.

4.1 BTEX

Petroleum distillation products are efficiently degraded by various individual microbial strains and their mixture. The mixture of the isolates enhanced slightly the degradation efficiency of crude oil and its distillation products.

Gasoline leaking from underground storage tanks, distribution facilities, and various industrial operations represents a prime source of soil and aquifer contamination (Swoboda-Colberg, 1995). Among the contaminants present in gasoline, benzene, toluene, ethylbenzene, and xylene

Table 10.2 Physical properties of BTEX and MTBE.

	Aromatics (BTEX)	Ether (MTBE)
Volume % in gasoline	Up to 25%	UP to 15%
Vapor pressure (mm Hg)	8–95	245
Specific gravity	0.88	0.744
Water solubility pure (ppm)	1780	43000
Water solubility in gasoline (ppm)	65	5100
Henry's law constant	0.22–0.39	0.018
Adsorption coefficient (Log K_{oc})	1.9	1.035–1.91
Taste thresholds in water (ppb)	>500	40–140
Odor thresholds in water (ppb)	NA	15–40
Toxicity	High	Low
Biodegradability	Medium	Low

(BTEX) are classified as priority pollutants because of their high mobility and toxicity (Mehlman, 1992).

The biodegradation of a mixture of benzene, toluene, ethylbenzene, xylene (BTEX), and methyl-*tert*-butyl ether (MTBE) was studied in soil microcosms (Prenafeta-Boldú et al., 2004). Soil inoculation with the toluene-metabolizing fungus *Cladophialophora* sp. strain T1 was evaluated in sterile and non-sterile soil. Induction of biodegradation capacity following BTEX addition was faster in the soil's native microflora than in axenic soil cultures of the fungus. Toluene, ethylbenzene, and the xylenes were metabolized by the fungus, but biodegradation of benzene required the activity of the indigenous soil microorganisms. MTBE was not biodegraded under the tested environmental conditions. Biodegradation profiles were also examined under two pH conditions after a long term exposure to BTEX. At neutral conditions, the presence of the fungus had little effect on the intrinsic soil biodegradation capacity. At an acidic pH, however, the activity of the indigenous degraders was inhibited, and the presence of *Cladophialophora* sp. increased significantly the biodegradation rates of toluene and ethylbenzene. Comparison of the BTEX biodegradation rates measured in soil batches combining presence and absence of indigenous degraders and the fungal inoculum indicated that no severe antagonism occurred between the indigenous bacteria and *Cladophialophora* sp. The presence of the fungal inoculum at the end of the experiments was confirmed by PCR-TGGE analysis of small subunits of 18S rDNA.

The techniques that refine or enhance the basic technologies briefly discussed here reflect an adaptation to unique hydrogeological site conditions, specific cleanup objectives, physical characteristics of the contaminant, and specific risk concerns. Before any technology or refinement thereof is selected, specific information must be obtained on site-specific hydrogeologic conditions, engineering design, data needs, performance data, pilot testing, duration of operation, sequencing with other technologies, case studies, and anticipated cost. While this is beyond the scope of this presentation, many of these data are readily available on the Internet at state and federal web sites, and from readily available, published sources, with the exception of site-specific hydrogeologic conditions, which must be characterized at the outset of the project.

Refinements and enhancement of remediation technologies continue to produce a variety of exciting techniques for optimizing in-situ destruction or physical extraction for ex-situ destruction. Development and implementation of these techniques require flexible designs and flexible, responsive management styles that can adapt to new information to expedite the remediation process.

Bioremediation of hydrocarbon pollution relies on the biodegradation activity of soil microorganisms. Bacteria and fungi capable of degrading BTEX and MTBE have been isolated from soil (Yadav and Reddy, 1993;

Hardison et al., 1997; Deeb et al., 2001). However, while most of the studies have focused on bacteria, little is known on the contribution of fungi to bioremediation of BTEX and MTBE. Mediation of the biodegradation of soil pollutants with fungi has mainly been assayed with white-rot fungi (Pointing, 2001).

These organisms oxidize aromatic hydrocarbons by co-metabolism, and significant mineralization can only be achieved through the synergic interaction of fungi and bacteria (Kotterman et al., 1998; Boonchan et al., 2000). Interestingly, fungi capable of growing on volatile aromatic hydrocarbons as their sole source of carbon and energy have recently been isolated from soil (Cox et al., 1993; Prenafeta Boldú et al., 2001; Woertz et al., 2001). These fungal isolates were found to be tolerant to acidic and dry conditions, characteristics that make them very suitable as biocatalyst in air bio-filters treating BTEX vapors (van Groenestijn et al., 2001).

An oil-refining plant site located in southern Taiwan has been identified as a petroleum hydrocarbon spill site: the spillages mainly involved benzene, toluene, ethylbenzene, and xylenes (BTEX) and methyl tert-butyl ether (MTBE) (Chen et al., 2006). Groundwater samples collected from the site were analyzed to assess the occurrence of intrinsic MTBE biodegradation. Microcosm experiments were conducted to evaluate the feasibility of biodegrading MTBE by indigenous microorganisms under aerobic, co-metabolic, iron reducing, and methanogenic conditions. Results from the field investigation and microbial enumeration indicate that the intrinsic biodegradation of MTBE and BTEX is occurring and causing the decrease in MTBE and BTEX concentrations. Microcosm results showed that the indigenous microorganisms were able to biodegrade MTBE under aerobic conditions, using MTBE as the sole primary substrate.

The detected biodegradation byproduct, tri-butyl alcohol (TBA), can also be biodegraded by the indigenous microorganisms. In addition, microcosms with site groundwater as the medium solution show higher MTBE biodegradation rate. This indicates that the site groundwater might contain some trace minerals or organics, which could enhance the MTBE biodegradation. Results showed that the addition of BTEX at low levels could also enhance the MTBE removal. No MTBE removal was detected in iron reducing and methanogenic microcosms. This might be due to the effects of low dissolved oxygen within the plume. The low iron reducers and methanogens ($<1.8 \cdot 103$ cell/g of soil) observed in the aquifer also indicate that the iron reduction and methanogenesis are not the dominant biodegradation patterns in the contaminant plume. Results from the microcosm study reveal that preliminary laboratory study is required to determine the appropriate substrates and oxidation-reduction conditions to enhance the biodegradation of MTBE. Results suggested that in situ or on-site aerobic bioremediation using indigenous microorganisms would be a feasible technology to clean up this MTBE-contaminated site.

4.2 Methyl Tert-Butyl Ether

Fuel oxygenates, mainly methyl tert-butyl ether (MTBE), are added to gasoline in replacement of lead tetraethyl to enhance its octane index. Their addition also improves the combustion efficiency, and therefore decreases the emission of pollutants (CO and hydrocarbons). On the other hand, MTBE, being highly soluble in water and resistant to biodegradation, is a major pollutant of water in aquifers contaminated by MTBE-supplemented gasoline during accidental release (John et al., 2000; Lopes Ferreira et al., 2006).

Methyl tertiary butyl ether (MTBE), used in gasoline to enhance the octane rating of the fuel (Speight, 2007), is difficult to remediate because it is very soluble in water, and is hard to break down using microbes normally present in soil. In vegetation-based bioremediation, MTBE is moved from the soil to the atmosphere, along with the water that plants take up from soil and release to the air. The MTBE breaks down rapidly in the atmosphere. Benzotriazole derivatives, used as corrosion inhibitors in antifreeze and aircraft deicer fluids, are treated by plant-based bioremediation. The benzotriazole adsorbs or sticks to the plant roots, and ends up as part of the plant biomass.

MTBE is generally 30 times more soluble in water than the hydrocarbon constituents in gasoline, and once MTBE is dissolved in water, it is difficult to separate (Johnson et al., 2000). Dispersion and dilution reduce the concentration of MTBE by mixing the contaminated water with cleaner water, but the total mass of MTBE present in situ does not change. While naturally-occurring biodegradation offers an alternative for in situ remediation of MTBE, it may be limited by the long delay times needed before degradation occurs consistently. In addition, because of the slow growth achieved, and perhaps because MTBE is not similar to petroleum hydrocarbons or other organic contaminants, few microbes seem to degrade MTBE.

The challenge of remediating MTBE in groundwater is one of the most difficult subsurface contamination problems, because of the compound's propensity to dissolve into water at high concentrations and to move with the water with little to no retardation. While in situ bioremediation offers the best hope for addressing MTBE contamination, natural processes will likely have to be accelerated using innovative engineering approaches (Steffan et al., 1997), such as the use of co-metabolic oxidation, or if it is used as a carbon and energy source by a few microorganisms (Lopes Ferreira et al., 2006).

Degradation of light boiling fractions achieved 80–100%. On the other hand, degradation removal of heavy, high boiling fractions was lower, and reached between 10–81%. The mixture of the isolates enhanced slightly the degradation efficiency of crude oil and its distillation products. Changes in crude oil toxicity as a function of biodegradation activity were also undertaken. Slight luminescence inhibition as a measure of toxicity was found during the experiment time. The samples were not found to be toxic to *Vibrio fisheri* during the crude oil biodegradation. The luminescence inhibition could be due to the presence of toxic intermediates produced during biodegradation (Plaza et al., 2008).

The biodegradation of methyl tert-butyl ether (MTBE) was included as part of an investigation that focused on determination of the kinetic rate laws for the aerobic biodegradation of a mixture of twelve volatile petroleum hydrocarbons (typical of constituents of the naphtha fraction and gasoline product) and in unsaturated alluvial sand. Laboratory column and batch experiments were performed at room temperature under aerobic conditions, and a reactive transport model for vapors of volatile organic compounds in soil gas coupled to degradation kinetics was used for data interpretation (Höhener et al., 2003).

After an acclimatization period of 23 days, during which time steady-state diffusive vapor transport through the horizontal column was achieved, kinetic parameters were derived. While the removal of cyclic alkanes, isooctane, and 1,2,4-trimethylbenzene followed first-order kinetics over the whole concentration range applied, MTBE and chlorofluorocarbons (CFCs) were not visibly degraded. For many compounds, including MTBE, disappearance rates in abiotic batch experiments were as high as in live batches, indicating sorption.

In summary, investigating microorganisms that are able to grow on MTBE remains a challenge, due to: (1) the low biomass produced from MTBE by microorganisms (Fayolle et al., 2003; Fortin et al., 2001), and (2) the lack of efficient genetic tools for working with environmental bacteria.

For the remediation of MTBE-contaminated aquifers, the monitored natural attenuation could benefit very efficiently from the use of genetic tools for the characterization of the biodegradation potential towards MTBE in contaminated environments. Microarrays could be built, including genes involved in the MTBE biodegradation pathways, as well as sequences specifically designed from the genes from the different groups of strains involved in the biodegradation. Their use could help in directing the remediation process in such contaminated sites.

Finally, while gasoline hydrocarbons also biodegrade under denitrifying conditions, there are indications that MTBE and ETBE (ethyl tert-butyl ether) biodegrade under denitrifying conditions in the laboratory. Depending on the aquifer, supplemental nitrate may be required to achieve this. After the demand for electron acceptors is satisfied, essential nutrients like phosphorus and nitrogen become growth and activity limiting. Both nutrients can be delivered as a liquid or gas for stimulating biological activity in the saturated or vadose zone, respectively.

5 References

Adly Ibrahim, M., Mizuno, H., Yasuda, Y., Fukunaga, K., and Nakao, K. 2001. "Removal of Mixtures of Acetaldehyde and Propionaldehyde from Waste Gas in Packed Column with Immobilized Activated Sludge Gel Beads." *Biochem. Eng. J.*, 8: 9–18.

Boonchan, S., Britz, M.L., and Stanley, G.A. 2000. "Degradation and Mineralization of High-Molecular-Weight Polycyclic Aromatic Hydrocarbons by Defined Fungal-Bacterial Co-cultures." *Appl. Environ. Microbiol.*, 66: 1007–1019.

Chen, K.F., Kao, C., Chen, T.Y., Weng, C.H., and Tsai, C.T. 2006. "Intrinsic Bioremediation of MTBE-Contaminated Groundwater at a Petroleum-Hydrocarbon Spill Site." *Environ. Geol.*, 50: 439–445.

Cherry, J.A. 1987. "Groundwater Occurrence and Contamination in Canada." *Can. Aquat. Resour.*, 215: 387–426.

Cox, H.H.J., Houtman, J.H.M., Doddema, H.J., and Harder, W. 1993. "Enrichment of Fungi and Degradation of Styrene in Biofilters." *Biotechnol. Lett.*, 15: 737–742.

Day, M.J., Reinke, R.F., and Thomson, J.A.M. 2001. "Fate and Transport of Fuel Components below Slightly Leaking Underground Storage Tanks." *Environ. Forensics*, 2: 21–28.

Deeb, R.A., Hu, H.Y., Hanson, J.R., Scow, K.M., and Alvarez, C.L. 2001. "Substrate Interactions in BTEX and MTBE Mixtures by an MTBE-Degrading Isolate." *Environ. Sci. Tech.*, 35: 312–317.

Fayolle, F., François, A., Garnier, L., Godefroy, D., Mathis, H., Piveteau, P., and Monot, F. 2003. "Limitations in MTBE biodegradation." *Oil Gas Sci. Technol.*, 58: 497–504.

Fortin, N., Morales, M., Nakagawa, Y., Focht, D.D., and Deshusses, M.A. 2001. "Methyl Tert-Butyl Ether (MTBE) Degradation by a Microbial Consortia." *Environ. Microbiol.*, 3: 407–416.

George, S.C., Boreham, C.J., Minifie, S.A., and Teerman, S.C. 2002. "The Effect of Minor to Moderate Biodegradation on C5 to C9 Hydrocarbons in Crude Oils." *Organic Geochemistry*, 33: 1293–1317.

Höhener, P., Duwig, C., Pasteris, G., Kaufmann, K., Dakhel, N., and Harms, H. 2003. "Biodegradation of Petroleum Hydrocarbon Vapors: Laboratory Studies on Rates and Kinetics in Unsaturated Alluvial Sand." *Journal of Contaminant Hydrology*, 66: 93–115.

Hardison, L.K., Curry, S.S., Ciuffetti, L.M., and Hyman, M.R. 1997. "Metabolism of Diethyl Ether and Co-metabolism of Methyl *tert*-Butyl Ether by a Filamentous Fungus, A *Graphium* Sp." *Appl. Environ. Microbiol.*, 63: 3059–3067.

Johnson, R., Pankow, J., Bender, D., Price, C., and Zogorsky, J. 2000. "MTBE. To What Extent Will Past Releases Contaminate Community Water Supply Wells?" *Environ. Sci. Technol.*, 34: 210A–217A.

Kotterman, M.J.J., Vis, E.H., and Field, J.A. 1998. "Successive Mineralization Detoxification of Benzo(A)Pyrene by the White Rot Fungus *Bjerkandera* Sp. Strain BOS55 and Indigenous Microflora." *Appl. Environ. Microbiol.*, 64: 2853–2858.

Mays, M.A. 1989. "The Use of Oxygenates Hydrocarbons in Gasoline and their Contribution to Reducing Urban Air Pollution." *Pure Appl. Chem.*, 61: 1373–1378.

Mehlman, M.A. 1992. "Dangerous and Cancer-Causing Properties of Products and Chemicals in the Oil Refining and Petrochemical Industry: VIII. Health Effects of Motor Fuels: Carcinogenicity Of Gasoline: Scientific Update." *Environ. Res.*, 59: 238–249.

Obire, O., and Nwaubeta, O. 2001. "Biodegradation of Refined Petroleum Hydrocarbons in Soil." *J. Appl. Sci. Environ. Mgt.*, 5(1): 43–46.

Płaza, G.A., Łukasik, K., Wypych, J., Nał cz-Jawecki, G., Berry, C., and Brigmon, R.L. 2008. "Biodegradation of Crude Oil and Distillation Products by Biosurfactant-Producing Bacteria." *Polish J. of Environ. Stud.*, 17(1): 87–94.

Pointing, S.B. 2001. "Feasibility of Bioremediation by White-Rot Fungi." *Appl. Microbiol. Biotech.*, 57: 20–33.

Prenafeta-Boldú, F.X., Ballerstedt, H., Gerritse, J., and Grotenhuis, J.T.C. 2004. "Bioremediation of BTEX Hydrocarbons: Effect of Soil Inoculation with the Toluene-Growing Fungus *Cladophialophora* sp. strain T1." *Biodegradation*, 15: 59–65.

Prince, R.C., and Suflita, J.M. 2007. "Anaerobic Biodegradation of Natural Gas Condensate can be Stimulated by the Addition of Gasoline." *Biodegradation*, 18: 515–523.

Plaza, G.A., Łukasik, K., Wypych, J., Nał cz-Jawecki, G., Berry, C., and Brigmon, R.L. 2008. "Biodegradation of Crude Oil and Distillation Products by Biosurfactant-Producing Bacteria." *Polish J. of Environ. Stud.*, 17(1): 87–94.

Razavi-Shirazi, F., and Veenstra, J.N. 2000. "Development of a Biological Permeable Barrier to Remove 2,4,6-Trichlorophenol from Groundwater using Immobilized Cells." *Water Environ. Res.*, 72: 460–468.

Shishido, M., Kojima, T., Araike, Y.K., and Toda, M. 1995. "Biological Phenol Degradation by Immobilized Activated Sludge in Gel Bead with Three-Phase Fluidized-Bed Bioreactor." *Trans. Inst. Chem. Eng.*, 73(A): 719–726.

Solano-Serena, F., Marchal, R., Ropars, M., Lebeault, J-M., and Vandecasteele, J-P. 1999. "Biodegradation of Gasoline: Kinetics, Mass Balance and Fate of Individual Hydrocarbons." *J Microbiol.*, 86: 1008–1016.

Solano-Serena, F., Marchal, R., Huet, T., Lebeault, J-M., and Vandecasteele, J.P. 2000. "Biodegradability of Volatile Hydrocarbons of Gasoline." *Appl. Microbiol. Biotechnol.*, 54: 121–125.

Speight, J.G. 1996. *Environmental Technology Handbook.* Taylor & Francis, Washington, DC.

Speight, J.G., and Lee, S. 2000. *Environmental Technology Handbook.* 2nd Edition. Taylor & Francis, New York.

Speight, J.G. 2005. *Environmental Analysis and Technology for the Refining Industry.* John Wiley & Sons Inc., Hoboken, New Jersey.

Speight, J.G. 2007. *The Chemistry and Technology of Petroleum.* 4th Edition. CRC Press, Taylor & Francis Group, Boca Raton, Florida.

Steffan, R.J., Mccloy, K., Vainberg, S., Condee, C.W., and Zhang, D. 1997. "Biodegradation of the Gasoline Oxygenates Methyl tert-Butyl Ether, Ethyl tert-Butyl Ether and tert – Amyl Methyl Ether by Propane-Oxidizing Bacteria." *Appl. Environ. Microbiol.*, 63(11): 4216–4222.

Swoboda-Colberg, N.G. 1995. C"hemical Contamination of the Environment: Sources, Types, and Fate of Synthetic Organic Chemicals." In *Microbial Transformations and Degradation of Toxic Organic Chemicals.* L.Y. Young and C.E. Cerniglia (Editors). Wiley-Liss, New York. pp. 27–74).

Van Groenestijn, J.W., van Heiningen, W.N.M., and Kraakman, N.J.R. 2001. "Biofilters Based on the Action of Fungi." *Water Sci. Tech.*, 44: 227–232.

Woertz, J.R., Gunsch, C.K., Czerwinski, R.M., Kinney, K.A., Szaniszlo, P.J., and Whitman, C.P. 2001. "Toluene and Ethylbenzene Biodegradation by *Exophiala Lecani-Cornii* in Biofilters." *Proceedings.* American Society of Microbiology 101st General Meeting. American Society of Microbiology, Orlando, FL. pp. 638.

Yadav, J.S., and Reddy, C.A. 1993. "Degradation of Benzene, Toluene, Ethylbenzene and Xylenes (BTEX) by The Lignin-Degrading Basidiomycete *Phanerochaete chrysosporium.*" *Appl. Environ. Microbiol.*, 59: 756–762.

Yadav, B.K., and Hassanizadeh, S.M. 2011. "An Overview of Biodegradation of LNAPLs in Coastal (Semi)-arid Environment." *Water Air Soil Pollut.*, 220: 225–239.

11

Biodegradation of Kerosene and Diesel

1 Introduction

The extent of hydrocarbon biodegradation in contaminated soils is critically dependent upon four factors, namely the creation of optimal environmental conditions to stimulate biodegradative activity, the predominant petroleum hydrocarbon types in the contaminated matrix, and the bioavailavility of the contaminants to microorganisms.

The petroleum hydrocarbon degradation is also affected by the molecular composition of the hydrocarbons, a characteristic that is directly related with the bioavailability of these compounds, and, as a consequence, the biodegradation rate may be altered (Huesemann, 1995).

Bioremediation is being used or proposed as a treatment option at many hydrocarbon-contaminated sites (Braddock et al., 1997). The effectiveness of bioremediation is often a function of the microbial population or consortium, and how it can be enriched and maintained in an environment. Microorganisms with the ability to degrade crude oil are ubiquitously distributed in soil and marine environments (Venkateswaran and Harayama, 1995). However, when few or no indigenous degradative microorganisms exist in a contaminated area, or when there is not time for the natural enrichment of a suitable population, inoculation (bioaugmentation) can be a realistic option. Inoculation of bacteria with hydrocarbon biodegradation capabilities shorten the time of the treatment. The same ecological principles that influence biodegradation with the native microorganism population, in general, will also govern the effectiveness of the inoculation, regardless of whether they are natural isolates or genetically engineered microorganisms (Liu and Suflita, 1993).

Biodegradation experiments for diesel/biodiesel blends in liquid cultures by-petroleum degrading microbial consortium showed that for low amendments of biodiesel (10%), the overall biodegradation efficiency of the mixture after seven days was lower than for petroleum diesel fuel (Owsianiak et al., 2009). Preferential usage of methyl esters in the broad biodiesel concentration range and diminished biodegradation of petroleum hydrocarbons for 10% biodiesel blend was confirmed. Rhamnolipids improved biodegradation efficiency only for blends with low content of biodiesel. Emulsion formation experiments showed that biodiesel amendments significantly affected dispersion of fuel mixtures in water. The presence of rhamnolipids biosurfactant affected stability of such emulsions, and altered cell surface properties of tested consortium.

Some researchers have reported that inoculation had no positive, or only marginal, effects on oil biodegradation rates. Microorganisms able to degrade organic pollutants in cultures may fail to function when inoculated into natural environments, because they may be susceptible to toxins or predators in the environment. They may use other organic compounds in preference to the pollutant, or they may be unable to move through the soil to the contaminated site. The successful use of microbial inocula in soils requires that the microorganisms contact the contaminant.

Physical adsorption to soil particles or filtration through small pores may limit the transport of organisms (Margesin and Schinner, 1997). Most physical-chemical technologies have the ability to be scaled up with predictive reliability from laboratory or pilot experiments to full-scale, a normally *ex situ* operation. This is true for the processing industries as well as for non-biological environmental remediation applications.

Bioremediation, especially *in situ* bioremediation, has been identified as one technology in which accurate scale-up can be problematic, and the prediction of how long a remediation treatment will require and how well that treatment will work can at times be uncertain (Blackburn, 1998; Márquez-Rocha et al., 2001).

Petroleum hydrocarbons present in spills undergo mineralization (conversion to carbon dioxide and water) or are used as a primary food source by microorganisms, which use the energy to generate new cells. The most frequently used isolated hydrocarbon degraders in the bacterial genera are *Pseudomonas*, *Acinetobacter*, *Flavobacterium*, *Brevibacterium*, *Corynebacterium*, *Arthrobacter*, *Mycobacterium*, and *Nocardia*, while the fungal genera are *Candida*, *Cladosporium*, *Rhodotorula*, *Torulopsis*, and *Trichosporium*.

2 Identity and Origin of Kerosene and Diesel

Kerosene (*kerosine*), also called paraffin or paraffin oil, is a flammable, pale-yellow or colorless, oily liquid with a characteristic odor. Also called paraffin oil, it should not be confused with the much more viscous paraffin oil used as a laxative: use of this type of paraffin oil was used by miners to lubricate the alimentary tract that would contain coal dust residues inhaled during working hours in the mine.

The term *kerosene*, like many petroleum fractions, is not used consistently: several sources list Jet Fuel 1 (JP-1) as an exact synonym for kerosene, while other sources refer to kerosene as a distillation fraction or group that includes kerosene as fuel oil No. 1. The spelling is also inconsistent: most industry sources use the spelling as *kerosene*, while others use the spelling *kerosine*. Throughout this text, the spelling will be *kerosene*, and the fraction is defined by distillation range and other physical properties, as presented below.

Kerosene originated as a straight-run petroleum fraction that boiled between approximately 205 and 260°C (400 to 500°F) (Speight, 2007). Kerosene was the major refinery product before the onset of the *automobile age*, but now kerosene can be termed one of several secondary petroleum products after the primary refinery product, gasoline.

Kerosene is intermediate in volatility between gasoline and gas oil, and has a flash point on the order of 25°C (77°F). The term *kerosene* is also too often incorrectly applied to various types of fuel oil (Chapter 3, Chapter 13), but fuel oil is actually any liquid or liquid petroleum product that produces heat when burned in a suitable container or that produces power when burned in an engine (Speight, 2007).

Jet fuel comprises both gasoline and kerosene type jet fuels meeting specifications for use in aviation turbine power units, and is often referred to as *gasoline-type jet fuel* and *kerosene-type jet fuel*.

Chemically, kerosene is a mixture of hydrocarbons; the chemical composition depends on its source, but it usually consists of about 10 different hydrocarbons, each containing 10 to 16 carbon atoms per molecule; the constituents include n-dodecane ($C_{12}H_{26}$) and alkyl benzenes, as well as naphthalene and its derivatives.

Kerosene is lighter than water (density: 0.78–0.81 g/cm^3), with a flash point between 37 and 65°C (100 and 150°F): the autoignition temperature is 220°C (428°F). The heat of combustion of kerosene is similar to that of diesel fuel: the lower heating value is approximately is around 18,500 Btu/lb, or 43.1 MJ/kg, and the higher heating value is approximately 19,800 Btu/lb. Kerosene is immiscible with water (cold or hot), but (as expected) miscible with petroleum solvents.

Diesel fuel is a complex fuel mixture primarily consisting of paraffin, olefin, and aromatic hydrocarbons, and smaller quantities of substances containing sulfur, nitrogen, metals, and oxygen. The hydrocarbon molecules contain from 8 to 40 atoms of carbon, and are generally heavier than those found in gasoline.

However, the diesel fuel produced by a refinery is a blend of all the appropriate available streams: (1) straight-run product, (2) FCC light cycle oil, and (3) hydrocracked gas oil. The straight-run diesel may be acceptable as is, or may need minor upgrading for use in diesel fuel prepared for off-road use. To meet the 15 ppm sulfur limit, all the streams used to prepare diesel fuel need hydrotreating to lower the sulfur concentration.

Diesel oil leakages from underground storage tanks, distribution facilities, and various industrial operations represent an important source of soil and aquifer contamination. This fuel is a complex mixture of normal, branched and cyclic alkanes, and aromatic compounds obtained from the middle-distillate fraction during petroleum separation (Speight, 2007).

Diesel oil consists mostly of linear and branched alkanes with different chain lengths, and contains a variety of aromatic compounds. Many of these compounds, especially linear alkanes, are known to be easily biodegradable. However, due to their low water solubility, the biodegradation of these compounds is often limited by slow rates of dissolution, desorption, or transport. In general, the bioavailability of hydrophobic compounds is determined by their sorption characteristics and dissolution or partitioning rates, and by transport process to microbial cell (Sticher et al., 1997).

Kerosene and diesel fuel are middle distillate petroleum hydrocarbon products of intermediate volatility and mobility. As intermediate products, kerosene has a combination of (mostly) lighter, less persistent, and more mobile compounds, as well as (some) heavier, more persistent, and less mobile compounds. These two different groups are associated with two distinctly different patterns of fate/pathway concerns.

The relatively lighter, more volatile, mobile, and water soluble compounds in kerosene will tend to fairly quickly evaporate into the atmosphere, or migrate to groundwater. When exposed to oxygen and sunlight, most of these compounds will tend to break down relatively quickly. In groundwater, many of these compounds tend to be more persistent than in surface water, and readily partition on an equilibria basis back and forth between water and solids (soil and sediment) media. Cleaning up groundwater without cleaning up soil contamination will usually result in a rebound of higher concentrations of these compounds partitioning from contaminated soils into groundwater.

Thus, the environmental effects of kerosene and diesel fuel can be easily categorized and used to identify cleanup options:

- Moderately volatile: refined products can evaporate with no residue.
- Low and moderate viscosity: spreads rapidly into thin slicks.

- Can form stable emulsions.
- Moderate to high acute toxicity to biota; product specific toxicity related to type and concentration of aromatic compounds.
- Has potential for subtidal impacts (dissolution, mixing, sorption onto suspended sediments).

The impact of kerosene of diesel fuel leaking from a source can vary site to site. The source must be identified, and where the leaked fuel impacts building structures and buried services (such as water supply pipes), there can be a significant impact, and the site occupants may be required to move out of the property while cleanup is undertaken. Where the leaking fuel may impact controlled waters (streams or underground water), action must be quickly taken to prevent the spread of the contamination.

3 Bioremediation

The effectiveness of bioremediation is often a function of the microbial population and how they can be enriched and maintained in an environment. Strategies for inexpensive in situ bioremediation of soil contaminated with petroleum hydrocarbons include stimulation of the indigenous microorganisms by introduction of nutrients (biostimulation) and/or through inoculation of an enriched mixed microbial culture into soil (bioaugmentation).

3.1 Kerosene

There are numerous literature reports on biodegradation or bioremediation of kerosene in contaminated environmental media (Song and Bartha, 1990; Song et al., 1990); many of the environmental factors affecting the biodegradability of petroleum hydrocarbons are described in the context of applied bioremediation (Hinchee et al., 1995).

Generally, sources of kerosene pollution include accidental spills, pipe leakage, deliberate disposal of oily wastes, corrosion of pipes, kerosene seeps, and other operational deficiencies. Apart from the aesthetic and economic damage caused by kerosene spills, plant, vertebrates, invertebrates, and microorganisms in both the terrestrial and aquatic environments are adversely affected (Atlas and Bertha, 1975; Toogood, 1977; Odu, 1987; Guzman et al., 2004).

In Nigeria, the Niger Delta Region constantly experiences high levels of crude oil refining activities and transportation to other regions of the country. Studies on bioremediation of coastal environments have been restricted to crude oil contamination with respect to individual treatment, with little emphasis on the effects of combined treatment processes on microbial activities (Chikere and Okpokwasili, 2003; Chukwara et al., 2005; Itah and Essien, 2005; Asitok and Antai, 2006; Ekpo and Ekpo, 2006). Kerosene serves as a major source of energy to all sectors of the society for cooking and lighting, hence, spills arising from tank overflow, bunkering, poor vending facilities, and so forth are inevitable. It is also logical that a study of bioremediation of kerosene polluted soils would substantially provide the necessary guide for studying bioremediation effects on other forms of pollutants arising from crude oil refining.

In general, kerosene/jet fuel components biodegrade significantly under aerobic conditions, provided sufficient nutrients are present for conversion of the hydrocarbons to microbial biomass. There is a complex interplay between partitioning, and thus bioavailability, and biodegradation in the various media in the environment. Lower molecular-weight, normal hydrocarbons are most readily biodegraded, but tend to partition to air rather than water, while more complex, higher molecular weight polynuclear aromatics and substituted aromatics tend to sorb to soil or sediment; both processes limit bioavailability, and can slow biodegradation. The hydrocarbons in kerosene are generally not inhibitory to microbial activity (Aelion and Bradley, 1991; Song and Bartha, 1990), though changes in microbial community composition may occur in spill or impacted areas, due to the proliferation of species that can biodegrade the compounds (Song and Bartha, 1990).

A study on the effect of bioremediation treatments on the reclamation of kerosene spills on various soil samples was conducted around Calabar Petroleum Depot within the Niger Delta Region of Nigeria (Ikpeme et al., 2007). Soil samples subjected to different levels of kerosene concentrations ranging between 5 mL and 50 mL/100 g (v/w) of soil were analyzed, and a total of four bacterial isolates, *Bacillus subtilis*, *Pseudomonas acidovorans*, *Serratia marcescens*, and *Micrococcus sp.*, were identified in decreasing order of kerosene utilizing potential. The mean hydrocarbon utilizing bacteria counts increased significantly at higher incubation periods, though no significant difference was observed with depth of polluted soil. A combination of all treatments (tilling, fertilizing with nitrogen-phosphorus-potassium, and microbial seeding with *Bacillus subtilis*) significantly increased the amount of reclaimed kerosene. The amount of reclaimed kerosene differed significantly with the various levels of kerosene pollution.

Kerosene contaminated clay results in large amounts from the treatment of jet fuel-type kerosene produced from the Merox process, and represents a great environmental pollution problem. In order to investigate the mitigation of this phenomenon, treatment of the clay was performed with natural attenuation, biostimulation, and bioaugmentation in lab and field-scale microcosms (Gouda et al., 2007, 2008). More than 90% of the kerosene was biodegraded in bioaugmentation and biostimulation processes, while only 50% was obtained by natural attenuation after seven weeks. Urea and superphosphate were used as nitrogen and phosphorus sources, due to their low cost and local availability. The immobilized cells enhanced the biodegradation processes and reduced the time. Dehydrogenase activity was affected by the time and type of the treatment. In addition, the laboratory-scale work was successfully scaled up to field work.

To demonstrate the potential use of bioremediation in soil contaminated with kerosene, a laboratory study with the objective of evaluating and comparing the effects of bioattenuation, biostimulation, bioaugmentation, and combined biostimulation and bioaugmentation was performed (Agarry et al., 2010). The investigation dealt with the biodegradation of kerosene in soil under different bioremediation treatment strategies: bioattenuation, biostimulation, bioaugmentation, and combined biostimulation and bioaugmentation, respectively. Each treatment strategy contained 10% (w/w) kerosene in soil as a sole source of carbon and energy. After five weeks of remediation, the results revealed that bioattenuation, bioaugmentation, biostimulation, and combined biostimulation and bioaugmentation exhibited 44.1%, 67.8%, 83.1%, and 87.3% kerosene degradation, respectively.

Also, the total hydrocarbon-degrading bacteria (THDB) count in all the treatments increased with time up until the second week, after which it decreased. The highest bacterial growth was observed for combined biostimulation and bioaugmentation treatment strategy. A first-order kinetic model equation was fitted to the biodegradation data to further evaluate the rate of biodegradation, and the results showed that the specific degradation rate constant (k) value was comparatively higher for combined biostimulation and bioaugmentation treatment strategy than the values for other treatments.

The value of the kinetic parameter showed that the degree of effectiveness of these bioremediation strategies in the cleanup of soil contaminated with kerosene is in the following order: bioattenuation < bioaugmentation < biostimulation < combined biostimulation and bioaugmentation. This work defined combined biostimulation and bioaugmentation treatment strategy requirements for kerosene oil degradation, and thus opened an avenue for its remediation from contaminated soil.

Specifically, as with many hydrocarbon fractions, the degree of degradation of kerosene contaminants depends to some extent on soil depth, period of incubation, degree of pollution, and remediation treatment.

3.2 Diesel Fuel

Kerosene-range and diesel-range hydrocarbons are degraded primarily by bacteria and fungi, and adaptation by prior exposure of microbial communities to hydrocarbons increases hydrocarbon degradation rates.

Diesel oil, which is one of the major products of crude oil, constitutes a major source of environmental pollution. With the combined dependence on diesel oil by some vehicles and generators, greater quantities are being transported over long distances. Therefore, diesel oil can enter into the environment through wrecks of oil tankers carrying diesel oil, cleaning of diesel tanks by merchants, and war ships carrying diesel oil (Hill and Moxey, 1960; Speight, 2007). Diesel oil spills on agricultural land generally reduce plant growth. Suggested reasons for the reduced plant growth in diesel oil contaminated soils range from direct toxic effect on plants (Baker, 1982) and reduced germination (Udo and Fayemi, 1975) to unsatisfactory soil condition, due to insufficient aeration of the soil because of the displacement of air from the space between the soil particles by diesel oil.

A soil remediation project performed at the U.S. Army Cold Regions Research and Engineering Laboratory showed that after a period of treatment time, land farming was an effective method in reducing the concentration of total petroleum hydrocarbons in diesel-contaminated soil, as was bioventing. Nutrient additions had a significant effect on reducing diesel concentrations for both land farming and bioventing. Therefore, land farming has potential as an effective bioremediation method for heavy oils cleanup in remote or cold areas (Reynolds et al., 1997; Maila and Cloete, 2004).

The biodegradability of various types of diesel oil, such as straight-run diesel oil, light-cycle diesel oil, hydrocracking diesel oil, Fischer–Tropsch diesel oil, and commercial diesel oil, has been investigated in biodegradation tests performed in closed-batch systems using two microflorae (Penet et al., 2004). The first microflora was an activated sludge from an urban wastewater treatment plant, as commonly used in biodegradability tests of commercial products, and the second was a microflora from a hydrocarbon-polluted soil with possible specific capacities for hydrocarbon degradation. The kinetics of carbon dioxide production and extent of DO biodegradation were obtained by chromatographic procedures.

Under optimized conditions, the polluted-soil microflora was found to extensively degrade all the diesel oil types tested, the degradation efficiencies being higher than 88%. For all of the diesel tested, the biodegradation capacities of the soil microflora were significantly higher than those of the activated sludge. Using both microflora, the extent of biodegradation was highly dependent upon the type of diesel oil used, especially its hydrocarbon composition. Linear alkanes were completely degraded in each test, whereas identifiable branched alkanes, such as farnesane, pristane, or phytane, were degraded to variable extents. Among the aromatics, substituted mono-aromatics were also variably biodegraded.

In a follow-up study (Penet et al., 2006), it was recognized that in contaminated soils, efficiency of natural attenuation or engineered bioremediation largely depends on biodegradation capacities of the local microflorae. In the present study, the biodegradation capacities of various microflorae toward diesel oil were determined in laboratory conditions. Microflorae were collected from 9 contaminated and 10 uncontaminated soil samples, and were compared to urban wastewater activated sludge. The recalcitrance of hydrocarbons in tests was characterized using both gas chromatography (GC) and comprehensive two-dimensional gas chromatography (GC-GC). The microflorae from contaminated soils were found to exhibit higher degradation capacities than those from uncontaminated soil and activated sludge. In cultures inoculated by contaminated-soil microflorae, 80% of diesel oil on an average was consumed over a four-week incubation, compared to only 64% in uncontaminated soil and 60% in activated sludge cultures. As shown by GC, n-alkanes of diesel oil were totally utilized by each microflora, but differentiated degradation extents were observed for cyclic and branched hydrocarbons. The enhanced degradation capacities of impacted-soil microflorae resulted probably from an adaptation to the hydrocarbon contaminants, but a similar adaptation was noted in uncontaminated soils when conifer trees might have released natural hydrocarbons. GC-GC showed that a contaminated-soil microflora removed all aromatics and all branched alkanes containing less than C15. The most recalcitrant compounds were the branched and cyclic alkanes with 15–23 atoms of carbon.

The biodegradation rate of hydrocarbons is an important consideration in determining the time scale for bioremediation in oil-contaminated environments. Two naturally occurring bacterial cultures, *Exiguobacterium aurantiacum* and *Burkholderia cepacia*, were capable of utilizing diesel oil as the sole source of carbon and energy by induction of hydrophobic cell surfaces (Mohanty and Mukerji, 2008). The cultures demonstrated good degradation characteristics for diesel range n-alkanes (C_9 to C_{26}), and were also capable of degrading pristane. Biodegradation altered the relative abundance of n-alkanes in diesel, and resulted in a loss of symmetry in n-alkane distribution: C_9, C_{17} to C_{19}, and C_{26} were completely degraded by both of the cultures. In *B. cepacia*, the residual diesel was enriched in the higher carbon number n-alkanes C_{20} to C_{25}.

Biodegradation of petroleum hydrocarbon contamination, including contamination by diesel fuel, is a common method for remediating soils and groundwater. Due to complexities with field-scale studies, biodegradation rates are typically evaluated at the bench-scale in laboratory studies. However, important field conditions can be difficult to mimic in the laboratory. Three scaling factors can impact laboratory biodegradation rates that are frequently unaccounted for in typical laboratory experimental procedures. These factors are soil heterogeneity, morphology of petroleum hydrocarbon non-aqueous phase liquids (NAPLs), and soil moisture distribution. The effects of these factors on the biodegradation rate indicate that a high degree of variability results from even small-scale heterogeneities (Davis et al., 2003). In addition, it appears that as the experimental scale increases, the measured biodegradation rates slow. Most importantly, the results indicate that diesel biodegradation rates derived from small-scale experiments are not necessarily representative of field-scale biodegradation rates. There may be environmental adaptation after the spill that does not, or cannot, occur in the laboratory.

The high degree of variability observed at the various experimental scales indicates that small-scale soil and microbial heterogeneities significantly impact biodegradation rates. In addition, the biodegradation rates determined in small-scale batch or column systems are enhanced by more efficient diffusion rates of substrate and electron acceptors, and will be susceptible to variability introduced by small-scale heterogeneities (Davis et al., 2003).

Adaptation is brought about by selective enrichment of hydrocarbon-utilizing microorganisms and amplification of the pool of hydrocarbon-catabolizing genes (Jackson et al., 1996). Biodegradation of hydrocarbons usually requires the cooperation of more than a single species (Ghazali et al., 2004). This is particularly true in pollutants that are made up of many different compounds, such as crude oil or petroleum. Individual microorganisms can metabolize only a limited range of hydrocarbon substrates, so assemblages of mixed populations with overall broad enzymatic capacities are required to increase the rate and extent of petroleum biodegradation.

Microbial populations, which consist of strains that belong to various genera, have been detected in petroleum-contaminated soil or water (Sorkhoh et al., 1995). The intensity of hydrocarbon biodegradation in soil is influenced by several environmental factors, such as quality and quantity of contaminants, indigenous microbial populations, soil properties, pH, temperature, water content, and nutrient availability (Walworth et al., 2003).

Usually, the bacterial community is adapted to the presence of the contaminant, but other environmental conditions, such as nutrient availability and oxygen concentration, may be unfavorable, and thus, the microbial degradation of the contaminant is slow; bioremediation techniques accelerate the naturally occurring biodegradation by optimizing conditions with the addition of nutrients, organic matter, microbial inocula, and surfactants, and also by aeration, control of temperature, and water content (Jørgensen et al., 2000; Riffaldi et al., 2006).

The use of surfactants represents a valuable method to enhance the access of the microorganisms to low-soluble and recalcitrant compounds in bioremediation techniques (Franzetti et al., 2008; Whang et al., 2008; Wyrwas et al., 2011). The choice of surfactants is the first step of feasibility studies for this application. So far, no defined procedures are present in literature to select the most suitable surfactant for the treatment of a specific contaminated site. Furthermore, the characterization of physico-chemical parameters is important to understand the reason of successes and failures. In this paper, a step procedure to select and characterize a commercial surfactant to be used in bioremediation enhancement of hydrocarbon-contaminated media was developed. Among the commercial surfactants, the procedure was applied to alkyl polyethoxylates (Brij family) and sorbitan derivatives (Tween family). The selection resulted in the application of Brij 56 and Tween 80 as biodegradation-enhancer in different lab scale systems for remediation of diesel contamination. In liquid systems, Tween 80 greatly increased biodegradation of highly branched and high-molecular weight hydrocarbons, while Brij 56 enhanced degradation of highly branched hydrocarbons. Based on these results, the potential applications and the limitations of these surfactants at full scale level were estimated.

The biodegradation of diesel oil is more difficult that the lower boiling petroleum products, debatably because it is a middle distillate composed of hydrocarbons ranging from C_{11} to C_{25}. Mineralization of diesel oil has been reported to be complete in air-sparged liquid cultures (Geerdink et al., 1996) and incomplete in agitated flasks (Olson et al., 1999), and it was generally accepted that degradation of diesel oil hydrocarbons in soil microcosms was incomplete (Chaîneau et al., 1995; Gallego et al., 2001; Seklemova et al., 2001).

This partial recalcitrance possibly resulted from the low intrinsic biodegradability of diesel oil, a limitation in hydrocarbon transfer, or an insufficient amount of available oxygen. Furthermore, numerous degradation tests were performed in open systems where substantial amounts of substrate might disappear by volatilization or irreversible adsorption to soil (Song et al., 1990).

In fact, commercial diesel oil is a complex hydrocarbon mixture of thousands of individual components. The hydrocarbon composition of diesel oil depends on the origin of the crude oil used for the distillation process (straight-run diesel oil), on the refining processes of crude oil, and on the mixtures added by the refiner for final formulation.

Bioremediation of diesel oil in soil can occur by natural attenuation, or treated by biostimulation or bioaugmentation. In this study, all three technologies were evaluated for the degradation of total petroleum hydrocarbons (TPH) in soil (Bento et al., 2005). In addition, the number of diesel-degrading microorganisms present and microbial activity as indexed by the dehydrogenase assay were monitored. Soils contaminated with diesel oil in the field were collected from Long Beach, California, USA and Hong Kong, China. After 12 weeks of incubation, all three treatments showed differing effects on the degradation of light (C_{12}–C_{23}) and heavy (C_{23}–C_{40}) fractions of total petroleum hydrocarbons in the soil samples.

Bioaugmentation of the Long Beach soil showed the greatest degradation in the light (72.7%) and heavy (75.2%) fractions of total petroleum hydrocarbons. Natural attenuation was more effective than biostimulation (addition of nutrients), most notably in the Hong Kong soil. The greatest microbial activity (dehydrogenase activity) was observed with bioaugmentation of the Long Beach soil (3.3-fold), and upon natural attenuation of the Hong Kong sample (4.0-fold). The number of diesel-degrading microorganisms and heterotrophic population was not influenced by the bioremediation treatments. Soil properties and the indigenous soil microbial population affect the degree of biodegradation; hence, detailed site specific characterization studies are needed prior to deciding on the proper bioremediation method (Bento et al., 2005).

The bioremediation processes were applied to a sandy soil with a high level of contamination originating from the leakage of a diesel oil underground storage tank at a service station (Pinto Mariano et al., 2007). Enhancement of biodegradation was carried out through biostimulation (addition of nitrogen and phosphorus solutions or Tween 80 surfactant) and bioaugmentation (bacterial consortium isolated from a land farming system). To investigate interactions between optimizing factors, and to find the right combination of these agents, the study was based on full factorial experimental design. Efficiency of biodegradation was simultaneously measured by two methods: respirometric (microbial production of carbon dioxide) and gas chromatography. Acute toxicity tests with *Daphnia similis* were applied for examination of the efficiency of the processes in terms of the generation of less toxic products. Results showed that all bioremediation strategies enhanced the natural bioremediation of the contaminated soil, and the best results were obtained when treatments had nutritional amendment. Respirometric data indicated a maximum hydrocarbon mineralization of 19.8%, obtained through the combination of the three agents, with a total petroleum hydrocarbons (TPH) removal of 45.5% in 55 days of treatment. At the end of the experiments, two predominant bacteria species were isolated and identified (*Staphylococcus hominis* and *Kocuria palustris*).

Diesel biodegradation does not appear to be limited by the metabolic capabilities of the bacterial consortium, since a significant fraction of high molecular weight hydrocarbons ($>C_{12}$) were biodegraded at the same rate (Nwaogu et al., 2008). However, bacterial inability for total hydrocarbon utilization seems to be due to a mass transfer limitation, in which hydrocarbons are not available to microorganisms. *P. ostreatus* achieved more efficiently the removal of diesel, because this white rot fungus grows well through the soil (Márquez-Rocha et al., 2001).

A recent study (Piehler et al., 1999) has explored the potential role of the microbial community in the biodegradation of petroleum pollution by indigenous N2 fixing marine microbial consortia. The study also sought to explore potential bioremedial methods with minimal environmental impacts on coastal diesel fuel pollution for areas in which nitrogen has been found to be the primary nutrient limiting biodegradation of petroleum (Piehler and Paerl, 1996).

Successful stimulation of nitrogen fixation and petroleum hydrocarbon degradation in indigenous microbial consortia may decrease exogenous N requirements and reduce environmental impacts of bioremediation following petroleum pollution. The biodegradation of petroleum pollution by indigenous nitrogen-2 fixing marine microbial consortia was also investigated. Particulate organic carbon in the form of ground, sterile corn-slash (post-harvest leaves and stems) was added to diesel fuel amended coastal water samples to stimulate biodegradation of petroleum hydrocarbons by native microorganisms capable of supplying a portion of their own nitrogen. It was hypothesized that addition of particulate organic carbon to petroleum amended water samples from nitrogen-limited coastal waters would promote the growth of nitrogen fixing consortia and enhance biodegradation of petroleum. Throughout this study, there was a consistent enhancement of petroleum hydrocarbon degradation in response to the addition of particulate organic carbon.

Stimulation of diesel fuel biodegradation following the addition of particulate organic carbon was likely attributable to increases in bacterial nitrogen fixation, diesel fuel bioavailability, bacterial biomass, and metabolic activity. Toxicity of the bulk phase water did not appear to be a factor affecting biodegradation of diesel fuel following particulate organic carbon addition. These results indicate that the addition of particulate organic carbon to diesel fuel-polluted systems stimulated indigenous nitrogen-fixing microbial consortia to degrade petroleum hydrocarbons.

More recently, the rapid development of ionic liquids is gaining more and more attention, as these valuable chemicals will undoubtedly lead to environmental pollution. New formulations and application of ionic liquids may result in contamination in the presence of hydrophobic compounds, such as petroleum mixtures.

It might be hypothesized that that in the presence of diesel fuel, low-water-soluble ionic liquids may become more toxic to hydrocarbon-degrading microorganisms. Accordingly, the influence of 1-alkoxymethyl-2-methyl-5-hydroxypyridinium chloride homologues (side chain length from C3 to C18) on biodegradation of diesel fuel by a bacterial consortium has been investigated (Chrzanowski et al., 2009).

Tests performed for the bacterial consortium cultivated on disodium succinate showed that toxicity of the investigated ionic liquids decreased with increase in side-chain length, while only higher homologues (C_8 to C_{18}) caused a decrease in diesel fuel biodegradation. As a result of exposure to toxic compounds, modification in cell surface hydrophobicity was also observed. Disulphine blue active substances method was employed to determine partitioning index of ionic liquids between water and diesel fuel phase, which varied from 1.1 to 51% for C_3 and C_{18} homologues, respectively. Therefore, in the presence of hydrocarbons acting as a solvent, the increased bioavailability of hydrophobic homologues is responsible for the decrease in biodegradation efficiency of diesel fuel (Chrzanowski et al., 2009).

The number of studies on composting of petroleum-contaminated soil and petroleum-based oil wastes is increasing (Beaudin et al., 1996, 1999; Al-Daher et al., 1998; Kirchmann and Ewnetu, 1998; Milne et al., 1998; Jørgensen et al., 2000; Chaw and Stoklas, 2001; Namkoong et al., 2002). Elevated temperatures stimulate hydrocarbon degradation (Atlas, 1975), and enhance the contaminant availability by increased solubility and mass transfer (Pignatello and Xing, 1996). However, in the study of Namkoong et al. (2002), the temperature was controlled at 20°C (68°F), while in some other studies, the maximum composting temperature reached only 30°C (68°F) (Kirchmann and Ewnetu, 1998) or 40°C (104°F) (Jørgensen et al., 2000). In most studies, the organic amendment was added at a percentage ranging from 33 to 75%. However, in all of these studies, the main emphasis was laid on the degradation of the petroleum products, and not on the composting of the organic amendment.

In another study (Van Gestel et al., 2003), soil spiked with diesel oil was mixed with biowaste (vegetable, fruit and garden waste) at a 1:10 ratio (fresh weight) and composted in a monitored composting bin system for 12 weeks. Pure biowaste was composted in parallel. In order to discern the temperature effect from the additional biowaste effect on diesel degradation, one recipient with contaminated soil was hold at room temperature, while another was kept at the actual composting temperature. Measurements of composting parameters, together with enumerations and identifications of microorganisms, demonstrate that the addition of the contaminated soil had a minor impact on the composting process. The first-order rate constant of diesel degradation in the biowaste mixture was four times higher than in the soil at room temperature, and 1.2 times higher than in the soil at composting temperature.

In cold regions, biodegradation of fuel spills can take a prolonged period of time. Conventional fuels and crude oil contain contaminants, such as aromatics and polynuclear aromatic hydrocarbons, which can pose risks to humans and the environment. The goal of one study (Horel and Schiewer, 2009) was to investigate the biological degradation of an alternative synthetic fuel, Syntroleum, which is less toxic and, as shown in this study, more easily biodegradable than conventional diesel fuel. Use of alternative fuels such as Syntroleum would be especially beneficial in sensitive regions, where spills of conventional fuel are highly undesirable.

Both fuel types were biodegraded, with up to 75% mineralization after 17 weeks. The faster degradation rate was achieved in

Syntroleum-contaminated soils, while diesel fuel showed minimal degradation during several short-term studies (4–6 weeks), less than 5% total mineralization of the hydrocarbons in the fuel. The different moisture contents in the sandy soil showed no significant impact on respiration. The addition of fertilizer was essential to achieve good degradation rates. After the end of the 17-week experiment, the recovered contaminant was approximately 50% less in the case of Syntroleum when nutrients were added to the soil, as compared with nutrient-deficient conditions. Respiration rates were higher in sand than in gravel, which may be due to differences in soil porosity and the available surface area for more even hydrocarbon distribution.

Degradation rates varied significantly over time. A first-order model, which used different rate constants for three growth phases, was able to model cumulative carbon dioxide production quite well over a period of four months. In the carbon mass balance, the sum of the diesel range organics recovered from the soil plus the produced carbon dioxide accounted for approximately 30–85%. The remaining amount of carbon either was incorporated into biomass, degraded incompletely, or evaporated.

The full scale in situ remediation of diesel-contaminated soils has only recently been used in Antarctica, partly because it has been assumed that temperatures are too low for effective biodegradation (Delille et al., 2007). To challenge this idea, the effects of temperature on the hydrocarbon mineralization rate have been quantified during mesocosms, biopiles, and field pilot studies carried out on an artificially contaminated sub-Antarctic soil.

Field studies were initiated in December 2000 in a selected soil of The Grande Terre (Kerguelen Archipelago, 698 42VE–498 19VS). Four experimental plots (0.75_0.75 m) were settled firmly into the studied soil. Each plot received 500 mL of diesel fuel, and two of them were covered with a black plastic sheet. All plots were regularly sampled over a one-year period. Under natural sub-Antarctic conditions, the field tests revealed that up to 95% of the contaminants were degraded within one year, indicating that low temperatures (0–7°C) can still accommodate oil biodegradation by indigenous microorganisms. Covering the soil with plastic sheets induced a small but permanent increase of the temperature in the surface soil (annual mean of +2.2°C). The microbial response was improved by this bioremediation treatment. Mesocosm studies and pilot biopile experiments confirmed that a constant heating of soil could be an effective mean to accelerate bioremediation of diesel-contaminated sub-Antarctic soils. However, the microbial response was always improved by a complementary fertilizer addition.

Microbial fuel cell technology can potentially be applied to enhance subsurface bioremediation of contaminants such as petroleum hydrocarbons by providing an inexhaustible source of terminal electron acceptors to a groundwater environment that is likely depleted in thermodynamically favorable electron acceptors, such as oxygen and nitrate (Morris et al., 2009). Results indicate that anaerobic biodegradation of diesel range organics (compounds eluting with *n*-alkane markers ranging in size from C_8 to C_{25}) was significantly enhanced (82% removal) by microbial fuel cell technology, as compared to an anaerobically incubated control cell (31% removal) over 21 days at 30°C (86°F), meanwhile. The microbial consortium on the anode of a diesel-degrading microbial fuel cell were cloned sequences that showed >98% similarity to bacteria capable of denitrification, such as *Citrobacter* sp., *Pseudomonas* sp., and *Stenotrophomonas* sp. This study suggests that microbial fuel cell technology may be used for enhancing biodegradation of petroleum contaminants in anoxic environments, thus eliminating the need to amend terminal electron acceptors, such as oxygen.

The intrinsic biodegradability of fuels, such as diesel oil, has been determined by using a reference aerobic microflora taken from an urban waste water treatment plant (Marchal et al., 2003). As expected, diesel has a lower intrinsic biodegradability than gasoline. The recalcitrant hydrocarbons of fuels were cycloalkanes and branched alkanes, in particular those having quaternary carbon atoms or consecutive substituting groups on the main carbon chain. In the case of various types of diesel oil, the composition in terms of hydrocarbon structural classes accounted for the diverse biodegradation rates observed. In particular, the biodegradation rate was close to 100% when linear alkanes were most abundant (Fischer-Tropsch diesel oil). The fuel degradation capability was widespread among the environment microflorae tested. Microflorae from polluted soils displayed in general a slightly higher degradation capacity than that of non-polluted soils. Several mechanisms are involved in the efficiency of microflorae taken from polluted environments: (1) the presence of microorganisms with specialized metabolic capacities, (2) the occurrence of co-metabolism, and (3) positive interactions between strains (cooperation).

A comparative assessment of the biodegradative abilities of fungal species isolated from Nigerian soils contaminated with engine oil, diesel, and petrol was carried out. Microbiological and physicochemical properties such as moisture content, pH, particle size, total hydrocarbon (THC), organic carbon, nitrogen, and phosphorus contents of the contaminated soils were also investigated, and the biodegradative potentials of the isolated fungal species were determined by measuring the optical densities of the fungal cultures spectrophotometrically (Eja et al., 2006).

Incidence of fungal growth was highest in diesel-contaminated soil (41%) after an initial response to the diesel toxic effect. Among the fungal isolates, *Saccharomyces*, *Aspergillus*, *Cladosporium*, *Rhizopus*, *Mucor*, and *Penicillium* species, *Cladosporium* species from diesel-contaminated soil had the highest growth turbidity, while *Penicillium* from petrol-contaminated soil showed the highest growth turbidity. All of the fungal isolates showed low ability to degrade (used) engine oil, with the conclusion that *Penicillium* and *Cladosporium* were the microorganisms of choice for the degradation of gasoline and diesel, respectively (Eja et al., 2006).

3.3 Diesel Fuel Additives

The addition of oxygen-bearing compounds to diesel fuel considerably reduces particulate emissions. However, these additives add another dimension in terms of soil (or water) contamination and cleanup (bioremediation) technologies.

The 2-Ethyhexyl nitrate (2-EHN), the nitric acid ester of 2-ethyl-1-hexanol, is currently added in significant amounts (0.05%–0.4%) to diesel oil, to improve ignition and boost cetane number (Guibet and Faure, 1999; Bornemann et al., 2002). 2-EHN is a large-scale commodity, the worldwide production of which is estimated to be about 100,000 tons per year.

Although 2-EHN is a large-scale commodity that can be considered as a potential pollutant for soils and aquifers, little information on its biodegradability is available, but it is generally considered as being not readily biodegradable. Recent work has shown that 2-EHN is indeed biodegradable. This can be achieved using a NAPL system involving HMN or silicone oil and microbial communities (Solano-Serena et al., 2009). Most strains of *M. austroafricanum* were found capable of extensive 2-EHN biodegradation, thereby highlighting the versatility of those strains for the biodegradation of hydrophobic compounds to which 2-EHN belongs.

Tripropylene glycol methyl ether (TGME) and dibutyl maleate (DBM) (Figure 11.1) have been identified as possible diesel additives, based on their physicochemical characteristics and performance in engine tests. Although these compounds will reduce particulate emissions, their potential environmental impacts are unknown. As a means of characterizing their persistence in environmental media, such as soil and groundwater, a series of biodegradation tests was conducted of DBM and TGME (Marchetti et al., 2003). Benzene and methyl tertiary butyl ether (MTBE; Chapter 11) were also tested as reference compounds.

Primary degradation of DBM fully occurred within three days, while TGME presented a lag phase of approximately eight days and was not completely degraded by day 28. Benzene primary degradation occurred completely by day three, and MTBE did not degrade at all. The total mineralized fractions of DBM and TGME achieved constant values as a function of time of _65% and _40%, respectively. Transport predictions show that, released to the environment, DBM and TGME would concentrate mostly in soils and waters with minimal impact to air. From an environmental standpoint, these results, combined with the transport predictions, indicate that DBM is a better choice than TGME as a diesel additive.

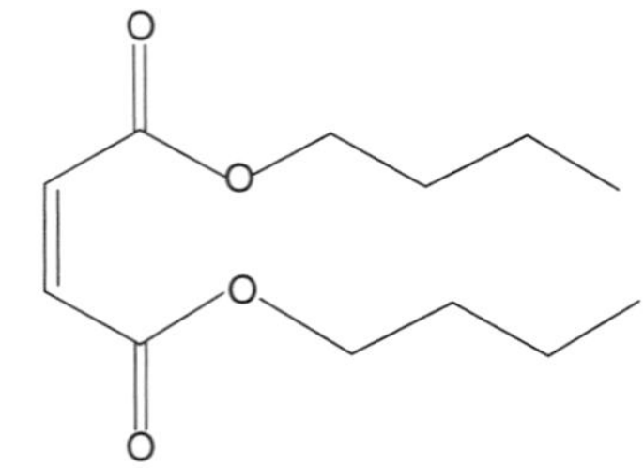

DBM

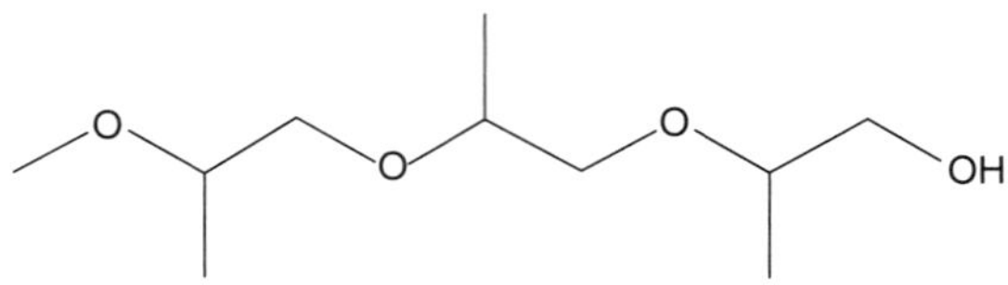

TGME

Figure 11.1 Chemical structures of dibutyl maleate (DBM) and tripropylene glycol monomethyl ether (TGME).

4 Jet Fuel

The bioremediation of jet fuel, a mixture of volatile and semi-volatile compounds at room temperature that is usually obtained as a fractional distillation of petroleum between 150 and 275°C (300 and 525°F), has been reported (Gouda et al., 2007, 2008; Machackova et al., 2008; Tzovolou et al., 2009).

In fact, bioremediation of this fuel is facilitated because volatile and semi-volatile linear alkanes, branched alkanes, and cycloalkanes predominate, while aromatics are less abundant. However, the high biodegradability of most components of jet fuel is not the only factor to be considered when determining the feasibility of bioremediation. In fact, the decrease in pollutant concentration may be due to both biotic and abiotic (mainly volatilization) processes. Indeed, the distinction between degradation processes may be complicated when evaluating the performance of bioremediation. This complexity has evident economic implications for decisions concerning the appropriate amendments and dosages, and predictions of the time required in the treatments. In this context, gas chromatography–mass spectrometry (GC-MS) analytical techniques aim not only to quantify a range of organic compounds, but also to differentiate between contaminant sources with similar chain length molecules, or provide available estimates of changes associated with dilution, evaporation, photo-oxidation, and biodegradation (Fingas, 1995; Garrett et al., 1998; Wang and Fingas, 2003).

In regard to biodegradation, among the useful procedures available to measure depletion (isotopes, respirometry, microbial growth, etc.), the use of the GC-MS to analyze non-biodegradable compounds as a standard reference presents some advantages, particularly when dealing with a mixture of contaminants, such as oil and its derivatives. For instance, in crude oil spills, hopane normalization has been reported as the best way to measure hydrocarbon biodegradation (Gallego et al., 2006). Nevertheless, light fuel has no high molecular weight (high-boiling) biomarkers, such as hopanes or steranes (Gagni and Camm, 2007); thus, C17/pristane and/or C18/phytane molecular ratios can be used to determine the progress of chemical breakdown processes (Bragg et al., 1994; Pollard et al., 1999; Gallego et al., 2007).

Concerning microbiology, the presence of bacteria with degrading capacity is a critical aspect of soil bioremediation; however, it is well known that hydrocarbon degraders are ubiquitous in natural soils (Gallego et al., 2002; Christopher et al., 2004), and hence, the limitations of soil bioremediation are usually more related to bioavailability than to the metabolic capacity of soil microorganisms (Huesemann, 1997; Trindade et al., 2005). Bioavailability (referred to that fraction available to microbial attack) restrictions are thought to be especially significant in aged soils, where lipophilic contaminant molecules have had sufficient time to diffuse into soil micropores, thus becoming inaccessible to hydrocarbon-degrading microorganisms (Ortega et al., 2007). This is a factor that limits the success of bioremediation, and it is directly associated with a high proportion of clay/silt-sized particles.

Following all of these premises, a complete biotreatability study for soil in a polluted emplacement affected by jet fuel spills has been described (Gallego et al., 2011). The work included the evaluation of habitual parameters considered in the selection of bioremediation strategies, i.e., the chemical, microbiological, and agronomical characteristics of the soil; nevertheless, in order to obtain useful information about bioavailability, a special focus was given to the textural properties of the soil. Subsequently, real-scale biopiles were implemented onsite by combining several biostimulation strategies. One of these strategies consisted of the use of a slow-release fertilizer (SRF) to add nitrogen and phosphorus, and another of the use of the SRF plus a surfactant (Volkering et al., 1998) to improve bioavailability; finally, a complementary approach, based on an oleophilic fertilizer (Santas et al., 1999), was also implemented.

The main objective of this full-scale site remediation, to reduce the total petroleum hydrocarbons (TPHs) to an environmental target, was defined in coherence with a previous risk assessment. Furthermore, the extent of degradation using geochemical biomarkers was examined to distinguish between biodegradation and volatilization (the principal objective being the identification of biodegradation losses), and textural data was also studied in order to predict the bioavailability of hydrocarbons. Finally, on the basis of the results, we draw several pertinent conclusions about the utility of oleophilic fertilizers and surfactants in other remediation cases.

JP-4 jet fuel constituents have been observed to be biodegraded, but only in the presence of nitric oxide (NO) as the electron acceptor (Hutchins et al., 1991).

Remediation of jet fuel at a former Soviet Hradcany Air Force Base in the Czech Republic (Machackova et al., 2008) involved a combination of technologies, including removal of volatile organic compounds using soil vapor extraction and air sparging, free product vacuum recovery, and aerobic biodegradation of organics with oxygen supplied by the air sparging system, along with nutrient addition.

The primary remedial method was found to be biodegradation, which has removed 93% of the contaminants from the site to date. A significant aspect of the remedial action was performance monitoring, including documentation of remediation efficiency.

The goal of the research was to assess the relative accuracy of methods commonly used for monitoring in situ remediation of total petroleum hydrocarbons. Two such methods were selected for the research: monitoring change in total petroleum hydrocarbons concentration in the soil (specified as nonpolar extractable substances) and monitoring respiration activity in soil with a subsequent stoichiometric mass balance to estimate the mass of total petroleum hydrocarbons destroyed. The study demonstrated that both of the methods provided comparable results regarding the effectiveness of in situ petroleum hydrocarbons destroyed remediation, despite the fact that their methodologies are very different.

5 References

Aelion, C.M., and Bradley, P.M. 1991. "Aerobic Biodegradation Potential of Subsurface Microorganisms from a Jet Fuel-Contaminated Aquifer." *Appl. and Envir. Microbiol.*, 57(1): 57–63.

Agarry, S.E., Owabor, C.N., and Yusuf, R.O. 2010. "Studies on Biodegradation of Kerosene in Soil under Different Bioremediation Strategies." *Bioremediation Journal*, 14(3): 135–141.

Al-Daher, R., Al-Awadhi, N., and El-Nawawy, A. 1998. "Bioremediation of Damaged Desert Environment Using Windrow Soil Pile System in Kuwait." *Environment International*, 24: 175–180.

Asitok, A.D., and S.P. Antai, 2006. "Petroleum Hydrocarbon Utilization and Biosurfactant Production by *Pseudomonas* and *Bacillus* species." *Nigerian J. Microbial.*, 20: 824–831.

Atlas, R.M. 1975. "Effects of Temperature and Crude Oil Composition on Petroleum Biodegradation." *Applied Microbiology*, 30: 396–403.

Atlas, R., and Bartha, R. 1975. "Biodegradation of Petroleum in Raw Water at Low Temperature." *Can. J. Microbial.*, 18: 1851–1855.

Baker, J.M. 1982. "Mangrove Swamps and The Oil Industry." *Environ. Pollut. Bull.*, p. 12.

Beaudin, N., Caron, R.F., Legros, R., Ramsay, J., Lawlor, L., and Ramsay, B. 1996. "Co-composting of Weathered Hydrocarbon-Contaminated Soil." *Compost Science and Utilization*, 4: 37–45.

Beaudin, N., Caron, R.F., Legros, R., Ramsay, J., and Ramsay, B. 1999. "Identification of the Key Factors Affecting Composting of a Weathered Hydrocarbon-Contaminated Soil." *Biodegradation*, 10: 127–133.

Bento, F.M., Camargo, F.A.O., Okeke, B.C., and Frankenberger, W.T. 2005. "Comparative Bioremediation of Soils Contaminated with Diesel Oil by Natural Attenuation, Biostimulation And Bioaugmentation." *Bioresource Technology*, 96: 1049–1055.

Braddock, J.F., Ruth, M.L., Walworth, J.L., and McCarthy, K.A. 1997. "Enhancement and Inhibition of Microbial Activity in Hydrocarbon-Contaminated Arctic Soils: Implications for Utrientamended Bioremediation." *Environ Sci. Technol.*, 31: 2078–2084.

Bragg, J.R., Prince, R.C., Harner, E.J., and Atlas, R.M. 1994. "Effectiveness of Bioremediation for the Exxon Valdez Oil Spill." *Nature*, 368: 413–418.

Chaîneau, C.H., Morel, J.L., and Oudot, J. 1995. "Microbial Degradation in Soils Microcosms of Fuel Oil Hydrocarbons from Drilling Cuttings." *Environ Sci. Technol.*, 29: 1615–1621.

Chaw, D., and Stoklas, U. 2001. "Co-composting of Cattle Manure and Hydrocarbon Contaminated Flare Pit Soil." *Compost Science and Utilization*, 9: 322–335.

Chikere, B.O., and Okpokwasili, G.C. 2003. "Enhancement of Biodegradation Of Petrochemicals By Nutrient Supplementation." *Nigerian J. Microbial.*, 17: 130–135.

Christopher, W., Kaplan, R., and Christopher, L. 2004. "Bacterial Succession in a Petroleum Land Treatment Unit." *Applied and Environmental Microbiology*, 70: 1777–1786.

Chrzanowski, L., Stasiewicz, M., Owsianiak, M., Szulc, A., Piotrowska-Cyplik, A., Olejnik-Schmidt, A.K., and Wyrwas, B. 2009. "Biodegradation of Diesel Fuel by a Microbial Consortium in the Presence of 1-Alkoxymethyl-2-Methyl-5-Hydroxypyridinium Chloride Homologs." *Biodegradation*, 20: 661–671.

Chukwura, E.I., Nwokolo, C.I., and Nwachukwu, S.C.U. 2005. "Bioremediation of Crude Oil-Polluted Escravos River using *Candida utilis*." *Nigerian J. Microbial.*, 19: 623–630.

Davis, C., Cort, T., Dai, D., Illangasekare, T.H., and Munakata-Marr, J. 2003. "Effects of Heterogeneity and Experimental Scale on the Biodegradation of Diesel." *Biodegradation*, 14: 373–384.

Delille, D., Pelletier, E., and Coulon, F. 2007. "The Influence of Temperature on Bacterial Assemblages during Bioremediation of a

Diesel Fuel-Contaminated Sub-Antarctic Soil." *Cold Regions Science and Technology*, 48: 74–83.

Ekpo, M.A., and Ekpo, E.I. 2006. "Utilization of Bonny Light and Bonny Medium Crude Oils by Microorganisms Isolated from Qua Iboe River Estuary." *Nigerian J. Microbial.*, 20: 832–839.

Fingas, M. 1995. "A Literature Review of the Physics and Predictive Modelling of Oil Spill Evaporation." *Journal of Hazardous Materials*, 42: 157–175.

Franzetti, A., Di Gennaro, P., Bestetti, G., Lasagni, M., Pitea, D., and Collina, E. 2008. "Selection of Surfactants for Enhancing Diesel Hydrocarbons-Contaminated Media Bioremediation." *Journal of Hazardous Materials*, 152: 1309–1316.

Gallego, J.L., Loredo, J., Llamas, J.F., Vazquez, F., and Sanchez, J. 2001. "Bioremediation of Diesel-Contaminated Soils: Evaluation of Potential In Situ Techniques by Study of Bacterial Degradation." *Biodegradation*, 12: 325–335.

Gallego, J.R., Loredo, J., Llamas, J.F., Vázquez, F., and Sánchez, J. (2002). "Bioremediation of Diesel-Contaminated Soils: Evaluation of Potential in Situ Techniques by Study of Bacterial Degradation." *Biodegradation*, 12: 325–335.

Gallego, J.R., González-Rojas, E., Peláez, A.I., Sánchez, J., García-Martínez, M.J., and Ortiz, J.E. 2006. "Natural Attenuation and Bioremediation of Prestige Fuel Oil in Atlantic Coasts of Galicia (Spain)." *Organic Geochemistry*, 37: 1869–1884.

Gallego, J.R., García-Martínez, M.J., Llamas, J.F., Belloch, C., Peláez, A.I., and Sánchez, J. 2007. "Biodegradation of Oil Tank Bottom Sludge using Microbial Consortia." *Biodegradation*, 18: 269–281.

Gallego, J.R., Sierra, C., Permanyer, A., Peláez, A.I., Menéndez-Vega, D., and Sánchez, J. 2011. "Full-Scale Remediation of a Jet Fuel-Contaminated Soil: Assessment of Biodegradation, Volatilization, and Bioavailability." *Water Air Soil Pollut.*, 217: 197–211.

Garrett, P.M., Pickerring, I.J., Haith, C.E., and Prince, R.C. 1998. "Photooxidation of Crude Oils." *Environmental Science and Technology*, 32: 3719–3723.

Geerdink, M.J., Van Loosdrecht, M.C.M., and Luyben, K.C.A.M. 1996. "Biodegradability of Diesel Oil." *Biodegradation*, 7: 73–81.

Ghazali, F., Zaliha, R., Salleh, A., and Basri, M. 2004. "Biodegradation of Hydrocarbons in Soil by Microbial Consortium." *International Biodeterioration & Biodegradation*, 54: 61–67.

Gouda, M.K., Omar, S.H., Chekrouda, Z.A., and Nour, H.M. 2007. "Bioremediation of Kerosene I: A Case Study in Liquid Media." *Chemosphere*, 69: 1807–1814.

Gouda, M.K., Omar, S.H., Chekrouda, Z.A., and Nour, H.M. 2008. "Bioremediation of Kerosene II: A Case Study in Contaminated

Clay (Laboratory and Field: Scale Microcosms)." *World Journal of Microbiology and Biotechnology*, 24: 1451–1460.

Guzman, J.A., Warrick, A.W., Myers, D.E., Musil, S.A., and Artiola, J.F. 2004. "Statistics and Geostatistics in Environmental Monitoring." *Environmental Monitoring and Characteristics*. Elsevier Academic Press, Burlington, Massachusetts. pp. 29–48.

Horel, A., and Schiewer, S. 2009. "Investigation of the Physical and Chemical Parameters Affecting Biodegradation of Diesel and Synthetic Diesel Fuel Contaminating Alaskan Soils." *Cold Regions Science and Technology*, 58: 113–119.

Huesemann, M.H. 1997. "Incomplete Hydrocarbon Biodegradation in Contaminated Soils: Limitations in Bioavailability or Inherent Recalcitrance?" *Bioremediation Journal*, 1: 27–39.

Hill, G.B., and Moxey, J.G. 1960. "Gasoline and Diesel Oil." In *Petroleum Product Handbook*. V.B. Gathee (Editor). McGraw Hill, New York.

Hinchee, R.E., Kittel, J.A., and Reisinger, H.J. 1995. *Applied Bioremediation of Petroleum Hydrocarbons*. Battelle Press, Columbus, Ohio.

Hutchins, S.R., Sewell, G.W., Kovacs, D.A., and Smith, G.A. 1991. "Biodegradation of Aromatic Hydrocarbons by Aquifer Micro-Organisms under Denitrifying Conditions." *Environ. Sci. Technol.*, 25: 68–76.

Huesemann, M.H. 1995. "Predictive Model for Estimating the Extent of Petroleum Hydrocarbon Biodegradation in Contaminated Soils." *Environmental Science and Technology*, 29(1): 7–18.

Ikpeme, E.M., Nfongeh, J.F., and Etim, L. 2007. "Comparative Bioremediation Enhancement Procedures on Kerosine Polluted Ultisol from a Niger Delta Region, Southern Nigeria." *Research Journal of Microbiology*, 2: 856–860.

Itah, A.Y., and Essien, J.P. 2005. "Growth Profile and Hydrocarbonoclastic Potential of Micro-Organisms Isolated from Tarballs in the Bight of Bonny, Nigeria." *World J. Microbial. Biotechnol.*, 21: 1317–1322.

Jackson, A.W., Pardue, J.H., and Araujo, R. 1996. "Monitoring Crude Oil Mineralization in Salt Marshes: Use of Stable Carbon Isotope Ratios." *Environ. Sci. Techn.*, 30; 1139–1144.

Jørgensen, K.S., Puustinen, J., and Suortti, A.M. 2000. "Bioremediation of Petroleum Hydrocarbon-contaminated Soil by Composting in Biopiles." *Environ. Pollut.*, 107: 245–254.

Kirchmann, H., and Ewnetu, W. 1998. "Biodegradation of Petroleum-Based Oil Wastes through Composting." *Biodegradation*, 9: 151–156.

Liu, S. and Suflita, J.M. 1993. "Ecology and Evolution of Microbial Populations for Bioremediation." *Trends Biotechnol.*, 11(8): 344–52.

Machackova, J., Wittlingerova, Z., Vlk, K., Zima, J., and Linka, A. 2008. "Comparison of Two Methods for Assessment of In Situ Jet-fuel Remediation Efficiency." *Water Air Soil Pollut.*, 187: 181–194.

Maila, M.P., and Cloete, T.E. 2004. "Bioremediation of Petroleum Hydrocarbons through Land Farming: Are Simplicity and Cost-Effectiveness the Only Advantages?" *Review in Environmental Science and Biotechnology*, 3: 349–360.

Marchal, R., Penet, S., Solano-Serena, F., and Vandecasteele, J.P. 2003. "Gasoline and Diesel Oil Biodegradation." *Oil & Gas Science and Technology - Rev. IFP*, 58(4): 441–448.

Marchetti, A.A., Knize, M.G., Chiarappa-Zucca, M.L., Pletcher, R.J., and Layton, D.W. 2003. "Biodegradation of Potential Diesel Oxygenate Additives: Dibutyl Maleate (DBM) and Tripropylene Glycol Methyl Ether (TGME)." *Chemosphere*, 52: 861–868.

Margesin, R., and Schinner, F. 1997. "Efficiency of Indigenous and Inoculated Cold-Adapted Soil Microorganisms for Biodegradation of Diesel Oil in Alpine Soils." *Appl. Environ. Microbiol.*, 63: 2660–2664.

Márquez-Rocha, F.J., Hernández-Rodríguez, V., and Lamela, Ma.T. 2001. "Biodegradation of Diesel Oil in Soil by a Microbial Consortium." *Water, Air, and Soil Pollution*, 128: 313–320.

Milne, B.J., Baheri, H.R., and Hill, G.A. 1998. "Composting of a Heavy Oil Refinery Sludge." *Environmental Progress*, 17: 24–27.

Mohanty, G., and Mukherji, S. 2008. "Biodegradation Rate of Diesel Range N-Alkanes by Bacterial Cultures *Exiguobacterium aurantiacum* and *Burkholderia cepacia*." *International Biodeterioration & Biodegradation*, 61: 240–250.

Morris, J.M., Jin, S., Crimi B., and Pruden, A. 2009. "Microbial Fuel Cell in Enhancing Anaerobic Biodegradation of Diesel." *Chemical Engineering Journal*, 146: 161–167.

Namkoong, W., Hwang, E.-Y., Park, J.-S., and Choi, J.-Y. 2002. "Bioremediation of Diesel-Contaminated Soil with Composting." *Environmental Pollution*, 119: 23–31.

Nwaogu, L.A., Onyeze, G.O.C., and Nwabueze, R.N. 2008. "Degradation of Diesel Oil in a Polluted Soil using *Bacillus subtili*." *African Journal of Biotechnology*, 7(12): 1939–1943.

Odu, C.T. 1987. "The Effect of Nutrient Application and Aeration on Oil Degradation in Soil." *Environ. Pollut.*, 15: 235–240.

Olson, J.J., Mills, G., Herbert, B.E., and Morris, P.J. 1999. "Biodegradation Rates of Separated Diesel Components." *Environ. Toxicol. Chem.*, 18: 2448–2453.

Ortega, J.J., Ball, W.P., Schulin, R., Semple, K.T., and Wick, L.Y. 2007. "Bioavailability of Pollutants and Soil Remediation." *Journal of Environmental Quality*, 36: 1383–1384.

Owsianiak, M., Chrzanowski, Ł., Szulc, A., Staniewski, J., Olszanowski, A., Olejnik-Schmidt, A.K., and Heipieper, H.J. 2009. "Biodegradation of Diesel/Biodiesel Blends by a Consortium of Hydrocarbon Degraders:

Effect of the Type of Blend and the Addition of Biosurfactants." *Bioresource Technology*, 100: 1497–1500.

Penet, S., Marchal, R., Sghir, A., and Monot, F. 2004. "Biodegradation of Hydrocarbon Cuts Used for Diesel Oil Formulation." *Appl. Microbiol. Biotechnol.*, 66: 40–47.

Penet, S., Vendeuvre, C., Fabrice Bertoncini, F., Marchal, R., and Monot, F. 2006. "Characterization of Biodegradation Capacities of Environmental Microflorae for Diesel Oil by Comprehensive Two-Dimensional Gas Chromatography." *Biodegradation*, 17: 577–585.

Piehler, M.F., and Paerl, H.W. 1996. "Enhanced Biodegradation of Diesel Fuel through the Addition of Particulate Organic Carbon and Inorganic Nutrients in Coastal Marine Waters." *Biodegradation*, 7: 239–247.

Piehler, M.F., Swistak, J.G., Pinckney, J.L., and Paerl, H.W. 1999. "Stimulation of Diesel uel Biodegradation by Indigenous Nitrogen Fixing Bacterial Consortia." *Microb. Ecol.*, 38: 69–78.

Pignatello, J.J., and Xing, B. 1996. "Mechanisms of Slow Sorption of Organic Chemicals to Natural Particles." *Environmental Science and Technology*, 30: 1–11.

Pinto Mariano, A., de Arruda Geraldes Kataoka, A.P., de Franceschi de Angelis, D., and Marcos Bonotto, D. 2007. "Laboratory Study on the Bioremediation of Diesel Oil Contaminated Soil from a Petrol Station." *Brazilian Journal of Microbiology*, 38: 346–353.

Pollard, S.J., Whittaker, M., and Risden, G.C. 1999. "The Fate of Heavy Oil Wastes in Soil Microcosms I: A Performance Assessment of Biotransformation Indices." *The Science of the Total Environment*, 226: 1–22.

Reynolds, C.M., Bhunia, P., and Koenen, B.A. 1997. *Soil Remediation Demonstration Project: Biodegradation of Heavy Fuel Oils*. Special Report 97–20. Cold Regions Research & Engineering Laboratory US Army Corps of Engineers, Washington, DC. August.

Riffaldi, R., Levi-Minzi, R. Cardelli, R., Palumbo, S., and Saviozzi, A. 2006. "Soil Biological Activities in Monitoring the Bioremediation of Diesel Oil-Contaminated Soil." *Water, Air, and Soil Pollution*, 170: 3–15.

Santas, R., Korda, A., Tenente, A., Buchholz, K., and Santas, P.H. 1999. "Mesocosm Assays of Oil Spill Bioremediation with Oleophilic Fertilizers: Inipol, F1 or Both?" *Marine Pollution Bulletin*, 38: 44–48.

Seklemova, E., Pavlova, A., and Kovacheva, K. 2001. "Biostimulation-Based Bioremediation of Diesel Fuel: Field Demonstration." *Biodegradation*, 12: 311–316.

Solano-Serena, F., Nicolau, E., Favreau, G., Jouanneau, Y., and Marchal, R. 2009. "Biodegradability of 2-Ethylhexyl Nitrate (2-EHN), a Cetane Improver of Diesel Oil." *Biodegradation*, 20: 85–94.

Song, H-G., and Bartha, R. 1990. "Effects of Jet Fuel Spills on The Microbial Community of Soil." *Appl. Environ Microbiol.*, 56(3): 646–651.

Song, H-G., Wang, X., and Bartha, R. 1990. "Bioremediation Potential of Terrestrial Fuel Spills." *Appl. Environ. Microbiol.*, 56(3): 652–656.

Sorkhoh, N.A., Al-Hasan, R.H., Khanafer, M., and Radwan, S.S. 1995. "Establishment of Oil-Degrading Bacteria Associated with Cyanobacteria in Oil-Polluted Soil." *J. Appl. Bact.*, 78: 194–199.

Speight, J.G. 2007. *The Chemistry and Technology of Petroleum*. 4th Edition. CRC Press, Taylor & Francis Group, Boca Raton, Florida.

Sticher, P., Jaspers, M.C., Stemmler, K., Harms, H., Zehnder, A.J., and van der Meer, J.R., 1997. "Development and characterization of a whole-cell bioluminescent sensor for bioavailable middle-chain alkanes in contaminated groundwater samples." *Appl. Environ. Microbiol.*, 63: 4053–4060.

Toogood, J.A. 1977. "Effects of Oil Spills on Physical Properties of Soil." *Reclamation of Agricultural Soil after Oil Spills*. Institute Petrology, University, Edmonton, Alberta, Canada. pp. 108–118.

Trindade, P.V.O., Sobral, L.G., Rizzo, A.C.L., Leite, S.G.F., and Soriano, A.U. 2005. "Bioremediation of a Weathered and Recently Oil-Contaminated Soils from Brazil: A Comparison Study." *Chemosphere*, 58: 515–522.

Tzovolou, D.N., Benoit, Y., Haeseler, F., Klint, K.E., and Tsakiroglou, C.D. 2009. "Spatial Distribution of Jet Fuel in the Vadose Zone of a Heterogeneous and Fractured Soil." *Science of the Total Environment*, 407: 3044–3054.

Udo, E.J., and Fayemi, A.A. 1975. "The Effects of Oil Pollution of Soil on Germination, Growth and Nutrient Uptake of Corn." *J. Environ Qual.*, 4: 537–540.

Van Gestel, K., Mergaert, J., Swings, J., Coosemans, J., and Ryckeboer, J. 2003. "Bioremediation of Diesel Oil-Contaminated Soil by Composting With Biowaste." *Environmental Pollution*, 125: 361–368.

Venkateswaran, K., and Harayama, S. 1995. "Sequential enrichment of microbial populations exhibiting enhanced biodegradation of crude oil." *Can. J. Microb.*, 41: 767–775.

Volkering, F., Breure, A.M., and Rulkens, W.H. 1998. "Microbiological Aspects of Surfactant Use for Biological Soil Remediation." *Biodegradation*, 8: 401–417.

Walworth, J.L., Woolard, C.R., and Harris, K.C. 2003. "Nutrient Amendments for Contaminated Peri-Glacial Soils: Use of Cod Bone Meal as a Controlled Release Nutrient Source." *Cold Regions Science and Technology*, 37: 81–88.

Wang, Z.D., and Fingas, M. 2003. "Development of Oil Hydrocarbon Fingerprinting and Identification Techniques." *Marine Pollution Bulletin*, 47: 423–452.

Whang, L-M., Liu, P-W.G., Ma, C-C., and Cheng, S-S. 2008. "Application of Biosurfactants, Rhamnolipid, and Surfactin, for Enhanced

Biodegradation of Diesel-Contaminated Water and Soil." *Journal of Hazardous Materials*, 151: 155–163.
Wyrwasa, B., Chrzanowskib, Ł., Ławniczakb, Ł., Szulcb, A., Cyplikc, P., Białas, W., Szymański, A., and Hołderna-Odachowskaa, A. 2011. "Utilization of Triton X-100 and Polyethylene Glycols during Surfactant-Mediated Biodegradation of Diesel Fuel." *Journal of Hazardous Materials*, 197: 97–103.

12

Biodegradation of Fuel Oil

1 Introduction

Fuel oil is a fraction obtained from petroleum distillation, either as a distillate or a residue. The term *fuel oil* is also used in a stricter sense to refer only to the heaviest commercial fuel that can be obtained from crude oil, but higher boiling and having a higher density than naphtha and gasoline. More generally, fuel oil is any liquid petroleum product that is burned in a furnace or boiler for the generation of heat, or used in an engine for the generation of power, except oils having a flash point on the order of 40°C (104°F) and oils burned in cotton or wool-wick burners. In this sense, diesel (Chapter 11) is a type of fuel oil. Fuel oil consists of alkane constituents, cycloalkane constituents, and aromatic constituents.

Fuel oil that is used for heating is graded from No. 1 Fuel Oil to No. 6 Fuel Oil (Table 12.1 and Table 12.2), and covers relatively low-boiling distillate oils, medium-boiling distillate, high-boiling distillate, a blend of distillate and residue, and residue oil (Chapter 3).

Heavy fuel oil consists primarily of the residue from distillation or cracking units in the refinery. Historically, fuel oils were based on long residues (see section 2.2) from the atmospheric distillation column, and were known as straight run fuels. However, the increasing demand for transportation fuels such as gasoline, kerosene, and diesel has led to an increased value for the atmospheric residue as a feedstock for vacuum distillation and for cracking processes. As a consequence, most heavy fuel oils are currently based on atmospheric residua and residua from thermal

Table 12.1 Different grades of fuel oil.

Name	Alias	Alias	Type	Chain Length
No. 1 fuel oil	No. 1 distillate	No. 1 diesel fuel	Distillate	9–16
No. 2 fuel oil	No. 2 distillate	No. 2 diesel fuel	Distillate	10–20
No. 3 fuel oil	No. 3 distillate	No. 3 diesel fuel	Distillate	10–26
No. 4 fuel oil	No. 4 distillate	No. 4 residual fuel oil	Distillate/ Residual	12–70
No. 5 fuel oil	No. 5 residual fuel oil	Heavy fuel oil	Residual	12–70
No. 6 fuel oil	No. 6 residual fuel oil	Heavy fuel oil	Residual	20–70

and catalytic cracking operations. These fuels differ in character from straight run fuels in that the density is higher and the hydrogen/carbon atomic ration is lower. The density of some heavy fuel oils can be above that of water (1.000 kg/m^3), which has environmental implications in the event of a spillage into fresh water.

Heavy fuel oils are composed of residual materials blended with middle distillates, or cutter stocks of lower molecular weight, and their composition is highly variable. As a result, the ecotoxicity associated with these oils varies depending on the nature and proportion of the added middle distillates. The following review collects together the available data on the environmental effects of heavy fuel oils, based both on laboratory studies and observations following accidental spillages.

However, fuel oil, like other petroleum-derived fuels, is rarely produced by a single process.

While not part of the actual refining process, on-site refinery fuel blending, handling, and storage can have a significant impact on fuel quality as bunkered. By blending, less desirable cracked residual oil can be made more attractive as fuel oil. This is accomplished by adding to a given quantity of residual oil a small amount of lighter distillate or cutter stock. It is not uncommon for these fuels to have been produced in different refinery units and even from different crude stocks. Thus, the potential for incompatibility problems to occur, such as sludge formation or stratification, exists.

Table 12.2 Properties of the various grades of fuel oil.

No. 1 fuel oil	Similar to kerosene or range oil (fuel used in stoves for cooking). A distillate intended for vaporizing in pot-type burners and other burners where a clean flame is required.
No. 2 fuel oil	Often called domestic heating oil with properties similar to diesel and higher-boiling jet fuel. Defined as a distillate for general purpose heating in which the burners do not require the fuel to be completely vaporized before burning.
No. 3 fuel oil	A light industrial heating oil that is intended for use where preheating is not required for handling or burning. Two grades that differ primarily in safety (flash) and flow (viscosity) properties.
No. 4 fuel oil	A heavy industrial oil that often requires preheating for burning and, in cold climates, for handling.
No. 5 fuel oil	A more viscous oil usually containing residuum. Commonly referred to as *Bunker C oil* when it is used to fuel ocean-going vessels. Preheating is required for both handling and burning this grade oil.

Currently, marine fuels bunkered as residual oils are in fact blends of deep cracked residuum and lighter distillate cutter stock. These have, as a result of increasing refinery yield demands on crude supplies, all but replaced the straight-run residual fuels obtained from atmospheric distillation, which characterized marine residual fuels of the past.

While the most significant portion of fuel contamination by water and debris normally occurs during barge transport, pipeline transport and tank farm storage at the refinery also can increase contaminant levels prior to transport for shipboard use.

There have been numerous petroleum releases in cold, remote regions. Alternatives for cleanup are limited by the remote locations, difficulties in mobilization of heavy equipment, inability to effectively monitor the biotreatment processes, and the costs of site operation. To remediate petroleum-contaminated sites, we need a low-cost, low input treatment alternative to use in conjunction with existing methods.

An assessment of the effects of biodegradation on the composition of a spilled fuel oil will be useful not only for the feasibility studies regarding the application of bioremediation techniques, but also for providing monitoring indicators of the natural evolution of the spill, which will be particularly valuable for the source recognition of oil samples collected a certain time after the incident.

The particular characteristics of fuel oil residue, devoid of the more labile fractions (b.p. <300°C, <570°F) and with high content of aromatic hydrocarbons (on the order of 50%) and resin constituents and asphaltene constituents (on the order of 30%), advocate for an assessment of the biodegradation process to better understand the ultimate fate of the spilled and stranded heavy fuel oil. A laboratory assay needs to be performed to provide preliminary information on the potential biodegradability of the fuel oil. However, it must be recognized that weathering will decrease the rate of degradation of the fuel oil unless the rate of biodegradation of recalcitrant compounds can be increased.

2 Identity and Origin of Fuel Oil

Heavy fuel oils are blended products based on the residues from various refinery distillation and cracking processes (Chapter 3). They are viscous liquids with a characteristic odor and require heating for storage and combustion. Heavy fuel oils are used in medium to large industrial plants, marine applications, and power stations in combustion equipment such as boilers, furnaces, and diesel engines.

Heavy fuel oil is a general term, and other names commonly used to describe this range of products include: residual fuel oil, bunker fuel, bunker C, fuel oil No 6, industrial fuel oil, marine fuel oil, and black oil. In addition, terms such as heavy fuel oil, medium fuel oil, and light fuel oil are used to describe products for industrial applications to give a general indication of the viscosity and density of the product.

Heavy fuel oils are blended products based on the residues from various refinery distillation and cracking processes. It is a general term used to describe the range of products that includes residual fuel oil, bunker fuel, bunker C, fuel oil No 6, industrial fuel oil, marine fuel oil and black oil. Heavy fuel oils also contain organo-metallic compounds from their presence in the original crude oils. Other elements that occur in heavy fuel oils include nickel, iron, potassium, sodium, aluminum, and silicon.

Residual fuel oils are complex mixtures of high molecular weight compounds having a typical boiling range from 350 to 650°C (660 to 1200°F), and consist of aromatic, aliphatic, and naphthenic hydrocarbons, typically having carbon numbers from C_{20} to C_{50}, together with asphaltene constituents and smaller amounts of heteronuclear compounds containing sulfur, nitrogen, and oxygen. These high molecular weight petroleum distillates have a density (specific gravity) approaching or even exceeding that of water, and are usually extremely viscous. These properties present serious environmental challenges when discharged into the surroundings.

Numerous small releases of petroleum products, including heavier grades, such as #6 fuel oil and bunker C, have occurred in cold regions, and many of these releases have occurred in remote sites, where alternatives for cleanup are few, and mobilization of heavy equipment is difficult to justify. Because of the large number of these sites, low-cost, low-input systems are needed to remediate these soils. A cost-effective bioremediation strategy applicable to remote sites would both address the need for soil cleanup at these sites and allow limited resources to be used to treat a larger number of sites.

2.1 Production

Because the process history of a refinery stream determines its chemical composition, it is expected that streams that have undergone similar processing will have similar physical/chemical/biologic properties and environmental fate and transport characteristics. The carbon ranges for streams in this category are directly related to physical/chemical properties and the potential for environmental effects. See Appendix A for a more detailed description of each of these streams. Knowledge of refining processes, in addition to carbon range and physical/chemical properties, coupled with tests of representative substances, can be useful in evaluating human health effects. As shown in Figure 1, the major processes used to produce the refinery streams included in the heavy fuel oils category are: (1) atmospheric distillation, (2) vacuum distillation, (3) cracking, and (4) reforming.

Heavy fuel oil related streams produced by *atmospheric distillation* comprise fractions of crude oil separated by heating (346 to 374°C, 650 to 700°F) at atmospheric pressure. They include atmospheric distillates (heavy gas oils) and the heavier residual materials. Some of these streams may be further hydrotreated or desulfurized to remove sulfur, nitrogen, and other impurities. Most atmospheric distillates undergo further processing in order to convert them into higher value fuels (diesel, kerosene).

The residuum from the atmospheric distillation unit is distilled in the *vacuum distillation* unit to further separate heavier molecules without the use of high temperatures. This is done under reduced pressure to prevent thermal cracking. In addition to producing lube oils, various vacuum distillates (vacuum gas oils) and vacuum residuals are produced. Similar to the atmospheric distillates, some of the vacuum distillates may be hydrotreated or desulfurized to remove sulfur, nitrogen, and other impurities. Most vacuum distillates undergo further processing in order to convert them into higher value fuels (diesel, kerosene).

Portions of the heavier atmospheric or vacuum distillate streams may be used as blending stocks to reduce the viscosity of other residual streams. The atmospheric and vacuum residual refinery streams each comprise a heterogeneous group of poorly defined, viscous, high boiling hydrocarbon streams that usually contain suspensions of resin/asphaltene complexes. These streams often have high levels of heterocyclic aromatic and naphthenic compounds. Varying percentages of sulfur, nitrogen, oxygen, and other elements are present as heterocyclic inclusions, primarily in the aromatics fraction. These residual streams often have a polynuclear aromatic hydrocarbon content in excess of 5% w/w.

Many of the distillate and residual streams used to blend heavy fuel oils are derived from *cracking processes* (thermal or catalytic), which convert higher boiling petroleum streams produced by atmospheric or vacuum distillation into lighter molecular weight materials, such as gasoline, diesel fuel, jet fuel, and kerosene.

The refinery streams produced by the various cracking processes represent a continuum in the severity of the cracking process. All of the cracking processes produce refinery streams that are similar from a physical-chemical perspective, being differentiated from each other primarily by the ratio of their unsaturated and saturated hydrocarbon content. The saturated and aromatic hydrocarbons species are similar, but may vary in ratio between streams. For instance, refinery streams that are produced by catalytic cracking have high levels of aromatics. In contrast, hydrocracked streams have relatively low amounts of aromatics, since hydrocracking introduces hydrogen into the cracking process resulting in saturation of aromatic compounds. As shown in Figure 1, there are eleven refinery streams in the heavy fuel oils category that are produced by cracking (five distillate and six residual streams).

Catalytic reforming employs a catalyst to facilitate the structural rearrangement of hydrocarbon molecules in order to increase the aromatic content of a refinery stream, ultimately producing higher octane gasoline blending stocks. During reforming, olefins are saturated to form paraffins, which are then converted to shorter paraffins, isoparaffins, and naphthenes. The naphthenes are converted to aromatics by dehydrogenation (Speight, 2007).

In addition to the process streams discussed above, the heavy fuel oil category also includes two blended residual fuel oils: (1) residual fuel oil and (2) No. 6 fuel oil. These two fuel oils are most often produced by blending any combination of the distillate and residual streams so that the finished fuel meets the appropriate product specifications. The residual fuels can also be blended with petroleum distillates (cutter stocks).

Using distillation endpoints, Bunker C fuel oil has been cited and used as a supporting material that is representative of a No. 6 fuel oil. Bunker fuel gets its name from the containers on ships and in ports in which it is stored (storage bunkers). Therefore, the composition of Bunker C fuels is expected to be similar to other substances in this category, and any differences may be explained by the variability in the streams from which these products are made and the characteristics of the original crude oil. For this reason, Bunker C fuel oil is a valid supporting substance to this category that provides valuable data for characterizing SIDS endpoints. Furthermore, much of the data on the fate and effects of heavy fuel oils are derived from studies on oil spilled at sea, of which Bunker C fuel has reported in a number of studies (Jézèquel et al., 2003; Lee et al., 2003).

2.2 Properties

The heavy fuel oil category includes both finished products (residual fuels) and the primary refinery streams from which they are blended. The residual fuels are primarily used in industrial boilers and other direct source heating applications (e.g., blast furnaces), and as a fuel for large marine diesel engines. The residual fuels are products that consist primarily of the residuum of the refining process after virtually all of the higher quality hydrocarbons have been removed from crude oil feedstock.

Historically, fuel oils were based on residuals from atmospheric distillation. However, the increasing demand for transportation fuels, such as gasoline, kerosene, and diesel, and lubricants, has led to an increased value for the atmospheric residuum as a feedstock for vacuum distillation and for cracking processes. As a consequence, most heavy fuel oils are currently based on vacuum residuum and residua from thermal and catalytic cracking operations (Speight, 2007). These high viscosity residual streams may in turn be diluted with lower quality, lighter weight distillates to produce a finished residual fuel of a specified viscosity.

The choice of the distillate diluent (cutter stock) is itself variable, and largely a function of availability at any given time within the refinery and the viscosity specifications of the fuel being manufactured. While materials frequently used as cutter stocks are included in this category, other streams may be used, e.g., kerosene and gas oils. The exact blend used for a specific residual fuel is determined largely by the desired viscosity of the finished fuel and applicable technical specifications. Some of the refinery streams in the heavy fuel oil category that have lower viscosities and lower molecular weight polynuclear aromatic compound levels have low-volume, specialty applications, such as cutter stock in cutback asphalt, and carbon electrode production.

Members of the heavy fuel oils category are a diverse group of substances that encompass hydrocarbons with a wide range of molecular weights, with carbon numbers ranging from C_7 to $\geq C_{50}$ and boiling points between 120 to 600°C (250 to 1110°F), but higher and lower boiling blending components may be present, depending on either the manufacturing processes or the technical requirements of the final product. However, typical heavy fuel oils are C_{20} to $\geq C_{50}$, with the low carbon numbers and boiling temperatures being associated with lighter weight diluent streams.

All of the category members are complex substances, containing variable amounts of alkanes, cycloalkanes, aromatics, olefins, asphaltene constituents, and heteroatom molecules containing sulfur, oxygen, nitrogen, and organo-metals. Because they are complex substances composed of relatively high molecular weight compounds, the materials in this category are difficult to characterize in detail. Consequently, they are typically not defined by detailed compositional information, but instead by process history, physical properties, and product use specifications. Since viscosity and sulfur are often the controlling specifications, and other limiting requirements, notably boiling point ranges, are unspecified, there is significant variation in the chemical composition of the resulting commercial residual fuel products.

No. 6 fuel oil is highly viscous, and is produced by blending heavy residual oils with a lighter oil (often No. 2 fuel oil) to meet specifications for viscosity and pour point. The specific gravity of a particular No. 6 fuel oil can vary from 0.95 to greater than 1.03. Thus, spilled oil can float, suspend in the water column, or sink. Small changes in water density may dictate whether the oil will sink or float.

When spilled on water, No. 6 fuel usually spreads into thick, dark colored slicks, which can contain large amounts of oil. The most viscous No. 6 oils will often breakup into discrete patches and tar balls when spilled, instead of forming slicks. Oil recovery by skimmers and vacuum pumps can be very effective, early in the spill. Very little of this viscous oil is likely to disperse into the water column.

No. 6 fuel oil is persistent: typically 5 to 10% is expected to evaporate within the first hours of a spill. Consequently, the oil can be carried hundreds of miles in the form of scattered tar balls by winds and currents. The tar balls will vary in diameter from several meters to a few centimeters, and may be very difficult to detect visually or with remote sensing techniques.

Floating oil in a high sediment environment (rivers, beaches) could potentially sink once it picks up sediment, resulting in subsurface tar balls or tar mats. These oils can occasionally form an emulsion, but usually only slowly, and after a period of days. Because of its high viscosity, beached oil tends to remain on the surface rather than penetrate sediments. Light accumulations usually form a line at the high-tide mark; heavy accumulations can pool on the beach.

Shoreline cleanup can be very effective before the oil weathers, becoming stickier and even more viscous. Natural degradation rates for these heavy oils are very slow. The oil may persist on beaches for months to years. Adverse effects of floating No. 6 fuel oil are related primarily to coating of wildlife dwelling on the water surface, smothering of intertidal organisms, and long-term sediment contamination. No. 6 fuel oil is not expected to be as acutely toxic to water column organisms as lighter oils, such as No. 2 fuel oil.

Direct mortality rates can be high for seabirds, waterfowl, and fur-bearing marine mammals, especially where populations are concentrated in small areas, such as during bird migrations or marine mammal haul outs. Direct mortality rates are generally less for shorebirds, because they rarely enter the water. Shorebirds, which feed in intertidal habitats where oil strands persist, are at higher risk of sublethal effects from either contaminated or reduced population of prey. The most important factors determining the impacts of No. 6 fuel oil contamination on marshes are the extent of oiling on the vegetation and the degree of sediment contamination from the spill or disturbance from the cleanup. Many plants can survive partial oiling; fewer survive when all or most of the above-ground vegetation is coated with heavy oil. However, unless the substrate is heavily oiled, the roots often survive, and the plants can re-grow.

3 Biodegradation

The enrichment of hydrocarbon-degrading bacteria and the persistence of petroleum hydrocarbons on an estuarine beach after a spill of residual fuel oil on 11 April 1973 in Upper Narragansett Bay, Rhode Island was the subject of an early investigation (Pierce et al., 1975). A rapid enrichment occurred during days 4 to 16 after the oil spill, and a significant population of hydrocarbon-degrading bacteria was maintained in the beach sand for at least a year. The concentration of petroleum hydrocarbons in the mid-tide area declined rapidly during the bacterial enrichment period, remained fairly constant throughout the summer, and then declined to a low concentration after one year. An increased concentration of branched and cyclic aliphatic hydrocarbons in the low-tide sediment 128 days after the spill suggested a migration of hydrocarbons during the summer. Hydrocarbon biodegradation was apparent during the winter months, but at a reduced rate.

This investigation has demonstrated an enrichment of a population of hydrocarbon-degrading bacteria in beach sediments. This population would apparently play a significant role in the degradation, and perhaps translocation, of petroleum hydrocarbons in response to a spill of residual fuel oil. After one year, the original fuel oil was still recognizable in the low-tide sediment, and although the normal aliphatic and aromatic hydrocarbon fractions were considerably reduced, the branched and cyclic aliphatic hydrocarbons persisted, presumably as a part of the organic material sequestered in the sediment.

Upon release into the environment, heavy fuel oil will break into small masses, and will not spread as rapidly as less viscous oil (Petersen, 1978; Fremling, 1981; Wolfe, 1977; Grose et al., 1979; Mehrasbi et al., 2003). The density of some heavy fuel oils means that they may sink on release to water, rather than float on the surface like other petroleum fuels. Loss of the lower molecular weight components due to volatility and dissolution will increase the density of the floating oil, causing it to sink.

This heavy fraction will assume a tar-like consistency and stick to exposed substrates, or become adsorbed to particulates. Weather conditions and temperature during the period after the spill will significantly influence the rate of dispersion; wind and wave action will tend to disperse oil into the water column, while higher temperatures will increase the rate of evaporation of lighter hydrocarbons. Water temperature is a major factor in determining the extent of the environmental impact following a heavy fuel oil spill, since higher temperatures will enhance loss of lighter constituents by evaporation as well as degradation processes.

In November 2002, the oil tanker *Prestige* sank into 12000 feet of water at about 130 nautical miles off the Spanish coasts, west-southwest of Cape Finisterre, after releasing an estimated 120,000 barrels of heavy fuel oil into the sea. Several hundred kilometers of coastline were coated in oil sludge after the disaster. Extensive hand and mechanically assisted removal was the chosen procedure for the initial cleanup of the shore. High pressure water jets were used to ease oil detachment from the stones. The *Prestige* oil spill was determined as a complex mixture of hydrocarbons made of saturated (23%) and aromatic (53%) hydrocarbons and resin constituents, plus asphaltene constituents (24%). The aromatic fraction was mainly composed of naphthalene, phenanthrene, and alkyl derivatives, and the saturated fraction was composed of lineal and cyclic hydrocarbons of variable length (Alzaga et al., 2004).

Microbial biodegradation is known to be an efficient process in the decontamination of oil-polluted environments. Common steps in establishing *in situ* bioremediation potential include search for hydrocarbon degrading indigenous microorganisms and optimization of conditions promoting biodegradation *in situ* by indigenous microbiota. In this process, it is important to assess whether intrinsic bioremediation, i.e., removal of the contaminant by indigenous bacterial populations, is taking place at the contaminated site. Many biochemical reactions result in carbon isotope fractionation, and molecules containing lighter 12C-isotope are used preferentially. Early studies already suggested that microbial degradation may lead to the 13C/12C isotope fractionation of compounds with high molecular masses, such as alkanes, in marine environments (Stahl, 1980). Isotopic fractionation was also observed in aromatic compound degradation under different conditions, both in the laboratory and in aquifers (Dempster et al., 1997; Meckenstock et al., 1999; Ahad et al., 2000).

In these examples, labeled aromatic substrates were added, and isotopic fractionation was followed in the remaining substrates after degradation. The fractionation levels obtained were generally below 3‰. On the other hand, crude oil, a natural product of biological origin, is depleted in 13C. Degradation of this oil by seawater microorganisms should render a local relatively ^{13}C-poor carbon dioxide and, as a consequence, a decrease in the $^{13}C/^{12}C$ ratio of the water dissolved inorganic carbon.

Stable carbon isotope ratios of inorganic carbon have been used in the past years to verify petroleum hydrocarbon mineralization (Van De Velde et al., 1995; Aggarwal et al., 1997; Bolliger et al., 1999). Recently, it has been demonstrated (Bugna et al., 2004) that the products of hydrocarbon respiration reflected the ^{13}C value of the substrate, and thus, could be used *in situ* to trace aerobic hydrocarbon degradation, provided that local dissolved inorganic carbon had a value significantly different from that of the contaminant.

In another study (Medina-Bellver et al., 2005), it was decided to use the carbon isotope ratio ($\delta^{13}C$) of the dissolved inorganic carbon in contaminated water samples taken from the shore of the Ons Island in the north of Spain as an easy and rapid approach to assess *in situ* microbial activity on crude oil at the *Prestige* spill area. On the basis of the results obtained, laboratory experiments were carried out to estimate crude oil biodegradation capability of the natural population under different conditions, in order to determine possible limiting factors to this process. The presence of bacterial populations with specific degradation traits was tested in samples collected at several sites in the Galician shore.

Putative biodegradation processes taking place at areas with high income of fresh seawater could not be detected with this technique. Laboratory-scale biostimulation processes carried out in samples with the highest oil biodegradation activity showed that N/P deficiency in seawater is a limiting factor for crude oil degradation. The most probable number (MPN) of crude oil component degraders was estimated for several aromatic compounds (naphthalene, anthracene, phenanthrene, and pyrene) and for undecane. Our results clearly show that bacteria present in the contaminated water are readily able to transform components of the crude oil into inorganic carbon.

Sedimentation has been shown to be an important mechanism for removal of heavy fuel oil from the water column following some spills. During the second week following a spill of No. 5 fuel oil from the Soviet tanker Tsesis in 1977, 0.7% of sediment collected at the site was fuel oil adsorbed to detritus or clay particles (Johansson et al., 1980; Linden et al., 1979). In all, 30 to 60 tons of the oil remaining after cleanup at this site was transported to the bottom by sedimentation.

The composition of heavy fuel oil will change drastically during weathering. Initially, a rapid loss of the low molecular weight components by evaporation, dissolution, and biodegradation will occur, resulting in almost complete removal of n-alkanes up to C_{17} in the first year following a spill (Guard and Cobet, 1972; Petersen, 1978; Vandermeulen, 1977; Rashid, 1974). The more recalcitrant pristane and phytane will be biodegraded more slowly, but should be degraded in a few years. The branched chain alkane components will be the most resistant to degradation, and will persist for many years (Song et al., 1990). In 28 day laboratory studies with crude and refined oils, Bunker C was the least degraded (11%, compared to 51–82% for crudes), due to its higher proportion of high molecular weight aromatics (Walker et al., 1976; Geyer, 1980; Mulkins-Phillips and Stewart, 1974). This slow rate of degradation has also been observed in the field (Petersen, 1978; Wolfe, 1977; Rashid, 1974; Geyer, 1980).

Bunker C from the Arrow spill could still be found in the subsurface sediments and on some shorelines up to seven years after the spill (Rashid, 1974; Vandermeulen and Gordon, 1976). Oil removal was occurring fastest in areas with heavy wave action, and slowest at high tidal areas in sheltered locations (Wolfe, 1977; Rashid, 1974). Covering of oil through bioturbation and sediment accumulation were more effective removal mechanisms for this tar-like residue than biodegradation (Baker et al., 1993). Re-suspension of these residues could lead to contaminant spread and continued impact on benthic and interstitial organisms.

In general, indigenous soil microbiota can degrade petroleum compounds. When soils fail to bioremediate at optimum rates, it is often a function of the water solubility of the compound and environmental limitations imposed on the microbes. Major limitations to the microbiota are temperatures that are too high or too low, excess or deficient water, insufficient or excessive nutrients, insufficient carbon in a form that microorganisms can use, poor mixing or distribution of the petroleum in the soil, and, for aerobic microorganisms, lack of oxygen.

The relative effectiveness of different treatment systems will vary over time. For example, lack of oxygen has been believed to be the primary limitation at depth; thus, air-injection technologies are commonly employed to overcome this. However, it is now well established that subsurface (relatively deep) microbial activity is common, and anaerobic biodegradation of water-soluble petroleum takes place without the need to inject air if alternate electron acceptors, such as oxidized species of iron or nitrogen (such as nitrate), are available. Bioremediation treatment is successful when limitations are overcome. The key problem, however, is identifying and implementing the most cost-effective means of doing this at isolated, cold sites.

In comparing treatment methods, it is important to measure effectiveness over time. In applying a bioremediation technology to a particular site, it is also important to consider contaminant transport, leaching, and the location of the problem. The rate of treatment must be compared to the rate of leaching and the distance between the contaminated soil and the area where it might cause harm: the potential receptor. For many sites, transport rates and distance to receptors are such that remediation need not be immediate. If a longer time is acceptable for remediation, more treatment options and lower costs can be considered.

A final difficulty in comparing the effectiveness of treatments is that of obtaining accurate measurements in the field. Limitations to effective bioremediation have been identified primarily through laboratory studies. Measuring rates in the field is much more difficult, because field conditions cannot be controlled effectively. In one field study, where attempts were made to measure bioremediation rates in a land farm that had been treated uniformly, the bioremediation rates varied up to seven-fold (Reynolds, 1993b).

One study has shown that land farming can be a reasonable alternative to bioventing for both heavy-oil- and diesel-contaminated soils at remote sites typical of cold regions (Reynolds, 1993a, 1993b; Reynolds et al., 1994; Reynolds et al., 1997). For comparison, in diesel-contaminated soil, reduction in the concentration of total petroleum hydrocarbons was greatest using land farming with nutrient amendments. Land farming with nutrient additions should be considered as a treatment option at sites where it can be easily used. Annual reductions in the concentration of total petroleum hydrocarbons for heavy fuel oil after approximately one year were not significantly different for any of the treatments.

For fuel oil-contaminated soil, land farming was more effective than bioventing during the initial phase of treatment, and this effect was more pronounced when nutrients were added. The lag period that occurs prior to rapid microbial growth and increased activity appeared to have a greater effect in the heavy-oil-contaminated soil, compared to the diesel-contaminated soil. It is possible that the measured increase in the concentration of total petroleum hydrocarbons is an analytical artifact of microbial fatty acids and related microbial biomass that are produced during adaptation of the soil microbiota to new conditions and carbon sources.

If this is the cause, it is reasonable to predict that, given favorable incubation conditions that are sufficiently long, there may be a period of rapid heavy-oil bioremediation following the temporary increase in measured concentration of total petroleum hydrocarbons. However, if the favorable incubation time is too short owing to a brief summer season, the same cycle may occur in subsequent years.

Other studies have suggested that plant-based systems may be useful (Nichols et al., 1997). These data demonstrate that, depending on the nature of the contaminant and nutrient status of the soil, the brevity of the summer season, when temperatures are favorable for microbial activity, may prevent either unamended or nutrient-amended biotreatment from attaining sufficient microbial numbers and activity to substantially reduce the concentration of total petroleum hydrocarbons during that season.

This may result from microbial population cycling, from low populations at early summer to high populations at late summer, with little metabolization of the total petroleum hydrocarbons.

The persistence of the heavier fraction of older oil spills in cold regions suggests that this may be happening (Collins et al., 1993). By comparison, for diesel spills, the effect is not as pronounced, and many of the compounds are more readily degraded; hence, nutrient amendments appeared to provide sufficient stimulation to significantly reduce the concentration of the total petroleum hydrocarbons. Despite both treatments being done ex situ (requiring excavation and transportation of soil), costs for the land farming treatment were less than for bioventing.

Bioventing of stockpiled, diesel-contaminated soil has also been used in northern climates, but owing to low temperatures and the distance of many sites from readily accessible highways, logistical, construction, and maintenance requirements and costs are greater than in more temperate climates.

Studies have shown that stimulating soil microbial activity in diesel-contaminated soils in bio-piles may raise soil temperatures by 1 to 3°C (2 to 5°F) (Brar et al., 1993). Also, nutrient amendments to diesel-contaminated soil can enhance bioremediation rates in both land farming and bioventing (Reynolds, 1993b; Brar et al., 1993; Walworth and Reynolds, 1995).

Biotreatment of soil contaminated with heavier oils, such as No. 6 heating oil, Bunker C fuel oil, oil sludge (Dibble and Bartha, 1979), and crude oil (Chapter 10), have not been evaluated to the same degree as diesel-fuel-contaminated soil. And, because a greater proportion of heavy fuel oil is composed of recalcitrant compounds that have limited water solubility, these contaminants are less readily bioremediated than diesel fuel.

Measuring field bioremediation rates is difficult because of the nonuniformity of contaminant distribution; it is also costly. Yet, bioremediation rate comparisons, based on direct field measurements, are the true measure of treatment effects. To effectively compare land farming to bioventing, we must either develop a better system for field monitoring, or wait a longer time before measuring, so that differences would be more pronounced, or know that the effects of the treatments would be vastly different. Because the effect of the treatments is our unknown, and a comparison of rates over time is important, the only acceptable option was to develop a better system for field monitoring.

To determine a bioremediation rate at any given time, it is necessary to compare contaminant concentrations to the initial concentration of the contaminant in the soil. To make valid rate comparisons, it is essential either to provide known contaminant concentrations at the start (*to*), or take an extremely large number of samples to quantify the random error caused by spatial variability. In addition, the *destructive nature* of taking samples (because resampling of the same volume of soil either cannot be done or cannot be reproduced accurately) and the difficulty in determining exactly how many samples would be required to sufficiently reduce the variability make the process even more difficult.

During the initial phases of degradation, the microorganisms use bioavailable forms of native soil carbon rather than contaminant carbon. Native soil carbon forms are less likely to be included in total petroleum hydrocarbon extraction or measurement (Chapter 8). However, the metabolites produced by the rapidly expanding population do appear as part of the measurement of the in total petroleum hydrocarbons.

Low-cost strategies that further stimulate the microbial population may be needed to effectively treat these soils. Recent studies suggest that plant-based systems may be useful (Nichols et al., 1997).

The reduced requirements for infrastructure also make land farming attractive in remote sites typical of cold regions. Landform installation and operation in a remote area is less costly than bioventing. Similarly, the remoteness of many contaminated sites may allow more time for the biomediation process, which would favor a low-input, low-maintenance land farming system.

4 References

Aggarwal, P.K., Fuller, M.E., Gurgas, M.M., Manning, J.F., and Dillon, M.A. 1997. "Use of Stable Oxygen and Carbon Isotope Analyses for Monitoring the Pathways and Rates of Intrinsic and Enhanced *In Situ* Biodegradation." *Environ. Sci. Technol.*, 31: 590–596.

Ahad, J.M.E., Lollar, B.S., Edwards, E.A., Slater, G.F., and Sleep, B. 2000. "Carbon Isotope Fractionation during Anaerobic Degradation of Toluene: Implications for Intrinsic Bioremediation." *Environ. Sci. Technol.*, 34: 892–896.

Alzaga, R., Montuori, P., Ortiz, L., Bayona, J.M., and Albaigés, J. 2004. "Fast Solid-Phase Extraction-Gas Chromatography-Mass Spectrometry Procedure for Oil Fingerprinting. Application to the Prestige Oil Spill." *J. Chromatogr. A*, 1025: 133–138.

Bolliger, C., Hohener, P., Hunkeler, D., Haberli, K., and Zeyer, J. 1999. "Intrinsic Bioremediation of a Petroleum Hydrocarbon-Contaminated Aquifer and Assessment of Mineralization Based on Stable Carbon Isotopes." *Biodegradation*, 10: 201–217.

Brar, G.S., Currier, P.M., and Reynolds, C.M. 1993. "Remediation of Petroleum Contaminated Soils with Bioventing in Cold Regions." Superfluid XIV Conference, 29 November–1 December, Washington, DC.

Bugna, G.C., Chanton, J.P., Kelley, C.A., Stauffer, T.B., MacIntyre, W.G., and Libelo, E.L. 2004. "A Field Test of ^{13}C as a Tracer of Aerobic Hydrocarbon Degradation." *Org. Geochem.*, 35: 123–135.

Collins, C.M., Racine, C.H., and Walsh, M.E. 1993. Fate and Effects of Crude Oil Spilled on Subarctic Permafrost Terrain in Interior Alaska: Fifteen Years Later. CRREL Report 93–13. Cold Regions Research and Engineering Laboratory, US Army Corps of Engineers, Washington, DC.

Dempster, H., Lollar, B.S., and Feenstra, S. 1997. "Tracing Organic Contaminants in Groundwater: A New Methodology using Compound-Specific Isotopic Analysis." *Environ. Sci. Technol.*, 31: 3193–3197.

Dibble, J.T., and Bartha, R. 1979. "Effect of Environmental Parameters on the Biodegradation of Oil Sludge." *Applied and Environmental Microbiology*, 37(4): 729–739.

Fremling, C.R. 1981. "Impacts of a Spill of No. 6 Fuel Oil on Lake Winona." In *Proceedings of the 1981 oil spill conference*, pp. 419–421. American Petroleum Institute, Washington, DC.

Geyer, R.A. 1980. *Marine Environmental Pollution*, "1. Hydrocarbons." Elsevier, Amsterdam. Scientific Publishing Company.

Grose, P.L., et al. 1979. USNS Potomac Oil Spill Melville Bay, Greenland, August 5, 1977. NOAA-80031106. National Oceanic and Atmospheric Administration, Washington, DC.

Guard, H.E. and Cobet, A. 1972. Fate of Petroleum Hydrocarbons in Beach Sand. Final Report No. CG-D-8–75. US Coast Guard, Washington, DC.

Jézèquel, R., Menot, L., Merlin, F.X., and Prince, R.C. 2003. "Natural Cleanup of Heavy Fuel Oil on Rocks: An In Situ Experiment." *Mar. Pollut. Bull.*, 46: 983–990.

Johansson, S., et al. 1980. "The Tsesis Oil Spill: Impact on The Pelagic Ecosystem." *Mar Poll Bull.*, 11: 284–293.

Lee, K., Prince, R.C., Greer, C.W., Doe, K.G., Wilson, J.E.H., Cobanli, S.E., Wohlgeschaffen, G.D., Alroumi, D., King, T., and Tremblay, G.H. 2003. "Composition and Toxicity of Residual Bunker C Fuel Oil in Intertidal Sediments after 30 Years." *Spill Sci. Technol. Bull.*, 8(2): 187–199.

Linden, O., et al. 1979. "The Tsesis Oil Spill: It's Impact on the Coastal Ecosystem of the Baltic Sea." *Ambio*, 8(6): 244–253.

Meckenstock, R.U., Morasch, B., Warthmann, R., Schink, B., Annweiler, E., Michaelis, W., and Richnow, H.H. 1999. "$^{13}C/^{12}C$ Isotope Fractionation of Aromatic Hydrocarbons during Microbial Degradation." *Environ. Microbiol.*, 1: 409–414.

Medina-Bellver, J.I., Marín, P., Delgado, A., Rodríguez-Sánchez, A., Reyes, E., Ramos, J.L., and Marqués, S. 2005. "Evidence for *In Situ* Crude Oil Biodegradation after the *Prestige* Oil Spill." *Environmental Microbiology*, 7(6), 773–779.

Mehrasbi, M.R., Haghighi, B., Shariat, M., Naseri, S., and Naddafi, K. 2003. "Biodegradation of Petroleum Hydrocarbons in Soil." *Iranian J. Publ. Health*, 32(3): 28–32.

Mulkins-Phillips, G.J., and Stewart, J.E. 1974. "Effect of Environmental Parameters on Bacterial Degradation of Bunker C Oil, Crude Oils and Hydrocarbons." *Appl Microbiol.*, 28(6): 915–922.

Nichols, T.D., Wolf, D.C., Rogers, H.B., Beyrouty, C.A., and Reynolds, C.M. 1997. "Rhizosphere Microbial Populations in Contaminated Soils." *Water, Air, and Soil Pollution*, 95: 165–178.

Petersen, H.K. 1978. "Fate and Effect of Bunker C Oil Spilled By the USNS Potomac in Melville Bay, Greenland, 1977." *Proceedings*. Conference on Assessment of Ecological Impacts of Oil Spills. American Institute of Biological Sciences. pp. 331–343.

Pierce, R.H. Jr., Cundell, A.M., and Traxler, R.W. 1975. "Persistence and Biodegradation of Spilled Residual Fuel Oil on an Estuarine Beach." *Applied Microbiology*, May: 646–652.

Rashid, M.A. 1974. "Degradation of Bunker C Oil under Different Coastal Environments Of Chedabucto Bay, Nova Scotia." *Estuarine Coastal Mar Sci.*, 2: 137–144.

Reynolds, C.M. 1993a. "Contaminated Soil Treatment in Cold Regions." *American Public Works Association Reporter*. July, p. 26.

Reynolds, C.M. 1993b. "Field Measured Bioremediation Rates in a Cold Region Land Farm: Spatial Variability Relationships." *Proceedings*. Seventh Annual East Coast Conference on Hydrocarbon Contaminated Soils. P.T. Kostecki and E.J. Calabrese (Editors). Lewis Publishers, Chelsea, Michigan. pp. 487–499.

Reynolds, C.M., Travis, M., Braley, W.A., and Scholze, R.J. 1994. "Applying Field Expedient Bioreactors and Land Farming In Cold Climates." *Proceedings*. Second International Symposium, In-Situ and Onsite Bioreclamation. Lewis Publishers, Chelsea, Michigan. pp. 100–106.

Reynolds, C.M., Bhunia, P., and Koenen, B.A. 1997. Soil Remediation Demonstration Project: Biodegradation of Heavy Fuel Oils. Special Report 97–20. Cold Regions Research & Engineering Laboratory US Army Corps of Engineers, Washington, DC. August.

Song, H-G., Wang, X., and Bartha R. 1990. "Bioremediation Potential of Terrestrial Fuel Spills." *Appl. Environ. Microbiol.*, 56(3): 652–656.

Vandermeulen, J.H., and Gordon, D.C. Jr. 1976. "Reentry of 5-Year Old Stranded Bunker C Fuel Oil from a Low-Energy Beach into The Water, Sediments and Biota of Chedabucto Bay, Nova Scotia." *J. Fish Res. Board Can.*, 33: 2002–2010.

Vandermeulen, J.H., et al. 1977. "Persistence of Non-Alkane Components of Bunker C Oil in Beach Sediments of Chedabucto Bay, and Lack of Their Metabolism by Molluscs." *Proceedings*. 1977 Oil Spill Conference, American Petroleum Institute, Washington, DC. pp. 469–473.

Van De Velde, K.D., Marley, M.C., Studer, J., and Wagner, D.M. 1995. "Stable carbon isotope analysis to verify bioremediation and bioattenuation." In *Monitoring and Verification of Bioremediation*. R.E. Hinchee,

G.S. Douglas, and S.K. Ong (Editors). Batelle Press, Columbus, Ohio. pp. 241–257.

Walker, J.D., et al. 1976. "Comparison of the Biodegradability of Crude and Fuel Oils." *Can J Microbiol.*, 22: 598–602.

Walworth, J.L., and Reynolds, C.M. 1995. "Bioremediation of a Petroleum Contaminated Soil: Effects of Phosphorus, Nitrogen, and Temperature." *Journal of Soil Contamination*, 4(3): 299–310.

Wolfe, D.A. 1977. *Fate and Effects of Petroleum Hydrocarbons in Marine Ecosystems and Organisms*. Pergamon Press, New York.

13

Biodegradation of Lubricating Oil

1 Introduction

The economic importance of crude oil is to be found in the numerous possible products obtainable from crude oil through refining: the uses of these products justify the resources committed to its exploration and production. Motor fuels, domestic fuels, industrial fuels for heating and power generation, and lubricants are among the products derivable from crude oil. Of particular interest to this chapter is lubricating oil, which may also be related (but not always in the true chemical sense, because of the additives that are included in lubricating oil) to atmospheric and vacuum gas oil.

The uses of the various kinds of lubricating oils (also called *mineral oils*) in various industrial situations have made them an indispensable ingredient of the industrialization and development that has characterized the past century. The class of petroleum products known as mineral oils can be generally understood to include a variety of products that go by different names, such as white oils, lubricating oils, light fuel oils, residual fuel oils, as well as transformer and cable oils (Gary and Handwerk, 2001; Speight, 2007).

Moreover, the term *mineral oil* tends to be non-specific, and its usage in the literature could give rise to some confusion, as it may be found to refer to various substances. However, in this context, it is taken to refer to all oils that are made from dewaxed paraffin-based crude oils that are blended with additives to impart particular properties for specific uses. This classification is as against other oils, such as organophosphate esters and poly-alpha-olefin fluids. Mineral oil also includes the base oils from which lubricating oils are manufactured.

Hence, for the purposes of this text, petroleum is the origin of lubricating oil, which is formulated from base petroleum stocks with approximately 10 to 25% additives. These additives consist of antioxidants and metal deactivators, detergents, dispersants, corrosion inhibitors, and rust inhibitors, which also have a deleterious effect on the environment (Farrington and Slater, 1997; Herdan, 1997). The carbon number distribution and, by extension, the viscosity in hydraulic fluids varies depending upon the anticipated application of the fluid. Generally, however, C_{15} to C_{50} is a reasonable range. Also, the higher the carbon number, the higher the viscosity of the fluid in question.

Recent focus on health, safety, and the preservation of the environment has turned the searchlight to the effects of lubricating oils on the environment when they are accidentally spilled, or intentionally disposed. Biodegradation, which provides an indication of the persistence of a particular substance in the environment, is the yardstick for assessing the eco-friendliness of substances. Due to the poor biodegradation rates observable for oils, various methods of bioremediation are currently being researched, with the isolation of various microbial species with the ability to use up lubricating oils as a carbon and energy source. Ultimately, vegetable oil based hydraulic fluids will come to the fore as a suitable and more environmentally friendly solution to the demand for hydraulic fluids.

The materials that fall into the lubricating oil category are complex petroleum mixtures, composed primarily of saturated hydrocarbons with carbon numbers ranging from C_{15} to C_{50} (some ranges give C_{15} to C_{40}). At ambient temperatures, lubricating base oils are liquids of varying viscosities, with negligible vapor pressures. Base oils are produced by first distilling crude oil at atmospheric pressure to remove lighter components (e.g., gasoline and distillate fuel components), leaving a residue (residuum) that contains base oil precursors. This atmospheric residuum is then distilled under vacuum to yield a range of distillate fractions (unrefined distillate base oils) and a vacuum residuum (Chapter 1). Removal of the asphalt components of the vacuum residuum results in unrefined residual base oils. These distillate and residual base oil fractions may then undergo a series of extractive or transforming processes that improve the base oils' performance characteristics and reduce or eliminate undesirable components.

Distillate base oils contain components whose boiling points typically range from 300 to 600°C (570 to 1110°F), which are often described as either *naphthenic* (saturated ring hydrocarbons) or *paraffinic* (straight or branched chain hydrocarbons), depending on the original petroleum from which they are derived and/or the dominant hydrocarbons present. The difference between naphthenic and paraffinic base oils is one of relative percentage, since naphthenes and paraffins are present in both types of oils.

Thus, base oil might be called *paraffinic base oil* if it is 60% w/w paraffins and 30% w/w naphthenes, or *naphthenic base oil* if it is 60% naphthenes w/w and 30% paraffins w/w. Base oils are also often described as either *light* (viscosity less than 19 cSt at 40°C, 104°F) or *heavy* (viscosity greater than 19 cSt at 40°C, 104°F). The naphthenic/paraffinic and light/heavy nomenclatures are primarily used to distinguish product application and lubricant quality parameters, rather than health and safety characteristics, since a significant amount of toxicology data exists that shows little differentiation between these four classifications.

Refined distillate base oils are produced from unrefined base oil fractions by undergoing additional processing designed to reduce or transform the undesirable components. In general, each additional step of processing (increasingly severe processing) results in: (1) lower levels of unwanted components, including aromatics, metals, waxes, and trace components causing unwanted colors or odors, (2) a narrower range of hydrocarbon molecules (increasing concentration of paraffins and naphthenes), and (3) lower, if any, carcinogenic or mutagenic activity. Some distillate base oils are destined for use in food, food contact, cosmetic, pharmaceutical, and related applications. Known as white oils, these very severely refined distillate base oils undergo numerous processing steps that essentially eliminate or transform all undesired components, including unsaturated hydrocarbons and aromatics.

Synthetic base oils can be substituted for conventional petroleum-based oils. Most synthetic motor oils are fabricated by polymerizing short chain hydrocarbon molecules, called alpha-olefins, into longer chain hydrocarbon polymers, called poly-alpha-olefins. The degree of variation in molecular size, chain length, and branching in synthetically produced fluids is much less than occurs in base stocks extracted from crude oil (OECD, 2004). While they appear chemically similar to mineral oils refined from crude oil, poly-alpha-olefins do not contain the impurities or waxes inherent in conventional mineral oils. Poly-alpha-olefins constitute the most widely used synthetic motor oil in the United States and Europe.

These various products have been applied ever since the discovery of crude oil. The recent worldwide emphasis on environmental preservation now raises questions which concern the long term effects of the use, accidental spills, and intentional disposal of industrial products, such as lubricating oils. Pressure is mounting to use more environmentally friendly products, especially in ecologically sensitive situations (Khawaja and

Aban, 1996; Lea, 2005). Hence, the biodegradability of lubricating oil is of interest, since that the environmental persistence of possibly toxic constituents of these lubricating oils is unacceptable.

2 Identity and Origin of Lubricating Oil

Lubricating oil is a complex mixture composed of straight and branched chain paraffinic, naphthenic, and aromatic hydrocarbons with more than fifteen carbon atoms. Generally, the physical properties of lubricating oils depend on their composition in terms of carbon number distribution, and this is defined by the source of crude oil. Accordingly, these physical properties vary widely: boiling points generally range from 300 to 600°C, while specific gravities range from 0.820 for light paraffin-base/process oils to just over 1.0 for high aromatic-base/process oils. They are all generally insoluble in water and alcohol, but are soluble in benzene, chloroform, ether, carbon disulfide, and petroleum ether.

In most crude oil refineries, lubricating oil base stocks are obtained from the heavy vacuum gas oil fraction and reduced crude from paraffinic or naphthenic base crude oils. Reduced crude (bottoms from the atmospheric distillation unit) is processed in a propane deasphalting unit, where asphaltene constituents are removed; next, it is processed in a blocked operation with the vacuum gas oils to give rise to lubricating oils base stocks, which form the basis for the wide range of lubricating oils. Vacuum gas oils and deasphalted reduced crude stocks are first solvent extracted to remove the aromatic compounds and the dewaxed to improve the pour point. Then, they are treated with special clays or high severity hydrotreating to improve their color and stability before they are blended into lubricating oils (Gary and Handwerk, 2001; Lea, 2005; Speight, 2007).

The uses of lubricating oils includes applications such as engine oils, automotive and industrial gear oils, transmission fluids, hydraulic fluids, circulating and hydraulic oils, bearing oils, and machine oils. Other uses are as machine-tool oils, compressor and refrigerator oils, textile machine oils, air tool oils, steam engine oils, metalworking oils, rust prevention oils, and transformer oils (Lea, 2005). Another use is found for mineral oils in medical applications: oil such as medicinal white oils are mixtures of highly refined paraffinic and naphthenic liquid hydrocarbons obtained from crude oil through various refining steps including distillation, extraction, and crystallization, with subsequent purification by acid and catalytic hydro treatment (Speight, 2007). In addition to being used as laxatives, skin care products, suspending agents in sulfur based ointments, and in livestock vaccines, white mineral oils are also applied as release agents, as well as glazing agents.

Lubricating oils are generally recognized to cause pollution of soil and water when spilled accidentally or when disposed, affecting plant and animal life (Khawaja and Aban, 1996). They cannot be regarded as readily biodegradable, and hence, their harmful effects often persist in the environment. In fact, petroleum-based lubricating oils, greases, and hydraulic fluids are usually toxic and not readily biodegradable; because of these characteristics, if these materials escape to the environment, the impacts tend to be cumulative and consequently harmful to plants, fish, and wildlife (Herdan, 1997; US Army Corps and Engineers, 1999).

The overall environmental hazard of lubricating oils is not entirely determined by the lubricating oil based base fluids, but the several additives used in hydraulic fluids that contribute substantially to the toxicity of the spill. This increased toxicity is coupled with the physical effect of staining essential organs of animal life in aquatic environments (Broekhuizen et al., 2003). Additionally, by means of spills and improper disposal of lubricating oils, the general population, if potentially exposed to unused and used lubricating oils, which are present as environmental hazardous pollutants, can be exposed to cancer-causing chemicals.

Finally, lubricating oil is classified on the basis of properties and use. For example, the American Petroleum Institute (API) classifies base oils under five categories, which assist in identifying the base oils for finished oil formulations to ensure that lubricating engine specifications and performance are met.

Group I: Conventional mineral oils derived from refining selected crude oil fractions. Solvent processing, first used in the 1930s to improve base oil performance, is still used today to produce Group I lubricant stock. While some automotive oils on the market use Group I stocks, they are generally used in less demanding applications.

Group II: The development of hydrotreating and hydrocracking technology led to the introduction of highly refined, low wax base stocks, or oils (Chapter 3). These hydroprocessed base oils still have performance limitations, due to the presence of undesirable impurities. Group II base oils are common in mineral based motor oils currently available on the market, and have good performance in lubricating properties.

Group III: These base oils are also manufactured using the hydrotreatment process, but are subjected to higher temperatures or processing times. These highly hydroprocessed products have greatly improved oxidation stability and low temperature performance, but still contain some undesirable impurities that cannot be removed.

Group IV: These oils, produced from poly-alpha-olefins, are a special class of base stock. The term *synthetic* was originally used to refer to Group IV and Group V base stocks.

Group V: These specialized base oils, with few exceptions, are chemically engineered stocks that do not fall into any of the categories previously mentioned. Typical examples of Group V stocks are esters, poly glycols, and silicone oils.

3 Composition and Properties of Lubricating Oil

Biodegradation of complex hydrocarbons, particularly pollutants that are made up of many different compounds, such as lubricating oil, usually requires the cooperation of more than one single species. Individual microorganisms can metabolize only a limited range of hydrocarbon substrates, so the mixed cultures with overall broad enzymatic capacities are required to increase the rate of petroleum biodegradation. Several microbial populations have been detected in petroleum-contaminated soils or water. This phenomenon strongly suggests that each strain or genera has its role in hydrocarbon transformation processes.

Large amounts of lubricating oils, composing long-chain saturated hydrocarbons (base oil), and additives are used in car engines. The main components of the base oil are cyclic alkanes (c-alkanes). Long-chain hydrocarbons and c-alkanes are known as recalcitrant to microbial degradation. The base oil contains C_{16}–C_{36} hydrocarbons, and more than 75% c-alkanes. The rings number of c-alkanes in the base oil is from one to three, and any ring contains five or six members.

Most of the c-alkanes in the base oil have long alkyl side chains. Susceptibility of a hydrocarbon to microbial degradation varies with type and size of the hydrocarbon molecule. Alkanes of intermediate chain length (C_{10}–C_{24}) are often degraded rapidly, while very long chain alkanes are increasingly resistant to microbial degradation [12].

It must be noted that, apart from base petroleum stocks, lubricating oils in their commercial applications contain about 10–25% of substances known as additives (Farrington and Slater, 1997). These additives are necessary to impart properties that are considered appropriate to specific applications, and constitute a very important part of the composition of lubricating oils, which include, among others: (1) extreme pressure additives, which help prevent surface damage under severe loading (organic sulfur-, phosphorus-, and chlorine-containing compounds), (2) anti-wear additives, which prevent wearing under light loads (fatty acids and derivatives, organo- phosphate esters), (3) corrosion inhibitors, which prevent corrosion by oxygen and water (fatty acids, sulfonates, and salts of fatty acids), (4) oxidation inhibitors, which inhibit oxidation of the hydraulic fluid (phenols, amines, and sulfides), (5) de-foaming agents, which prevent foam formation (silicone oils), (6) viscosity index improvers, which reduce the dependence of viscosity on temperature (poly alpha-olefins, poly-methacrylates, and poly-alkylstyrenes), (7) pour point depressants, which lower the pour point temperature (poly-methacrylates and condensation products), (8) demulsifying agents, which allow separation of oil and water (ionogenic and non-ionogenic polar compounds), and (9) dispersants, which prevent unwanted deposits (sulfonates and amides).

Considering the variety of additives and their chemical constitution, it is not surprising that lubricating oil additives, although these may be used in lower amounts, also cause environmental problems (Herdan, 1997). They may promote ecological injury, but there are only limited data concerning the biodegradability of the additives used. Used oils are generally collected and recovered, but a significant quantity of used oil is directly burned. Burning of used lubricating oil in incorrect equipment leads to environmentally-harmful emissions.

3.1 Thermal Stability of Lubricating Oils

The term *thermal stability* when used in reference to lubricating oil is the resistance posed by the lubricant to either molecular breakdown, or some form of rearrangement at elevated temperatures in the absence of oxygen (Booser, 1984). It is a well observed fact that all oils will start to decompose even in the absence of oxygen when they are heated above a certain temperature. When heated, lubricating oils break down to yield basically methane, ethane, and ethylene (Stachowiak and Batchelor, 2005). More moles of methane are liberated than of ethane and ethylene (Booser, 1984).

Thermal stability of crude oil products, which includes lubricating oil, determines their use and applicability in crude oil industry (Rychla et al., 2001). The improvement of the thermal stability of lubricating oils is achieved right from the refining process. It is known that organo-nitrogen molecules are largely responsible for poor color and stability of base oils; hence, the finishing processes undertaken during refining must necessarily involve operations that are effective for removing organo-nitrogen molecules (Mortier and Orszulik, 1997).

Thermal stability may not be improved via the means of additives; rather, usually, additives used with lubricant base stocks have lower thermal stability than the base stocks themselves (Stachowiak and Batchelor, 2005). Thermal stability is also dependent on the length of usage (Karacan et al., 1999). Carbon to carbon single bonds are the most vulnerable points for thermal breakdown. Hence, fluids, such as lubricating oils, that are constituted of a substantial percentage of C-C single bonds exhibit a thermal stability of about 650 to 700°F. Synthetic oils, in general, exhibit better thermal stability than lubricating oils (Stachowiak and Batchelor, 2005).

3.2 Environmental Aspects of Lubricating Oil Use

As specified elsewhere in this chapter, used lubricating oil refers to a petroleum- based product that is used as a lubricating agent or in other protective applications. In terms of environmental effects of lubricating oils, each type of crude oil or refined product has distinct physical properties, which affect the way the oil spreads and breaks down, and ultimately determine the hazard it may pose to marine and human life, and the likelihood that it would pose a threat to natural and manmade resources. Lubricating oils are persistent, and present a greater remediation challenge than lighter petroleum products (Khawaja and Aban, 1996).

This prospective hazard is recognized by many regulatory bodies, which stipulate, for example, that used oils in liquid form cannot be disposed of by any of the following methods: (1) disposed of into a drain or sewer or into surface water or groundwater, (2) disposed of in a landfill, (3) burned in a municipal solid waste incinerator without any form of gas cleaning technology, and (4) used as dust control or weed control by spraying directly (or indirectly) on the ground.

The United States Environmental Protection Agency has laid down guidelines that stipulate that accessible oil be recovered in the event of a spill, and all contaminated soil be removed. Additionally, the EPA also requires that no visible oil sheen be evident downstream from facilities close to waterways. Another regulation requires that point discharges into waterways should not exceed 10 ppm of lubricating-based oils (US Army Corps and Engineers, 1999).

Residual base oils are derived from the residuum of the vacuum distillation tower, and may contain components boiling as high as 800°C (1470°F). The residual oils have molecular weights that are much higher than the distillate base oils. Residual base oils are primarily used in situations requiring oils with a high viscosity, e.g., gear oils. Detailed analytical testing indicated that the polynuclear aromatic constituents they contain are predominantly one-to-three rings that are highly alkylated (paraffinic and naphthenic). Because they are found in such a high boiling material (>575°C, >1070°F), it is estimated that the alkyl side-chains of the ring aromatics would be approximately 13 to 25 carbons in length. These highly alkylated aromatic ring materials are either devoid of the biological activity necessary to cause mutagenesis and carcinogenesis, or are largely non-bioavailable to the organisms (Roy et al., 1988).

4 Biodegradation of Lubricating Oils

It has been estimated that about 1% of total lubricating oil consumption worldwide is used to formulate lubricants, and significant amounts of lubricants were lost into the environment by improper disposal or leakage (Bartz, 1998). Petroleum-based lubricants, which consist predominantly of hydrocarbons, such as alkanes and aromatic compounds, and additives, such as antioxidants, are often environmentally persistent and hazardous: often chemical additives prevent or reduce the oxidative reactions thereby, and most additives are dangerous and can impact human health and the environment, due to their inherent toxicity and non-biodegradable nature (Haus et al., 2001).

For these reasons, the biodegradability and toxicity of petroleum based lubricants are important factors with respect to environmental management. Although during the last decades, the increased public attention of protection of the environment has stimulated the development of lubricants that show more or less compatibility with the environment, environmental compliance of lubricants and bioremediation of lubricant-polluted sites have become topics of interested research (Erhan and Asadauskas, 2000; Boyde, 2002; Willing, 2001; Lee et al., 2007; Cecutti and Agius, 2008; Matos Lopes and Bidoia, 2009).

Petroleum hydrocarbons are among the most ubiquitous contaminants in the environment (Watanabe, 2001; Margesin et al., 2003). There has been extensive evidence of microbial degradation for petroleum products (Atlas, 1981). Since biodegradation of petroleum products is a natural process limited by temperature, pH, and scarcity of nutrients, such as N and P (Ladousse and Tramier, 1991; Leahy and Colwell, 1990), bioremediation, a process whereby the natural biodegradation capacities are enhanced by nutrient addition and/or culture microorganisms, is a viable and promising method for cleanup and remedy of petroleum polluted environments.

In spite of the large number of works on petroleum, petroleum derivatives, and hydrocarbon biodegradation, little information is available on petroleum lubricant oils biodegradation by microbial inoculation (Rhykerd et al., 1995; Nocentini et al., 2000). In practice, an effective bacterial inoculum should be able to tolerate high levels of petroleum lubricant while maintaining a level of activity to provide efficient biodegradation. Understanding the physiological and biochemical properties of lubricant oil-degrading bacteria is required before optimum use of bacteria in environmental applications. Intensive investigation of microbial degradation of petroleum lubricants is thus imperative, and of important practical significance. In the present paper, the lubricating oil-degrading abilities of three bacterial strains previously isolated from oil-contaminated soils by the authors were investigated. Effect of pH on degrading ability of the bacterial isolates and impact of temperature on microbial growth were also explored.

The large variation in the composition of used oil relates to its crude oil source, chemical additives, fuel type, and combustion and mechanical properties of the engine (Delistraty and Stone, 2007). Furthermore, due to its chemical composition, global distribution, and potentially toxic effects, used motor oil can be a serious environmental problem (Vazquez-Duhalt, 1989). The environmental pollution by hydrocarbons can cause enormous damage to biota. The evaluation of long term impact and risk of oil spill is a complex process involving chemical analyses and development of the ecosystem-based toxicology (Matos Lopes et al., 2010).

4.1 Petroleum-Based Oils

The notion of biodegradability conveys the meaning of the breaking down of complex, and possibly toxic, material into simple and common forms, in which the elements carbon, hydrogen, and oxygen exist, or are simply associated. Whenever any substance is placed in the environment, it begins to undergo degradation due to the action of relevant substrate utilizing microorganisms. Every substance, irrespective of complexity and toxicity, is subject to this process. When used in the present context, it is referenced to a time frame. That is, for a given substance to be referred to as biodegradable, a specific time limit must be used as a reference.

By definition, biodegradation is the chemical transformation of a substance caused by organisms or their enzymes. There are two major types of biodegradation. Primary biodegradation refers to the modification of a substance by microorganisms such that a change is caused in some specific measurable property of the substance (US Army Corps and Engineers, 1999). When the term primary biodegradation is used, it refers to minimal transformation that alters the physical characteristics of a compound, while leaving the molecule largely intact. Intermediary metabolites produced may, however, be more toxic than the original substrate. Thus, mineralization is the true aim. When this happens, it is referred to as ultimate or complete biodegradation, which is the degradation achieved when a substance is totally utilized by microorganisms, resulting in the production of carbon dioxide, methane, water, mineral salts, and new microbial cellular constituents (US Army Corps and Engineers, 1999).

Various methods exist for the testing of biodegradability of substances (Table 13.1) (Aluyor and Ori-jesu, 2009). Biodegradability is assessed by following certain parameters which are considered to be indicative of the consumption of the test substance by microorganisms, or the production of simple basic compounds which indicate the mineralization of the test substance. Hence, there are various biodegradability testing methods which measure the amount of carbon dioxide (or methane, for anaerobic cases) produced during a specified period; there are those which measure the loss of dissolved organic carbon for substances which are water soluble; those that measure the loss of hydrocarbon infrared bands; and there are yet others which measure the uptake of oxygen by the activities of microorganisms (the Biochemical Oxygen Demand, BOD).

Table 13.1 Tests for ultimate biodegradability (Aluyor and Ori-jesu, 2009).

Ultimate Biodegradation Tests: Test title	Measurements	Sponsoring Organization
D5864, test method for the aerobic aquatic biodegradation of lubricants	Theoretical CO_2 (%)	ASTM
9429:1990, Technical corrigendum 1, water quality-evaluation in an aqueous medium of the "ultimate" biodegradation of organic compounds-method by analysis of released carbon dioxide	Theoretical CO_2 (%)	ISO
301B, CO_2 evolution test (modified strum test)	Theoretical CO_2 (%)	OECD
Aerobic aquatic biodegradation test	Theoretical CO_2 (%)	US EPA
C.4-C: Carbon dioxide (CO_2) evolution	Theoretical CO_2 (%)	EUC
L-33-A-934, biodegradability of two-stroke cycle outboard engine oils in CH_2 groups water (formerly L-33-T-82)	Loss of extractable (%)	CEC

However, when the reference is specifically to lubricants, there are two major methods of biodegradability testing, and these are outlined in brief below:

1. ASTM D5864 is used to determine the biodegradation of lubricating oil. Specifically, the test is used to determine the rate and extent of aerobic aquatic biodegradation of lubricants when exposed to an inoculum under laboratory conditions. The inoculum may be the activated sewage-sludge from a domestic sewage-treatment plant, or it may be derived from soil or natural surface waters, or any combination of the three sources. The degree of biodegradability is measured by calculating the rate of conversion of the

lubricant to carbon dioxide. A lubricant, hydraulic fluid, or grease is classified as readily biodegradable when 60% or more of the test material carbon is converted to carbon dioxide within a 28-day period.

2. The most established test methods used by the lubricant industry for evaluating the biodegradability of their products are Method CEC-L-33-A-94, developed by the Coordinating European Council (CEC; Method OECD 301B); the modified Sturm test, developed by the Organization for Economic Cooperation and Development (OECD); and Method EPA 560/6–82–003, number CG-2000, the Shake Flask Test, adapted by the U.S. Environmental Protection Agency (EPA). These tests can also be used to determine the rate and extent of aerobic aquatic biodegradation under laboratory conditions. The modified Sturm test and shake flask test can also be used to calculate the rate of conversion of the lubricant to carbon dioxide. The Coordinating European Council (CEC; Method OECD 301B) test measures the disappearance of the lubricant by analyzing test material at various incubation times through infrared spectroscopy. Laboratory tests have shown that the degradation rates may vary widely among the various test methods indicated above (US Army Corps of Engineers, 1999).

Biodegradability tests based on the CEC method described above have certain trends which indicate that lubricating oils, along with alkylated benzenes and polyalkeleneglycols, among others, generally have poor biodegradability: between 0–40%. Specifically, lubricating oil biodegradability varies from 15 – 35%, as conducted by the CEC biodegradability test.

4.2 Synthetic Lubricating Oil

Synthetic ester lubricants are degraded more rapidly in soil and in aquatic systems than traditional mineral oil-based products (Haigh, 1995). Poly-alpha-olefins show higher biodegradability than mineral oils of equivalent viscosity, because of their higher degree of hydrocarbon chain linearity (Boyde, 2002). Within a class of synthetic lubricants, the percent of material biodegraded within a prescribed time period can cover a large range, and different biodegradability tests can give different results for the same lubricant type.

For example, biodegradability of poly-alpha-olefins can range from 20 to 80% w/w after 21 days using a primary biodegradability test, which measures the initial transformation from the parent material (Boyde, 2002). Using this same test, biodegradation of mineral-based oils ranged from 10 to 45% w/w. Rates and extents of biodegradation vary considerably between laboratory and field situations, largely due to the influence of factors such as temperature, the types and number of microbes, and the availability of oxygen and water (Haigh, 1995).

4.3 Used Lubricating Oil

Used lubricating oil is of variable composition, depending on the composition of the original oil and the degree of degradation. This material is a complex mixture of paraffinic, naphthenic, and aromatic petroleum hydrocarbons that may contain one or more of the following: carbon deposits, sludge, aromatic and non-aromatic solvents, water (as a water-in-oil emulsion), glycols, wear metals and metallic salts, silicon-based antifoaming compounds, fuels, polynuclear aromatics, aromatic hydrocarbons, and miscellaneous lubricating oil additive materials. In the unlikely event that used transformer oils are mixed with other waste oil, then polychlorinated biphenyl derivatives (PCBs) and polychlorinated terphenyl derivatives (PCTs) may also be present.

Thus, it is not surprising that used lubricating oil contains several toxic components, including up to 30% aromatic hydrocarbons, with as much as 22 ppm benzo(a)pyrene, and has a density on the order of 0.828 g/ml (Upshall et al., 1992).

Used lubricating oil is typically (but not always) used lubricating oils removed from the crankcase of internal combustion engines. Before they are used, crankcase oils consist of a base lubricating oil (a complex mixture of hydrocarbons, 80 to 90% by volume) and performance-enhancing additives (10 to 20% v/v). Lubricating oil in the crankcase is altered during use because of the breakdown of additives, contamination with the products of combustion, and the addition of metals from the wear and tear of the engine. Therefore, the composition of used lubricating oil is difficult to generalize in exact chemical terms. It is recognized that the major components consist of aliphatic and aromatic hydrocarbons (such as phenol, naphthalene, benz(a)anthracene, benzo(a)pyrene, and fluoranthene).

Used lubricating oils are divided into two categories: (1) clear oils, of industrial origin and little deteriorated in use, which can easily be regenerated through a simple purification process, such as filtering and/or centrifugation, and (2) dark oils, which typically originate from automotive lubrication, and which have been subject to severe thermal and mechanical conditions, under which they have been loaded with metals and combustion residues and oxidized. Neither of these oils should be confused with used soluble oils and other aqueous machining fluids, or vegetable frying oils, or water-hydrocarbon mixtures, for which entirely different methods for collection and disposal are used.

The land farming of oily wastes (for the most part, these wastes are lubricating oils or lubricating oil derivatives, or fuel oil derivatives or difficult-to-identify oily sludge) has been used by the petroleum industry (Cansfield and Racz, 1978; Morgan and Watkinson, 1989) since it has been recognized that hydrocarbons can be metabolized by the indigenous microbial population of the soil (Raymond, 1976; Bossert and Bartha, 1984; Oudot et al., 1989). The biodegradation of such waste hydrocarbons has been studied in soil microcosms (Chaîneau et al., 1995) and in a field experiment (Chaîneau et al., 1996.)

Numerous microorganisms are known for their ability to degrade hydrocarbons (Atlas, 1984), but most work focuses on the biodegradation of petroleum hydrocarbons by bacteria, whereas relatively few papers deal with the degradation of fuel oil by fungi (Llanos and Kjoller, 1976; Davies and Westlake, 1979; Oudot et al., 1987, 1993). The biodegradation potential of strains of bacteria and fungi that were isolated from an unpolluted and hydrocarbon treated temperate soil has been estimated (Chaîneau et al., 1999).

Thus, three months after spreading the hydrocarbon, hydrocarbon-adapted bacteria and fungi were isolated at different soil depths in the two plots, and identified. The biodegradation potential of the isolated strains was monitored by measuring the degradation rate of total hydrocarbons, saturated hydrocarbons, aromatic hydrocarbons and resins of the fuel. Bacteria of the genera *Pseudomonas*, *Brevundimonas* (a genus of proteo-bacteria), *Sphingomonas*, *Acinetobacter*, *Rhodococcus*, *Arthrobacter*, *Corynebacterium*, and fungi belonging to *Aspergillus*, *Penicillium*, *Beaver* (a genus of asexually-reproducing fungi), *Acremonium*, *Cladosporium*, *Fusarium*, and *Trichoderma* were identified. The most active strains in the assimilation of saturate constituents and aromatic constituents were *Arthrobacter* sp., *Sphingomonas spiritivorum*, *Acinetobacter baumanii*, *Beaveria alba*, and *Penicillium simplicissimum*. The biodegradation potential of the hydrocarbon utilizing microorganisms isolated from polluted or unpolluted soils were similar. In laboratory pure cultures, saturated hydrocarbons were more degraded than aromatic HC, whereas resins were resistant to microbial attack. On average, individual bacterial strains were more active than fungi in hydrocarbon biodegradation.

In addition to being a complex mixture of hydrocarbons and other organic compounds, including some organometallic constituents (Hagwell et al., 1992; Butler and Mason, 1997), used lubricating oil contains metals and polycyclic aromatic hydrocarbons that could contribute to chronic hazards, including mutagenicity and carcinogenicity (Keith and Telliard, 1979; Hagwell et al., 1992; Boonchan et al., 2000). Prolonged exposure to high oil concentration may cause the development of liver or kidney disease, possible damage to the bone marrow, and an increased risk of cancer (Propst et al., 1999; Lloyd and Cackette, 2001; Mishra et al., 2001). In many developing countries, about 20 million gallons of waste engine oil are generated annually from mechanic workshops and discharged into the environment (Adegoroye, 1997; Faboya, 1997). Used engine oil also renders the environment unsightly, and constitutes a potential threat to humans, animals, and vegetation (Edewor et al., 2004; Adelowo et al., 2006).

Environmental pollution by petroleum and petroleum products has attracted much attention in recent times. The presence of various kinds of automobiles and machinery vehicles has caused an increase in the use of motor oil. Spillages of used motor oils, such as diesels or jet fuels, contaminate our natural environment with hydrocarbon (Husaini et al., 2008).

As the use of petroleum and petroleum products increase, soil contamination with diesel and engine oils is becoming one of the major environmental problems. To investigate the countermeasure to remediate soils contaminated with oils, bioremediation provides an effective and efficient strategy to speed up the clean-up processes (Lin and Mandri, 2007). Various factors may limit the rate of petroleum hydrocarbon degradation, including lack of essential nutrients, such as nitrogen. Therefore, the addition of inorganic or organic nitrogen-rich nutrients (biostimulation) is an effective approach to enhance the bioremediation process (Hollender et al., 2003; Semple et al., 2006; Walworth et al., 2007). Positive effects of nitrogen amendment on microbial activity and/or petroleum hydrocarbon degradation have been widely demonstrated (Jørgensen et al., 2000; Margesin et al., 2000; Brook et al., 2001; Margesin and Schinner, 2001; Riffaldi et al., 2006; Margesin et al., 2007).

The illegal dumping of used engine oil is an environmental hazard with global ramification. Most mechanical methods to reduce hydrocarbon pollution are expensive, time consuming, and depend mainly on excavating of these soils, treating in separate areas, or better treatment facilities. These treatments include, for example, incineration and/or burial in secure landfills (Udeani et al., 2009). These are effective treatments, but after burning, this soil loses most of its nutritional value and structure. These methods do not remove the contamination, but only relocate the problem (Lageman et al., 2005).

Twenty-five bacterial strains capable of utilizing used engine oil as a carbon source were isolated from the contaminated soils (Thenmozhi et al., 2011). The three isolates with best oil degradation ability were identified as *Serratia marcescens* PDKT-1 (Accession No: HM 998315), *Pseudomonas aeruginosa* PDKT-2, and *Bacillus licheniformis* PDKT-5, using the biochemical tests. The genetic diversity assessment by RFLP shows that genetic similarity between the isolates was varied. The results indicate all three isolates are capable of utilizing engine oil as the nutrient source. *Pseudomonas aeruginosa* was found to be the best oil degrading isolate in this study, with 81% degradation after 30 days incubation period, while 72% and 60%

degradation were observed using *Serratia marcescens* and *Bacillus licheniformis*, respectively, under the standard assay conditions.

All three isolates utilized engine oil as a sole carbon source with a higher degradation rate. The peak reduction is a strong reason for degradation, as it shows a significant reduction in aliphatic compounds under equal condition. The most significant reduction is related to the sample seeded by the consortium culture, which could degrade 90% of aliphatic compounds during 30 days. More reduction of the hydrocarbons was seen with *Pseudomonas aeruginosa*, among 3 pure cultures that were investigated. It is obvious that consortium cultures are more effective. Also, the biodegradation of short- and middle-chain aliphatic compounds were more extensive, compared to the long chain hydrocarbons. The discovery of *Pseudomonas* was not surprising, based on their frequency in the soil, as well as their frequent biodegradability. More reduction of engine oil hydrocarbons by *Pseudomonas* sp has also been reported (Shojaosadati, et al., 2008). Nevertheless, isolation and apparently a good growth on hydrocarbon substrates are surprising for *Serratia marcescens*. This bacteria has been not previously considered be a strong hydrocarbon degrader. It has showed very good crude oil and gasoline degradability (Balows et al., 1992; Ijah, 1998). The presence of *Bacillus sp* in engine oil contaminated sites was reported by Udeani et al., 2008. Other workers (Mandri and Lin, 2007) reported that there was an increase in cell number of *B. stearothemophilus* during the degradation process, demonstrating the ability of utilizing engine oil as energy source for this organism, which supports the present study results. The consortia proved to be a better degrader compared to individual isolates, with degradation rates of 90% in 4 weeks.

Different nutrients sources and environmental conditions, such as pH and temperature, were substituted in the growth pattern assay. The results showed that *Serratia marcesens* isolated in this study utilized ammonium nitrate in engine oil degradation metabolic mechanism. The addition of urea or nitrite reduced the cell growth. *P. aeruginosa* and *Bacillus sp*, on another hand, were capable of utilizing different nitrogen sources, including nitrite. However, individual organisms often prefer to metabolize a limited range of hydrocarbon substrates (Marquez-Rocha et al., 2001).

Consequently, a mixed population of bacteria and fungi is usually required to provide all metabolic capabilities for complete degradation of complex mixture of hydrocarbons (Leahy and Cowell, 1990). Microorganisms able to degrade organic pollutants in culture have failed to function when inoculated into natural environments, so further research is required in this field. These results showed that each member in a microbial community has a significant role, and may be dependent on the presence of the other species or strains for surviving. Furthermore, this study has demonstrated a very good biodegradation capability of engine oil hydrocarbons by bacterial consortium (Thenmozhi et al., 2011).

The use of fungi as a method of bioremediation provides an option to clean up environmental pollutants. Bioremediation using fungi has drawn little attention in the past two decades, since most of the bioremediation researches focused mainly on the use of bacteria. Nevertheless, recently, fungi have received considerable attention for their bioremediation potential that is attributed to the enzymes they produced that are involved in lignin breakdown, which degrade a wide range of recalcitrant pollutants, such as polyaromatic hydrocarbons, chlorophenols, and pesticides (Bumpus et al., 1985). In addition, fungi have advantages over bacteria, such as fungal hyphae, that can penetrate contaminated soil to reach the PAHs that have spread beyond the top layer of the soil (Novotny et al., 1999; April et al., 2000).

Penicillium janthinellum was also previously reported as having the ability to transform pyrene and benzo[a]pyrene into phenols, diols, and quinones (Launen et al., 1995). It has also been reported (Saraswathy and Hallberg, 2005) that *Penicillium ochrochloron* was capable in degrading and utilizing pyrene as sole carbon source. *Aspergillus terreus*, which was isolated from oil-polluted areas, was reported capable of mineralizing the aliphatic and aromatic hydrocarbons efficiently (Colombo et al., 1996).

Recently, the isolation, identification, and characterization of a potential indigenous hydrocarbon degrading fungi, *Penicillium* species (P1), for possible use and application in the bioremediation of used motor oil has been reported (Husaini et al., 2008). This research demonstrated the potential application of the bioremediation oil-contaminated soil and water using *Penicillium* species, an indigenous fungus, as a prospective and better way to resolve the problem of oil pollution.

Phytoremediation is a non-destructive *in situ* technology that can be used for the cleanup of contaminated soils. The potential for this technology in the tropics is high, due to the prevailing climatic conditions which favor plant growth and stimulate microbial activity. In one investigation (Erute Magdalene et al., 2009), the use of four common weeds in Nigeria (*Phyllanthus amarus* Schum and Thonn., *Hyptis spicigera* Lam., *Sida rhombifolia* L., and *Mariscus alternifolius* Vahl.) were tested for their reaction to spent lubricating oil contamination and subsequent reduction of the contaminant by the plants. Shoot length, leaf area, root length, and chlorophyll contents were determined for these plants grown in spent lubricating oil contaminated soils. The residual hydrocarbons were extracted from soil, and percentage degradation was gravimetrically determined for the total hydrocarbons and saturated hydrocarbons present in the spent lubricating oil. The contamination caused a reduction in the shoot length, leaf area, root length, and total chlorophyll content of the test plants used.

A statistically significant influence of spent lubricating oil on the test plants could not, however, be established. The degradation of total petroleum content was low, as the highest degradation recorded was 35.30%

for the plant *P. amarus*; however, appreciable degradation of the saturated hydrocarbons was recorded, as the plants *S. rhombifolia* and *M. alternifolia* removed over 60% of the saturated hydrocarbons present. *H. spicigera* recorded the least degradation for the saturated hydrocarbons (39.04%). The growth of the plants *S. rhombifolia* and *M. alternifolius*, which caused a reduction of over 60% of the saturated hydrocarbons, makes these two plants choice plants for the remediation of spent lubricating oil from contaminated soils.

Phytoremediation of soil polluted with used lubricating oil using nonedible plants, like *Jatropha curcas*, offers an environmental friendly and cost-effective method for remediating the polluted soil. In a recent investigation (Park et al., 2010), phytoremediation of soil contaminated with 2.5 and 1% (w/w) waste lubricating oil using *Jatropha curcas* and enhancement with organic wastes, banana skin, brewery spent grain, and spent mushroom compost was undertaken for a period of 180 days under room condition. 56.6% and 67.3% loss of waste lubricating oil was recorded in *Jatropha*-remediated soil without organic amendment for 2.5% and 1% contamination, respectively. However, addition of brewery spent grain to *Jatropha* remediation rapidly increases the removal of waste lubricating oil, to 89.6% and 96.6% in soil contaminated with 2.5% and 1% oil, respectively. Jatropha root did not accumulate hydrocarbons from the soil, but the number of hydrocarbon utilizing bacteria was high in the rhizosphere of the Jatropha plant, thus suggesting that the mechanism of the oil degradation was via rhizodegradation (Park et al., 2010).

In a similar unusual approach, three organic wastes, banana skin, brewery spent grain, and spent mushroom compost, were used for bioremediation of soil spiked with used engine oil to determine the potential of these organic wastes in enhancing biodegradation of used oil in soil (Abioye et al., 2010). The rates of biodegradation of the oil were studied for a period of 84 days under laboratory conditions. Hydrocarbon-utilizing bacterial counts were high in all the organic waste-amended soil compared to unamended control soil throughout the 84 days of study. Oil-contaminated soil amended with brewer spent grain showed the highest reduction in total petroleum hydrocarbon, with net loss of 26.76% in 84 days, compared to other treatments. A first-order kinetic model revealed that brewery spent grain was the best of the three organic wastes used, and the results obtained demonstrated the potential of organic wastes for oil bioremediation.

4.4 Vegetable Oils

The application of vegetable oils and animal fats for industrial purposes and, specifically, lubrication has been in practice for many years. Inherent disadvantages and the availability of inexpensive options have, however, brought about low utilization of vegetable oils for industrial lubrication (Honary, 2004). When applied in the science of tribology, vegetable oils fall under the class known as fixed oils (Gunther, 1971). They are so named because they do not volatilize without decomposing. Prior to recent developments, vegetable and animal oils in tribology have functioned mainly as additives to lubricating oil formulations, although in some cases they are applied exclusively, or in blends. For instance, tallow (free from acid) has been used as an emulsifying agent for steam cylinder oils, while castor, peanut, and rapeseed oils have been used in blends with lubricating oils to improve lubrication performance. Palm oil has been used in isolation as a fluxing dip in the tin plating of steel, while olive oil has applications as a yarn lubricant (Gunther, 1971).

Reasons for the use of vegetable oils in the science of lubrication abound (Aluyor et al., 2009). Their superior lubricity and emulsifying characteristics increase their desirability as additives to the cheaper but less effective petroleum-oil based lubricants. Their superior lubricity in industrial and machinery lubrication sometimes even necessitates the addition of friction materials in tractor transmissions in order to reduce clutch slippage (Honary, 2004).

Other advantages that encourage the use of vegetable oils include their relatively low viscosity-temperature variation, that is, their high viscosity indices, which are about twice those of lubricating oils (Honary, 2004). Additionally, they have low volatilities, as manifested by their high flash points (Honary, 2004). Significantly, they are environmentally friendly: renewable, non-toxic, and biodegradable (Howell, 2007). In summary, engine lubricants formulated from vegetable oils have the following advantages deriving from the base stock chemistry:

1. Higher Lubricity, resulting in lower friction losses, and hence, more power and better fuel economy.
2. Lower volatility, resulting in decreased exhaust emissions.
3. Higher viscosity indices.
4. Higher shear stability.
5. Higher detergency, eliminating the need for detergent additives.
6. Higher dispersion.
7. Rapid biodegradation, and hence, decreased environmental/toxicological hazards (Erhan and Perez, 2002).

In a comparison of palm oil and petroleum-based lubricants, palm oil based lubricants were found to be more effective in reducing the hydrocarbon and carbon monoxide emission levels, among other things (Masjuki et al., 1999). Vegetable oils have also been identified as having a lot of potential as alternative diesel engine fuels (Kayisoglu et al., 2006).

This is supported by an interest in a cleaner environment, as well as the increasing cost of mineral deposit based energy (Howell, 2007). Based on availability to meet demand, soybean, peanut, and sunflower oils have been identified as the most promising fuel sources (Kayisoglu et al., 2006). When used as a fuel, the term "biodiesel" is applicable.

Biodiesel is the mono alkyl ester (Howell, 2007), and consists primarily of long chain fatty acid esters, produced by the transesterification reaction of vegetable oils with short chain alcohols. Distinct advantages of biodiesel include a high flash point of over 100°C, excellent lubricity, a BTU content comparable to that of petro diesel, and virtually no sulfur or aromatic content. Above all, biodiesel is non-toxic and biodegradable (Howell, 2007).

Results from investigating performance of vegetable oils in blends with diesel indicate that blending up to 25% biodiesel (sunflower) with lubricating-based diesel has no adverse effect on performance (Kayisoglu et al., 2006). Vegetable oils have also been applied as transformer coolant oils, and have been found to conform to all industry standards, with performances and cost profiles comparable to the conventional lubricating oils applied in transformer cooling (ABB Inc., 2002; McShane, 2002). Transformer oil products have been produced from soybean oils, as well as castor oils (Honary, 2004).

The major disadvantage militating against the use of vegetable oils in industrial applications is its oxidative stability. This factor has been most researched (Nedyalka, 2001; Bahruddin et al., 2008), particularly as biodiesel (Kapilani et al., 2009). Several proposals on how to tackle this problem have been investigated (Anderson, 2007).

Ways of evaluating the oxidative stability of oils have occupied several authors (Tan et al., 2001; International Organization for Standards, 1996; Gertz et al., 2000). Several oils have been proposed for industrial uses, primarily due to their recognized high oxidative stability compared to other oils (Brimberg and Afaf, 1994; Gordon and Lenka, 1995; Nedyalka, 2001; Ruger et al., 2002; Schober and Mittelbach, 2004; Ghazalia et al., 2006).

The notion of biodegradability conveys the meaning of the breaking down of complex, and possibly toxic, material into simple and common forms, in which the elements carbon, hydrogen, and oxygen exist, or are simply associated. Whenever any substance is placed in the environment, it begins to undergo degradation due to the action of relevant substrate utilizing microorganisms.

Every substance, irrespective of its complexity and toxicity, is subject to this process. When used in the present context, it is referenced to a time frame. That is, for a given substance to be referred to as biodegradable, a specific time limit must be used as a reference.

According to the California advertising statute, a manufacturer cannot claim his product is biodegradable unless it has the proven capability to decompose in the most common environment where the product may be disposed within three years through natural biological processes into nontoxic carbonaceous soil, water, carbon dioxide, or methane. Biodegradation is important for the natural and industrial cycling of environmental chemicals. Government regulators and industries are always interested in the fate of industrial chemicals and waste products when discarded to the environment, either intentionally or unintentionally. This is because toxic substances in the environment eventually affect the ecosystem and humans adversely (Wackett et al., 1999).

Lipids (fats, oils, and greases) form a major part of domestic and industrial waste; hence, they contribute their fair share towards environmental pollution. Sources include waste water from the edible oil refinery, slaughter houses, and dairy products. These waste products are responsible for clogging sewer networks and unsettling the balance of waste water treatment plants (Saifudin et al., 2006).

The first step in the degradation of vegetable based oils is the enzyme-catalyzed cleavage of the ester bond to fatty acids. The enzymes that catalyze this biodegradation reaction include esterases and lipases, and these are synthesized by a wide range of microorganisms (Broekhuizen et al., 2003). "Lipases are hydrophobic proteins that catalyze the cleavage of carboxyl ester bonds in tri-, di-, and mono-acylglycerol derivatives (the major constituents of animal, plant and microbial fats and oils)" (Saifudin et al., 2006). Due to their degradative effects, lipases are applied in remediation efforts to degrade lipid rich waste.

Drawbacks, however, include thermal instability of the enzyme and the high cost of the single use of the enzyme (Saifudin et al., 2006). Following the first step of degradation, both saturated and unsaturated fatty acids biodegrade via a process of oxidation.

Indeed, the biodegradation of materials also is dependent on the nature of the environment, where pH adjustments of soils, for instance, was found to have a noticeable effect on the biodegradation of certain compounds (Kaakinen et al., 2007). Microbial communities are also prone to adapt to a substrate when it is a regular contaminant, that is, increases in rates of transformation of hydrocarbons associated with oil contaminated environments has been identified and documented (Khalida et al., 2006). Modified strains of bacteria emerge, which are characterized by the ability to degrade the substances which induce the modification (Abrashev et al., 2002; Mansee et al., 2004). However, biodegradability primarily is a function of the chemical nature of the substance in question.

That natural and vegetable oils are biodegradable is not in doubt. "Vegetable oils and synthetic esters have a much better biodegradation capacity than lubricating oil under aerobic as well as anaerobic conditions" (Broekhuizen et al., 2003). Tests carried out severally indicate that vegetable oils undergo about 70–100% biodegradation in a period of 28 days.

In a specific comparative study (Mecurio et al., 2003), vegetable-derived lubricants were established to be in fact more biodegradable than comparable petroleum-derived lubricating oil in the presence of tropical mangrove or coral reef microbial communities (Mecurio et al., 2003), while some others have examined the biodegradation of vegetable oils under spill conditions (Pereira et al., 1998; Campo et al., 2007; Al-Darbi et al., 2005).

Several authors have examined the biodegradability and oxidative stability of industrial fluids obtained from vegetable oils. These include as methyl esters (Demirbaş, 2009); as hydraulic fluids (Abdalla and Patel, 2006; Petlyuk et al., 2004); as lubricant (Fernando and Hanna, 2002), and as dielectric fluid (McShane, 2002).

5 Bioremediation – The Challenge

Consequent on the poor biodegradability of lubricating oils, remediation activities are required when spills occur. It has been demonstrated that a variety of contaminants can be degraded by microbes, and that such biological processes are cost and cleanup effective, especially when compared to conventional pump and treat processes applied for the remediation of contaminated groundwater. Cleanup efforts were carried out between 1997 and 2000 in Berlin for a 500 m^2 area, 10 m deep region contaminated by cable insulating oil. Bioremediation techniques were applied, and an impressive biodegradation rate of 86% in one thousand days was recorded (Zittwitz et al., 2000).

A similar study to demonstrate the potential of bioremediation of contaminated soils was carried out in Boucherville, Quebec, where a transformer oil spill had been recorded. The summary of the results is that on bioremediation, the site recorded a removal rate of 72% within 48 weeks (Biogenie, 1995).

In keeping with recorded successes on bioremediation efforts, research to optimize the process is ongoing, with the isolation of strains of microorganisms that are found to have hydrocarbon degradative capabilities. Such microorganisms have enzyme systems that degrade and utilize oils as sources of carbon and hydrogen. In research carried out by Nwaogu et al. (2008), diesel oil was used as a source for the isolation of certain bacteria strains, including *Bacillus subtilis* and *Bacillus cereus*, which are found to possess hydrocarbon degrading capability (Nwaogu et al., 2008).

Yet other studies have identified strains belonging to genus *Bacillus* as having impressive biodegradative ability, utilizing crude oil as the sole source of carbon and energy, even under stressed environmental conditions (Khalida et al., 2006). Further studies (Muratovba and Turkovskaia, 2001) reported isolation of a series of microbial associations capable of degrading various petroleum oils, emulsols, and crude oil. Oil product degradation by these microbial associations was identified to be most efficient during aerobic flow cultivation. Under these conditions, oils were degraded by about 92%; also, certain parameters, such as the brand of oil, the concentration of the oil, its degree of emulsification, and aeration, were found to have an effect on the microbial degradation of the oil (Muratovba et al., 2001). In another related study, surfactants belonging to the fatty acid-acylated amino acids were found to accelerate the microbial degradation of lubricating oils.

The faster, more complete degradation of lubricating oils is identified as being caused solely by interfacial activity (Riis et al., 2000).

6 Conclusion

Like fuels (Chapter 10, Chapter 11, Chapter 12), lubricating oils are very important products in the industrialized 21st century, with such numerous applications that they may be described as *indispensable*. However, increasing environmental awareness has placed the spotlight on the persistence of the harmful components of lubricating oils in the environment, and their cumulative effects on ecosystems, especially in the light of their established poor biodegradability.

Spills are inevitable, and the development of bioremediation processes utilizing isolated hydrocarbon degrading bacteria has gone a long way in improving the efficiency of spill cleanup. The poor biodegradability of lubricating oils, coupled with the fact that remediation efforts always tend to be expensive, has made the option of the development of environmentally friendly oils based primarily on vegetable oils an attractive one. The future will bring with it the improvement of the efficiency of bio-based fluids in various applications, leading to the widespread use of these products, a situation which will no doubt be in the best interests of the global ecosystem to which we belong.

7 References

Abdalla, H.S., and Patel, S. 2006. "The Performance And Oxidation Stability Of Sustainable Metalworking Fluid Derived From Vegetable Extracts." *Proceedings*. Institution of Mechanical Engineers, Part B: J. Engine. Manufacture, 220: 12.

Abioye, P.O., Abdul Aziz, A., and Agamuthu, P. 2010. "Enhanced Biodegradation of Used Engine Oil in Soil Amended with Organic Wastes." *Water Air Soil Pollut.*, 209: 173–179.

Abrashev, I., Rachev, R., Donev, T., and Petrova, M. 2002. "A Biodegradation Activity of Microbial Associations." *J. Cult. Collect.*, 3(1): 43–47.

Adegoroye, G. 1997. "Environmental Considerations in Property Design, Urban Development and Renewal." In *Dimensions of Environmental Problems in Nigeria*. O. Akinjide (Editor). Friedrich Ebert Foundation, Washington, DC. pp. 12–25.

Adelowo, O.O., Alagbe, S.O., and Ayandele, A.A. 2006. "Time-Dependent Stability of Used Engine Oil Degradation by Cultures of *Pseudomonas fragi* and *Achromobacter aerogens*." *African Journal of Biotechnology*, 5(24), 2476–2479.

Al-Darbi, M.M., Saeed, N.O., and Islam, M.R. 2005. "Biodegradation of Natural Oils in Seawater." *Energy Sources*, 27: 19–34.

Aluyor, E.O., and Ori-jesu, M. 2009. "Biodegradation of Mineral Oils – A Review." *African Journal of Biotechnology*, 8(6): 915–920.

Aluyor, E.O., Obahiagbon, K.O., and Ori-jesu, M. 2009. "Biodegradation of Vegetable Oils: A Review." *Scientific Research and Essay*, 4 (6): 543–548.

Anderson, K.D., Emir, B., Saad, W.W.D., Vechiatto, H.M., Wilhelm, L.P., and Ramos, J.B. 2007. "The Influence of BHA, BHT and TBHQ on the Oxidation Stability of Soybean Oil Ethyl Esters." *Biod. Chem. Soc.*, 18: 2.

April, T.M., Fought, J.M., and Currah, R.S. 2000. "Hydrocarbon-Degrading Filamentous Fungi Isolated from Flare Pit Soils in Northern and Western Canada." *Can. J. Microbiol.*, 46: 38–49.

Atlas, R.M. 1981. "Microbial Degradation of Petroleum Hydrocarbons: An Environmental Perspective." *Microbiol. Rev.*, 45: 180–209.

Atlas, R.M. 1984. *Petroleum Microbiology*. Macmillan Publishing Co., New York.

Bahruddin, S., Wan, T.W., and Boey, P.L. 2008. "Comparative Study on Oxidative Decomposition Behavior of Vegetable Oils and Its Correlation with Iodine Value Using Thermogravimetric Analysis." *J. Oleo Sci.*, 57(4): 257–266.

Balows, A., Truper, H.G., Dworkin, M., Harder, W., and Schleifer, K.H. 1992. *The Prokaryotes: A Handbook on the Biology of Bacteria*. Springer Verlag, Heidelberg, Germany.

Bartz, W.J. 1998. "Lubricants and the Environment." *Tribology International*, 31(1–3): 35–47.

Biogenie Inc. 1995. "Project Summary: Bioremediation of Soil Contaminated by Transformer Oil (Boucherville) and Diesel Fuel (Jonquiere)." Technology Development Section, Environment Canada. www.qc.ec.gc.ca/dpe.

Boonchan, S., Britz, M.L., and Stanley, G.A. 2000. "Degradation and Mineralization of High-Molecular Weight Polycyclic Aromatic Hydrocarbons by Defined Fungal–Bacterial Co-cultures." *Appl. Environ. Microbiol.*, 66(3): 1007–1019.

Booser, E.R. 1984. *CRC Handbook of Lubrication: Theory and Practice of Tribology*. CRC Press, Boca Raton, Florida.

Bossert, I., and Bartha, R. 1984. "The Fate of Petroleum in Soil Ecosystem." In *Petroleum Microbiology*. R.M. Atlas (Editor). Macmillan Publishing Co., New York. pp. 435–476.

Boyde, S. 2002. "Green Lubricants: Environmental Benefits and Impacts on Lubrication." *Green Chemistry*, 12(4): 293–307.

Brimberg, U.I., and Afaf, K-E. 1994. "On the Kinetics of the Autoxidation of Fats: Influence of Pro-Oxidants, Antioxidants and Synergists." *European. J. Lipid Sci. Technol.*, 105(2): 83–91.

Broekhuizen, P., Theodori, D., Le Blanch, K., and Ullmer, S. 2003. *Lubrication in Inland and Coastal Water Activities*. A. A. Balkema Publishers, Tokyo.

Brook, T.R., Stiver, W.H., and Zytner, R.G. 2001. "Biodegradation of Diesel Fuel in Soil under Various Nitrogen Addition Regimes." *Soil Sediment Contamination*, 10: 539–553.

Bumpus, J.A., Tien, M., Wright, D., and Aust, S.D. 1985. "Oxidation of Persistent Environmental Pollutants by a White Rot Fungus." *Science*, 228: 1434–1436.

Butler, C.S., and Mason, J.R. 1997. "Structure-Function Analysis of the Bacterial Aromatic Ring-Hydroxylating Dioxygenases." *Advanced Microbial Physiology*, 38: 47–84.

Campo, P.Z.Y., Suidan, M.T., Venosa, A.D., and Sorial, G.A. 2007. "Biodegradation Kinetics and Toxicity of Vegetable Oil Triacylglycerols under Aerobic Conditions." *Chemosphere*, 68(11): 2054–2062.

Cansfield, P.E., and Racz, G.J. 1978. "Degradation of Hydrocarbon Sludge in the Soil." *Can. J. Soil Sci.*, 3: 339–345.

Cecutti, C., and Agius, D. 2008. "Ecotoxicity and Biodegradability in Soil and Aqueous Media of Lubricants used in Forestry Applications." *Bioresource Technology*, 99: 8492–8496.

Chaîneau, C.H., Morel, J.L., and Oudot, J. 1995. "Microbial Degradation in Soil Microcosms of Fuel Oil Hydrocarbons from Drilling Cuttings." *Environ, Sci. Technol.*, 29: 1615–1621.

Chaîneau, C.H., Morel, J.L., and Oudot, J. 1996. "Land Treatment of Oil-Based Drill Cuttings in an Agricultural Soil." *J. Environ. Qual.*, 4: 858–867.

Chaîneau, C.H., Morel, J., DuPont, J., Bury, E., and Oudot, J. 1999. "Comparison of The Fuel Oil Biodegradation Potential of Hydrocarbon-Assimilating Microorganisms Isolated from a Temperate Agricultural Soil." *The Science of the Total Environment*, 227: 237–247.

Colombo, J.C., Cabello, M., and Arambarri, M. 1996. "Biodegradation of Aliphatic and Aromatic Hydrocarbons by Natural Soil Microflora and Pure Cultures of Imperfect and Ligninolytic Fungi." *Environ. Pollut.*, 94: 355–362.

Davies, J.S., and Westlake, D.W.S. 1979. "Crude Oil Utilization by Fungi." *Can. J. Microbiol.*, 25: 146–156.

Delistraty, D., and Stone, A. 2007. "Dioxins, Metals, and Fish Toxicity in Ash Residue from Space Heaters Burning used Motor Oil." *Chemosphere*, 68: 907–914.

Edewor, T.I., Adelowo, O.O., and Afolabi, T.J. 2004. "Preliminary Studies into the Biological Activities of a Broad Spectrum Disinfectant Formulated from Used Engine Oil." *Pollution Research*, 23(4): 581–586.

Erhan, S.Z., and Asadauskas, S. 2000. "Lubricant Basestocks from Vegetable Oils." *Industrial Crops and Products*, 11(2): 277–282.

Erhan, S.Z., and Perez, J.M. 2002. *Bio-based Industrial Fluids and Lubricants*. The American Oil Chemists' Society, Urbana, Illinois.

Erute Magdalene, O., Ufuoma, A., and Gloria, O. 2009. "Screening of Four Common Nigerian Weeds for use in Phytoremediation of Soil Contaminated with Spent Lubricating Oil." *African Journal of Plant Science*, Vol. 3 (5), pp. 102–106.

Faboya, O.O.P. 1997. "Industrial Pollution and Waste Management." In *Dimensions of Environmental Problems in Nigeria*. O. Akinjide (Editor). Friedrich Ebert Foundation, Washington, DC. pp. 12–25.

Farrington, A.M., and Slater, J.M. 1997. "Monitoring of Engine Oil Degradation by Voltammetric Methods Utilizing Disposable Solid Wire Microelectrodes." *Analyst*, 122: 593–596.

Fernando, S., and Hanna, M. 2002. "Oxidation Characteristics of Soybean Oils as Water Pump Lubricants." *Transactions of the ASAE*, 45(6): 1715–1719.

Gary J.H., and Handwerk, G.E. 2001. *Petroleum Refining Technology and Economics*. 4th Edition. Marcel Dekker Inc. New York, USA.

Gertz, C., Klostermann, S., and Kochhar, S.P. 2000. "Testing and Comparing Oxidative Stability of Vegetable Oils and Fats at Frying Temperature." *Eur. J. Lipid Sci. Technol.*, 102(8–9): 543–541.

Ghazalia, Z., Wan, N.W.B., Ku, B.KH, Ani, F., and Xiand, L.F. 2006. "The Effect of Light on the Oxidative Stability of Palm Olein." *Proceedings*. 1st International Conference on Natural Resources Engineering and Technology. Putrajaya, Malaysia. pp. 631–637.

Gordon, M.H., and Lenka, K. 1995. "The Effects of Antioxidants on Changes in Oils during Heating and Deep Frying." *J. Sci. Food Agric.*, 68 (3): 347–353.

Gunther, R.C. 1971. *Lubrication*. Bailey Brothers and Swinfen Ltd., Folkestone, Kent, United Kingdom.

Hagwell, I.S., Delfino, L.M., and Rao, J.J. 1992. "Partitioning of Polycyclic Aromatic Hydrocarbons from Oil into Water." *Environ. Sci. Technol.*, 26: 2104–2110.

Haigh, S.D. 1995. "Fate and Effects of Synthetic Lubricants in Soil: Biodegradation and Effect on Crops in Field Studies." *Science of the Total Environment*, 168: 71–83.

Haus, F., Jean, G., and Junter, G.A. 2001. "Primary Biodegradability of Mineral Base Oils in Relation to their Chemical and Physical Characteristics." *Chemosphere*, 45(6): 983–990.

Herdan, J.M. 1997. "Lubricating Oil Additives and The Environment — An Overview." *Lubrication Science*, Volume 9, Issue 2, 9(2): 161–172.

Hollender, J., Althoff, K., Mundt, M., and Dott, W. 2003. "Assessing the Microbial Activity of Soil Samples, Its Nutrient Limitation and Toxic Effects on Contaminants using a Simple Respiration Test." *Chemosphere*, 53: 269–275.

Honary, L.A.T. 2004. "Biodegradable/Biobased Lubricants and Greases." *Machinery Lubrication Magazine*, Issue Number 200109, Noria Corporation, Tulsa, Oklahoma.

Howell, S. 2007. "Promising Industrial Applications for Soybean Oil in the US." In International Organization for Standards (1996). *Animal and Vegetable Fats and Oils – Determination of Oxidative Stability (Accelerated Oxidation Test)*, ISO 6886: 200.

Husaini, A., Roslan, H.A., Hii, K.S.Y., and Ang, C.H. 2008. "Biodegradation of Aliphatic Hydrocarbons by Indigenous Fungi Isolated from Used Motor Oil Contaminated Sites." *World Journal of Microbiology and Biotechnology*, 24: 2789–2797.

Ijah, U.J.J. 1998. "Studies on Relative Capabilities of Bacterial and Yeast Isolates from Tropical Soil in Degrading Crude Oil." *Waste Management*, 18: 293–299.

Jørgensen, K.S., Puustinen, J., and Suortti, A.M. 2000. "Bioremediation of Petroleum Hydrocarbon-Contaminated Soil by Composting In Biopiles." *Environmental Pollution*, 107: 245–254.

Kaakinen, J., Vahaoja, P., Kuokkanen, T., and Roppola, K. 2007. "Studies on the Effects of Certain Soil Properties on the Biodegradation of Oils Determined by the Manometric Respirometric Method." *J. Automated Methods and Management in Chemistry*, 1–7.

Kapilani, N., Ashok, B.T.P., and Reddy, R.P. 2009. "Technical Aspects of Biodiesel and its Oxidation Stability." *Int. J. Chem. Tech. Res.*, 1(2): 278–282.

Karacan, O., Kok, M., Karaaslan U. 1999. "Dependence of Thermal Stability of an Engine Lubricating Oil on Usage Period." *J. Thermal Anal. Calorimetry*, 55(1): 109–114.

Kayisoglu, B., Ulger, P., Akdemir, B., and Aytac, S. 2006. "A Research on Determining Some Performance Values by Using Proportional Mixture of Vegetable Oils and Diesel Fuel at a Diesel Engine." *J. Tekirdag Agric. Faculty*, 3(1): 16–24.

Keith, L.H., and Telliard, W.A. 1979. "Priority Pollutants 1: A Perspective View." *Environ. Sci. Technol.*, 13: 416–423.

Khalida, K., Naeem, M., Arshed, M.J., and Asif, M. 2006. "Extraction and Characterization of Oil Degrading Bacteria." *J. Appl. Sci.*, 6(10): 2302–2306.

Khawaja, M.A., and Aban, M.M. 1996. "Characteristics of Used Lubricating Oils: Their Environmental Impact and Survey of Disposal Methods." *Environmental Management and Health*, 7(1): 23–32.

Ladousse, A., and Tramier, B. 1991. "Results of 12 Years of Research in Spilled Oil Bioremediation. Inipole EAP 22." *Proceedings*. 1991 Oil Spill Conf., American Petroleum Institue, Washington, DC.

Lageman, R., Clarke, R., and Pool, W. 2005. "Electro-reclamation: A Versatile Soil Remediation Solution." *Eng. Geol.*, 77: 191–201.
Launen, L., Pinto, L., Wiebe, C., Kiehlmann, E., and Moore, M. 1995. "The Oxidation of Pyrene and Benzo[A]Pyrene by Nonbasidiomycete Soil Fungi." *Can. J. Microbiol.*, 41: 477–488.
Lea, C. 2005. *Use of Crop-derived Raw Materials as Replacement for Mineral Oils in the Production of Metalworking Fluids: Challenges for the Lubricants Industry*. Fuchs Lubricants (UK) Plc. New Century Street, Hanley, Stoke-on-Trent, Staffs. www.bbsrc.ac.uk/businesscollaborative_researchlink.
Leahy, J.G., and Cowell, R.R. 1990. "Microbial Degradation of Hydrocarbons in the Environment." *Microbio. Rev.*, 54: 305–315.
Lee, S.H., Lee, S., and Kim, D.Y. 2007. "Degradation Characteristics of Waste Lubricants Under Different Nutrient Conditions." *Journal of Hazardous Materials*, (143): 65–72.
Llanos, C., and Kjoller, A. 1976. "Changes in the Flora of Soil Fungi Following Oil Waste Application." *Oikos*, 27: 377–382.
Lloyd, C.A., and Cackette, T.A. 2001. "Diesel Engines: Environmental Impact and Control." *Air and Waste Management Association*, 51: 805–847.
Mandri, T., and Lin, J. 2007. "Isolation and Characterization of Engine Oil Degrading Indigenous Microorganisms in Kwazulu-Natal, South Africa." *African Journal of Biotechnology*, 6(1): 23–27.
Mansee, A.H., Montasser, M.R., and Shanab, A.S.A. 2004. "Decontamination of Pollutants in Aquatic System: 1. Biodegradation Efficiency of Isolated Bacteria Strains from Certain Contaminated Areas." *Pakistan J. Biol. Sci.*, 7(7): 1202–1207.
Margesin, R., Zimmerbauer, A., and Schinner, F. 2000. "Monitoring of Bioremediation by Soil Biological Activities." *Chemosphere*, 40: 339–346.
Margesin, R., and Schinner, F. 2001. "Bioremediation (Natural Attenuation and Biostimulation) of Diesel-Oil-Contaminated Soil in an Alpine Glacier Skiing Area." *Applied and Environmental Microbiology*, 67: 3127–3133.
Margesin, R., Hammerle, M., and Tscherko, D. 2007. "Microbial Activity and Community Composition during Bioremediation of Diesel-Oil-Contaminated Soil: Effects of Hydrocarbon Concentration, Fertilizers, and Incubation Time." *Microbiol. Ecol.*, 53: 259–269.
Marquez-Rocha, F.J., Rodriguez, H.V., and Lamela, M.T. 2001. "Biodegradation of Engine and Diesel Oil in Soil by a Microbial Consortium." *Water, Air and Soil Pollution*, 128: 313–320.
Masjuki, H.H., Maleque, M.A., Kubo, A., and Nonaka, T. 1999. "Palm Oil and Mineral Oil Based Lubricants – their Tribological and Emission Performance." *Tribol. Int.*, 32(6): 305–314.

Matos Lopes, P.R., Nallin Montagnolli, R., de Fátima Domingues, R., and Bidoia, E.D. 2010. "Toxicity and Biodegradation in Sandy Soil Contaminated by Lubricant Oils." *Bull. Environ. Contam. Toxicol.*, 84: 454–458.

Matos Lopes, P.R., and Bidoia, E.D. 2009. "Evaluation of the Biodegradation of Different Types of Lubricant Oils in Liquid Medium." *Braz. arch. biol. technol.*, 52(5): 1285–1290.

Mecurio, P., Burns, K.A., and Negri, A. 2003. "Testing the Ecotoxicology of Vegetable Versus Mineral Based Lubricating Oils: 1. Degradation Rates Using Tropical Marine Microbes." *Environ. Pollution*, 129(2): 165–173.

Mishra, S.J., Jyot, R.C., and Kuhad, B.L. 2001. "Evaluation of Inoculum Addition to Stimulate In Situ Bioremediation of Oily–Sludge-Contaminated Soil." *Applied Environmental Microbiology*, 67(4): 1675–1681.

Morgan, P., and Watkinson, R.J. 1989. "Hydrocarbon Degradation in Soils and Methods for Soil Biotreatment." *CRC Crit. Rev. Biotechnol.*, 4: 305–333.

Mortier, R.M., and Orszulik, S.T. 1997. *Chemistry and Technology of Lubricants*. Springer, USA.

Muratovba, A., and Turkovskaia, O. 2001. "Degradation of Mineral Oils by a Selected Microbial Association." *Prikl. Biokhim. Mikrobiol.*, 37(2): 175–180.

Nedyalka, V.Y., and Emma, M.M. 2001. "Stabilization of Edible Oils with Natural Antioxidants." *Eur. J. Lipid Sci. Technol.*, 103(1): 1752–1767.

Nocentini, M., Pinelli, D., and Fava, F. 2000. "Bioremediation of a Soil Contaminated by Hydrocarbon Mixtures: The Residual Concentration Problem." *Chemosphere*, 41: 1115–1123.

Novotny, C., Erbanova, P., Sasek, V., Kubatova, A., Cajthaml, T., and Lang, E. 1999. "Extracellular Oxidative Enzyme Production And PAH Removal in Soil by Exploratory Mycelium of White Rot Fungi." *Biodegradation*, 10: 159–168.

Nwaogu, L.A., Onyeze, G.O.C., and Nwabueze, R.N. 2008. "Degradation of Diesel Oil in a Polluted Soil Using Bacillus subtilis." *Afr. J. Biotechnol.*, 7(12): 1939–1943.

OECD. 2004. Emission Scenario Document on Lubricants and Lubricant Additives. Report No. ENV/JM/MONO(2004)21. Organization for Economic Cooperation and Development, Paris, France.

Oudot, J., Fusey, P., Abdelouahid, D.E., Haloui, S., and Roquebert, M.F. 1987. "Capacités Dégradatives de Bactéries et de Champignons Isolés d'un Sol Contaminé par un Fuel." *Can. J. Microbiol.*, 33: 232–243.

Oudot, J., Amblès, A., Bourgeois, S., Gatellier, C., and Sebyera, N. 1989. "Hydrocarbon Infiltration and Biodegradation in a Land Farming Experiment." *Environ. Pollut.*, 59: 17–40.

Oudot, J., Dupont, J., Haloui, S., and Roquebert, M.F. 1993. "Biodegradation Potential of Hydrocarbon-Assimilating Tropical Fungi." *Soil Biol. Biochem.*, 25: 1167–1173.

Park, S-W., Lee, J-Y., Yang, J-S., Kim, K-J., and Baek, K. 2010. "Phytoremediation of Soil Contaminated with Used Lubricating Oil using *Jatropha curcas*." *Journal of Hazardous Materials*, 179(1–3): 1168–1172.

Pereira, M.G., Mudge, S.M., and Latchford, J. 1998. "Bacterial Degradation of Vegetable Oils." *Chemistry and Ecology*, 14 (3–4): 291–303.

Petlyuk, A., Adams, M., and Richard, J. 2004. "Oxidation Stability and Tribological Behavior of Vegetable Oil Hydraulic Fluids." *Tribology Transactions*, 47(2): 182–187.

Probst, T.L., Lochmiller, R.L., Qualis, C.W., and McBee, K. 1999. "In Situ (Mesocosm) Assessment of Immunotoxicity Risks to Small Mammals Inhabiting Petrochemical Waste Sites." *Chemosphere*, 38: 1049–1067.

Raymond, R.L. 1976. "Oil Degradation in Soil." *Appl. Env. Microbiol.*, 31: 522–535.

Rhykerd, R.L., Weaver, R.W., and McInnes, K.J. 1995. "Influence of Salinity on Bioremediation of Oil in Soil." *Environ. Pollut.*, 90: 127–130.

Riffaldi, R., Levi-Minzi, R., Cardelli, R., Palumbo, S., and Saviozzi, A. 2006. "Soil Biological Activities in Monitoring the Bioremediation of Diesel Oil-Contaminated Soil." *Water, Air & Soil Pollution*, 170: 3–15.

Riis, V., Brandt, M., Miethe, D., and Babel, W. 2000. "Influence of Special Surfactants on the Microbial Degradation of Mineral Oils." *Chemosphere*, 41(7): 1001–1006.

Roy, T.A., Johnson, S.W., Blackburn, G.R., and Mackerer, C.R. 1988. "Correlation of Mutagenic and Dermal Carcinogenic Activities of Mineral Oils with Polycyclic Aromatic Compound Content." *Fund. Appl. Toxicol.*, 10: 466–476.

Ruger, C.W., Klinker, E.J., and Hammond, E.G. 2002. "Abilities of Some Antioxidants to Stabilize Soybean Oil in Industrial Use Conditions." *J. Am. Oil Chemists' Soc.*, 79: 734–736.

Rychla, L., Rychly, J., and Srivastava, S.P. 2001. "Determination of Stability of Mineral Oils Crude Oils and Waxes by Chemiluminescence." *Petroleum Coal*, 43(1): 35–40.

Saifudin, N., and Chua, K.H. 2006. "Biodegradation of Lipid-rich Waste Water by Combination of Microwave Irradiation and Lipase Immobilized on Chitosan." *Biotechnol.*, 5 (3): 315–323.

Saraswathy, A., and Hallberg, R. 2005. "Mycelial Pellet Formation by *Penicillium ochrochloron* Species due to Exposure to Pyrene." *Microbiol. Res.*, 160: 375–383.

Schober, S., and Mittelbach, M. 2004. "The Impact Of Antioxidants On Biodiesel Oxidation Stability." *European J. Lipid Sci. Technol.*, 106(6): 382–389.

Semple, K.T., Dew, N.M., Doick, K.J., and Rhodes, A.H. 2006. "Can Mineralization be used to Estimate Microbial Availability of Organic Contaminants In Soil?" *Environmental Pollution*, 140: 164–172.

Shojaosadati, S.A., Kolaei, S.M.V., Babaeipour, V., and Farnoud, A.M. 2008. "Recent advances in high cell density cultivation for production of recombinant protein." *Iranian Journal of Biotechnology*, 6(2): 63–84.

Stachowiak, G.W., and Batchelor, A.W. 2005. *Engineering Tribology*. Butterworth – Heinemann, USA.

Swannell, R.P.J., Lee, K., and Donagh, M.M. 1996. "Field Evaluation of Marine Oil Spill Bioremediation." *Microbiol Rev.*, 60: 342–365.

Thenmozhi, R., Nagasathya, A., and Thajuddin, N. 2011. "Studies on Biodegradation of Used Engine Oil by Consortium Cultures." *Advances in Environmental Biology*, 5(6): 1051–1057.

Udeani, T.K.C., Obroh, A.A., Okwuosa, C.N., Achukwu, P.U., and Azubike, N. 2009. "Isolation of Bacteria from Mechanic Workshops' Soil Environment Contaminated with Used Engine Oil." *African J. Microbiol.*, 8(22): 6301–6303.

Upshall, C., Payne, J.F., and Hellou, J. 1992. "Induction of MFO Enzymes and Production of Bile Metabolites in Rainbow Trout (*Oncorhynchus mykiss*) Exposed to Waste Crankcase Oil." *Environmental Toxicology and Chemistry*, 12: 2105–2112.

US Army Corps of Engineers. 1999. *US Army Manual EM1110–2–1424*. Chapter 8.

US EPA Emergency Response Program. 2008. "Physical Properties of Petroleum Based Oil." www.epa.gov.

US Army Corps of Engineers. 1999. *US army Manual EM1110–2–1424* (chapter 8). Viewed from www.usace.army.mil/usace-docs/eng-manuals/em1110–2–1424/c-8.pdf.

Vazquez-Duhalt, R. 1989. "Environmental Impact of used Motor Oil." *Sci. Total Environ.*, 79(1): 1–23.

Wackett, L.P., and Ellis, L.B.M. 1999. "Predicting Biodegradation." *Environmental Microbiology*, 1 (2): 119–124.

Walworth, J., Pond, A., Snape, I., Rayner, J., Ferguson, S., and Harvey, P. 2007. "Nitrogen Requirements for Maximizing Petroleum Bioremediation in a Sub-Antarctic Soil." *Cold Regions Science and Technology*, 48: 84–91.

Watanabe, K. 2001. "Microorganism relevant to bioremediation." *Curr. Opin. Biotechnol.*, 12: 237–241.

Willing, A. 2001. "Lubricants Based on Renewable Resources: An Environmentally Compatible Alternative to Mineral Oil Products." *Chemosphere*, 43: 89–98.

Zittwitz, M., Gerhardt, M., and Ringpfeil, M. 2000. "Practical Experience from Commercial In-situ Bioremediation in Cases of Cable Insulating Oil and Tri-/Perchlorethylene." Biopract GmbH, Rudower Chaussee, Berlin, Germany. www.biopract.de.

14

Biodegradation of Residua and Asphalt

1 Introduction

The residuum or asphaltic fraction of petroleum is the fraction that is either is (1) the residue remaining after completion of atmospheric distillation, or (2) the residue remaining after the completion of vacuum distillation. On the other hand, asphalt is the product produced from residua that is used, among other uses, for road and highway surfacing (Speight, 2007).

A *residuum* (*pl. residua*, also shortened to *resid, pl. resids*) is the residue obtained from petroleum after nondestructive distillation has removed all of the volatile materials. The temperature of the distillation is usually maintained below 350°C (660°F), since the rate of thermal decomposition of petroleum constituents is minimal below this temperature, but the rate of thermal decomposition of petroleum constituents is substantial above 350°C (660°F). *Residua* are black, viscous materials, and are obtained by distillation of a crude oil under atmospheric pressure (atmospheric residuum) or under reduced pressure (vacuum residuum) (Chapter 16). They may be liquid at room temperature (generally atmospheric residua), or almost solid (generally vacuum residua), depending upon the nature of the crude oil (Table 14.1) (Speight, 2007).

The chemical composition of a residuum from an asphaltic crude oil is complex. Physical methods of fractionation usually indicate high

Table 14.1 Properties of various residua.

Residuum	Gravity API	Sulfur % w/w	Nitrogen % w/w	Nickel ppm	Vanadium ppm	Asphaltene fraction (heptane)	Carbon residue (Conradson, % w/w
Arabian Light. >650 F	17.7	3.0	0.2	10.0	26.0	1.8	7.5
Arabian Light, >1050 F	8.5	4.4	0.5	24.0	66.0	4.3	14.2
Arabian Heavy, > 650 F	11.9	4.4	0.3	27.0	103.0	8.0	14.0
Arabian Heavy, >1050 F	7.3	5.1	0.3	40.0	174.0	10.0	19.0
Alaska, North Slope, >650 F	15.2	1.6	0.4	18.0	30.0	2.0	8.5
Alaska, North Slope, >1050 F	8.2	2.2	0.6	47.0	82.0	4.0	18.0
Lloydminster (Canada), >650 F	10.3	4.1	0.3	65.0	141.0	14.0	12.1
Lloydminster (Canada), >1050 F	8.5	4.4	0.6	115.0	252.0	18.0	21.4
Kuwait, >650 F	13.9	4.4	0.3	14.0	50.0	2.4	12.2
Kuwait, >1050 F	5.5	5.5	0.4	32.0	102.0	7.1	23.1
Tia Juana, >650 F	17.3	1.8	0.3	25.0	185.0		9.3
Tia Juana, >1050 F	7.1	2.6	0.6	64.0	450.0		21.6
Taching, >650 F	27.3	0.2	0.2	5.0	1.0	4.4	3.8
Taching, >1050 F	21.5	0.3	0.4	9.0	2.0	7.6	7.9
Maya, >650 F	10.5	4.4	0.5	70.0	370.0	16.0	15.0

Source: Speight, J.G. 2007. *The Chemistry and Technology of Petroleum*. 4th Edition. CRC Press, Taylor & Francis Group, Boca Raton, Florida.

proportions of asphaltene constituents and resin constituents, even in amounts up to 50% (or higher) of the residuum. In addition, the presence of ash-forming metallic constituents, including such organometallic compounds as those of vanadium and nickel, is also a distinguishing feature of residua and the heavier oils. Furthermore, the deeper the *cut* into the crude oil, the greater is the concentration of sulfur and metals in the residuum, and the greater the deterioration in physical properties (Speight, 2007).

On the other hand, *asphalt* is manufactured from petroleum, and is a black or brown material that has a consistency varying from a viscous liquid to a glassy solid. To a point, asphalt can resemble bitumen, hence the tendency to refer to bitumen (incorrectly) as *native asphalt*. It is recommended that there be differentiation between asphalt (manufactured) and bitumen (naturally-occurring) other than by use of the qualifying terms *petroleum* and *native*, since the origins of the materials may be reflected in the resulting physicochemical properties of the two types of materials. It is also necessary to distinguish between the asphalt which originates from petroleum by refining and the product in which the source of the asphalt is a material other than petroleum, e.g., *wurtzilite asphalt*. In the absence of a qualifying word, it is assumed that the term *asphalt* refers to the product manufactured from petroleum.

When the asphalt is produced simply by distillation of an asphaltic crude oil, the product can be referred to as *residual asphalt* or *straight run asphalt*. If the asphalt is prepared by solvent extraction of residua or by light hydrocarbon (propane) precipitation, or if blown or otherwise treated, the term should be modified accordingly to qualify the product (e.g., *propane asphalt*, *blown asphalt*).

Asphalt softens when heated, and is elastic under certain conditions. The mechanical properties of asphalt are of particular significance when it is used as a binder or adhesive. The principal application of asphalt is in road surfacing, which may be done in a variety of ways. Light oil *dust layer* treatments may be built up by repetition to form a hard surface, or a granular aggregate may be added to an asphalt coat, or earth materials from the road surface itself may be mixed with the asphalt.

Other important applications of asphalt include canal and reservoir linings, dam facings, and sea works. The asphalt so used may be a thin, sprayed membrane, covered with earth for protection against weathering and mechanical damage, or thicker surfaces, often including riprap (crushed rock). Asphalt is also used for roofs, coatings, floor tiles, soundproofing, waterproofing, and other building-construction elements, and in a number of industrial products, such as batteries. For certain applications, an asphaltic emulsion is prepared, in which fine globules of asphalt are suspended in water.

Other terminologies (often used incorrectly) that the investigator should be made aware of are the terms *tar* and *pitch*. These terms are often applied indiscriminately to any material that is black and viscous, no matter what the source. Confusion or misinterpretation of these terms with the terms *resid* and *asphalt* can lead any environmental investigator to a situation where inappropriate methods are applied to remediation of spills of such materials.

Tar is a product of the destructive distillation of many bituminous or other organic materials, and is a brown to black, oily, viscous, liquid to semi-solid material. *Tar* is most commonly produced from *bituminous coal*, and is generally understood to refer to the product from coal, although it is advisable to specify *coal tar* if there is the possibility of ambiguity. The most important factor in determining the yield and character of the coal tar is the carbonizing temperature. Three general temperature ranges are recognized, and the products have acquired the designations: *low-temperature tar* (approximately 450 to 700°C; 540 to 1290°F), *mid-temperature tar* (approximately 700 to 900°C; 1290 to 1650°F), and *high-temperature tar* (approximately 900 to 1200°C; 1650 to 2190°F). Tar released during the early stages of the decomposition of the organic material is called *primary tar*, since it represents a product that has been recovered without the secondary alteration that results from prolonged residence of the vapor in the heated zone.

Treatment of the distillate (boiling up to 250°C, 480°F) of the tar with caustic soda causes separation of a fraction known as *tar acids*; acid treatment of the distillate produces a variety of organic nitrogen compounds known as *tar bases*. The residue left following removal of the heavy oil, or distillate, is *pitch*, a black, hard, and highly ductile material.

In the chemical-process industries, pitch is the black or dark brown residue obtained by distilling coal tar, wood tar, fats, fatty acids, or fatty oils. *Coal tar pitch* is a soft to hard and brittle substance containing chiefly aromatic resinous compounds, along with aromatic and other hydrocarbons and their derivatives; it is used chiefly as road tar, in waterproofing roofs and other structures, and to make electrodes. *Wood tar pitch* is a bright, lustrous substance containing resin acids; it is used chiefly in the manufacture of plastics and insulating materials, and in caulking seams. *Pitch* derived from fats, fatty acids, or fatty oils by distillation are usually soft substances containing polymers and decomposition products; they are used chiefly in varnishes and paints, and in floor coverings.

Bioremediation of petroleum residua and asphalt has becomes increasingly important, due to the damage caused by the spills of such materials. To lay a solid foundation of bioremediation processes, it is a necessary (as with all petroleum products) that there is an understanding to the composition and properties of residua and asphalt.

Both the atmospheric residuum (which also contains a fraction known as vacuum gas oil) and the vacuum residuum residue (which is the atmospheric residuum without the vacuum gas oil) contain compounds that either are (1) non-biodegradable, or (2) degrade very slowly. One of the reasons that residuum, which is high in asphaltene constituents, makes an excellent road paving material is because it is slow to degrade. Tar balls, like mousse, are difficult to degrade, because their low surface area restricts the availability of oxygen and other nutrients.

Resin constituents include petroleum compounds containing nitrogen, sulfur, and/or oxygen as constituents. If not highly condensed, they may be subject to limited microbial degradation. Asphaltene constituents and resin constituents are difficult to analyze, and, to date, little information is available on the biodegradability of most compounds in these groups. Light oils may contain about 1 to 5 percent of both asphaltene constituents and resin constituents, whereas heavy oil or weathered oil may have up to 25% w/w asphaltene constituents and 20% w/w resin constituents.

2 Identity and Origin of Residua and Asphalt

2.1 Residua

When a residuum is obtained from a crude oil and thermal decomposition has commenced, it is more usual to refer to this product as *pitch*. The differences between the parent petroleum and the residua are due to the relative amounts of various constituents present, which are removed or remain by virtue of their relative volatility.

The environmental impact and toxicological profile of petroleum residua is high because of the high content of polynuclear aromatic hydrocarbon derivatives (typically in the asphaltene fraction), some of which contain heteroaromatic ring systems. Soil contamination has been a growing concern, because it can be a source of groundwater (drinking water) contamination, contaminated soils can reduce the usability of land for development, and weathered petroleum residua are particularly recalcitrant, and may stay bound to soil for years, if not decades.

2.2 Asphalt

Asphalt is a major product of many petroleum refineries. Asphalt may be residual (straight-run) asphalt, which is made up of the nonvolatile hydrocarbons in the feedstock, along with similar materials produced by thermal alteration during the distillation sequences, or they may be produced by air blowing residua. Alternatively, asphalt may be the residuum from a vacuum distillation unit. In either case, the properties of the asphalt are, essentially, the properties of the residuum (Speight, 2007). If the properties are not suitable for the asphalt product to meet specifications, changing the properties by, for example, blowing is necessary.

Asphalt manufacture is, in essence, a matter of distilling everything possible from crude petroleum until a residue with the desired properties is obtained. This is usually done by stages (Figure 26–3); crude distillation at atmospheric pressure removes the lower boiling fractions and yields reduced crude that may contain higher boiling (lubricating) oils, asphalt, and even wax. Distillation of the reduced crude under vacuum removes the oils (and wax) as overhead products, and the asphalt remains as a bottom (or residual) product. The majority of the polar functionality in the original crude oil end to be non-volatile and concentrate in the vacuum residuum (Figure 26–4). It is this concentration effect that confers upon asphalt some of its unique properties. At this stage, the asphalt is frequently and incorrectly referred to as pitch, and has a softening point related to the amount of oil removed: the more oil distilled from the residue, the higher the softening point (Corbett and Petrossi, 1978).

However, since there are wide variations in refinery operations and crude petroleum, asphalt with softening points ranging from 25 to 55°C (80 to 130°F) may be produced. Blending with higher and lower softening point asphalt may make asphalt of intermediate softening points. If lubricating oils are not required, the reduced crude may be distilled in a flash drum, which is similar to a bubble tower, but has few, if any, trays. Asphalt descends to the base of the drum as the oil components pass out of the top of the drum. If the asphalt has a relatively low softening point, it can be hardened by further distillation with steam or by oxidation.

Asphalt is also produced by propane deasphalting (Chapter 19), and the asphalt so produced may have a softening point of about 95°C (200°F). Softer grades are made by blending *hard asphalt* with the extract obtained in the solvent treatment of lubricating oils.

Soft asphalt can be converted into harder asphalt by oxidation, which promotes the formation of resins and asphaltene constituents from the lower molecular weight constituents:

$$\text{Oil constituents} \rightarrow \text{resin constituents} \rightarrow \text{asphaltene constituents}$$

The increase in the proportion of semisolid (resins) and solid (asphaltene constituents) constituents as a result of air blowing is accompanied by an increase in softening point, with only a small loss in volume. A similar increase in softening point by removing oily constituents would cause a considerable decrease in volume. Oxidation is carried out by blowing air through asphalt heated to about 260°C (500°F), and is usually done in a tower (an oxidizer) equipped with a perforated pipe at the bottom through which the air is blown. The asphalt, in the batch mode or continuous mode, is heated until the oxidation reaction starts, but the reaction is exothermic, and the temperature is controlled by regulating the amount of air and by circulating oil or water through cooling coils within the oxidizer. Asphalt with softening points as high as 180°C (350°F) may be produced.

Asphalt, normally a liquid when applied at higher temperature, may be referred to as *nonvolatile liquid asphalt*, but semisolid or solid asphalt may be made liquid for easier handling by dissolving them in a solvent, and is referred to as *cutback asphalt*. The asphalt and solvent (naphtha, kerosene, or gas oil) are heated to about 105°C (225°F), and passed together through a mechanical mixer, and the effluent then enters a horizontal tank, which is used for further mixing. Liquid asphalt is pumped from the top of the tank into a perforated pipe lying at the bottom of the tank, and circulation is continued until mixing is complete.

Asphalt paving materials typically are composed of aggregate and/or sand (90–95% by weight) and asphalt (5 to 10% by weight). The aggregate and/or sand is responsible for the primary load-bearing properties, while asphalt serves as the binder, and as a protective coating. The asphalt binder functions best when the aggregate/sand particles are *wetted* by the asphalt. To ensure good adhesion between the asphalt and the aggregate or sand particles, moisture and clays are to be avoided.

Similar to petroleum residua, the spillage of asphalt (other than for its intended purpose of road use as part of an asphalt-aggregate mixture) onto soil and waterways is disadvantageous.

The main hazard associated with asphalt is from the polynuclear aromatic hydrocarbon derivatives that can move into the ecosystem from the breakdown of asphalt. Since asphalt contains so many toxic and carcinogenic compounds, and since leaching of harmful polynuclear aromatic hydrocarbon derivatives has been documented even in water pipe use, asphalt should be kept out of rivers, streams, and other natural waters to the extent possible.

3 Biodegradation of Residua and Asphalt

In residua and asphalt, which are characterized by high viscosity and high density or low API gravity (Speight, 2007), asphaltene constituents, resin constituents, and the matrix compose a dynamic stable system, in which the non-asphaltene and non-resin fraction act as the medium for dispersion in which the asphaltene constituents are stabilized by the resin constituents (Speight, 2007).

The inherent biodegradability of these individual high-boiling constituents is a reflection of their chemical structure, but is also strongly influenced by the physical state and toxicity of the compounds. While n-alkanes as a structural group are the most biodegradable petroleum hydrocarbons, the higher molecular weight constituents become increasingly recalcitrant, because not only the chemical structure, but also the physical state of the constituents, strongly influences biodegradation (Bartha and Atlas, 1977).

In fact, asphaltene and resin constituents tend to increase during biodegradation in relative, and sometimes absolute, amounts (Okoh, 2006). This suggests that they not only tend to resist biodegradation, but may also be formed by condensation reactions of biodegradation and photodegradation intermediates.

3.1 Residua

Worldwide higher-quality lighter crude has been depleting during the last decades. Future demands for petroleum is focused on the use of heavy and extra-heavy crude oil, which is difficult to produce, transport, and refine; therefore, the interest in transportation and conversion of the high-molecular weight fractions of these materials into refined fuels and petrochemicals has increased (Gray, 1994). There are several known strategies for future heavy crude oil production (the up-stream) and upgrading (the down-stream) (Speight, 2007). However, when heavy crudes enter a refinery, their processing, as well as processing of bitumens, requires conversion of the vacuum residue components, including the asphaltene constituents, resin constituents, long chain aliphatic compounds, and waxes into distillable fractions. This upgrading has typically been accomplished with either thermal conversion (cracking or coking) (Twerdok, 2004; Rana et al., 2007) or by catalytic hydroconversion (Kirkwood et al., 2004). These processings require expensive investment in process equipment and supporting infrastructure (Kirkwood et al., 2004).

Heavy crude oil and residue have many similarities in composition. Hence, vacuum residua could be a net sample of problematic components of heavy crudes. In spite of some similarities between residue and heavy oil, residue feedstock differs in several ways, such as its higher concentration of asphaltene, sulfur, nitrogen and metal compounds, which depend on the origin of heavy or extra-heavy crudes (bitumen) (Gray, 1994). Since biological processing may offer less severe process conditions and higher selectivity to specific reactions to increase net distillates, it is proposed that the bacteria with the ability to biodegrade heavy fractions of vacuum residua could be a useful apparatus for upgrading heavy crude oils.

The biodegradation of residua requires a complex metabolic pathway, which usually can be observed in a microbial community. Many studies have been carried out on the biodegradation of petroleum hydrocarbons using a consortium of microorganisms (Morais and Tauk-Tornisielo, 2009). In literature, it is revealed that in soils that are permeated with the heavy hydrocarbons, bacteria, including indigenous ones, would survive and function after contaminations (Whyte et al., 1997; Foght et al., 1990) seeped through the soil. Selection of bacterial communities for petroleum substances occurs rapidly after even short-term exposures of soil to petroleum hydrocarbons (Van der Meer et al., 1992, 1994). During adaptation of microbial communities to hydrocarbon components, particularly complex ones, genes for hydrocarbon-degrading enzymes that are carried on plasmids or transposons may be exchanged between species, and new catabolic pathways eventually may be assembled and modified for efficient regulation (Rabus et al., 2005). Other cell adaptations leading to new ecotypes may include modifications of the cell envelope to tolerate solvents (Ramos et al., 2002) and development of community level interactions that facilitate cooperation within consortia (Kim et al., 2007).

One of the limitation factors for biodegradation of petroleum hydrocarbons in oil is the bioavailability (Banat, 1995; Kim, et al., 2001) of these components due to their chemical structure (Harvey, 1997; Cerniglia, 1992; Makkar and Rockne, 2003). One of the options to increase bioavailability of PAHs is the use of surfactants to increase desorption and apparent solubility in the aqueous phase (Makkar and Rockne, 2003), and consequently enhance the oil mobility, improving the biodegradation rates (Barathi et al., 2001; Burd and Ward, 1996a, 1996b; Burd and Ward, 1997; Rosenberg et al., 1988; Sar and Rosenberg, 1983).

3.2 Asphalt

In the United States alone, several million tons of asphalt are used annually as cement in the construction of roads. Asphalt is the unrefined residue of the fractional distillation of petroleum crude oil. The chemical composition of asphalt depends not only on the source of the petroleum crude oil from which it is derived, but also on the method of its production (Speight, 2007).

An early study (Phillips and Traxler, 1963) indicated that organisms of the genus *Pseudomonas* vary, to a greater or lesser extent, in the mechanisms by which they oxidize an asphalt substrate. All of these studies were made with a single asphalt cement (135 penetration; American Society for Testing Materials), designated asphalt 1-A. The use of a single asphalt rules out the possibility that differences observed were due to differences in the chemical nature of the substrate.

Previous investigations (Ramamurti et al., 1984) have established that some bacteria can utilize asphalt. However, the complexity of asphalt has limited the study of its biodegradation (Atlas, 1981). Petroleum-degrading bacteria excrete biosurfactants, which emulsify hydrocarbons of petroleum by reducing the interfacial tension between the hydrocarbons and water (Cooper and Zajic, 1980).

One petroleum-degrading bacterium, *Acinetobacter alcoaceticus*, excretes an emulsifier, emulsan (Goldman et al., 1982; Gutnick, 1987). Emulsan is an extracellular polysaccharide polymer with fatty acid side chains

(Shoham and Rosenberg, 1983; Shoham et al., 1983) (Chapter 10). This extracellular molecule forms a complex with proteins. These proteins bind non-covalently to the polysaccharide backbone. This emulsifier loses most of its activity when these proteins are removed from the complex (Zosim et al., 1987; Foght et al., 1989).

Biodegradation of asphalt occurs mostly on the surface of the asphalt exposed to oxygen, where the primary nutrients (asphalt, 02, and other trace minerals) are all present in adequate levels. Surface degradation is limited by oxygen diffusion and diffusion or presence of needed minerals/ions (i.e., phosphorous and nitrogen) (Pendrys, 1989). As time progresses, degradation should diminish, as all of these become limited. In addition, the mechanism for biodegradation may be such that a residue that is resistant to biodegradation remains on the surface, protecting the underlying asphalt from further biodegradation. As the saturate and naphthene aromatic hydrocarbons are emulsified, a residue remains on the surface that is resistant to biodegradation, and protects the underlying asphalt from biodegradation. The most potent asphalt-degrading bacterium, *Acinetobacter calcoaceticus* NAV2, excretes an emulsifier that is capable of emulsifying the saturate and naphthene aromatic fractions of asphalt cement-20 (Pendrys, 1989).

In another study (Luey and Li, 1993), testing was initiated in March 1991, and completed in November 1992, to determine the rate at which asphalt is biodegraded by microorganisms native to the Hanford Site soils. The asphalt tested (AR-6000, U.S. Oil, Tacoma, Washington) is to be used in the construction of a diffusion barrier for the Hanford grout vaults. Experiments to determine asphalt biodegradation rates were conducted using three separate test sets. These test sets were initiated in March 1991, January 1992, and June 1992, and ran for periods of 6 months, 11 months, and 6 months, respectively.

The experimental method used was one originally developed by Bartha and Pramer (1965), and further refined by Bowerman et al. (1985), that determined the asphalt biodegradation rate through the measurement of carbon dioxide evolved.

Using data from the January and June 1992 test sets, asphalt biodegradation rates were determined for test flasks containing soil and irradiated asphalt sample incubated at ambient and elevated (–35°C, –31°F) temperature. The average asphalt biodegradation rate determined at ambient temperature for test flasks incubated for less than 160 days was 1.6×10^{-4} cm/yr, with a high endpoint at the 95% confidence interval of 1.9×10^{-4} cm/yr, and a low endpoint at the 95% confidence interval of 1.4×10^{-4} cm/yr. For ambient temperature test flasks incubated up to 310 days, the average asphalt biodegradation rate was determined to be 1.0×10^{-4} cm/yr, indicating that the rate of biodegradation decreases as a function of time.

For test flasks at elevated temperature, the average asphalt biodegradation rate was determined to be 1.1×10^{-3} cm/yr with a high endpoint of 1.3×10^{-3} cm/yr, and a low endpoint of 9.9×10^{-4} cm/yr. Conservatively, the asphalt diffusion barrier will degrade at 1.3×10^{-3} cm/yr during elevated temperatures, with the rate decreasing to 1.9×10^{-4} cm/yr at lower, ambient temperature.

This is a conservative estimate of the long-term degradation rates. Parameters such as available water, available oxygen, and available nutrients will affect the long-term biodegradation rate. Since the environment surrounding the asphalt diffusion barrier will likely be deficient in these parameters, the long-term biodegradation rate is predicted to be lower than determined by this study. However, additional work is needed to investigate the influence of these parameters, and others, on the asphalt biodegradation rate and refine the estimated long-term rate.

Extrapolating the biodegradation rate to 10,000 years indicates that approximately 3 cm of asphalt would degrade, assuming that the asphalt diffusion barrier remained at elevated (35°C, 95°F) temperature for 1,000 years, and then was at ambient temperature for 9,000 years. This amount of degradation is within the 10-acre allowance in the design criteria for the asphalt diffusion barrier. This allowance would be approached only if the barrier remained at elevated temperature during the entire 10,000-year performance period (based upon the conservative asphalt biodegradation rate).

3.3 Asphaltene Constituents

The asphaltene constituents have drawn considerable attention due to problems caused by their detrimental effects in the extraction, transportation, and processing of residua because of their viscous and flocculating nature and their relative resistance to biodegradation following spills (Speight, 2007).

The asphaltene constituents are the highest molecular weight and most polar fraction of crude oil. Despite that the structure of asphaltene constituents structure has not been fully elucidated, it is widely accepted that they are constituted by interacting systems of polyaromatic sheets bearing alkyl side-chains. Asphaltene molecules have a high content of O, N, and S heteroatoms, as well as metals (V, Ni, and Fe) (Speight, 2007). The problems associated with asphaltene constituents have increased due to the need to extract heavier crude oils, as well as the trend to extract larger amounts of light fractions out of crude oil by cracking and visbreaking.

The asphaltenic fraction is recognized as the most recalcitrant oil fraction. There is no clear evidence that asphaltene constituents can be degraded or transformed by microbial activity. Microorganisms have been found associated with bitumens (Wyndham and Costerton, 1981) and natural asphalt lake (Naranjo et al., 2007), which contain high amounts of asphaltene constituents. A molecular study (Kim and Crowley, 2007) revealed a wide range of phylogenetic groups within the Archaea and Bacteria domains in natural asphalt-rich tar pits; interestingly, genes encoding novel oxygenases were also detected in such samples.

On the other hand, an extensive screening, involving more than 750 strains of filamentous fungi, was carried out to select strains able to modify untreated hard coal (Bublitz et al., 1994; Hofrichter et al., 1997). Only six of the 750 strains tested exhibited some activity, from which the most active fungi, *Panus tigrinus*, growing on wood shavings coated with coal asphaltene constituents, led to a decrease of the average molecular weight (Hofrichter et al., 1997), although the average molecular weight of any complex mixture is not a measure of the constituents that were actually biodegraded.

Furthermore, most of the studies on asphaltene constituents biodegradation should be considered cautiously, as the asphaltene content was usually determined gravimetrically after n-alkane precipitation, and thus, the reported changes may be attributed to the disruption of the asphaltenic matrix by the production of surfactants during bacterial growth, liberating trapped hydrocarbons. Other studies have reported that the asphaltene fraction does not support bacterial growth, and no changes in asphaltene content are found after bioconversion of heavy oil and asphaltene constituents (Lacotte et al., 1996; Thouand et al., 1999).

There have been claims of the biodegradation of asphaltene constituents by mixed bacteria (Bertrand et al., 1983; Rontani et al., 1985). However, none of these reports described the analytical results of extractable materials recovered from appropriate sterile controls. Therefore, most of the asphaltene losses during microbial activity could be considered abiotic losses (Lacotte et al., 1996).

A study (Pineda et al., 2004) reported a bacterial consortium able to grow in the asphaltene fraction as the sole carbon source. Mineralization of the asphaltene constituents was estimated by measuring CO2 production. The authors found in two control experiments (inoculum without asphaltene constituents and non-inoculated asphaltene constituents), a CO2 production equivalent to 39% and 26%, respectively, of that found in the consortium growing in the asphaltene fraction.

The microbial inoculum for consortium stabilization contained 1% of crude oil, which could serve as carbon source. Thus, it is not possible to distinguish the origin of the CO2 production. The first clear experimental evidence that enzymes are able to modify asphaltene molecules has been reported (Fedorak et al., 1993). Chloroperoxidase from the fungus *Caldariomyces fumago* and a chemically modified *cytochrome c* were able to transform petroporphyrin derivatives and asphaltene constituents in reaction mixtures containing organic solvents (Fedorak et al., 1993; Mogollon et al., 1998; Garcia-Arellano et al., 2004). Notable spectral changes in the petroporphyrin-rich fraction of asphaltene constituents were observed, and the enzymatic oxidation of petroporphyrin derivatives led to the removal up to 74% of Ni and 95% of V.

According to Fourier Transform Infrared Spectroscopy (FTIR) spectra, the chemically modified *cytochrome c* catalyzed the oxidation of sulfur and carbon atoms in asphaltene molecules (Garcia-Arellano et al., 2004). The enzymatic treatment of asphaltene constituents is an interesting alternative for the removal of heavy metals. It would result in reduced catalyst poisoning during hydrotreatment and cracking processes. On the other hand, the introduction of polar groups in asphaltene molecules could positively affect their sedimentation properties and improve their behavior.

It has also been reported (Uribe-Alvarez et al., 2011) that a fungus isolated from a natural asphalt lake is able to grow using asphaltene constituents as the sole source of carbon and energy.

Thus, a fungal strain isolated from a microbial consortium growing in a natural asphalt lake is able to grow in purified asphaltene constituents as the only source of carbon and energy. The asphaltene constituents were rigorously purified in order to avoid contamination from other petroleum fractions. In addition, most of petroporphyrin derivatives were removed. The 18S rRNA and b-tubulin genomic sequences, as well as some morphologic characteristics, indicate that the isolate is *Neosartorya fischeri*. After 11 weeks of growth, the fungus is able to metabolize 15.5% of the asphaltenic carbon, including 13.2% transformed to CO2. In a medium containing asphaltene constituents as the sole source of carbon and energy, the fungal isolate produces extracellular lactase activity, which is not detected when the fungus grows in a rich medium. The results obtained in this work clearly demonstrate that there are microorganisms able to metabolize and mineralize asphaltene constituents, which is considered the most recalcitrant petroleum fraction.

To overcome the shortcomings of conventional methods, microbial degradation of asphaltene has been accepted worldwide as the most promising environmentally sound technology for remediation of spills and discharges related to petroleum and petroleum products.

Furthermore, bacterial metabolites (especially polysaccharides) are of great value as enhancers of oil recovery, due to their surfactant activity and bio-emulsifying properties (Banat, 1995). Because the conditions in oil deposits are often saline, the use of salt-resistant metabolites may be advantageous to the recovery of oil. Furthermore, hypersaline water and soil are often contaminated with crude oils, heavy metals, or other toxic compounds from anthropogenic sources. However, conventional microbiological treatment processes do not function at high salt concentrations; therefore, the use of moderately halophilic bacteria should be considered (Hao and Lu, 2009).

In the past, biodegradation of asphaltene constituents, through the use of a microbial consortium or mixed cultures isolated from soil samples, sediments contaminated with hydrocarbons, and oil wells, have taken place, but in low proportions of 0.55–3.5% (Venkateswaran et al., 1995; Thouand et al., 1999). This is most likely due to the complex molecular structure of asphaltene constituents (Speight, 2007), which makes these molecular species resistant to biodegradation, thereby causing their accumulation in ecosystems where petroleum and its refining byproducts are spilled in either accidental or purposeful ways (Guiliano et al., 2000).

The focus of many studies has generally been bioremediation of sites contaminated by total petroleum hydrocarbons (Iturbe et al., 2007; Machackova et al., 2008), and there is a general lack of detailed work that has been done on the biodegradation of asphaltene constituents.

However, more recently, viscosity reduction by asphaltene degradation has been structurally characterized by Fourier transform infrared spectroscopy (FTIR). The work was focused on reduction of viscosity of heavy oil in order to improve enhanced recovery from the reservoir or deposit. The bacterium (*Garciaella petrolearia* TERIG02) also showed an additional preference to degrade toxic asphalt and aromatics compounds first, unlike the other known strains. Furthermore, these characteristics make the species *Garciaella petrolearia* TERIG02 a potential candidate for residua and asphalt biodegradation and a solution to degrading toxic aromatic compounds (Lavania et al., 2012).

In contrast to low-molecular-weight hydrocarbons, polycyclic aromatics and hydrocarbons included in the asphaltene fraction are usually considered as being only slightly biodegradable, because of their insufficient availability to microbial attack (Gibson and Subramanian, 1984; Cerniglia, 1992; Kanaly and Harayama, 2000). Among the pentacyclic triterpane derivatives, the hopane constituents are so stable that they are commonly used as ubiquitous biomarkers for the assessment of biodegradation levels of crude oil (Ourisson et al., 1979). They were shown to be only slightly biodegraded by specialized microflorae under laboratory conditions (Frontera-Suau et al., 2002).

The mechanism of the degradation is complex, but is believed to be a sequential process, in which n-alkanes moieties are generally removed first, followed by the degradation of iso-alkanes moieties, cycloalkanes moieties, one-to-three ring aromatics, and finally, polyaromatics (Greenwood et al., 2008). However, the typical pattern of degradation varies with different bacteria, as well as type and composition of oil (Greenwood et al., 2008; Diaz-Ramirez et al., 2008; Zrafi-Nouira et al., 2009).

Microbes reduce the viscosity by degrading high molecular weight constituents into lower molecular weight constituents, such as biological surface active substances, acids, and gases. In addition, anaerobic fermentation leads to the production of acids, carbon dioxide, hydrogen, and alcohols. Anaerobic bacteria produce acetate and butyrate during the initial growth phase (acidogenic phase) of the fermentation process.

As the culture moves to the second phase of fermentation, the stationary growth phase, there is a shift in the metabolism of the cells to solvent production (solvent-generation phase). These gaseous and liquid metabolites dissolve into the oil, resulting in reduced viscosity (Bryant et al., 1998). Moreover, the reaction of asphalt degradation within an acidic background is preferable, as the proton (H^+) effectively interacts with the polar functionalities in the asphaltene constituents and resin constituents, thereby efficiently reducing the polar interactions, which results in breaking the intermolecular associations that exist in the raw residuum or asphalt.

Furthermore, when each of the fractions, namely aliphatic, aromatic, and asphaltic (asphaltene and resin) fractions, were treated with *Garciaella petrolearia* TERIG02 for 30 days, it was found that maximum degradation was in the case of asphalt followed by aromatic fraction (Lavania et al., 2012). Indications were that *Garciaella petrolearia* TERIG02 could tolerate the toxicity of these compounds, and was capable in utilizing them as a carbon and energy source.

In addition, five asphaltene degrading bacterial strains were obtained from crude oil and polluted soil samples of Dorood oilfield in the south of Iran. Maximum degradation ability of 46% and 48% were observed by *B. lentus* and the mixed culture of five selected isolates at 28 _C, respectively. Statistical optimization of asphaltene biodegradation was successfully carried out. and the optimum values of pH, salinity. and asphaltene concentration for asphaltene biodegradation at 40°C (104°F) were obtained for pure cultures of *B. lentus*. The kinetic study showed that a good model fit the biodegradation of asphaltene constituents.

4 References

Atlas, R.M. 1981. "Microbial Degradation of Petroleum Hydrocarbons: An Environmental Perspective." *Microbiol. Rev.*, 45: 180–209.

Banat, I.M. 1995. "Biosurfactants Production and Possible Uses in Microbial Enhanced Oil. Recovery and Oil Pollution Remediation: A Review." *Biores. Tech.*, 51: 1–12.

Barathi, S., and Vasudevan, N. 2001. "Utilization of Petroleum Hydrocarbons by *Pseudomonas fluorescens* Isolated from a Petroleum-Contaminated Soil." *Environment International*, 26: 413–416.

Bartha, R., and Pramer, D. 1965. "Features of a Flask and Method for Measuring the Persistence and Biological Effects of Pesticides in Soil." *Soil Science*, 100: 68–70.

Bartha, R., and Atlas, R.M. 1977. "The Microbiology of Aquatic Oil Spills." *Adv. Appl. Microb.*, 22: 225–266.

Bertrand, J.C., Rambeloarisoa, E., Rontani, J.F., Giusti, G., and Mattei, G. 1983. "Microbial Degradation of Crude Oil in Sea Water in Continuous Culture." *Biotechnol. Lett.*, 5: 567–572.

Bowerman, B.S., Swyler, K.J., Dougherty, D.R., Davis, R.E., Siskind, B., and Barletta, R.E. 1985. An Evaluation of the Stability Tests Recommended in the Branch Technical Position on Waste Forms and Container Materials. Report No. TI85–008726. Prepared for Division of Waste Management, Office of Nuclear Material Safety and Safeguards, U.S. Nuclear Regulatory Commission, Washington, DC.

Bryant, R.S., Bailey, S.A., Step, A.K., Evans, D.B., and Parli, J.A. 1998. "Biotechnology for heavy oil recovery." *SPE Paper No. 36767 Proceedings*. SPE/DOE improved oil recovery symposium. pp. 1–7.

Bublitz, F., Guenther, T., and Fritsche, W. 1994. "Screening of Fungi for the Biological Modification of Hard Coal and Coal Derivatives." *Fuel Proc. Technol.*, 40: 347–354.

Burd, G., and Ward, O.P. 1996a. "Physico-Chemical Properties of PM-Factor, a Surface Active Agent Produced by *Pseudomonas marginalis*." *Canadian journal of Microbiology*, 42: 243–251.

Burd, G., and Ward, O.P. 1996b. "Bacterial Degradation of Polycyclic Aromatic Hydrocarbons on Agar Plates: The Role of Biosurfactants." *Biotechnology Techniques*, 10: 371–374.

Cerniglia, C.E. 1992. "Biodegradation of Polycyclic Aromatic Hydrocarbons." *Biodegradation*, 3: 351–368.

Cooper, D.G., and Zajic, J.E. 1980. "Surface-Active Compounds from Microorganisms." *Adv. Appl. Microbiol.*, 26: 229–253.

Fedorak, P.M., Semple, K.M., Vazquez-Duhalt, R., and Westlake, D.W.S. 1993. "Chloroperoxidase Mediated Modifications of Petroporphyrins and Asphaltenes." *Enzyme Microb. Technol.*, 15: 429–437.

Foght, J.M., Gutnick, D.L., and Westlake, D.W.S. 1989. "Effect of Emulsan on Biodegradation of Crude Oil by Pure and Mixed Bacterial Cultures." *Applied and Environmental Microbiology*, January: 36–42.

Foght, J.M., Fedorak, P.M., and Westlake, D.W.S. 1990. "Mineralization of [^{14}C] Hexadecane and [^{14}C] Phenanthrene in Crude Oil: Specificity among Bacterial Isolates." *Can. J. Microbiol.*, 36: 169–175.

Frontera-Suau, R., Bost, F., McDonald, T., and Morris, P.J. 2002. "Aerobic Biodegradation of Hopanes and Other Biomarkers by Crude Oil Degrading Enrichment Cultures." *Environ. Sci. Technol.*, 36: 4585–4592.

Garcia-Arellano, H., Buenrostro-Gonzalez, E., and Vazquez-Duhalt, R. 2004. "Biocatalytic Transformation of Petroporphyrins by Chemical Modified Cytochrome c." *Biotechnol. Bioeng.*, 85: 790–798.

Gibson, D.T., and Subramanian, V. 1984. "Microbial Degradation of Aromatic Hydrocarbons." In *Microbial Degradation of Organic Compounds*. D.T. Gibson (Editor). McGraw–Hill, New York. pp. 181–252.

Goldman, S., Shabtai, Y., Rubinovitz, C., Rosenberg, E., and Gutnick, D.L. 1982. "Emulsan in Acinetobacter Calcoaceticus RAG-1: Distribution of Cell-Free and Cell-Associated Cross-Reacting Material." *Appl. Environ. Microbiol.*, 44: 165–170.

Greenwood, P.F., Wibrow, S., George, S.J., and Tibbett, M. 2008. "Sequential Hydrocarbon Biodegradation in a Soil from Arid Coastal Australia, Treated with Oil Under Laboratory-Controlled Conditions." *Org. Geochem.*, 39: 1336–1346.

Guiliano, M., Boukir, A., Doumenq, P., and Mille, G. 2000. "Supercritical Fluid Extraction of BAL 150 Crude Oil Asphaltenes." *Energy & Fuels*, 14: 89–94.

Gutnick, D.L. 1987. "The Emulsan Polymer: Perspectives on a Microbial Capsule as an Industrial Product." *Biopolymers*, 26: 223–240.

Hao, R., and Lu, A. 2009. "Biodegradation of Heavy Oils by Halophilic Bacterium." *Progress in Natural Science*, 19: 997–1001.

Harvey, R. 1997. *Polycyclic Aromatic Hydrocarbons*. Wiley-VCH, New York.

Hofrichter, M., Bublitz, F., and Fritsche, W. 1997. "Fungal Attack on Coal: I. Modification of Hard Coal by Fungi." *Fuel Proc. Technol.*, 52: 43–53.

Iturbe, R., Flores, C., Castro, A., and Torres, L.G. 2007. "Sub-Soil Contamination due to Oil Spills in Zones Surrounding Oil Pipeline-Pump Stations and Oil Pipeline Right-Of-Ways in Southwest-Mexico." *Environ. Monit. Assess.*, 133: 387–398.

Kanaly, R.A., and Harayama, S. 2000. "Biodegradation of High-Molecular Weight Polycyclic Aromatic Hydrocarbons by Bacteria." *J. Bacteriol.*, 182: 2059–2067.

Kim, I.S., Park, J., and Kim, K.W. 2001. "Enhanced Biodegradation of Polycyclic Aromatic Hydrocarbons using Nonionic Surfactants in Soil Slurry." *Applied Geochemistry*, 16: 1419–1428.

Kim, J-S., and Crowley, D.E. 2007. "Microbial Diversity in Natural Asphalts of the Rancho La Brea Tar Pits." *Appl. Environ. Microbiol.*, 73: 4579–4591.

Kirkwood, K.M., Foght, J.M., and Gray, M.R. 2004. "Prospects for Biological Upgrading of Heavy Oils and Asphaltenes." *Petroleum Biotechnology: Developments and Perspectives (Studies in Surface Science and Catalysis)*. Elsevier, Netherlands. Volume 151. Chapter 4. pp. 113–145.

Lacotte, D.J., Mille, G., Acquaviva, M., and Bertrand, J.C. 1996. "Arabian Light 150 Asphaltene Biotransformation with n-Alkanes as Co-Substrate." *Chemosphere*, 32: 1755–1761.

Lavania, M., Cheema, S., Sarma, P.M., Mandal, A.K., and Lal, B. 2012. "Biodegradation of Asphalt by *Garciaella petrolearia* TERIG02 for Viscosity Reduction of Heavy Oil." *Biodegradation*, 23: 15–24.

Luey, J., and Li, S.W. 1993. Determination of the Biodegradation Rate of Asphalt for the Hanford Grout Vaults. Report. Contract DE-AC06-76RLO 1830. Pacific Northwest Laboratory Richland, US Department of Energy, Richland, Washington. April.

Machackova, J., Wittlingerova, Z., Vlk, K., Zima, J., and Ales, L. 2008. "Comparison of Two Methods for Assessment of In Situ Jet Fuel Remediation Efficiency." *Water Air Soil Pollut.*, 187: 181–194.

Mogollon, L., Rodriguez, R., Larrota, W., Ortiz, C., and Torres, R. 1998. "Biocatalytic Removal of Nickel and Vanadium from Petroporphyrins and Asphaltenes." *Appl. Biochem. Biotechnol.*, 70–72: 765–777.

Morais, E.B. de, and Tauk-Tornisielo, S.M. 2009. "Biodegradation of Oil Refinery Residues using Mixed-Culture of Microorganisms Isolated from Land Farming." *Brazilian Archives of Biology and Technology*, 52(6): 1571–1578.

Naranjo, L., Urbina, H., De Sisto, A., and León, V. 2007. "Isolation of Autochthonous Non-White Rot Fungi with Potential for Enzymatic Upgrading of Venezuelan Extra-Heavy Crude Oil." *Biocatal. Biotransformation*, 25: 1–9.

Okoh, A.I. 2006. "Biodegradation Alternative in the Cleanup of Petroleum Hydrocarbon Pollutants." *Biotechnology and Molecular Biology Review*, 1(2): 38–50.

Ourisson, G., Albrecht, P., and Rohmer, M. 1979. "The Hopanoids. Paleochemistry and Biochemistry of a Group of Natural Products." *Pure Appl. Chem.*, 51: 709–729.

Pendrys, J.P. 1989. "Biodegradation of Asphalt Cement-20 by Aerobic Bacteria." *Applied and Environmental Microbiology*, 55(6): 1357–1362.

Phillips, U.A., and Traxler, R.W. 1963. "Microbial Degradation of Asphalt." *Appl. Microbiol.*, 11: 235–238. 1963.

Pineda, F.G., Mesta-Howard, A.M., Boll-Argúello, G., and Lira-Galeana, C. 2004. "A Microbial Consortium Isolated from a Crude Oil Sample that uses Asphaltenes as a Carbon and Energy Source." *Biodegradation*, 15: 145–151.

Rabus, R., Kube, M., Heider, J., Beck, A., Heitmann, K., Widdel, F., and Reinhardt, R. 2005. "The Genome Sequence of an Anaerobic Aromatic-degrading Denitrifying Bacterium, Strain Ebn1." *Archive of Microbiology*, 183: 27–36.

Ramos, J.L., Duque, E., Gallegos, M.T., Godyoy, P., Ramos-Gonzalez, M.I., Rojas, A., Teran, W., and Segura, A. 2002. "Mechanisms of Solvent Tolerance in Gram-Negative Bacteria." *Annual Review of Microbiology*, 56: 743–768.

Rana, M.S., Sámano, V., Ancheyta, J., and Diaz, J.A.I. 2007. "A Review of Recent Advances on Process Technologies for Upgrading Of Heavy Oils and Residua." *Fuel*, 86: 1216–1231.

Rosenberg, E., Rubinovitz, C., Gottlieb, A., Rosenhak, S., and Ron, E.Z. 1988. "Production of Biodispersan by *Acinetobacter calcoaceticus* A2." *Applied and Environmental Microbiology*, 54: 317–322.

Rontani, J.F., Bosser-Joulak, F., Rambeloarisoa, E., Bertrand, J.C., Giusti, G., and Faure, R. 1985. "Analytical Study of Asyhart Crude Oil Biodegradation." *Chemosphere*, 14: 1413–1422.

Sar, N., and Rosenberg, E. 1983. "Emulsifier Production by *Acinetobacter calcoaeticus* Strains." *Current Microbiology*, 9: 309–314.

Shoham, Y., and Rosenberg, E. 1983. "Enzymatic depolymerization of emulsan." *J. Bacteriol.*, 156: 1 61–167.

Shoham, Y., Rosenberg, M., and Rosenberg, E. 1983. "Bacterial Degradation of Emulsan." *Appl. Environ. Microbiol.*, 46: 573–579.

Speight, J.G. 2007. *The Chemistry and Technology of Petroleum*. 5th Edition. CRC Press, Taylor & Francis Group, Boca Raton, Florida.

Thouand, G., Bauda, P., Oudot, J., Kirsch, G., Sutton, C., and Vidalie, J.F. 1999. "Laboratory Evaluation of Crude Oil Biodegradation with Commercial or Natural Microbial Inocula." *Can. J. Microbiol.*, 45: 106–115.

Twerdok, L., 2004. Heavy Fuel Oils Test Plan and Robust Summary. The Petroleum HPV Testing Group, Heavy Fuel Oils Consortium Registration #201–15368A.

Uribe-Alvarez, C., Ayala, M., Perezgasga, L., Naranjo, L., Urbina, H., and Vazquez-Duhalt, R. 2011. "First Evidence of Mineralization of Petroleum Asphaltenes by a Strain of *Neosartorya fischeri*." *Microbial Biotechnology*, 4(5): 663–672.

Van Der Meer, J.R., Devos, W.M., Harayama, S., and Zehnder, A.J.B. 1992. "Molecular Mechanisms of Genetic Adaptation to Xenobiotic Compounds." *Microbiology Review*, 56: 677 694.

Van Der Meer, J.R. 1994. "Genetic Adaptation of Bacteria to Chlorinated Aromatic Compounds." *FEMS Microbiol. Rev.*, 15: 239–249.

Venkateswaran, K., Hoaki, T., Kato, M., and Maruyama, T. 1995. "Microbial Degradation of Resins Fractionated from Arabian Light Crude Oil." *Can. J. Microbiol.*, 41: 418–424.

Whyte, L.G., Bourbonnière, L., and Greer, C.W. 1997. "Biodegradation of Petroleum Hydrocarbons by Psychrotrophic *Pseudomonas* Strains Possessing Both Alkane (*alk*) and Naphthalene (*nah*) Catabolic Pathways." *Applied and Environmental Microbiology*, 63(9): 3719–3723.

Wyndham, R.C., and Costerton, J.W. 1981. "In Vitro Microbial Degradation of Bituminous Hydrocarbons and In Situ Colonization of Bitumen Surfaces within the Athabasca Oil Sands Deposit." *Appl. Environ. Microbiol.*, 41: 791–800.

Zosim, Z., Gutnick, D.L., and Rosenberg, E. 1987. "Effect of Protein Content on the Surface Activity and Viscosity of Emulsan." *Colloid Polymer Sci.*, 265: 442–447.

Zrafi-Nouira, I., Guermazi, S., Chouari, R., Safi, N.M.D., Pelletier, E., Backhrouf, A., Saidane-Moshabi, D., and Sghir, A. 2009. "Molecular Diversity Analysis and Bacterial Population Dynamics of an Adapted Seawater Microbiota during the Degradation of Tunisian Zarzantine Oil." *Biodegradation*, 20: 467–486.

15

Bioremediation Methods

1 Land Ecosystems

Bioremediation is a process by which chemical substances are degraded by bacteria and other microorganisms (Chapter 1). The use of these microorganisms has been successfully applied for the treatment of waste and wastewater in controlled systems.

Bioremediation is not a new concept. Microbiologists have studied the process since the 1940s. However, in the United States, bioremediation became known to a large public audience after the *Exxon Valdez* oil spill in 1989 in Prince William Sound, Alaska, and it is not surprising that in the years since 1989, bioremediation has become a technology that is discussed, applied, and considered in many different circumstances (Hoff, 1993).

The story of bioremediation and its recent development as an oil spill response technology provides an interesting example of how a new environmental technology comes into being. Based on the experience in the United States (highly influenced by the *Exxon Valdez* oil spill), the history of bioremediation in spill response can be divided into three development periods (Hoff, 1993): (1) the *proposal period*, which is pre-1990, (2) the *experimental period*, which ran from 1990 until approximately 1995, and (3) the *acceptance period*, which has extended from 1995 until the present time.

The *proposal period* was primarily an ideas and research period, when bioremediation was little known outside the microbiology of petroleum in the reservoir or the hazardous waste community. The microbial degradation of petroleum in the laboratory and in field trials had been documented, and the mechanisms of biodegradation were of increasing interest (Hoff, 1993). These investigations established an understanding of biodegradation of petroleum as the importance of the effect of microbial species on the process commonly referred to as *weathering*, which includes the microbial, chemical, and physical processes that promote chemical breakdown and/or chemical change of petroleum constituents.

During the *experimental period*, bioremediation received much attention, and during this time, the promise of the technology was not always borne out by its use in real situations. Reports of investigations catalyzed further interest in bioremediation technology as a means of dealing with spills related to petroleum and petroleum products. If anything, this period suffered from the *rush-to-apply-and-rush-to-publish syndrome*, in which few of the investigations were able to confirm the effectiveness of bioremediation applications in field tests. Several of these investigations suffered from poor design, were conducted over too short a time period, or had analytical difficulties detecting changes in oil concentrations.

In fact, it was during this time that the groundwork was laid for the test methods relating to the determination of *total petroleum hydrocarbons*. In the initial test method, and it is still advocated, hexane was the extraction solvent of choice. It is widely known in the petroleum industry that hexane is not fully effective in solubilizing all of the petroleum constituents (Speight, 2007), thereby leading to errors in the estimation of the petroleum-based molecular species in the land-based or water-based ecosystems (Speight, 2005).

During the *acceptance period*, bioremediation has indeed achieved a certain level of acceptance, with more realistic expectations than earlier, and there have been attempts to resolve issues related to: (1) the toxicity of microbial products, and (2) monitoring programs to determine whether or not the technique is accelerating petroleum degradation at rates in excess of background rates.

When a site is selected for cleanup, it is extremely important to link *biodegradation* with *bioremediation*. The former is a component of oil weathering, and is a natural process, whereby bacteria or other microorganisms alter and break down organic molecules into other substances, eventually producing fatty acids and carbon dioxide (Hoff, 1993). On the other hand, *bioremediation* is the acceleration of this process through the addition of exogenous microbial populations, through the stimulation of indigenous populations or through manipulation of the contaminated media using techniques such as aeration or temperature control (Atlas, 1995; Hoff, 1993; Swannell et al., 1996).

Many microorganisms possess the enzymatic capability to degrade petroleum hydrocarbons. Some microorganisms degrade alkane constituents, others degrade aromatic constituents, while others will degrade both alkane and aromatic constituents. Often the normal alkanes in the range C_{10} to C_{26} are viewed as the most readily degraded, but low-molecular-weight aromatics, such as benzene, toluene, and xylene, which are among the toxic compounds found in petroleum, are also biodegraded by many marine microorganisms. More complex structures are more resistant to biodegradation (Chapter 12, Chapter 13, Chapter 14), and the rates of biodegradation of these more complex structural entities are lower than biodegradation rates of the simpler hydrocarbon structures found in petroleum (Atlas, 1995).

In fact, it has been reported that biodegradation rates of hydrophobic organic compounds decreased in the presence of organic matrices. The biodegradation of n-alkanes in a contaminated soil is inhibited in the presence of unresolved complex mixture of hydrocarbons originating from fuel oil residue (Jonge et al., 1997). It was further suggested that inhibition of n-alkane dissolution into water by an unresolved complex mixture of hydrocarbons resulted in slow biodegradation. Therefore, oil residue in soil or sediment may decrease the biodegradation rate of hydrophobic organic compounds, such as polynuclear aromatic hydrocarbons.

In addition to the varying rates of biodegradation of organic molecules, a lag time (chemically: an induction period) occurs after spillage of petroleum or a petroleum product before indigenous microbes begin to degrade the petroleum-related constituents (Hoff, 1993). This lag time is related to presence of volatile constituents of the spilled material (that also may exhibit toxicity to the indigenous microbes), which evaporate in the first few days after the spill. In order to be effective, the indigenous microbial populations must begin to use oil and increase their population density before measurable degradation takes place. This period usually lasts several days, and is an important aspect when the appropriateness of application of a bioremediation technology is considered (Hoff, 1993).

There are several different bioremediation techniques, and the underlying idea of selecting a bioremediation technique is to accelerate the rates of *natural* hydrocarbon biodegradation by overcoming the rate-limiting factors.

Thus, several techniques can lead to the desired results: (1) indigenous populations of microbial bacteria can be stimulated through the addition of nutrients or other materials, (2) exogenous microbial populations can be introduced in the contaminated environment (bio-augmentation), (3) genetically altered bacteria can be used, and (3) the contaminated site (soil or water) can be manipulated by, for example, aeration or temperature control.

One approach often considered for the bioremediation of petroleum pollutants after an oil spill is the addition of microorganisms (seeding) that are able to degrade hydrocarbons. Most microorganisms considered for seeding are obtained by enrichment cultures from previously contaminated sites. However, because hydrocarbon-degrading bacteria and fungi are widely distributed in marine, freshwater, and soil habitats, adding seed cultures has proven less promising for treating oil spills than adding fertilizers and ensuring adequate aeration. Most tests have indicated that seed cultures are likely to be of little benefit over the naturally-occurring microorganisms at a contaminated site for the biodegradation of the bulk of petroleum contaminants (Atlas, 1995).

Hydrocarbon biodegradation in marine environments is often limited by abiotic environmental factors, such as molecular oxygen, phosphate, and nitrogen (ammonium, nitrate, and organic nitrogen) concentrations. Rates of degradation of petroleum constituents are negligible in anaerobic sediments, because molecular oxygen is required by most microorganisms for the initial step in hydrocarbon metabolism. Oxygen, however, is not limiting in well aerated (high energy) marine environments (Atlas, 1995; Stelmaszewski, 2009).

Furthermore, marine ecosystems typically have low (sometime zero) concentrations of nitrogen, phosphorus, and various mineral nutrients that are needed for the incorporation into cellular biomass, and the availability of these within the area of hydrocarbon degradation is critical.

1.1 Site Evaluation

The bioremediation process typically begins with an environmental site assessment, but may include many steps (Sweed et al., 1996).

A thorough site assessment that includes an evaluation of the current condition of the property, the operational history of the facility, and the processes used to discover this information should be conducted early in the revitalization process. Accurate characterization of environmental conditions is required to estimate potential costs associated with site cleanup and regulatory negotiation. Environmental costs can represent a significant financial issue impacting the success or start of a revitalization project. In the event that significant contamination exists, communication with local regulatory agencies is important, and should include discussion of impacts to project schedule.

Typically, site assessment (sometimes called environmental assessment) is a method used to identify any potential risk to the environment, and the resulting report is usually divided into three separate phases or types of work: (1) qualitative investigation and report, (2) quantitative investigation and report, and (3) remediation and environmental management.

In the first phase, a qualitative investigation and report, visual observations and site history are researched, and relevant regulatory records and databases are consulted to determine the reasonable probability of environmental risk. At this stage, sampling and testing are not usually performed, but a detailed waste classification and a determination of the origin of the waste assist in planning the removal activities and in utilizing the appropriate funding instrument. A thorough *paper review* and site history must be conducted to establish Oil Pollution Act (OPA) of 1990 or Comprehensive Environmental Response and Liability Act (CERCLA) authority. Typically, an oil refinery waste consists of both OPA wastes, CERCLA wastes, and Resource Conservation and Recovery Act (RCRA) wastes. A careful waste classification is required to use appropriate funding to remediate the site. If a risk assessment report is required, this may include some minor soil sampling and testing as a precautionary measure, but this does not necessarily qualify the report as an environmental assessment.

In the second phase, a quantitative investigation and report, sampling of materials (such petroleum and petroleum products) suspected of contamination are analyzed and characterized. It is also opportune to analyze unspilled sample to determine potential sources of the spilled material (if these sources have not already been identified). The object of this phase is to determine as much as reasonable the extent of any environmental contamination present at the site.

The third and final phase, the remediation and environmental management project, includes study of any reports from the two prior phases as well as an environmental site assessment in order to formulate an appropriate solution to the environmental problem. The remediation or bioremediation project is planned and estimated in terms of time, costs, and objectives.

Other phases may also be introduced, and these phases will include a timeline for site examination (say, after one year with continuous weekly/monthly monitoring) to determine the success/failure of the remediation effort. After the first year, the timeline may be shortened or lengthened appropriately, depending on the first-year data.

Reference was made above to the identification of the source of the petroleum-related pollutants, if the source was not identified at the time of the spill. This can be done through *finger printing* of the spilled petroleum or petroleum product, which can be used to identify the origin of the spill (Michelsen and Petito Boyce, 1993).

The finger-printing approach requires collection and analysis to search for specific chemical biomarkers in the petroleum or chemical markers specific to the petroleum product. All nearby potential sources of petroleum and petroleum products that might be responsible for the spill must be included in the collection and analysis. When a positive correlation of spilled material and the relevant facility has been identified, the origin of the spill has been determined. If the spilled oil has undergone weathering, water washing, and/or any level of biodegradation, then biomarker analysis and/or other techniques may be required to identify the spill source.

Biomarkers are a group of compounds, primarily hydrocarbons, found in oils, rock extracts, recent sediment extracts, and soil extracts. Biomarkers are distinguished from other compounds and are generally referred to as *molecular fossils*: they are structurally similar to specific natural products. Typically, even though diagenetic alteration of the original mileages has occurred, biomarkers retain all or most of the original carbon skeleton of the original natural product. Biomarkers reflect the type and age of the source rock that generated the petroleum.

Since different potential sources of a spill may involve oil derived from different basins, biomarker distribution (such as the sterane/hopane ratio or the ratio of steroid, diterpenoid, triterpenoid, and hopanoid biological molecules) can be used to either point to the source, or rule out the potential sources of a spill, and can be used to determine if oil in a contaminated area actually represents more than one spill (Stout et al., 2001, 2005; Speight, 2007). Specific chemical compounds (not necessarily biomarkers, since these may have been formed in the refining process) can also be used to assess the origin of some refined hydrocarbon products (Peters et al., 1992; Stout et al., 2005; Speight, 2007).

Polynuclear aromatic hydrocarbons are another group of compounds present in oil that are especially useful in identifying the source(s) of a spill. A subset of the polynuclear aromatic hydrocarbons in petroleum is products of the diagenesis of steroid, diterpenoid, triterpenoid, and hopanoid biological molecules originally deposited in sediments (biomarkers, see above). Several of these biomarkers are present as fully or partially aromatized compounds with multiple aromatic rings; therefore, they are polycyclic aromatic hydrocarbons. These polynuclear aromatic hydrocarbons are resistant to biodegradation conditions typically encountered in spill situations, and have proved useful for defining a unique *fingerprint* characteristic of a specific petroleum (or petroleum product), which can be used to correlate a biodegraded oil to a sample of its non-degraded equivalent, and hence, can be used to identify the source(s) of a petroleum-related release (Burns et al., 1997; Stout et al., 2001, 2005).

Many land-based environments are characterized by low or elevated temperatures, acidic or alkaline pH, high salt concentrations, or high pressure. Extremophile microorganisms (organisms that thrives in physically or geochemically extreme conditions that are generally detrimental to most terrestrial life forms) are adapted to grow and thrive under these adverse conditions. Hydrocarbon degrading extremophiles are thus ideal candidates for the biological treatment of polluted extreme habitats. In this review, we summarize the recent developments, obtained both in laboratory and field studies, in biodegradation and bioremediation of hydrocarbon contaminants that are of environmental concern in extreme habitats.

A wide range of hydrocarbons that contaminate the environment, mainly due to accidental release or industrial processes, has been shown to be biodegraded (mineralized or transformed) in various extreme environments characterized by low or elevated temperatures, acidic or alkaline pH, high salinity or high pressure, emphasizing the metabolic capacities of extremophiles (Margesin and Schinner, 2001b). Those adapted to more than one extreme offer a special potential for the biological decontamination of habitats where various different extreme conditions prevail simultaneously, and several examples of such behavior by microorganisms are presented below.

1.2 Soil Evaluation

When soil is the medium in which treatment will take place, it is of utmost importance to evaluate its properties, as well as the general properties of the contaminated site.

Soil is heterogeneous, and varies widely in physical, chemical, and biological properties. The soil may be acidic or basic, may have high or low nutrients, and may exhibit a different exchange capacity at different locations in the same area. The characteristics important in the design and operation of a land treatment site include the slope, the soil classification (texture and permeability), soil moisture content, pH, the cation exchange capacity (CEC), and salinity. If the initial soil properties are not ideal for the biodegradation of hydrocarbons, they can be optimized (USEPA, 2003).

1.2.1 Physical Properties

A gently sloped terrain can help minimize earthwork, but slopes in excess of 5% are not recommended for land treatment facilities, due to erosion problems and less than ideal surface drainage and run-off control capabilities. However, physical manipulation of the land may produce the appropriate slope incline.

A survey should be performed to classify the indigenous soil present on-site. A soil engineer or scientist may be consulted to perform soil classification. Soil particle analysis allows the identification of soil type, and is inexpensive to conduct. A general soil classification scheme based on standard sieve analysis is available, and provides the *unified soil classification system* (USCS) (Table 15.1).

Thus, if more than 50% of the soil is retained on a No. 200 sieve, it is considered coarse-grained soil; otherwise, it will be fine-grained soil. Coarse-grained soils permit rapid infiltration of liquids and allow good aeration; they are considered to be very permeable. However, they may not control containment of waste and nutrients added to the soil as well as fine-grained soils, which would be considered impermeable. The oxygen (air) transfer rate and substrate availability are greater in coarse-grained soils than in fine-grained soils, due to more air pore space, and thus favor

Table 15.1 Soil classification by particle size.

Soil Type		U.S. Sieve No.	Particle Size
Coarse-Grained	Gravely soil	Retained on No. 4	Larger than 4.75 mm
	Sandy soil	No. 4 through No. 200	From 4.75 to 0.075 mm
Fine-Grained	Clayey soil	Passing No. 200	Smaller than 0.075 mm
	Silty soil	Passing No. 200	Smaller than 0.075 mm

aerobic conditions desirable for biodegradation. Coarse-grained soils are also more desirable since they can be more favorably loaded with hydrocarbons. Fine-grained soils should be loaded more lightly in a shallower depth, and will generally require more tilling for equivalent performance.

Another important variable that should be assessed during soil characterization is the moisture content, or the amount of water the soil can retain (Figure 15.1). Saturation, field capacity, wilting point, and oven dry are the four conditions that will help evaluate the irrigation needs of the treated soil. Saturation is undesirable, as it decreases oxygen availability and limits site access for nutrient application and tilling. Approximately 50 to 70% of soil field capacity is ideal for microbial activities, and adequate drainage can help manage that range. Soil field capacity can be

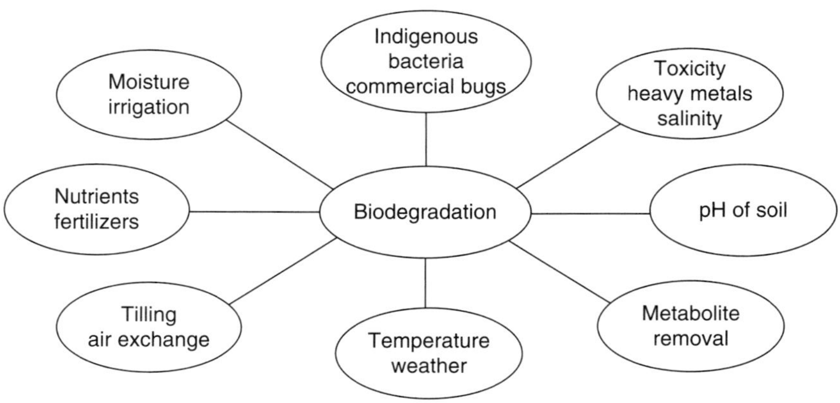

Figure 15.1 Factors requiring assessment prior to and during the bioremediation of petroleum-contaminated sites (US EPA, 2003).

determined by saturating the soil, draining it for 24 hours under gravity, then by weighing followed by oven drying at 105°C (221°F) to attain a constant weight. Thus:

$$W_{ds} - W_{ods} = W_{fc}$$
$$W_{\%fc} = W_w / W_s \times 100$$

W_{ds} is the weight of drained soil, W_{ods} is the weight of oven dry soil, W_{fc} is the weight of water in the soil at field capacity, $W_{\%fc}$ is the % of water in soil at field capacity, W_w is the weight of water, and W_s is the dry weight of soil.

The *infiltration rate* should also be assessed, because application of a liquid at a rate greater than that rate will result in flooding and erosion. At water levels greater than the field capacity, water may accumulate and result in flooding and erosion. Below the wilting point, the soil becomes too dry, slowing down microbial activities.

Soil moisture, pH, nutrients, oxygen transfer, presence of metals and toxics, and salinity are the utmost controlling factors that must be monitored, and can be optimized to achieve time-efficient biodegradation rates at a given site (Chapter 1). Another important factor is the climate, but it is beyond the control of the responder. Desirable soil parameters ranges that should be maintained to conduct a time-efficient bioremediation in the land treatment unit are: (1) moisture content, which is calculated as a percentage of the field capacity, and typically falls in the 50 to 70% range, (2) acidity or alkalinity, which is usually on the order of pH 6 to 8, but may be site specific, (3) temperature, typically 24 to 35°C (75 to 95°F), and (4) nutrient ratio, i.e., the ratio of carbon:nitrogen:phosphorus:potassium, which is generally on the order of 100:5:1:1.

The capacity of the soil to retain water at specific levels has been the most neglected area in bioremediation land-farming operations. Generally, a soil is at field capacity when soil micropores are filled with water and macropores are filled with air, but the water holding capacity depends upon the nature of the soil (Table 15.2). Too much water or too little water can be detrimental to an aerobic bioremediation operation. For example, saturation (too much water) will inhibit oxygen infiltration, and dry conditions (too little water) will decrease the rates of microbial activity or cause cessation of the biodegradation process if a wilting point is reached. A desirable range is between 70 to 80% of field capacity, thereby allowing the degrading bacteria access to both air and water, which are very much needed for microbial activity.

The pH of the soil not only affects the growth of microorganisms, but also has a noticeable effect on the availability of nutrients, mobility of metals, rate of abiotic transformation of organic waste constituents, and soil structure. Usually, a pH range of 6 to 8 units is considered optimum for

Table 15.2 Soil characteristics for effective bioremediation.

Soil Type[1]	Water Application Rate	Moisture Holding Capacity	Permeability	Field Capacity[2] (~ % by Weight)	Wilting Point[2] (~ % by Weight)
Sandy	10–12 inches	High	Low	9–25	3–10
Clayey[3]	8–9 inches	Low	High	38–43	25–28

[1] For detailed soil classification, refer to Hazardous Waste Land Treatment, SW-874, September 1980.

[2] Soil field capacity and wilting point are dependent upon silt and clay content. These numbers are approximate and proper evaluation should be conducted in the field.

[3] Provided the moisture content is maintained at optimum levels, studies have shown that generally clay soil biodegradation rates are higher than sandy soil.

biodegradation activities; the pH of the soil can be adjusted by addition of chemical reagents. For example, if the soil is too acidic (pH <6), agricultural lime may be used to raise the pH, while aluminum sulfate or ferrous sulfate may be used to lower the pH of alkaline soils. However, caution should be used to avoid *over correction* of the pH, and further consultation may be used to help calculate optimum quantities.

Biological activity is regulated by soil temperature, and an ideal temperature range is between 24 to 35°C (75 and 95°F). Since the soil temperature is difficult to control under field conditions, the waste loading rates should be adjusted according to temperatures. This adjustment should also be performed during the change in season, since the biodegradation rates are lower in the spring and the autumn compared to summer.

1.2.2 *Chemical Properties*

In addition to the pH of the soil, *cation exchange capacity* is an important variable that can be used to monitor in order to optimize the degradation process. The cation exchange capacity is an indication of the capacity of the soil to retain metallic ions (the cation exchange capacity is usually obtained through laboratory testing), and is measured in milli-equivalents per one hundred grams of dry soil (meq/100g). A cation exchange capacity value greater than 25 is an indication that the soil contains more nutrients and has high clay content, whereas values less than 5 indicate a sandy soil with little ion retention. Most metals found in oily wastes are not readily soluble in water; however, variations of pH may change that property, and when treating land where the soil has a low cation exchange capacity, care must be taken to manage subsurface of metal ions. With proper pH management, metals remain immobilized in the treatment zone, even with low cation exchange capacity values.

Soil salinity results from accumulation of neutral soluble salts (mainly due to neutral salts of sodium, calcium, magnesium, and potassium) in the upper soil horizon following capillary movement of the water, which evaporates and leaves the crystalline form of the salt, which is often indicated by a white crust. Elevated concentrations of the salts can be lethal to many microorganisms. Assessing the feasibility of biodegradation in relation to salinity is achieved by measuring electrical conductivity (deciSiemens/meter, (dS/m), which is a general measure of soil salinity. At conductivity values above 1 dS/m, biological growth is hindered, and values above 6 dS/m generally indicate that the soil is sterile.

Finally, the soil should be analyzed for heavy metal content, since a high metal concentration could be toxic to microbial survival and growth. Metals in the soil are difficult (if not impossible) to remediate by indigent soil bacteria. Therefore, if the heavy metal concentrations in soil exceed the acceptable residual levels as determined by federal and state regulations (http://www.cleanuplevels.com/), bioremediation is not a viable option.

1.2.3 Biological Properties

The action of indigenous bacteria in the soil can account for up to 80% of waste degradation in soil, the remainder being due to evaporation, photo-oxidation, and solubilization in water. This is true as long as environmental conditions, such as the presence of oxygen, adequate moisture, moderate temperatures, neutral pH, low to moderate salinity, and excess nutrients, are present to allow bacteria to grow exponentially. In addition, the spilled material should not be excessive insofar as it overwhelms the indigenous microbes.

While bacteria that degrade hydrocarbons are ubiquitous, the populations of hydrocarbon-degrading bacteria exposed to hydrocarbons increase rapidly when given adequate aeration, moisture, favorable pH, and excess nutrients.

Typically, hydrocarbon-degrading bacteria are found in the range of 10^5 to 10^6 bacteria per gram of soil under no oil spill conditions, and when exposed to crude oil, that number increases to 10^6 to 10^8 bacteria per gram of soil. For reference, one gram of rich agricultural soil generally contains 2.5×10^9 bacteria (heterotrophic count), 5×10^5 fungi, 5×10^4 algae, and 3×10^4 protozoa (US EPA, 2003). Soil samples should be analyzed for enumeration of both heterotrophic and hydrocarbon-utilizing bacteria population to verify population densities. The population of microorganisms could be assessed in soil by plate count, most probable number technique, phospholipid fatty acid analysis, or denaturing gradient gel electrophoresis (Zhu et al., 2001; Zhu et al., 2004). There are several methods that can be applied to the bioremediation of petroleum-contaminated sites (Table 15.3), but it is necessary to note that there is no single species of bacteria that can metabolize all of the components of crude oil.

However, the nature of the petroleum-related spill and the character of the site may require addition of non-indigenous microorganisms to enhance the native populations. Thus, whenever a spill has occurred and bioremediation is the best means to approach the cleanup after other options (Speight, 1996; Speight and Lee, 2000; Speight, 2005) have been dismissed, after suitable test methods have been applied to determine

Table 15.3 Methods for microbial analysis.

Plate Count	Most Probable Number	Phospholipids Fatty Acid (PLFA) Analysis	Denaturing Gradient Gel Electrophoresis (DGGE)
Provides a count of colonies formed on specific solid media.	Uses liquid media and hydrocarbons as the carbon source to evaluate microbial growth.	Can provide a quantitative assessment of viable biomass, community composition, and nutritional stature.	Identifies species distribution.
Inexpensive.	Simple field method, slightly more labor intensive and time consuming.	Requires specialized knowledge and expensive instrumentation.	Requires specialized knowledge and expensive instrumentation.
Does not differentiate between types of bacteria.	Specific to hydrocarbon-metabolizing bacteria.	Can be used to analyze culture-independent bacteria but does not identify species.	Species-specific, can provide fingerprint of bacterial community.

the physical and chemical character of the soil, investigators should consider one of two options, which include (1) bioaugmentation and (2) biostimulation.

In the *bioaugmentation approach* (Chapter 1), petroleum-degrading bacteria are added to the existing bacterial population in the soil to increase the rate of oil consumption. In the *biostimulation approach* (Chapter 1), nutrients are added and optimization of environmental conditions to improve the biodegradation efficiency of indigenous bacteria is practiced.

The nutrients nitrogen, phosphorus, and potassium (N, P, and K) are normally added during land treatment in order to enhance microbial activities, which decompose carbon (C) compounds in the soil. Nitrogen, when added through the ammonium salts, can be toxic to microorganisms, due to the possibility of generation of ammonia in the soil; the ammonium ion can also promote the increase of oxygen demand. A commonly used strategy is to add nutrients that provide a stoichiometric ratio of carbon:n itrogen:phosphorus:potassium of 100:5:1:1. However, when nitrogen and phosphorus are introduced in excessive quantities, biodegradation can be inhibited (Trindade et al., 2002).

For optimum biodegradation, nutrients can be added to the soil using organic or inorganic fertilizers, and their concentration should be closely monitored and supplemented as they are depleted during the biodegradation nitrate and process. Agriculture fertilizers, such as ammonium nitrate, urea, ammonium phosphate, and potassium phosphate, may be added to increase nutrient concentrations in the soil.

Fertilizers should be added gradually to the soil to minimize pH changes. The amount and frequency of fertilizer addition depend upon field conditions. However, evidence from documented land farming has shown that an appropriate fertilizer dosage that could be repeated, depending upon field conditions, is 500 pounds of nitrogen per acre, or 1,100 pounds of urea, or 1,500 pounds of ammonium nitrate per acre, and 250 pounds of phosphorus per acre. (McMillen et al., 2002).

Organic amendments, such as wood chips, sawdust, straw, hay, and animal manure, are used to improve soil structure and oxygen infiltration, and to increase the moisture-holding capacity of sandy soils. In general, animal manure should be applied at the rate of about 3 to 4% by weight of soil, and should be analyzed for nitrogen and phosphorus before its application. Bulking agents, like hay, palm husks, rice hulls, and straw, are added to clayey soils to increase pore space, and hence, air exchange. The bulking agent should be blended into the soil until a porous structure is obtained and visual evidence of oil is eliminated. A general rule to follow to add hay in contaminated media is 5 standard hay bales per 1,000 square feet of impacted soils. However, the source of bulking agent should be checked for residual substances (such as pesticides or heavy metals) for toxicity.

1.3 Effect of Temperature

Temperature plays a significant role in controlling the nature and the extent of microbial hydrocarbon metabolism, which is of special significance for in situ bioremediation (Margesin and Schinner, 2001a). Bioavailability and solubility of less soluble hydrophobic substances, such as aliphatic and polyaromatic hydrocarbons, are temperature dependent. Increase in temperature affects a decrease in viscosity, while a decrease in temperature affects an increase in viscosity, thereby affecting the degree of distribution, and an increase (decrease) in diffusion rates of organic compounds. Therefore, higher reaction rates due to smaller boundary layers are expected at elevated temperatures. The increased (decreased) volatilization and solubility of some hydrocarbons at elevated (low) temperature affect toxicity and allow bio-transformations with high (low) substrate concentrations (Müller et al., 1998; Whyte et al., 1998; Niehaus et al., 1999).

Cold habitats possess sufficient indigenous microorganisms, psychrotrophic bacteria being predominant, which adapt rapidly to the contamination, as demonstrated by significantly increased numbers of oil degraders shortly after a pollution event. In fact, certain types of (psychrophilic and psychrotrophic) microorganisms that have been adapted to cold temperature are able to grow at temperatures around 0°C (32°F). They are widely distributed in nature, since a large part of the Earth's biosphere is at temperatures below 5°C (Margesin and Schinner, 1999a, 1999b). Psychrophilic microorganisms (organisms that are capable of growth and reproduction in cold temperatures) have an optimum growth temperature of approximately 15°C (59°F), and do not grow above 20°C (68°F), whereas psychrotrophs (cold-tolerant) have optimum and maximum growth temperatures above 15°C (59oF) and 20°C (68°F), respectively (Morita, 1975). Cold-adapted indigenous microorganisms play a significant role in the in situ biodegradation of hydrocarbons in cold environments, where ambient temperatures often coincide with their growth temperature range.

Most studies showed that the indigenous microbial populations degrade hydrocarbons more efficiently than the strains introduced in an attempt at bioaugmentation (Margesin and Schinner, 2001a). However, bioaugmentation may result in a shorter hydrocarbon acclimation period. Inoculation of contaminated Arctic soils with consortia (Whyte et al., 1999) or with alkane-degrading *Rhodococcus* sp. (Whyte et al., 1998) decreased the induction period and increased the rate of C_{16} mineralization at 5°C (41°F). However, nitrogen-phosphorus-potassium fertilization alone had a comparable effect on hydrocarbon loss like fertilization plus bioaugmentation. This has been shown both in chronically oil-polluted Arctic soil (Whyte et al., 1999) and in artificially diesel oil-contaminated alpine soils (Margesin and Schinner, 1997a, 1997b).

The decreased bioavailability of the long-chain alkanes at low temperature (many form crystals at 0°C, 32°F) may be responsible for their increased recalcitrance, which affects in situ bioremediation in cold climates. The application of cold-active solubilizing agents could be useful for enhancing hydrocarbon bioavailability. For example, Antarctic marine bacteria produced bio-emulsifiers when grown with n-alkanes as the sole carbon source (Yakimov et al., 1999; Chugunov et al., 2000).

Benzene, toluene, ethyl benzene, xylene, and polynuclear aromatic hydrocarbons are frequent soil contaminants, but little is known about biodegradation of polynuclear aromatic hydrocarbons in cold climates. Cold-tolerant isolates (*Sphingomonas, Pseudomonas spp.*) from oil-polluted Antarctic soils utilized benzene, toluene, ethyl benzene, xylene, naphthalene, phenanthrene, and fluorene as the sole carbon and energy source (Aislabie et al., 1998, 2000). In a field study on the biodegradation of dispersed crude oil in cold and icy seawater (–1.8 to 5.5°C, 19 to 42°F), half-life times of polynuclear aromatic hydrocarbons ranged from two days (naphthalene) to eight days (phenanthrene) at a temperature above 0°C (32°F) and with effective chemical dispersion (Siron et al., 1995).

On the other hand, cold climates are not the only extreme environments, and a higher-temperature environment is also classed as an extreme environment. Microorganisms that grow optimally above 40°C are designated as thermophiles: most thermophiles known are moderate, and show an upper temperature border of growth between 50 and 70°C (122 to 158°F). Another class of thermophiles, the hyperthermophiles, experience optimal growth at temperatures in excess of 80°C (176°F) (Stetter, 1998). Thermophiles, predominantly bacilli, possess a substantial potential for the conversion of environmental pollutants, including all major classes (Müller et al., 1998).

1.4 Effect of pH

Hydrocarbon mineralization is favored by near neutral pH values (pH = 6 to 8), and it is an established practice to add lime to bioremediate acid soils containing hydrocarbon pollutants (Alexander, 1999). However, biodegradation has been reported to proceed in aquifers where the acidity is obvious (pH = <6) (Norris, 1994; Amadi et al., 1996).

Acidophiles (organisms that have their optimum growth rate at least 2 pH units below neutrality) are metabolically active in highly acidic environments, and often have a high heavy metal resistance (Norris and Johnson, 1998). Heterotrophic acidophilic microorganisms are an interesting potential for the bioremediation of acidic environments that contain both heavy metals and organic compounds, such as petroleum-polluted acidic drainage waters (Stapleton et al., 1998).

Alkaliphiles (organisms that have their optimum growth rate at least 2 pH units above neutrality) are able to grow or survive at pH values above 9, but their optimum growth rate is around neutrality or less (Kroll, 1990).

1.5 Effect of Salinity

Microorganisms requiring salt for growth are referred to as halophiles, whereas microorganisms that are able to grow in the absence as well as in the presence of salt are designated halo-tolerant (Kushner, 1978; Grant et al., 1998). A range of organic pollutants has been shown to be mineralized or transformed by microorganisms able to grow in the presence of salt (Oren et al., 1992; Margesin and Schinner, 2001b).

Halophilic archaea maintain an osmotic balance with the hypersaline environment by accumulating high salt concentrations, which requires salt adaptation of the intracellular enzymes. Eubacteria are more promising degraders than archaea, as they have a much greater metabolic diversity. Their intracellular salt concentration is low, and their enzymes involved in biodegradation may be conventional (i.e., not salt-requiring) enzymes similar to those of non-halophiles (Oren et al., 1992).

An inverse relationship between biodegradation of petroleum hydrocarbons and salinity has been assumed because enrichment cultures from the Great Salt Lake were not able to grow on mineral oil and to mineralize hexadecane in the presence of salt concentrations above 20% w/v (Ward and Brock, 1978; Whitehouse 1984; Means 1995). The inhibitory effect of salinity was found to be greater for the biodegradation of aromatic and polar fractions than of the saturated fraction of petroleum hydrocarbons (Mille et al., 1991), and there are reports of microorganisms able to oxidize petroleum hydrocarbons even in the presence of 30% w/v sodium chloride. Among such microorganisms are crude oil-degrading *Streptomyces albaxialis* (Kuznetsov et al., 1992), and an alkane-degrading member of the *Halobacterium* group (Kulichevskaya et al., 1992).

In terms of the bioremediation of petroleum-contaminated soil, an inhibitory effect of artificial salinity on mineralization of used lubricating oil has been reported (Rhykerd et al., 1995). Thus, the removal of salt from oil-contaminated soils may reduce the time required for bioremediation. However, different results may be obtained when investigating naturally salt-containing soils, since indigenous microorganisms in such environments are expected to be salt-adapted. In fact, seasonal variations of the mineralization potential of crude oil (520 g kg^{-1} soil) in soil samples from salt marshes have been reported to be high (Jackson and Pardue, 1997).

2 Water Ecosystems

In a water ecosystem, a spill of petroleum or a petroleum product represents the introduction of a non-aqueous phase liquid composed of a large amount of organic matrices, in addition to other toxic constituents, such as polynuclear aromatic hydrocarbons, as occur in resin constituents and asphaltene constituents.

Obviously, there are many different processes acting on a marine oil spill modifying its physical and chemical characteristics, such as biodegradation, scattering, evaporation, dissolution, dispersion, sedimentation, and photo-oxidation (Figure 15.1) (Prince, 1993). These processes occur simultaneously, but at different speeds, depending on the physical and chemical properties of the oil and the ambient conditions, like the environmental temperature and energy. On several occasions, however, these processes can be so slow that it is necessary to interfere in order to hasten degradation of the pollutants (Rosa and Triguis, 2007).

Oil type, weather, wind and wave conditions, as well as air and sea temperature all play important roles in ultimate fate of spilled oil in marine environment. After oil is discharged in the environment, a wide variety of physical, chemical, and biological processes begin to transform the discharged oil. These chemical and physical processes are collectively called weathering, and act to change the composition, behavior, route of exposure, and toxicity of discharged oil. The weathering processes are described as follows:

- *Spreading and Advection*: Spreading, which dominates the initial stages of the spill and involves the whole oil, is the movement of the entire oil slick horizontally on the surface of water due to effects of gravity, inertia, friction, viscosity, and surface tension: on calm water, spreading occurs in a circular pattern outward from the center of the release point. Advection is the movement of oil due to overlying winds or underlying currents, which increase the surface area of the oil, thereby increasing its exposure to air, sunlight, and underlying water. The advection effects are not uniform, and do not affect the chemical composition of the oil.
- *Evaporation*: Evaporation, which is subject to atmospheric conditions such as temperature, is the preferential transfer of lower-boiling constituents of the petroleum-related spill from the liquid phase to the vapor phase. The chemical composition of the spilled material is physically altered, and although the volume decreases, the remaining components have higher viscosity and specific gravity, leading to thickening of the slick and formation of entities such as tar balls.
- *Dissolution*: Dissolution is the transfer of constituents from the spilled material into the aqueous phase. Once dissolved, these constituents are bioavailable and, if exposed to marine life, can cause environmental impacts and injuries. The

largest concentration is found near the surface or the release point, and hence, the effect on marine life may be localized.

- *Natural Dispersion*: Natural dispersion is the process of forming small oil droplets that become incorporated in the water in the form of a dilute oil-in-water suspension, and occurs when breaking waves mix the oil in the water column. This phenomenon reduces the volume at the spill surface, but does not change the physicochemical properties of the spilled material. It is estimated to range from 10 to 60% per day for first three days of the spill, depending on the condition of the ecosystem, and may be independent of the type of petroleum or petroleum product.
- *Emulsification*: Emulsification is the mixing of sea water droplets into oil spilled on water surface, forming a water-in-oil emulsion.
- *Photo-oxidation*: Photo-oxidation occurs when sunlight in the presence of oxygen transforms hydrocarbons (by increasing the oxygen content of a hydrocarbon) into new by-products. This results in changing in the interfacial properties of the oil, affecting spreading and emulsion formation, and may result in transfer of toxic by-products into the water column, due to the by-product's enhanced water solubility.
- *Sedimentation and Shoreline Stranding*: Sedimentation is the incorporation of oil within both bottom and suspended sediments. Shoreline stranding is the visible accumulation of petroleum along the water's edge (shoreline) following a spill. It is affected by proximity of the spill to the affected shoreline, intensity of current and wave action on the affected shoreline, and persistence of the spilled product. Sedimentation can begin immediately after the spill, but increases and peaks after several weeks, whereas shoreline stranding is a function of the distance of the shoreline from the spill and the chemical nature of spilled oil.
- *Biodegradation*: Biodegradation occurs on the water surface, in the water, in the sediments, and at the shoreline. It can begin (depending on the lag time or induction period) after a spill, and can continue as long as degradable hydrocarbons are present.

Weathering processes occur simultaneously: one process does not stop before the other begins. The order in which these processes present is instantaneous, and the relative significance of these processes may change if the spill occurred below the water surface or in tropical or ice conditions. The spill chronology may also vary if the spill is near the shoreline, in which case it can contaminate the soil and groundwater even before the weathering or cleanup processes start. Also, there are many onshore and offshore operations in a petroleum industry that can cause soil pollution and aquifer contamination.

Aquifer contamination can also take place, because of the migration of the oil through the porous media and its subsequent adsorption on the rock surface. Many technologies have been proposed for the treatment of oil contaminated sites; these can be performed by two basic processes, in-situ and ex-situ treatment, using different cleaning technologies, such as thermal treatment, biological treatment, chemical extraction and soil washing, and aeration accumulation techniques (Speight, 1996; Speight and Lee, 2000).

2.1 Biodegradation

Hydrocarbon degradation rates in the water ecosystems generally follow the order (Perry, 1984):

n-Alkanes > branched alkanes > low molecular weight aromatics > cyclic alkanes

The process is usually aerobic, requiring terminal or sub-terminal oxidation of the alkanes (Harayama et al., 1999), while aromatic hydrocarbon ring structures are broken through hydroxylation and carboxylation processes (Cerniglia, 1992). Hydrocarbon biodegradation in water is associated both with water-soluble or -miscible compounds, and with the oil-water interphases, mainly on droplets and thin oil films with high surface/volume ratio (Bartha and Atlas, 1987; Button et al., 1992; Floodgate, 1984). A significant number of studies have shown that it is difficult to predict the extent and rates of hydrocarbon degradation processes in marine environments, due to the many factors involved in the process (Leahy and Colwell, 1990; Atlas and Bartha, 1992; Margesin and Schinner, 1999).

Biodegradation rates of polynuclear aromatic hydrocarbons (PAHs) in spilled oil stranded on tidal flats were evaluated using model reactors to clarify the effects of non-aqueous phase liquid (NAPL) on the biodegradation of polynuclear aromatic hydrocarbons in stranded oil on tidal flats, with special emphasis on the relationship between dissolution rates of polynuclear aromatic hydrocarbons into water and viscosity of the non-aqueous phase liquid. Biodegradation of polynuclear aromatic hydrocarbons in non-aqueous phase liquids was limited by the dissolution rates of polynuclear aromatic hydrocarbons into water. Biodegradation rate of chrysene was smaller than that of acenaphthene and phenanthrene, due to the smaller dissolution rates. Dissolution rates of polynuclear aromatic hydrocarbons in heavy fuel oil are typically lower than the dissolution rates of polynuclear aromatic hydrocarbons petroleum because of the higher viscosity of heavy fuel oil: the solubility of the fuel oil constituents (i.e., penetration of the solvent into the heavy fuel oil) is more than likely diffusion controlled. Hence, biodegradation rates of polynuclear aromatic hydrocarbons in heavy fuel oil are lower than the biodegradation rates of petroleum.

Biodegradation rates of polynuclear aromatic hydrocarbons in non-aqueous phase liquids with slow rate of decrease, like fuel oil C, was slower than those in non-aqueous phase liquids with rapid rate of decrease, like crude oil. The smaller rate of decrease of fuel oil C than crude oil was due to the higher viscosity of fuel oil C. Therefore, not only the dissolution rate of polynuclear aromatic hydrocarbons, but also the rates of decrease of non-aqueous phase liquids were important factors for the biodegradation of polynuclear aromatic hydrocarbons (Kose et al., 2003).

Generally, the data from the above study (Kose et al., 2003) show that the biodegradation of polynuclear aromatic hydrocarbons in non-aqueous phase liquids was limited by the dissolution rates of polynuclear aromatic hydrocarbons into water.

2.2 Bioremediation

The conditions of the contaminated area play a major role on whether bioremediation is the appropriate method of cleanup for the given oil spill. The success of bioremediation is dependent upon physical conditions and chemical conditions. Physical parameters include temperature, surface area of the oil, and the energy of the water. Chemical parameters include oxygen and nutrient content, pH, and the composition of the oil. Temperature affects bioremediation by changing the properties of the oil, and also by influencing the petroleum-degrading microbes (Nedwell, 1999). When the temperature is lowered, the viscosity of the petroleum is increased, which changes the toxicity and solubility of the oil, depending upon its composition (Zhu et al., 2001). Temperature also has an effect on the growth rate of the microorganisms, as well as the degradation rate of the hydrocarbons, depending upon their characteristics.

Biostimulation, the addition of nutrients, is practiced for cleanup of petroleum spills in seawater when there is an existing population of oil degrading microbes present. When an oil spill occurs, the result is a large increase in carbon, and this also stimulates the growth of the indigenous petroleum-degrading microorganisms. However, these microorganisms are limited in the amount of growth and remediation that can occur by the amount of available nitrogen and phosphorus.

By adding these supplemental nutrients in the proper concentrations, the hydrocarbon degrading microbes are capable of achieving their maximum growth rate, and hence, the maximum rate of pollutant uptake. It has been found that when using nitrogen for the supplemental nutrient, a maximum growth rate is achieved by the oil degrading microorganisms (Boufadel et al., 2006). Biostimulation has been proven to be an effective way of achieving increased hydrocarbon degradation by the indigenous microbial population (Coulon et al., 2006).

Supplemental nutrients tended to move downward during rising tides and seaward during falling tides (Boufadel et al., 2006). This is very useful information in determining the proper timing to add nutrients in order to allow for the maximum residence time of the nutrients in the contaminated areas. The results of this experiment concluded that the nutrients should be applied during low tide at the high tide line, which results in maximum contact time of the nutrients with the oil and hydrocarbon degrading microorganisms.

Waves also have an effect on the distribution and movement of the water and dissolved nutrients, which determine the residence time of the nutrients in the oil affected area. The role of waves on solute movement varies whether there is a tide or not. Research showed that when a wave is present, there is a sharp seaward hydraulic gradient in the backwash zone and a gentle gradient landward of this area (Boufadel et al., 2007). Furthermore, the contact time of the nutrients is increased when the waves break seaward of their location, the waves increased the dispersion and washout of the nutrients in the tidal zone, and residence time was approximately 75% when a wave was present with a tide, as compared to a tide with no waves.

The marine environment encompassing the vast majority of earth's surface is a repertoire of a large number of microorganisms. The environmental roles of the biosurfactants produced by many such marine microorganisms have been reported earlier (Poremba et al., 1991; Schulz et al., 1991; Abraham et al., 1998; Das et al., 2008).

The rate and extent of microbial biodegradation of petroleum-related spilled material are dependent on several factors.

2.2.1 *Effect of Temperature*

Water temperature affects the physical and chemical properties of oil and the rate of biodegradation. Colder temperatures slow the rate; warmer temperatures increase the rate.

The ubiquitous distribution of oil-degrading bacteria in the polar environment has already been reported (Leahy and Colwell, 1990). Before contamination, our counts indicated that hydrocarbon-degrading microorganisms comprised less than 0.01% of the total number of bacteria, but represented 10% of the cultivable heterotrophic microorganisms. These results are consistent with those that indicate that hydrocarbon-degrading microorganisms comprised 1–10% of the total number of saprophytic bacteria in marine bacterial communities (Wright et al., 1997).

In contrast to the results reported by others (Kennicutt, 1990), the introduction of oil into oil-free Antarctic seawater resulted in enrichment in hydrocarbon-degrading microorganisms by several orders of magnitude within a few days. Such a large enhancement of specific microorganisms has been previously reported (Delille and Vaillant, 1990; Wagner-Döbler et al., 1998; Delille and Delille, 2000).

The addition of a fertilizer had an immediate stimulation effect on both heterotrophic and hydrocarbon-degrading microorganisms. Several studies have reported favorable effects of fertilizers on oil biodegradation at low temperatures in Arctic (Braddock et al., 1997; Whyte et al., 1998, 1999; Mohn et al., 2001), alpine (Margesin and Schinner, 1997a, 1997b; Margesin and Schinner, 2001a; Margesin et al., 2007a), and Antarctic soils (Kerry, 1993; Wardell, 1995; Aislabie et al., 1998, 2006; Delille, 2000; Ferguson et al., 2003; Ruberto et al., 2003; Powell et al., 2006).

Temperature can affect the physical and chemical characteristics of the discharged oils: hydrocarbons with high pour points could be expected to show more temperature-related biodegradation than contaminants with low pour points. However, globally low temperatures have been reported to play a significant role in controlling the nature and extent of microbial hydrocarbon metabolism (Nedwell, 1999; Gerdes et al., 2005). Mineralization of isotopically-marked (^{14}C) hexadecane has been observed in Arctic soil at 5°C (41°F), but mineralization occurred faster and to a greater extent at 23°C (74°F).

Furthermore, and not surprisingly, during a two-month-long microcosms experiment in seawater, biodegradation efficiency was higher at 20°C (68°F) than at 4°C (39°F) for two psychrotrophic Antarctic marine isolated from Terra Nova Bay (Ross Sea) (Michaud et al., 2004). However, Mohn and Stewart (2000) reported that a change in incubation temperature from 7 to 22°C (13 to 72°F) did not affect the extent of mineralization of isotopically-marked (1^{4}C) dodecane. The behavior of isolated strains do not necessarily parallel the one of a much more complex natural consortium, as environmental factors characterizing natural polar ecosystems are impossible to replicate in laboratory.

Other studies indicated that heterotrophic and hydrocarbon-degrading bacteria counts are often similar in cold and warm climates (Aislabie, 1997; Eckford et al., 2002). These results seem in conflict with the typical temperature-related assumption predicting an increase of microbial metabolism with a temperature increase (Leahy and Colwell, 1990; Bossert and Bartha, 1984). According to the Arrhenius equation, any decrease in temperature should cause an exponential decrease of the reaction rates, the magnitude of which depends on the value of the activation energy (Margesin et al., 2007b).

Low temperatures may affect the utilization of substrates comprising a mixture of hydrocarbons (Atlas, 1986; Baraniecki et al., 2002). Altered growth responses might result from beneficial changes to membrane fluidity. The production and/or activity of cold-active enzymes are usually significantly below the "optimal" growth temperature (as determined from growth rate) of the enzyme producer, which reflects the thermal characteristics of the secretion process (Margesin et al., 2007b).

In another investigation (Coulon, 2006), it was found that when crude oil was added to the test, the presence of petroleum-degrading microbes increased by two orders of magnitude at 4°C (39°F), more than three orders of magnitude at 12°C (54°F), and more than four orders of magnitude at 20°C (68°F). The surface area of the oil is also a significant parameter in the success of bioremediation, because the growth of oil degrading microorganisms occurs at the interface of the water and oil. The larger the surface area of the oil results in a larger area for growth, and hence, larger numbers of microbes. The energy of the water is important, because rough waters will disperse and dilute essential nutrients for the microorganisms, and also spread the oil, contaminating more areas.

A more recent investigation (Delille et al., 2009) provides further evidence of the high potential of indigenous Antarctic bacterial communities for bioremediation action, even at low temperatures. Little difference in data obtained under three incubation temperatures and with two different concentrations of oil is clearly indicating that temperature had only a rather limited influence on petroleum degradation in the studied Antarctic seawater. This point is important when considering bioremediation as an efficient means to clean up contaminated soils in in remote polar locations.

2.2.2 *Effect of Oxygen*

In open water, oil hydrocarbons undergo aerobic biodegradation by bacteria that use oxygen dissolved in the water. Scientists have monitored dissolved oxygen levels around the spill since it occurred. Early measurements of the deep plume showed a rate of 30 percent oxygen depletion, which demonstrated the presence of biodegrading microbes. Sediment on the ocean floor and along the coast is, for the most part, *anoxic*: it does not contain oxygen. Hydrocarbons that settle into sediments on the ocean floor and along coasts undergo anaerobic biodegradation, a much slower process. Therefore, onshore oil lingers longer than oil at sea, and can become a chronic pollutant.

2.2.3 *Nutrients*

In addition to carbon and oxygen, bacteria need nitrogen and phosphorus to survive. These nutrients are found naturally in the ocean environment. Nitrogen and phosphorus-based fertilizers from farms and gardens on land also enter marine waters through storm water runoff.

2.2.4 Effect of Petroleum Characteristics

The type of oil, its concentration, and the types of hydrocarbons it contains influence the rate of biodegradation. The Gulf spill released light, sweet crude oil, which is more readily broken down than heavy, sour oil. *Mousse*, tar balls, or oil slicks that wash onshore are concentrated compared to dispersed oil, more protected from wind and wave action than oil in open water, and have less surface area for microbes to access. Smaller droplets of oil are more biodegradable.

2.2.5 Effect of Prior Exposure

Microbes adapt to gradual exposure to oil. The more oil a microbial community has been exposed to in the past, the greater its capacity and availability to biodegrade oil in the future. In one study, microbes from sediments previously contaminated with oil were able to metabolize oil 10 to 400 times faster than those from sediments that had never been contaminated. Once a species of bacteria is exposed to oil and metabolizes it, the next generations inherit that ability, a concept known as *genetic adaptation*. This has been studied in a particular species of *Vibrio* in the northwestern Gulf of Mexico.

2.2.6 Effect of Dispersants

Biodegradation occurs more readily when oil droplets are dispersed. Chemical dispersants break oil into smaller droplets, which increases the surface area available for bacteria to access. The exact effects of dispersants, such as Corexit, on the rate of biodegradation are unknown, especially in deep water. The effectiveness of dispersants also depends on the type and consistency of oil and the oil-dispersant ratio.

2.2.7 Effect of Flowing Water

Oil discharges generate a dynamic situation on the oil–water interphases, caused by a variety of physical and chemical processes, including wind, currents, oil weathering, film generation, and oil dispersion. Thus, immiscible oil surfaces are constantly washed with seawater, and new microbes are continuously contacting the oil film or droplet surfaces. Simulation of these dynamic conditions may be achieved in flow-through systems rather than by closed static experiments. Various controlled flow-through systems have been described that simulate biodegradation under marine conditions, including microcosms (Swannell and Daniel, 1999; MacNaughton et al., 2003; Röling et al., 2002; Xu et al., 2003), and large tank-based mesocosms (Wade and Quinn, 1979; Siron et al., 1993; Santas and Santas, 2000; Yamada et al., 2003). However, since dispersed oil was used in most of these systems, it was impossible to differentiate between HC processes in the oil and water phases.

More recent studies with immobilized oil in flow-through systems may give valuable contributions to the determination and understanding of various processes after oil discharges to aquatic environments (Brakstad et al., 2004). For instance, hydrocarbon dissolution and biodegradation from oil films, and the potential impacts in the water column, may be predicted after an oil spill in combination with remote sensing of oil films' thickness (Brown and Fingas, 2003). Further, if depletion from films and droplets are comparable, the system described here may be used for gaining new insights into dissolution, biodegradation, and microbiological processes of dispersed oil. This may also have implications for studies of oil spill treatment with chemical dispersants.

2.2.8 Effect of Deep-Sea Environments

The deep sea (as well as other habitats, such as deep groundwater, deep sediments, or oilfields) is influenced by high pressure. Barophiles (piezophiles) are microorganisms that require high pressure for growth, or grow better at pressures higher than atmospheric pressure (Prieur and Marteinsson, 1998).

Pollutants with densities greater than that of marine waters (such as oxidized petroleum-water emulsions) will be expected to sink to a level commensurate with the density, and even to the deep benthic zone, where the hydrostatic pressure is notably high. A combination of high pressure and low temperatures in the deep ocean results in low microbial activity (Alexander, 1999).

3 References

Abraham, W.R., Meyer, H., and Yakimov, M. 1998. "Novel Glycine Containing Glucose Lipids from the Alkane using Bacterium *Alcanivorax borkumensis*." *Biochim. Biophys. Acta*, 1393: 57–62.

Aislabie, J. 1997. "Hydrocarbon-Degrading Bacteria in Oil-Contaminated Soils near Scott Base, Antarctica." In *Ecosystem processes in Antarctica's ice-free landscape*. W.B. Lyons, C. Howard-Williams, and I. Hawes (Editors). Balkema Publishers Ltd, Rotterdam, Netherlands. pp. 253–258.

Aislabie, J., McLeod, M., and Fraser, R. 1998. "Potential for Biodegradation of Hydrocarbons in Soil from the Ross Dependency, Antarctica." *Appl. Microbiol. Biotechnol.*, 49: 210–214.

Aislabie, J., Foght, J., and Saul, D. 2000. "Aromatic Hydrocarbon-Degrading Bacteria from Soil near Scott Base, Antarctica." *Polar Biol.*, 23: 183–188.

Aislabie, J., Saul, D.J., and Foght, J.M. 2006. "Bioremediation of Hydrocarbon-Contaminated Polar Soils." *Extremophiles*, 10: 171–179.

Alexander, M. 1999. *Biodegradation and Bioremediation*. 2nd Edition. Academic Press, London, United Kingdom.

Amadi, A., Abbey, S.D., and Nma, A. 1996. "Chronic Effects of Oil Spill on Soil Properties and Microflora of a Rainforest Ecosystem in Nigeria." *Water Air Soil Pollut.*, 86: 1–11.

Atlas, R.M. 1995. "Petroleum Biodegradation and Oil Spill Bioremediation." *Marine Pollution Bulletin*, 31: 178–182.

Atlas, R.M., and Bartha, R. 1992. "Hydrocarbon Biodegradation and Oil-Spill Bioremediation." *Adv. Microb. Ecol.*, 12: 287–338.

Baraniecki, C.A., Aislabie, J., and Foght, J.M. 2002. "Characterization of *Sphingomonas sp. Ant 17*, an Aromatic Hydrocarbon-Degrading Bacterium Isolated from Antarctic Soil." *Microb. Ecol.*, 43: 44–54.

Bartha, R., and Atlas, R.M. 1987. "Transport and Transformation of Petroleum: Biological Processes." In *Long Term Environmental Effects of Offshore Oil and Gas Development*. Applied Science Publishers, London, Unted Kingdom. pp. 287–341.

Bossert, I., and Bartha, R. 1984. "The Fate of Petroleum in Soil Ecosystems." In *Petroleum Microbiology*. R.M. Atlas (Editor). MacMillan Publishing Co., New York. pp. 434–476.

Boufadel, M.C., Suidan, M.T., and Venosa, A.D. 2006. "Tracer Studies in Laboratory Beach Simulating Tidal Influences." *Journal of Environmental Engineering*, 132(6): 616.

Boufadel, M.C., Suidan, M.T., and Venosa, A.D. 2007. "Tracer Studies in a Laboratory Beach Subjected to Waves." *Journal of Environmental Engineering*, 133 (7), 722.

Braddock, J.F., Ruth, M.L., Catterall, P.H., Walworth, J.L., and McCarthy, K.A. 1997. "Enhancement and Inhibition of Microbial Activity in Hydrocarbon-Contaminated Artic Soils: Implications for Nutrient-Amended Bioremediation." *Environ. Sci. Technol.*, 31: 2078–2084.

Brakstad, O.G., Bonaunet, K., Nordtug, T., and Johansen, Ø. 2004. "Biotransformation and Dissolution of Petroleum Hydrocarbons in Natural Flowing Seawater at Low Temperature." *Biodegradation*, 15: 337–346.

Brown, C.E., and Fingas, M.F. 2003. "Development of Airborne Oil Thickness Measurements." *Mar. Pollut. Bull.*, 47: 485–492.

Burns, W.A., Mankiewicz, P.J., Bence, A.E., Page, D.S., and Parker, K.R. 1997. "A Principle Component and Least Squares Method for Allocating Polycyclic Aromatic Hydrocarbons in Sediment to Multiple Sources." *Environ. Toxicol. Chem.*, 16: 1119–1131.

Button, D.K., Robertson, B.B., McIntosh, D., and Jüttner, F. 1992. "Interactions between Marine Bacteria and Dissolved-Phase and

Beached Hydrocarbons after the Exxon Valdez Oil Spill." *Appl. Environ. Microbiol.*, 58: 243–251.

Cerniglia, C.E. 1992. "Biodegradation of Polycyclic Aromatic Hydrocarbons." *Biodegradation*, 3: 351–368.

Chugunov, V.A., Ermolenko, Z.M., Martovetskaya, I.I., Mironava, R.I., Zhirkova, N.A., Kholodenko, V.P., and Urakov, N.N. 2000. "Development and Application of a Liquid Preparation with Oil-Oxidizing Bacteria." *Appl. Biochem. Microbiol.*, 36: 577–581.

Coulon, F., McKew, B.A., Osborn, A.M., McGenity, T.J., and Timmis, K.N. 2006. "Effects of Temperature and Biostimulation on Oil-Degrading Microbial Communities in Temperate Estuarine Waters." *Environmental Microbiology*, 9 (1), 177–186.

Das, P., Mukherjee, S., and Sen, R. 2008. "Improved Bioavailability and Biodegradation of a Model Polyaromatic Hydrocarbon by a Biosurfactant Producing Bacterium of Marine Origin." *Chemosphere*, 72: 1229–1234.

Delille, D., and Vaillant, N. 1990. "The Influence of Crude Oil on the Growth of Sub-Antarctic Marine Bacteria." *Antarctic Sci.*, 2: 655–662.

Delille, D. 2000. "Response of Antarctic Soil Assemblages to Contamination by Diesel Fuel and Crude Oil." *Microb. Ecol.*, 40:159–168.

Delille, D., and Delille, B. 2000. "Field Observations on the Variability of Crude Oil Impact on Indigenous Hydrocarbon-Degrading Bacteria from Sub-Antarctic Intertidal Sediments." *Mar. Environ. Res.*, 49: 403–417.

Delille, D., Pelletier, E., Rodriguez-Blanco, A., and Ghiglione, J-F. 2009. "Effects of Nutrient and Temperature on Degradation of Petroleum Hydrocarbons in Sub-Antarctic Coastal Seawater." *Polar Biol.*, 32: 1521–1528.

Eckford, R., Cook, F.D., Saul, D., Aislabie, J., and Foght, J. 2002. "Free-Living Nitrogen-Fixing Bacteria from Antarctic Soils." *Appl. Environ. Microbiol.*, 68: 5181–5185.

Ferguson, S.H., Franzmann, P.D., Revill, A.T., Snape, I., and Rayner, J.L. 2003. "The Effects of Nitrogen and Water on Mineralization of Hydrocarbons in Diesel-Contaminated Terrestrial Antarctic Soils." *Cold Reg. Sci. Technol.*, 37: 197–212.

Gerdes, B., Brinkmeyer, R., Dieckmannm G., and Helmke, E. 2005. "Influence of Crude Oil on Changes of Bacterial Communities in Arctic Sea-Ice." *FEMS Microbiol Ecol.*, 53: 129–139.

Grant, W.D., Gemmell, R.T., and McGenity, T.J. 1998. "Halophiles." In *Extremophiles: Microbial Life in Extreme Environments*. K. Horikoshi and W.D. Grant (Editors). Wiley-Liss, New York. pp. 93–132.

Harayama, S., Kishira, H., Kasai, Y., and Shutsubo, K. 1999. "Petroleum Biodegradation in Marine Environments." *J. Molec. Microbiol. Biotechnol.*, 1(1): 63–70.

Hoff, R.Z. 1993. "Bioremediation: An Overview of Its Development and Use for Oil Spill Cleanup." *Marine Pollution Bulletin*, 29: 476–481.

Jackson, W.A., and Pardue, J.H. 1997. "Seasonal Variability of Crude Oil Respiration Potential in Salt and Fresh Marshes." *J. Environ. Qual.*, 26: 1140–1146.

Kennicutt, M.C. 1990. "Oil Spill in Antarctica." *Environ. Sci. Technol.*, 24: 620–624.

Kerry, E. 1993. "Bioremediation of Experimental Petroleum Spills on Mineral Soils in The Vesfold Hills, Antarctica." *Polar Biol.*, 13: 163–170.

Kose, T., Mukai, T., Takimoto K., and Okada, M. 2003. "Effect of Non-Aqueous Phase Liquid on Biodegradation of PAHs in Spilled Oil on Tidal Flat." *Water Research*, 37: 1729–1736.

Kroll, R.G. 1990. "Alkalophiles." In *Microbiology of extreme environments*. C. Edwards (Editor). Open University Press, Milton Keynes, United Kingdom. pp. 55–92.

Kulichevskaya, I.S., Milekhina, E.I., Borzenkov, I.A., Zvyagintseva, I.S., and Belyaev, S.S. 1992. "Oxidation of Petroleum Hydrocarbons by Extremely Halophilic Archaebacteria." *Microbiology*, 60: 596–601.

Kushner, D.J. 1978. "Life in High Salt and Solute Concentrations." In *Microbial Life in Extreme Environments*. D.J. Kushner (Editor). Academic Press, London, United Kingdom. pp. 317–368.

Kuznetsov, V.D., Zaitseva, T.A., Vakulenko, L.V., and Filippova, S.N. 1992. "Streptomyces *albiaxalis sp. nov.*: A New Petroleum Hydrocarbon-Degrading Species of Thermo- and Halotolerant Streptomyces." *Microbiology*, 61: 62–67.

Leahy, J.G., and Colwell, R.R. 1990. "Microbial Degradation of Hydrocarbons in the Environment." *Microbiol. Rev.*, 54: 305–315.

Margesin, R., and Schinner, F. 1997a. "Efficiency of Indigenous and Inoculated Cold-Adapted Soil Microorganisms for Biodegradation of Diesel Oil in Alpine Soils." *Appl. Environ. Microbiol.*, 63: 2660–2664.

Margesin, R., and Schinner, F. 1997b. "Bioremediation of Diesel-Oil Contaminated Alpine Soils at Low Temperatures." *Appl. Microbiol. Biotechnol.*, 47: 462–468.

Margesin, R., and Schinner, F. (Editors). 1999a. *Cold-Adapted Organisms*. Springer, New York.

Margesin, R., and Schinner, F. 1999b. "Biological Decontamination of Oil Spills In Cold Environments." *J Chem. Technol. Biotechnol.*, 74: 381–389.

Margesin, R., and Schinner, F. 2001a. "Biodegradation and Bioremediation of Hydrocarbons in Extreme Environments." *Appl. Environ. Microbiol.*, 56: 650–663.

Margesin, R., and Schinner, F. 2001b. "Potential of Halotolerant and Halophilic Microorganisms for Biotechnology." *Extremophiles*, 5: 73–83.

Margesin, R., Hämmerle, M., and Tscherko, D. 2007a. "Microbial Activity and Community Composition during Bioremediation of Diesel-Oil

Contaminated Soil: Effects of Hydrocarbon Concentration, Fertilizers and Incubation Time." *Microb. Ecol.*, 53: 259–269.

Margesin, R., Neuner, G., and Storey, K.B. 2007b. "Cold-Loving Microbes, Plants, and Animals – Fundamental and Applied Aspects." *Naturwissenschaften*, 94: 77–99.

Means, J.C. 1995. "Influence of Salinity upon Sediment-Water Partitioning of Aromatic Hydrocarbons." *Mar. Chem.*, 51: 3–16.

Michaud, L., Giudice, A.L., Saitta, M., De Domenico, M., and Bruni, V. 2004. "The Biodegradation Efficiency on Diesel Oil by Two Psychrotrophic Antarctic Marine Bacteria during a Two-Month-Long Experiment." *Mar. Pollut. Bull.*, 49: 405–409.

Michelsen, T.C., and Petito Boyce, C. 1993. "Cleanup Standards for Petroleum Hydrocarbons. Part 1. Review of Methods and Recent Developments." *Journal of Soil Contamination*, 2(2): 1–16.

Mohn, W.W., and Stewart, G.R. 2000. "Limiting Factors for Hydrocarbon Degradation at Low Temperature in Arctic Soils." *Soil Biol. Biochem.*, 32: 1161–1172.

Mohn, W., Radziminski, C., Fortin, M-C., and Reimer, K. 2001. "On Site Bioremediation of Hydrocarbon-Contaminated Arctic Tundra Soils in Inoculated Biopiles." *Appl. Microbiol. Biotechnol.*, 57: 242–247.

Morita, R.Y. 1975. "Psychrophilic Bacteria." *Bacteriol. Rev.*, 39: 144–167.

Müller, R., Antranikian, G., Maloney, S., and Sharp, R. 1998. "Thermophilic Degradation of Environmental Pollutants." In *Biotechnology of Extremophiles*. G. Antranikian (Editor). Advances in Biochemical Engineering/Biotechnology, Volume 61. Springer, New York. pp. 155–169.

Nedwell, D.B. 1999. "Effect of Low Temperature on Microbial Growth: Lowered Affinity for Substrates Limits Growth at Low Temperature." *FEMS Microbiol. Ecol.*, 30: 101–111.

Niehaus, F., Bertoldo, C., Kähler, M., and Antranikian, G. 1999. "Extremophiles as a Source of Novel Enzymes for Industrial Application." *Appl Microbiol Biotechnol*, 51: 711–729.

Norris, R.D. 1994. *Handbook of Bioremediation*. CRC Press, Boca Raton, Florida.

Norris, P.R., and Johnson, D.B. 1998. "Acidophilic Microorganisms." In *Extremophiles: Microbial Life in Extreme Environments*. K. Horikoshi and W.D. Grant (Editors). Wiley-Liss, New York. pp. 133–153.

Oren, A., Gurevich, P., Azachi, M., and Hents, Y. 1992. "Microbial Degradation of Pollutants at High Salt Concentrations." *Biodegradation*, 3: 387–398.

Peters, K.E., Scheuerman, G.L., Lee, C.Y., Moldovan, J.M., and Reynolds, R.N. 1992. "Effects of Refinery Processes on Biological Markers." *Energy and Fuels*, 6: 560–577.

Poremba, K., Gunkel, W., Lang, S., and Wagner, F. 1991. "Marine Biosurfactants, III. Toxicity Testing with Marine Microorganisms and Comparison with Synthetic Surfactants." *Z. Naturforsch.*, 46c: 210–216.

Powell, S.M., Harvey, P.McA., Stark, J.S., Snape, I., and Riddle, M.J. 2007. "Biodegradation of petroleum products in experimental plots in Antarctic marine sediments is location dependent." *Mar. Pollut. Bull.*, 54: 434–440.

Prieur, D., and Marteinsson, V.T. 1998. "Prokaryotes living under elevated hydrostatic pressure." In *Biotechnology of Extremophiles*. G. Antranikian (Editor). Advances in Biochemical Engineering/Biotechnology, Volume 61. Springer, New York. pp. 23–35.

Rhykerd, R.L., Weaver, R.W., and McInnes, K.J. 1995. "Influence of Salinity on Bioremediation of Oil in Soil." *Environ. Pollut.*, 90: 127–130.

Rosa, A.P., and Triguis, J.A. 2007. "Bioremediation Process on the Brazil Shoreline." *Env. Sci. Pollut. Res.*, 14(7): 470–476.

Ruberto, L., Vazquez, S.C., and MacCormarck, W.P. 2003. "Effectiveness of the Natural Bacteria *Xora*, Biostimulation, and Bioaugmentation on the Bioremediation of a Hydrocarbon-Contaminated Antarctic Soil." *Inter. Biodeter. Biodegrad.*, 52: 115–125.

Schulz, D., Passeri, A., Schmidt, M., Lang, S., Wagner, F., Wray, V., and Gunkel, W. 1991. "Marine Biosurfactants, I. Screening for Biosurfactants among Crude Oil Degrading Marine Microorganisms from the North Sea." *Z. Naturforsch.*, 46c: 197–203.

Siron. R., Pelletier, E., Delille, D., and Roy, S. 1993. "Fate and Effects of Dispersed Crude Oil under Icy Conditions Simulated in Mesocosms." *Mar. Environ. Res.*, 35: 273–302.

Siron, R., Pelletier, E., and Brochu, C. 1995. "Environmental Factors Influencing the Biodegradation of Petroleum Hydrocarbons in Cold Seawater." *Arch. Environ. Contam. Toxicol.*, 28: 406–416.

Speight, J.G. 1996. *Environmental Technology Handbook*. Taylor & Francis, Washington, DC.

Speight, J.G., and Lee, S. 2000. *Environmental Technology Handbook*. 2nd Edition. Taylor & Francis, New York.

Speight, J.G. 2005. *Environmental Analysis and Technology for the Refining Industry*. John Wiley & Sons Inc., Hoboken, New Jersey.

Stapleton, R.D., Savage, D.C., Sayler, G.S., and Stacey, G. 1998. "Biodegradation of Aromatic Hydrocarbons in an Extremely Acidic Environment." *Appl. Environ. Microbiol.*, 64: 4180–4184.

Stelmaszewski, A. 2009. "Determination of Petroleum Pollutants in Coastal Waters of the Gulf of Gda sk." *Oceanologia*, 51(1): 85–92.

Stetter, K.O. 1998. "Hyperthermophiles: Isolation, Classification, and Properties." In *Extremophiles: Microbial Life in Extreme Environments*. K. Horikoshi and W.D. Grant (Editors). Wiley-Liss, New York. pp. 1–24.

Stout, S.A., Uhler, A.D., and McCarthy, K.J. 2001. A Strategy and Methodology for Defensibly Correlating Spilled Oil to Source Candidates: Environmental Forensics, v. 2, pp. 87–98.

Stout, S.A., Uhler, A.D., and McCarthy, K.J. 2005. "Middle Distillate Fuel Fingerprinting using Drimane-Based Bicyclic Sesquiterpanes." *Environmental Forensics*, 6: 241–251.

Swannell, R.P.J., Lee, K., and McDonagh, M. 1996. "Field Evaluations of Marine Oil Spill Bioremediation." *Microbiological Reviews*, 60: 342–365.

Sweed, H.G., Bedient, P.B., and Hutchins, S.R. 1996. "Surface Application System for In Situ Groundwater Bioremediation Site Characterization and Modeling." *Ground Water*, 34(2): 211–222.

Trindade, P.V.O., Sobral, L.G., Rizzo, A.C.L., Leite, S.G.F., Lemos, J.L.S., Millioli, V.S., and Soriano, A.U. 2002. "Evaluation of the Biostimulation and Bioaugmentation Techniques in the Bioremediation Process of Petroleum Hydrocarbon Contaminated Soil." *Proceedings*. 9th Annual International Petroleum Environmental Conference, October. Albuquerque, New Mexico.

US EPA. 1990. *Interim Report. Oil Spill Bioremediation Project*. U.S. Environmental Protection Agency, Office of Research and Development, Washington, DC.

US EPA. 2003. *Aerobic Degradation of Oily Wastes: A Field Guidance Book for Federal On-Scene Coordinators*. Version 1.0. US Environmental Protection Agency Response and Prevention Branch, Region 6 South Central., Dallas Texas.

Venosa, A.D., Suidan, M.T., Wrenn, B.A., Strohmeier, K.L., Haines, J.R., Eberhart, B.L., King, D., and Holder, E. 1996. "Bioremediation of an Experimental Oil Spill on the Shoreline of Delaware Bay." *Environmental Science & Technology*, 30 (5), 1764.

Wagner-Döbler, I., Bennasar, A., Vancanneyt, M., Strömpl, C., Brümmer, I., Eichner, C., Grammel, I., and Moore, E.R.B. 1998. "Microcosm Enrichment of Biphenyl-Degrading Microbial Communities from Soils and Sediments." *Appl. Environ. Microbiol.*, 64: 3014–3022.

Ward, D.M., and Brock, T.D. 1978. "Hydrocarbon Degradation in Hypersaline Environments." *Appl. Environ. Microbiol.*, 35: 353–359.

Wardell, L.J. 1995. "Potential for Bioremediation of Fuel-Contaminated Soil in Antarctica." *J. Soil Contam.*, 4: 111–121.

Whitehouse, B.G. 1984. "The Effects of Temperature and Salinity on the Aqueous Solubility of Polynuclear Aromatic Hydrocarbons." *Mar. Chem.*, 14: 319–332.

Whyte, L.G., Hawari, J., Zhou, E., Bourbonniere, L., Inniss, W.E., and Greer, C.W. 1998. "Biodegradation of Variable-Chain-Length Alkanes at Low Temperature by a Psychrotrophic *Rhodococcus* sp." *Appl. Environ. Microbiol.*, 64: 2578–2584.

Whyte, L.G., Bourbonnière, L., Bellerose, C., and Greer, C.W. 1999. "Bioremediation Assessment of Hydrocarbon-Contaminated Soils from the High Arctic." *Biorem. J.*, 3: 69–79.

Wright, A.L., Weaver, R.W., and Webb, J.W. 1997. "Oil Bioremediation in Salt Marsh Mesocosms as Influenced by N and P Fertilization, Flooding and Season." *Water Air Soil Poll.*, 95: 179–191.

Yakimov, M.M., Giuliano, L., Bruni, V., Scarfi, S., and Golyshin, P.N. 1999. "Characterization of Antarctic Hydrocarbon-Degrading Bacteria Capable of Producing Bioemulsifiers." *Microbiologica*, 22: 249–256.

Zhu, X, Venosa, A.D., Suidan, M.T., and Lee, K. 2001. "Guidelines for the Bioremediation of Marine Shorelines and Freshwater Wetlands." US EPA Report, September.

Zhu, X., Venosa, A.D., and Suidan, M.T. 2004. Literature Review on the Use of Commercial Bioremediation Agents for Cleanup of Oil-Contaminated Environments. Report No. EPA/600/R-04/075. National Risk Management Research Laboratory, United States Environmental Protection Agency, Cincinnati, Ohio. July.

16

The Future of Bioremediation

1 Introduction

The increasingly long list of chemical contaminants released into the environment on a large scale includes numerous aliphatic and aromatic compounds, such as petroleum hydrocarbons. The local concentration of such contaminants depend on the amount present and the rate at which the compound is released, its stability in the environment under both aerobic and anaerobic conditions, the extent of its dilution in the environment, the mobility of the compound in a particular environment, and its rate of biological or non-biological degradation (Harayama, 1997; Ellis, 2000; Janssen et al., 2001; Dua et al., 2002).

Remediation technologies that can reduce or remove a contaminant can be classified into four categories based on the process acting on the contaminant. These categories are removal, separation, destruction, and containment, which can be either a physical, chemical, or biological process. Physical removal (Speight, 1996; Speight and Lee, 2000; Speight, 2005), isolation, and microbial remediation, which includes phytoremediation for the purposes of this text, are the most commonly used remediation techniques.

The physical removal of contaminated soil and groundwater is the most common form of remediation, but the process does not eliminate the contamination, but rather transfers it to another location. In ideal cases, the other location will be a facility that is specially designed to contain the contamination for a sufficient period of time, or treat it as necessary. In this way, proper removal reduces the risk of exposure to the environmental contaminants.

Isolation technology is typically carried out using clay, concrete, man-made liners, or a combination of these, and is often used when a contaminant is difficult or extremely expensive to remove or destroy. The process essentially isolates the contaminant from the affected ecosystem.

In terms of microbial remediation (bioremediation), which involves the breakdown of pollutants by microorganisms, aerobic processes are considered the most efficient and generally applicable: aerobic degradation is dependent on the presence of molecular oxygen, and is catalyzed by enzymes that have evolved for the catabolism of natural substrates and exhibit low specificities. Depending upon the type of enzyme catalyzing the reaction, either one (mono-oxygenase) or two (di-oxygenase) oxygen atoms are inserted into the molecule via an electrophilic attack on an unsubstituted carbon atom. Anaerobic degradation proceeds via reductive dehalogenation, wherein an electron transfer to the compound, resulting in hydrogenation.

Petroleum hydrocarbons are widespread common environmental pollutants (Megharaj et al., 2000). The search for an effective remediation for petroleum-contaminated soil is a huge challenge to environmental researchers. The requirement of bioremediation is highlighted after the case of Van Daze Oil Spills in 1989 (Bragg et al., 1994). Bioremediation research has recently attracted widespread attention (Atlas, 1995; Grishchenkov et al., 2000; Chen et al., 2008). It has been reported that suitable microbes are available and can be used to effectively remediate petroleum contamination even at low environmental temperatures (Rike et al., 2003; Sanscartier et al., 2009).

Although regulations are strictly enforced in developed countries like the United States and most of the European countries to meet the challenges of petroleum-related contamination, these regulations often remain unenforced in most of the developing countries. Cleaning up such sites is often not only technically challenging, but also very expensive. Considerable pressure encourages the adoption of waste management alternatives to burial, the traditional means of disposing of solid and liquid wastes.

Approaches such as air-stripping (to remove volatile compounds) and incineration have been used (Speight, 1996; Speight and Lee, 2000; Speight, 2005). However, where the contaminants infect a large area but are in low (albeit significant) concentration, such methods are either very costly or simply not feasible (Blackburn and Hafker, 1993; Singh et al., 2001). In such schemes, microorganisms can provide an effective alternative through the biodegradation of the contaminants.

Since most of the contaminants of concern in crude oil are readily biodegradable under the appropriate conditions, the success of oil-spill bioremediation depends mainly on the ability to establish these conditions in the contaminated environment using the above new developing technologies to optimize the microorganisms' total efficiency. The technologies used at various polluted sites depend on the limiting factors present at the location. For example, where there is insufficient dissolved oxygen, bioventing or sparging is applied, where biostimulation or bioaugmentation is suitable for instances where the biological count is low.

2 Status

Over the past decade, opportunities for applying bioremediation to a much broader set of contaminants have been identified. Indigenous and enhanced organisms have been shown to degrade industrial solvents, polychlorinated biphenyls (PCBs), explosives, and many different agricultural chemicals. Pilot, demonstration, and full-scale applications of bioremediation have been carried out on a continuing basis. However, the full benefits of bioremediation have not been realized, because processes and organisms that are effective in controlled laboratory tests are not always equally effective in full-scale applications. The failure to perform optimally in the field setting stems from a lack of predictability, due in part to inadequacies in the fundamental scientific understanding of how and why these processes work.

This, if bioremediation is to be effective, the microorganisms must enzymatically attack the pollutants and convert them to non-contaminating products: some microbes may produce products that are not only toxic to themselves, but also the ecosystem. Parameters that affect the bioremediation process include temperature, nutrients (fertilizers), and the amount of oxygen present in the soil and/or the affected water system (Chapter 1, Chapter 9). These conditions allow the microbes to grow and multiply and consume more of the contaminant. When conditions are adverse, microbes grow too slowly or die, or they can create more harmful chemicals. In addition, the application of any technology is dependent not only on the availability of the technology, but also on the reliability of the technology, as well as on the suitability of the technology for the specific site conditions, and whether the technology is readily available (i.e., emerging, developing, or proven).

2.1 Conventional Bioremediation

Conventional bioremediation methods used are composting, land farming, biopiling, and bioslurry reactors. Composting is a technique that involves combining contaminated soil with nonhazardous organic additives, such as manure or agricultural wastes; the presence of the organic materials allows the development of a rich microbial population and elevated temperature characteristic of composting. Land farming is a simple technique in which contaminated soil is excavated and spread over a prepared bed and periodically tilled until pollutants are degraded. While biopiling is a hybrid of land farming and composting, it is essentially engineered cells that are constructed as aerated composted piles.

Bioslurry reactors can provide rapid biodegradation of contaminants due to enhanced mass transfer rates and increased contaminant-to-microorganism contact. These units are capable of treating high concentrations of organic contaminants in soil and sludge. These reactors can aerobically biodegrade aqueous slurries created through the mixing of soils or sludge with water. The most common state of bioslurry treatment is batch; however, continuous-flow operation can be achieved.

Microorganisms excel at using organic substances, natural or synthetic, as sources of nutrients and energy. Indeed, the diversity of petroleum-related constituents for growth had led to the discovery of enzymes capable of transforming many unrelated natural organic compounds by many different catalytic mechanisms (Butler and Mason, 1997; Ellis, 2000).

However, depending on behavior in the environment, organic compounds are often classified as biodegradable, persistent, or recalcitrant. A biodegradable organic compound is one that undergoes a biological transformation (Blackburn and Hafker, 1993; Liu and Suflita, 1993). A persistent organic compound does not undergo biodegradation in certain environments, and a recalcitrant compound resists biodegradation in a wide variety of environments. While partial biodegradation is usually an alteration by a single reaction, primary biodegradation involves a more extensive chemical change. Mineralization is a parallel term to biodegradation, referring to complete degradation to the end products of carbon dioxide, water, and other inorganic compounds.

Biodegradation and its application in bioremediation of organic pollutants have benefited from the biochemical and molecular studies of microbial processes (Lal et al., 1986; Fewson, 1988; Sangodkar et al., 1989; Chaudhary and Chapalamadugu, 1991; Bollag and Bollag, 1992; Van der Meer et al., 1992; Dickel et al., 1993; Deo et al., 1994; Johri et al., 1996; Kumar et al., 1996; Johri et al., 1999; Faison, 2001; Janssen et al., 2001). Indeed, the biotransformation of organic contaminants in the natural environment has been extensively studied to understand microbial ecology, physiology, and evolution for their potential in bioremediation (Bouwer and Zehnder, 1993; Chen et al., 1999; Johan et al., 2001; Mishra et al., 2001; Watanabe, 2001).

As a result, there is a strong demand to increase the adoption of bioremediation as an effective technique for risk reduction on hydrocarbon impacted soils (Diplock et al., 2009). However, the biodegradation effectiveness diminishes with the time extension, and the inhibiting effect may become dominant with time. The key solution to bioremediation is to speed up the restoration process and eliminate or delay the inhibitory effect, such as through the selection of specifically targeted strains or microorganisms (Marijke and Vlerken, 1998), or through the alteration of microbial community structure changes (Antizar-Ladislao et al., 2008) during the treatment.

Like bioremediation, phytoremediation has recently been developed as a remedial strategy for organic contaminants. It is believed that phytoremediation is much less disruptive to the environment, and may have a high probability of public acceptance as a low-cost alternative (Alkorta and Garbisu, 2001). Furthermore, rhizosphere microbes can become contaminant degraders under stress condition (Wenzel, 2009; Gerhart et al., 2009; Korade and Fulekar, 2009).

The recent successful implementation of phytoremediation by a number of researchers (Escalante-Espinosa et al., 2005; Lin et al., 2005; Merkl et al., 2005; Erute Magdalene et al., 2009) has indicated that such an approach is quite promising, and can be a viable alternative to the conventional bioremediation. However, the use of phytoremediation is constrained by climate and geological conditions of the sites to be cleaned, such as temperature, altitude, soil type, and the accessibility by agricultural equipment (Macek et al., 2000). Plants growing in petroleum-contaminated soil have to cope with the nutrient deficiency and hydrocarbon toxicity. The complicated soil characteristics, such as the heterogeneous distribution of the petroleum, soil structure, nutrient shortage, and transplanting, may cause some unexpected consequences. Furthermore, the selection of proper plant species (Salt et al., 1998) is also crucial for the experiment. In general, phytoremediation can be a commercial strategy for low total petroleum hydrocarbon (TPH) contaminated soil, which could improve soil characteristics with minimal risks.

Pollutants sometimes cannot be removed completely by a single remediation process, especially when using biological methods. Therefore, multi-process bioremediation provides a promising and environmentally friendly solution, which is cost-effective and pollution-free during the active cleaning process. In addition, multi-process bioremediation may become an effective strategy for rapid biodegradation by altering microbial community structure. It is thus a challenging and rewarding research to search for an innovative solution to speed up TPH reduction for effective environmental cleaning.

2.2 Enhanced Bioremediation

The natural processes that drive bioremediation can be enhanced to increase the effectiveness and to reduce time required to meet cleanup objectives. Enhanced bioremediation involves the addition of microorganisms (e.g., fungi, bacteria, and other microbes) or nutrients (e.g. oxygen, nitrates) to the subsurface environment to accelerate the natural biodegradation process, a process in which indigenous microbes degrade (metabolize) organic contaminants found in soil and/or ground water, and convert them to innocuous end products. The process relies on general availability of naturally occurring microbes to consume contaminants as a food source (petroleum hydrocarbons in aerobic processes) or as an electron acceptor (chlorinated solvents). In addition to microbes being present, in order to be successful, these processes require nutrients (carbon: nitrogen: phosphorus).

The potential of microorganisms in the remediation of some of the compounds hitherto known to be non-biodegradable has been widely acknowledged globally. With advances in biotechnology, bioremediation has become a rapidly growing area, and has been commercially applied for the treatment of hazardous wastes and contaminated sites. Although a wide range of new microorganisms have been discovered that are able to degrade highly stable, toxic organic xenobiotics, still many pollutants persist in the environment.

Briefly, a xenobiotic is a chemical which is found in an organism, such as a bacterium, but which is not normally produced or expected to be present in the organism. The word is very often used in the context of a pollutant, and is a substance that is not indigenous to an ecosystem or a biological system, and which did not exist in nature before human intervention.

A number of reasons have been identified as challenges posed to the microorganisms working in contaminated sites. Such potential limitations to biological treatments include: poor bioavailability of chemicals, presence of other toxic compounds, inadequate supply of nutrients, and insufficient biochemical potential for effective biodegradation. A wide range of bioremediation strategies have been developed for the treatment of contaminated soils using natural and modified microorganisms.

Selecting the most appropriate strategy to treat a specific site can be guided by considering three basic principles: the amenability of the pollutant to biological transformation to less toxic products, the bioavailability of the contaminant to microorganisms, and the opportunity for bioprocess optimization. With the help of advances in bioinformatics, biotechnology holds a bright future for developing bioprocesses for environmental applications.

Biotechnological processes for the bioremediation of petroleum-related pollutants offer the possibility of in situ treatments, and are mostly based on the natural activities of microorganisms. Biotechnological processes to destroy contaminants of the type found in petroleum and petroleum products offer many advantages over physicochemical processes. When successfully operated, biotechnological processes may achieve complete destruction of petroleum-related pollutants. However, an important factor limiting the bioremediation of sites contaminated with such contaminants is the slow rate of degradation (Iwamoto and Nasu, 2001), which may limit the practicality of using microorganisms in remediating contaminated sites. This is an area where genetic engineering can make a marked improvement. Molecular techniques can be used to increase the level of a particular protein or enzyme or series of enzymes in bacteria with an increase in the reaction rate (Chakrabarty, 1986).

Biosurfactants are surface-active microbial products that have numerous industrial applications (Desai and Banat, 1997; Sullivan, 1998; Sekelsky and Shreve, 1999). Many microorganisms, especially bacteria, produce biosurfactants when grown on water-immiscible substrates. Most common biosurfactants are glycolipids, in which carbohydrates are attached to a long-chain aliphatic acid, while others, such as lipopeptides, lipoproteins, and hetero-polysaccharides, are more complex. The most promising applications of biosurfactants are in the cleaning of oil-contaminated tankers, oil-spill management, transportation of heavy crude, enhanced oil recovery, recovery of oil from sludge, and bioremediation of sites contaminated with hydrocarbons, heavy metals, and other petroleum-related pollutants.

Furthermore, the bioremediation of polynuclear aromatic hydrocarbons was designed by the addition of surfactants, and mathematical models were constructed to explain the effect of surfactants on biodegradation (Harayama, 1997). With the increasing awareness of the applicability of biosurfactants, the focus is now on the utilization of biosurfactants for the bioremediation of non-aqueous-phase liquids (NAPLs).

Advances in genetic and protein engineering techniques have opened up new avenues to move towards the goal of *genetically engineered microorganisms* to function as biocatalysts, in which certain desirable biodegradation pathways or enzymes from different organisms are brought together in a single host with the aim of performing specific reactions (Masai et al., 1995; Hauschild et al., 1996; Timmis and Piper, 1999; Sayler and Ripp, 2000). A strategy has also been suggested (Timmis and Piper, 1999) for designing organisms with novel pathways, and the creation of a bank of genetic modules encoding broad-specificity enzymes or pathway segments that can be combined at will to generate new or improved activities.

Petroleum and petroleum products as target pollutants are difficult to identify because of the complexity of the petroleum system (Chapter 4, Chapter 5, Chapter 6). The presence of metals and metallo-organic constituents, which are not destroyed biologically, but are only transformed from one oxidation state to another, interfere with the bioremediation processes. Genetic engineering allows transferring the heavy metal-resistance genes to identifiable microorganism hosts, that can then serve as excellent bases from which to construct recombinant strains to overcome the challenge of the metal constituents of petroleum.

Methods for the rapid and specific identification of microorganisms within their natural environments continue to be developed. Classic methods are time-consuming and only work for a limited number of microorganisms (Amann et al., 2001). An increasing need to develop new methods for characterization of microorganisms able to degrade the various types of petroleum-related pollutants has led to the use of molecular probes to identify, enumerate, and isolate microorganisms with degradative potential.

Through the genetic engineering of metabolic pathways, it is possible to extend the range of substrates that an organism can utilize. Aromatic hydrocarbon di-oxygenases have broad substrate specificity, and catalyze enantio-specific reactions with a wide range of substrates.

2.3 Bioremediation in Extreme Environments

The biodegradation of many components of petroleum hydrocarbons has been reported in a variety of terrestrial and marine cold ecosystems, extreme environments such as alpine soil (Margesin, 2000), Arctic soil (Braddock et al., 1997), Arctic seawater (Siron et al., 1995), and Antarctic soil (Aislabie et al., 1998; Aislabie et al., 2004), as well as Antarctic seawater and sediments (Delille et al., 1998; Delille and Delille, 2000; Al-Darbi et al., 2005).

Antarctic exploration and research have led to some significant, although localized, impacts on the environment. Human impacts occur around current or past scientific research stations, typically located on ice-free areas that are predominantly soils. Fuel spills, the most common occurrence, have the potential to cause the greatest environmental impact in the Antarctic through accumulation of aliphatic and aromatic compounds. Effective management of hydrocarbon spills is dependent on understanding how they impact soil properties such as moisture, hydrophobicity, soil temperature, and microbial activity. Numbers of hydrocarbon-degrading bacteria, typically *Rhodococcus*, *Sphingomonas*, and *Pseudomonas* species, for example, may become elevated in contaminated soils, but overall microbial diversity declines. Alternative management practices to the current approach of *dig it up and ship it out* are required, but must be based on sound information (Aislabie et al., 2004).

Cold-tolerant bacteria, isolated from oil-contaminated soils in Antarctica, were able to degrade *n*-alkanes (C_6 to C_{20}) typical of the hydrocarbon contaminants that persist in Antarctic soil (Bej et al., 2000). Representative isolates were identified as *Rhodococcus* species: they retained metabolic activity at sub-zero temperatures of –2°C. A psychrotrophic *Rhodococcus* sp. from Arctic soil (Whyte et al., 1998) utilized a broad range of aliphatic compounds (C_{10} to C_{21} alkanes, branched alkanes, and a substituted cyclohexane) present in diesel oil at 5°C (41°F). The strain mineralized the short-chain alkanes (C_{10} and C_{16}) to a significantly greater extent than the long-chain alkanes (C_{28} and C_{32}) at 0 and 5°C (32 and 41°F). The decreased bioavailability of the long-chain alkanes at low temperature (many form crystals at 0°C, 32°F) may be responsible for their increased recalcitrance, which affects in situ bioremediation in cold climates.

In addition, the physical environment is also important for hydrocarbon biodegradation. This has been demonstrated in sub-Antarctic intertidal beaches (Delille and Delille, 2000) and in Arctic soils (Mohn and Stewart, 2000). Soil nitrogen and concentrations of total petroleum hydrocarbons together accounted for 73% of the variability of the induction period (lag time) for dodecane mineralization at 7°C (45°F) in Arctic soils. High total carbon concentrations were associated with high mineralization rates; high sand content resulted in longer half-times for mineralization. Dodecane mineralization was limited by both N and P; mineralization kinetics varied greatly among different soils (Mohn and Stewart, 2000).

Cold habitats possess sufficient indigenous microorganisms, with psychrotrophic bacteria (bacteria capable of surviving or even thriving in a cold environment) being predominant. They adapt rapidly to the presence of the contaminants, as demonstrated by significantly increased numbers of oil degraders shortly after a pollution event, even in the most northerly areas of the world (Whyte et al., 1999). However, the temperature threshold for significant petroleum biodegradation is approximately 0°C (32°F) (Siron et al., 1995).

However, the bulk of information on hydrocarbon degradation borders on activities of mesophiles, although significant biodegradations of hydrocarbons have been reported in psychrophilic environments in temperate regions (Yumoto et al., 2002; Pelletier et al., 2004; Delille et al., 2004). Full-scale in situ remediation of petroleum contaminated soils has not yet been used in Antarctica, for example, partly because it has long been assumed that air and soil temperatures are too low for an effective biodegradation (Delille et al., 2004). Such omissions in research programs need to be corrected.

The technical feasibility of in situ bioremediation of hydrocarbons in cold groundwater systems has been demonstrated (Bradley and Chapelle, 1995). Rapid aerobic toluene mineralization was demonstrated in sediments from a cold (mean groundwater temperature 5°C/41°F) petroleum-contaminated aquifer in Alaska. The mineralization rate obtained at 5°C in this aquifer was comparable to that measured in sediments from a temperate aquifer (mean temperature 20°C). Rates of overall microbial metabolism in the two sediments were comparable at their respective in situ temperatures.

Bioaugmentation of contaminated cold sites with hydrocarbon-degrading bacteria has been tested as a bioremediation strategy. For the most part, indigenous microbial populations degrade hydrocarbons more efficiently than the introduced microbial strains. However, bioaugmentation (through the use of non-indigenous microbial species) may enhance biodegradation through the onset of a shorter hydrocarbon acclimation period (Mohammed et al., 2007).

Inoculation of contaminated Arctic soils with consortia (Whyte et al., 1999) or with alkane-degrading *Rhodococcus* sp. (Whyte et al., 1998) decreased the lag time and increased the rate of C_{16} mineralization at 5°C (41°F). However, nitrogen-phosphorus-potassium fertilization alone had a comparable effect on hydrocarbon loss like fertilization plus bioaugmentation; this has been shown both in chronically oil-polluted Arctic soil (Whyte et al., 1999) and in artificially diesel oil-contaminated alpine soils (Margesin and Schinner, 1997a, 1997b; Vieira et al., 2009).

Finally, the application of cold-active solubilizing agents could be useful for enhancing hydrocarbon bioavailability. Two hydrocarbon-degrading Antarctic marine bacteria, identified as *Rhodococcus fascians*, produced bioemulsifiers when grown with *n*-alkanes as the sole carbon source. The strains utilized hexadecane and biphenyl as sole carbon sources at temperatures ranging from 4 to 35°C (39 to 95°F); the optimum temperature was 15 to 20°C (59 to 68°F) (Yakimov et al., 1999; Chugunov et al., 2000).

In summary, hydrocarbon biodegradation rates in cold groundwater systems are not necessarily lower than in temperate systems, but activity measurements should be performed at the prevailing in situ temperature in order to obtain a realistic estimate of the naturally occurring biodegradation.

3 Advantages and Disadvantages

Bioremediation has advantages over traditional cleanup methods of petroleum-related spills (Table 16.1). One of the major advantages of bioremediation is the savings in cost, and there is also the savings in the time put forth by workers to clean a contaminated site. The financial savings of bioremediation, when used properly, have tremendous benefits compared to traditional cleanup processes. After the Exxon Valdez spill, the cost to clean 75 miles of shoreline by bioremediation was less than cost to provide physical washing of the shore for one day (Atlas, 1995; Zhu et al., 2004). In addition to traditional methods of site cleanup, bioremediation continues to clean the contaminated site without the constant need of workers.

Bioremediation is also advantageous due to its environmentally friendly approach, since no foreign or toxic chemicals are added to the site. It is also environmentally friendly because it does not require any disruption to the natural habitat, which often occurs from physical and chemical methods of cleanup. Bioremediation allows for natural organisms to degrade the

Table 16.1 Advantages and disadvantages of bioremediation.

Advantages
1) A natural process.
2) Toxic chemicals are destroyed or removed from environment.
3) Can affect complete destruction of a wide variety of contaminants.
4) Can transform contaminants to harmless products.
5) Can be carried out on site without causing a major disruption of normal activities.
6) If performed in situ, no excavation or transport costs.
7) Less energy is required as compared to other technologies.
8) Microbes do not need manual supervision.
Disadvantages
1) Not all contaminants are biodegradable.
2) The microbial action may be highly specific to one site only.
3) Heavy metals remain on the site.
4) The products may be more toxic than the original contaminant(s).
5) For in situ bioremediation, the site must have highly permeable soil.
6) Process may not remove all quantities of contaminants.
7) Extrapolation from laboratory scale to field scale my by subject to inaccuracies.
8) The contaminant may be a complex mixture that is not evenly dispersed.
9) The contaminant may be present as solids, liquids, and gases.

toxic hydrocarbons into simple compounds that pose no threat to the environment, and this also eliminates the need to remove and transport the toxic compounds to another site. This loss of a need to transport the oil and contaminated soils lowers further risk of additional oil spills, and also saves energy and money that would be put forth in the transportation process. These environmental benefits also make bioremediation a positively viewed method by the general public. With the limited resources in today's world, this is a very much supported technology that pleases the public, and hence, is given political support and funding for further research.

One of the greater downsides of bioremediation for petroleum-related spills is that it is a slow process. Such spills can pose a great threat to many different habitats, environments, and industries, and depending upon the urgency of cleanup, bioremediation may not always be the best available option. Also, there are many variables that affect whether bioremediation is capable and practical for the cleanup of different spills. Depending on where the spill takes place and the conditions of the soil or water, it may be very difficult to provide proper nutrient concentrations to the oil degrading microorganisms (Delille et al., 2009). If an oil spill occurs offshore, there is typically much more energy and waves, and this can cause for the quick loss and dilution of nutrients provided by biostimulation (Chapter 14). In the case of bioaugmentation, there are other issues, particularly the competition that will develop between the native and foreign microbes, which has the potential to render the bioremediation method unsuccessful.

However, it must be recognized that all of the spilled petroleum or petroleum product removed by microbes in a linear timeframe, since the residual fractions (especially those that contain high molecular eight polar constituents) that are not consumed initially will be more refractory to microbial attack (Speight, 2005, 2007). Conditions in any ecosystem are rarely favorable for maximum biodegradation.

Furthermore, a major issue relates to whether or not degradation that has been demonstrated to occur in the laboratory will occur in the soil or water ecosystems. Biodegradation phenomena reported in laboratory studies of pure cultures of microorganisms reflect only the *potential degradation* that *may* occur in any natural ecosystem. Physical-chemical properties of petroleum, petroleum products, and any of the related constituents, and the concentration of the petroleum related contaminant(s), as well as the concentration and diversity of the microbial flora of a specific ecosystem, are variable factors in the biodegradation process.

Furthermore, there have been various suggestions that when the concentration of the petroleum-related contaminants passes below a threshold level (which is typically not defined), biodegradation ceases, because bacterial growth is limited. Surely this relates to the amount of recalcitrant species (such as resin constituents and asphaltene constituents or related polynuclear aromatic systems) remaining in the non-biodegraded material.

The qualitative and quantitative differences in the hydrocarbon and the non-hydrocarbon content of petroleum (Speight, 2005, 2007) influence the susceptibility of petroleum and certain petroleum products to biodegradation. This must be acknowledged as a major consideration in determining eco-toxicological effects of petroleum constituents.

4 Conclusion

Biodegradation (with subsequent bioremediation) has shown great promise for cleanup of spills of petroleum-related contaminant, but caution is advised in extolling its virtues as the panacea for cleanup of all petroleum-related spills. It is, nevertheless, one of several viable options that are available as a single or piggy-back method for environmental cleanup, but considerations such as (1) site parameters, (2) the properties of the spilled material, (3) prevailing climatic conditions, and (4) the nature of the microorganisms that are available must be taken into account.

However, with advances in biotechnology, bioremediation has become one of the most rapidly developing fields of environmental restoration, utilizing microorganisms to reduce the concentration and toxicity of as petroleum-related hydrocarbons, polycyclic aromatic hydrocarbons, and other constituents, such as metals and metallo-organic compounds. However, although simple aromatic compounds are biodegradable by a variety of degradative pathways, their halogenated counterparts are more resistant to bacterial attacks, and often necessitate the evolution of novel pathways (Chakrabarty, 1982; Engasser et al., 1990).

A number of bioremediation strategies have been developed to treat petroleum-contaminated wastes and sites. However, selecting the most appropriate strategy to treat a specific site can be guided by considering three basic principles: the amenability of the pollutant to biological transformation to less toxic products (biochemistry), the accessibility of the contaminant to microorganisms (bioavailability), and the opportunity for optimization of biological activity (bioactivity).

Bioremediation is a preferred method for the long-term restoration of petroleum hydrocarbon polluted systems, with the added advantage of cost efficiency and environmental friendliness (Okoh, 2006). Although exhaustive investigations have been performed in relation to biodegradation of petroleum and petroleum products, these studies must continue. The identification of active microbial strains is not always ascertained to a sufficient degree, and misidentifications or incomplete identifications are sometimes reported. There is much to be done in terms of the optimization of the process conditions for more efficient application of biological degradation of petroleum and petroleum products under different climatic conditions.

Although a wide range of new microorganisms have been discovered that are able to degrade highly stable, toxic organic chemicals, many pollutants continue to persist in the environment. A number of reasons have been identified as challenges posed to the microorganisms working in contaminated sites, such as: (1) poor bioavailability of chemicals, (2) presence of non-petroleum compounds, (3) inadequate supply of nutrients, and (4) insufficient biochemical potential for effective biodegradation.

Thus, selecting the most appropriate strategy to treat a specific site can be guided by considering three basic principles: (1) the amenability of the pollutant to biological transformation to less toxic products, (2) the bioavailability of the contaminant to microorganisms, and (3) the opportunity for bioprocess optimization.

In summary, the potential and success of microorganisms in the remediation of spills of petroleum and petroleum-related products have been widely acknowledged. With advances in biotechnology, bioremediation has become a rapidly growing area, and has been commercially applied for the treatment of hazardous wastes and contaminated sites.

5 References

Aislabie, J.M., McLeod, M., and Fraser, R. 1998. "Potential for Biodegradation of Hydrocarbons in Soil from the Ross Dependency, Antarctica." *Appl. Microbiol. Biotechnol.*, 49: 210–214.

Aislabie, J.M., Balks, M.R., Foght, J.M., and Waterhouse, E.J. 2004. "Hydrocarbon Spills on Antarctic Soils: Effects and Management." *Environ. Sci. Technol.*, 38(5): 1265–74.

Al-Darbi, M.M., Saeed, N.O., Islam, M.R., and Lee, K. 2005. "Biodegradation of Natural Oils in Seawater." *Energy Sources*, 27:19–34.

Alkorta, I., and Garbisu, C. 2001. "Phytoremediation of organic contaminants in soils." *Bioresour. Technol.*, 79: 273–276.

Amann, R., Fuchs, B.M., and Behrens, S. 2001. "The Identification of Microorganisms by Fluorescence In Situ Hybridization." *Curr. Opin. Biotechnol.*, 12: 231–236.

Antizar-Ladislao, B., Spanova, K., Beck, A.J., and Russell, N.J. 2008. "Microbial Community Structure Changes during Bioremediation of PAHs in an Aged Coal Tar-Contaminated Soil by In-Vessel Composting." *Int. Biodeterior. Biodegrad.*, 61: 357–364.

Bej, A.K., Saul, D., and Aislabie, J. 2000. "Cold-tolerant Alkane-degrading *Rhodococcus* Species from Antarctica." *Polar Biol.*, 23: 100–105.

Blackburn, J.W., and Hafker, W.R. 1993. "The Impact of Biochemistry, Bioavailability and Bioactivity on the Selection of Bioremediation Techniques." *Trends Biotechnol.*, 11: 328–333.

Bollag, W.B., and Bollag, J.M. 1992. "Biodegradation." *Encycl. Microbiol.*, 1: 269–280.

Bouwer, E.J., and Zehnder, A.J.B. 1993. "Bioremediation of Organic Compounds: Microbial Metabolism to Work." *Trends Biotechnol.*, 11: 360–367.

Braddock, J.F., Ruth, M.L., Walworth, J.L., and McCarthy, K.A. 1997. "Enhancement and Inhibition of Microbial Activity in Hydrocarbon-Contaminated Arctic Soils: Implications for Utrientamended Bioremediation." *Environ Sci. Technol.*, 31: 2078–2084.

Bradley, P.M., and Chapelle, F.H. 1995. "Rapid Toluene Mineralization by Aquifer Microorganisms at Adak, Alaska: Implications for Intrinsic Bioremediation in Cold Environments." *Environ Sci. Technol.*, 29: 2778–2781.

Bragg, J.R., Prince, R.C., Harner, E.J., and Atlas, R.M. 1994. "Effectiveness of Bioremediation for the Exxon Valdez Oil Spill." *Nature*, 368: 413–418.

Butler, C.S., and Mason, J.R. 1997. "Structure, Function Analysis of the Bacterial Aromatic Ring Hydroxylating Dioxygenases." *Adv. Microb. Physiol.*, 38: 47–84.

Chakrabarty, A.M. 1982. "Genetic Mechanisms in the Dissimilation of Chlorinated Compounds." In *Biodegradation and Detoxification of Environmental Pollutants*. A.M. Chakrabarty (Editor). CRC Press, Boca Raton, Florida. pp. 127–139.

Chakrabarty, A.M. 1986. "Genetic Engineering and Problems of Environmental Pollution." *Biotechnology*, 8: 515–530.

Chaudhary, G.R., and Chapalamadugu, S. 1991. "Biodegradation of Halogenated Organic Compounds." *Microbiol, Rev.*, 55: 59–78.

Chen, W., Bruhlmann, F., Richnis, R.D., and Mulchandani, A. 1999. "Engineering of Improved Microbes and Enzymes for Bioremediation." *Curr. Opin. Biotechnol.*, 10: 137–141.

Chen, Y.D., Barker, J.F., and Gui, L. 2008. "A strategy for aromatic hydrocarbon bioremediation under anaerobic condition and the impacts of ethanol: a microcosm study." *J. Contam. Hydrol.*, 96: 17–31.

Chugunov, V.A., Ermolenko, Z.M., Martovetskaya, I.I., Mironava, R.I., Zhirkova, N.A., Kholodenko, V.P., and Urakov, N.N. 2000. "Development and Application Of A Liquid Preparation With Oil-Oxidzing Bacteria." *Appl. Biochem. Microbiol.*, 36: 577–581.

Delille, D., Bassères, A., and Dessommess, A. 1998. "Effectiveness of Bioremediation for Oil-Polluted Antarctic Seawater." *Polar Biol.*, 19: 237–241.

Delille, D., and Delille, B. 2000. "Field Observations on the Variability of Crude Oil Impact in Indigenous Hydrocarbon-Degrading Bacteria from Sub-Antarctic Intertidal Sediments." *Mar. Environ. Res.*, 49: 403–417.

Delille, D., Pelletier, E., Rodriguez-Blanco, A. and Ghiglione, J. 2009. "Effects of Nutrient and Temperature on Degradation of Petroleum Hydrocarbons in Sub-Antarctic Coastal Seawater." *Polar Biol.*, 32: 1521–1528.

Deo, P.G., Karanth, N.G., and Karanth, N.G.K. 1994. "Biodegradation of Hexachlorocyclohexane Isomers in Soil and Food Environment." *Crit. Rev. Microbiol.*, 20: 57–78.

Desai, J.D., and Banat, I.M. 1997. "Microbial Production of Surfactants and Their Commercial Potential." *Microbiol. Mol. Biol. Rev.*, 61: 47–64.

Dickel, O., Haug, W., and Knackmus, H-J. 1993. "Biodegradation of Nitrobenzene by a Sequential Anaerobic–Aerobic Process." *Biodegradation*, 4: 187–194.

Dua, M., Singh, A., Sethunathan, N., and Johri, A.K. 2002. "Biotechnology and Bioremediation: Successes and Limitations." *Appl. Microbiol. Biotechnol.*, 59: 143–152.

Ellis, B.M.L. 2000. "Environmental Biotechnology Informatics." *Curr. Opin. Biotechnol.*, 11: 232–235.

Engasser, K.H., Auling, G., Busse, J., and Knackmus, H-J. 1990. "3-Fluorobenzoate Enriched Bacterial Strain FLB 300 Degrades Benzoate and all Three Isomeric Monofluoro-Benzoates." *Arch. Microbiol.*, 153: 193–199.

Erute Magdalene, O., Ufuoma, A., and Gloria, O. 2009. "Screening of Four Common Nigerian Weeds for Use in Phytoremediation of Soil Contaminated with Spent Lubricating Oil." *African Journal of Plant Science*, Vol. 3 (5), pp. 102–106.

Escalante-Espinosa, E., Gallegos-Martinez, M.E., Favela-Torres, E., and Gutierrez-Rojas, M. 2005. "Improvement of the Hydrocarbon Phytoremediation Rate by Cyperus Laxus Lam. Inoculated with a Microbial Consortium in a Model System." *Chemosphere*, 59: 405–413.

Faison, B.D. 2001. "Hazardous Waste Treatment." *SIM News*, 51: 193–208.

Fewson, C.A. 1988. "Biodegradation of Xenobiotic and Other Persistent Compounds: The Causes of Recalcitrance." *Trends Biotechnol.*, 6: 148–153.

Gerhart, K.E., Huang, X., Glick, B.R., and Greenberg, B.M. 2009. "Phytoremediation and rhizoremediation of organic soil contaminants: potential and challenges." *Plant Sci.*, 176: 20–30.

Grishchenkov, V.G., Townsend, R.T., McDonald, T.J., Autenrieth, R.L., Bonner, J.S., and Boronin, A.M. 2000. "Degradation of Petroleum Hydrocarbons by Facultative Anaerobic Bacteria under Aerobic and Anaerobic Conditions." *Process Biochem.*, 35: 889–896.

Harayama, S. 1997. "Polycyclic Aromatic Hydrocarbon Bioremediation Design." *Curr. Opin. Biotechnol.*, 8: 268–273.

Hauschild, J.E., Masai, E., Sugiyama, K., Hatta, T., Kimbara, K., Fukuda, M., and Yano, K. 1996. "Identification of an Alternative 2,3-Dihydroxybiphenyl 1,2-Dioxygenase In *Rhodococcus sp.* Strain RHA1 and Cloning of the Gene." *Appl. Environ. Microbiol.*, 62: 2940–2946.

Iwamoto, T., and Nasu, M. 2001. "Current Bioremediation Practice and Perspective." *J. Biosci. Bioeng.*, 92: 1–8.

Janssen, D.B., Oppentocht, J.E., and Poelarends, G. 2001. "Microbial Dehalogenation." *Curr. Opin. Biotechnol.*, 12: 254–258.

Johan, E.T., Vlieg, V.H., and Janssen, D.B. 2001. "Formation and Detoxification of Reactive Intermediates in the Metabolites of Chlorinated Ethenes." *J. Biotechnol.*, 85: 81–102.

Johri, A.K., Dua, M., Tuteja, D., Saxena, R., Saxena, D.M., and Lal, R. 1996. "Genetic Manipulations of Microorganisms for the Degradation of Hexachlorocyclohexane." *FEMS Microbiol. Rev.*, 19: 69–84.

Johri, A.K., Dua, M., Singh, A., Sethunathan, N., and Legge, R.L. 1999. "Characterization and Regulation of Catabolic Genes." *Crit. Rev. Microbiol.*, 25: 245–273.

Korade, D.L., and Fulekar, M.H. 2009. "Development and Evaluation of Mycorrhiza for Rhizosphere Bioremediation." *Journal of Applied Biosciences*, 17: 922–929.

Kumar, S., Mukerjim, K.G., and Lal, R. 1996. "Molecular Aspects of Pesticide Degradation by Microorganisms." *Crit. Rev. Microbiol.*, 22: 1–26.

Lal, R., Lal, S., and Shivaji, S. 1986. "Use of Microbes for Detoxification of Pesticides." *Crit. Rev. Microbiol.*, 3: 1–14.

Lin, Q., and Mendelessohn, I.A. 2009. "Potential of Restoration and Phytoremediation with *Juncus roemerianus* for Diesel-Contaminated Coastal Wetlands." *Ecol. Eng.*, 35: 85–91.

Liu, S., and Suflita, J.M. 1993. "Ecology and Evolution of Microbial Populations for Bioremediation." *Trends Biotechnol.*, 11: 344–352.

Macek, T., Mackova, M., and Kas, J. 2000. "Exploitation of Plants for the Removal of Organics in Environmental Remediation." *Biotechnol. Adv.*, 18: 23–34.

Margesin, R., and Schinner, F. 1997a. "Efficiency of Indigenous and Inoculated Cold-Adapted Soil Microorganisms for Biodegradation of Diesel Oil in Alpine Soils." *Appl. Environ. Microbiol.*, 63: 2660–2664.

Margesin, R., and Schinner, F. 1997b. "Bioremediation of Diesel-Oil-Contaminated Alpine Soils at Low Temperatures." *Appl. Microbiol. Biotechnol.*, 47: 462–468.

Margesin, R. 2000. "Potential of Cold-Adapted Microorganisms for Bioremediation of Oil-Polluted Alpine Soils." *Int. Biodeterior. Biodegrad.*, 46: 3–10.

Marijke, M.A., and van Vlerken, F. 1998. "Chances for Biological Techniques in Sediment Remediation." *Water Sci. Tech.*, 37: 345–353.

Masai, E., Yamada, A., Healy, J.M., Hatta, T., Kimbara, K., Fukuda, M., and Yano, K. 1995. "Characterization of Biphenyl Catabolic Genes of Gram Positive Polychlorinated Biphenyl Degrader *Rhodococcus sp.* RH1." *Appl. Environ. Microbiol.*, 61: 2079–2085.

Megharaj, M., Singleton, I., McClture, N.C., and Naidu, R. 2000. "Influence of Petroleum Hydrocarbon Contamination on Microalgae and Microbial Activities in a Long-Term Contaminated Soil." *Arch. Environ. Contam. Toxicol.*, 38: 439–445.

Merkl, N., Schultze-Kraft, R., and Infante, C. 2005. "Phytoremediation in the tropics-influence of heavy crude oil on root morphological characteristics of graminoids." *Environ. Pollut.*, 138: 86–91.

Mishra, V., Lal, R., and Srinivasan, S. 2001. "Enzymes and Operons Mediating Xenobiotic Degradation in Bacteria." *Crit. Rev. Microbiol.*, 27: 133–166.

Mohammed, D., Ramsubhag, A.S., and Beckles, D.M. 2007. "An Assessment of the Biodegradation of Petroleum Hydrocarbons in Contaminated Soil Using Non-indigenous, Commercial Microbes." *Water Air Soil Pollut.*, 182: 349–356.

Mohn, W.W., and Stewart, G.R. 2000. "Limiting Factors for Hydrocarbon Biodegradation at Low Temperature in Arctic Soils." *Soil Biol. Biochem.*, 32: 1161–1172.

Pelletier, E., Delille, D., and Delille, B. 2004. "Crude Oil Bioremediation in Sub-Antarctic Intertidal Sediments: Chemistry and Toxicity of Oil Residues." *Mar. Environ. Res.*, 57: 311–327.

Rike, A.G., Haugen, K.B., Børresen, M., Engene, B., and Kolstad, P. 2003. "In Situ Biodegradation of Petroleum Hydrocarbons in Frozen Arctic Soils." *Cold Reg. Sci. Technol.*, 37: 97–120.

Salt, D.E., Smith, R.D., and Raskin, I. 1998. "Phytoremediation." *Annu. Rev. Plant Physiol. Plant Mol. Biol.*, 49: 643–668.

Sangodkar, U.M.X., Aldrich, T.L., Haugland, R.A., Johnson, J., Rothmel, R.K., Chapman, P.J., and Chakrabarty, A.M. 1989. "Molecular Basis of Biodegradation of Chloroaromatic Compounds." *Acta Biotechnol.*, 9: 301–316.

Sanscartier, D., Zeeb, B., Koch, I., and Reimer, K. 2009. "Bioremediation of Diesel-Contaminated Soil by Heated and Humidified Biopile System in Cold Climates." *Cold Reg. Sci. Technol.*, 55: 167–173.

Sayler, G.S., and Ripp, S. 2000. "Field Application of Genetically Engineered Microorganisms for Bioremediation Processes." *Curr. Opin. Biotechnol.*, 11: 286–289.

Sekelsky, A.M., and Shreve, G.S. 1999. "Kinetic Model of Biosurfactant-enhanced Hexadecane Biodegradation by *Pseudomonas aeruginosa*." *Biotechnol. Bioeng.*, 63: 401–409.

Singh, A., Mullin, B., and Ward, O.P. 2001. "Reactor-Based Process for the Biological Treatment of Petroleum Wastes." *Proceedings*. Middle East Petrotech 2001 Conference. Bahrain. pp. 1–13.

Siron, R., Pelletier, E., and Brochu, C. 1995. "Environmental Factors Influencing the Biodegradation of Petroleum Hydrocarbons in Cold Seawater." *Arch. Environ. Contam. Toxicol.*, 28: 406–416.

Speight, J.G. 1996. *Environmental Technology Handbook*. Taylor & Francis, Washington, DC.

Speight, J.G., and Lee, S. 2000. *Environmental Technology Handbook*. 2nd Edition. Taylor & Francis, New York.

Speight, J.G. 2005. *Environmental Analysis and Technology for the Refining Industry*. John Wiley & Sons Inc., Hoboken, New Jersey.

Speight, J.G. 2007. *The Chemistry and Technology of Petroleum*. 5th Edition. CRC Press, Taylor & Francis Group, Boca Raton, Florida.

Sullivan, E.R. 1998. "Molecular Genetics of Biosurfactant Production." *Curr. Opin. Biotechnol.*, 9: 263–269.

Timmis, K.N., and Piper, D.H. 1999. "Bacteria Designed for Bioremediation." *Trends Biotechnol.*, 17: 201–204.

Van der Meer, J.R. 1997. "Evolution of Novel Metabolic Pathways for the Degradation of Chloroaromatic Compounds." *Antonie Van Leeuwenhoek*, 71: 159–178.

Van der Meer, J.R., de Vos, W.M., Harayama, S., and Zehnder, A.Z.B. 1992. "Molecular Mechanisms Of Genetic Adaptation To Xenobiotic Compounds." *Microbiol. Rev.*, 56: 677–694.

Vieira, P.A., Faria, S.R., Vieira, B., De Franca, F.P., and Cardoso, V.L. 2009. "Statistical Analysis and Optimization of Nitrogen, Phosphorus, and Inoculum Concentrations for the Biodegradation of Petroleum Hydrocarbons by Response Surface Methodology." *J. Microbiol. Biotechnol.*, 25: 427–438.

Watanabe, K. 2001. "Microorganisms relevant to bioremediation." *Curr. Opin. Biotechnol.*, 12: 237–241.

Whyte, L.G., Hawari, J., Zhou, E., Bourbonnière, L., Inniss, W.E., and Greer, C.W. 1998. "Biodegradation of Variable-Chain-Length Alkanes at Low Temperatures by a Psychrotrophic *Rhodococcus* sp." *Appl. Environ. Microbiol.*, 64: 2578–2584.

Wenzel, W. 2009. "Rhizosphere Processes and Management in Plant-Assisted Bioremediation (Phytoremediation) of Soils." *Plant and Soil*, 321 (1–2): 385–408.

Whyte, L.G., Bourbonnière, L., Bellerose, C., and Greer, C.W. 1999. "Bioremediation Assessment of Hydrocarbon-Contaminated Soils from the High Arctic." *Bioremediation J.*, 3: 69–79.

Yakimov, M.M., Giuliano, L., Bruni, V., Scarfi, S., and Golyshin, P.N. 1999. "Characterization of Antarctic Hydrocarbon-Degrading Bacteria Capable of Producing Bioemulsifiers." *Microbiologica.*, 22: 249–256.

Yumoto, I., Nakamura, A., Iwata, H., Kojima, K., Kusumuto, K., Nodasaka, Y., and Matsuyama, H. 2002. "*Dietzia psychralcaliphila sp. nov.*, A Novel Facultatively Psychrophilic Alkaliphile that Grows On Hydrocarbons." *Int. J. Syst. Evol. Microbiol.*, 52: 85–90.

Zhu, X., Venosa, A.D., and Suidan, M.T. 2004. "Literature Review on the Use of Commercial Bioremediation Agents for Cleanup of Oil-Contaminated Environments." Report No. EPA/600/R-04/075. National Risk Management Research Laboratory, United States Environmental Protection Agency, Cincinnati, Ohio. July.

Glossary

Abiotic: not associated with living organisms; synonymous with abiological.

Abiotic transformation: the process in which a substance in the environment is modified by non-biological mechanisms.

Absorption: the penetration of atoms, ions, or molecules into the bulk mass of a substance.

Acidophiles: metabolically active in highly acidic environments, and often have a high heavy metal resistance.

Acyclic: a compound with straight or branched carbon-carbon linkages but without cyclic (ring) structures.

Adhesion: the degree to which oil will coat a surface, expressed as the mass of oil adhering per unit area. A test has been developed for a standard surface that gives a semi-quantitative measure of this property.

Adsorption: the retention of atoms, ions, or molecules on to the surface of another substance.

Aerobe: an organism that needs oxygen for respiration, and hence, for growth.

Aerobic: in the presence of, or requiring, oxygen; an environment or process that sustains biological life and growth, or occurs only when free (molecular) oxygen is present.

Aerobic bacteria: any bacteria requiring free oxygen for growth and cell division.

Aerobic conditions: conditions for growth or metabolism in which the organism is sufficiently supplied with oxygen.

Aerobic respiration: the process whereby microorganisms use oxygen as an electron acceptor.

Aliphatic compound: Any organic compound of hydrogen and carbon characterized by a linear-chain or branched-chain of carbon atoms; three subgroups of such compounds are alkanes, alkenes, and alkynes.

Alkalitolerants: organisms that are able to grow or survive at pH values above 9, but their optimum growth rate is around neutrality or less.

Alkaliphiles: organisms that have their optimum growth rate at least 2 pH units above neutrality.

Alkane (paraffin): a group of *hydrocarbons* composed of only carbon and hydrogen with no double bonds or aromaticity. They are said to be "saturated" with hydrogen. They may be straight-chain (normal), branched, or cyclic. The smallest alkane is methane (CH_4), the next, ethane (CH_3CH_3), then propane ($CH_3CH_2CH_3$), and so on.

Alkene (olefin): an unsaturated *hydrocarbon*, containing only hydrogen and carbon with one or more double bonds, but having no aromaticity. *Alkenes* are not typically found in crude oils, but can occur as a result of heating.

Alcohol: compounds in which a hydroxy group (–OH) is attached to a saturated carbon atom (e.g., ethyl alcohol, C_2H_5OH).

Aldehydes: compounds in which a carbonyl group is bonded to one hydrogen atom and to one alkyl group [RC(=O)H,].

Aliphatic compounds: a broad category of hydrocarbon compounds distinguished by a straight, or branched, open chain arrangement of the constituent carbon atoms, excluding aromatic compounds; the carbon-carbon bonds may be either single or multiple bonds; alkanes, alkenes, and alkynes are aliphatic hydrocarbons.

Alkanes: the homologous group of linear (acyclic) aliphatic hydrocarbons having the general formula C_nH_{2n+2}; alkanes can be straight chains (linear), branched chains, or ring structures; often referred to as paraffins.

Alkenes: acyclic branched or unbranched hydrocarbons having one carbon-carbon double bond (–C=C–) and the general formula C_nH_{2n}; often referred to as olefins.

Alkyl groups: a *hydrocarbon* functional group (C_nH_{2n+1}) obtained by dropping one hydrogen from fully saturated compound; e.g., methyl ($-CH_3$), ethyl ($-CH_2CH_3$), propyl ($-CH_2CH_2CH_3$), or isopropyl [$(CH_3)_2CH-$].

Alkyl radicals: carbon-centered radicals derived formally by removal of one hydrogen atom from an alkane, for example, the ethyl radical (CH_3CH_2).

Alkynes: the group of acyclic branched or unbranched hydrocarbons having a carbon-carbon triple bond (–C≡C–).

Ambient: the surrounding environment and prevailing conditions.

Anaerobe: an organism that does not need free-form oxygen for growth. Many anaerobes are even sensitive to free oxygen.

Anaerobic: a biologically-mediated process or condition not requiring molecular or free oxygen; relating to a process that occurs with little or no oxygen present.

Anaerobic bacteria: any bacteria that can grow and divide in the partial or complete absence of oxygen.

Anaerobic respiration: the process whereby microorganisms use a chemical other than oxygen as an electron acceptor; common substitutes for oxygen are nitrate, sulfate, and iron.

Analyte: the component of a system to be analyzed; for example, chemical elements or ions in groundwater sample.

Anoxic: an environment without oxygen.

API Gravity: An American Petroleum Institute measure of *density* for petroleum:

API Gravity = [141.5/(specific gravity at 15.6°C) –131.5]

Fresh water has a gravity of 10° API. The scale is commercially important for ranking oil quality; heavy oils are typically <20°API; medium oils are 20 to 35° API; light oils are 35 to 45°API.

Aquifer: a water-bearing layer of soil, sand, gravel, rock, or other geologic formation that will yield usable quantities of water to a well under normal hydraulic gradients or by pumping.

Aromatic: organic cyclic compounds that contain one or more benzene rings; these can be monocyclic, bicyclic, or polycyclic hydrocarbons and their substituted derivatives. In aromatic ring structures, every ring carbon atom possesses one double bond.

Asphaltene fraction: a complex mixture of heavy organic compounds precipitated from oils and *bitumens* by natural processes or in laboratory by addition of excess *n*-pentane, or *n*-heptane; after precipitation of the *asphaltene fraction*, the remaining oil or *bitumen* consists of *saturates*, *aromatics*, and *resins*.

Assay: qualitative or (more usually) quantitative determination of the components of a material or system.

ATSDR: Agency for Toxic Substances and Disease Registry.

Attenuation: the set of human-made or natural processes that either reduce or appear to reduce the amount of a chemical compound as it migrates away or is disposed from one specific point towards another point in space or time; for example, the apparent reduction in the amount of a chemical in a ground-water plume as it migrates away from its source; degradation, dilution, dispersion, sorption, or volatilization are common processes of attenuation.

Benthic zone: the ecological region at the lowest level of a body of water such as an ocean or a lake, including the sediment surface and some sub-surface layers; organisms living in this zone (benthos or benthic organisms) generally live in close relationship with the substrate bottom; many such organisms are permanently attached to the bottom; because light does not penetrate very deep ocean water, the energy source for the benthic ecosystem is often organic matter from higher up in the water column that sinks to the depths.

Bio-augmentation: a process in which acclimated microorganisms are added to soil and groundwater to increase biological activity. Spray irrigation is typically used for shallow contaminated soils, and injection wells are used for deeper contaminated soils.

Biodegradation: the natural process whereby bacteria or other microorganisms chemically alter and break down organic molecules; the breakdown or transformation of a chemical substance or substances by microorganisms using the substance as a carbon and/or energy source.

Biological marker (biomarker): complex organic compounds composed of carbon, hydrogen, and other elements which are found in oil, *bitumen*, rocks, and sediments and which have undergone little or no change in structure from their parent organic molecules in living organisms; typically, biomarkers are isoprenoids, composed of isoprene subunits. Biomarkers include pristane, phytane, *triterpanes*, steranes, porphyrins, and other compounds.

Bioremediation: a treatment technology that uses biological activity to reduce the concentration or toxicity of contaminants: materials are added to contaminated environments to accelerate natural biodegradation.

Biota: living organisms.

Bitumen: a complex mixture of *hydrocarbonaceous constituents* of natural or pyrogenous origin, or a combination of both.

Boiling point: the temperature at which a liquid begins to boil, that is, the temperature at which the vapor pressure of a liquid is equal to the atmospheric or external pressure. The boiling point distributions of crude oils and petroleum products may be in a range from 30 to in excess of 700°C (86 to 1290°F).

Breakdown product: a compound derived by chemical, biological, or physical action on a chemical compound; the breakdown is a process which may result in a more toxic or a less toxic compound, and a more persistent or less persistent compound, than the original compound.

BTEX: The collective name given to benzene, toluene, ethylbenzene, and the xylene isomers (*p*-, *m*-, and *o*-xylene); a group of volatile organic compounds (VOCs) found in petroleum hydrocarbons such as gasoline, and other common environmental contaminants.

Carbon: element number 6 in the periodic table of elements.

Carbon preference index (CPI): the ratio of odd to even *n*-alkanes; odd/even CPI *alkanes* are equally abundant in petroleum, but not in biological material; a CPI near 1 is an indication of petroleum.

Catabolism: the breakdown of complex molecules into simpler ones through the oxidation of organic substrates to *provide* biologically available energy; ATP (adenosine triphosphate) is an example of such a molecule.

Catalysis: the process where a catalyst increases the rate of a chemical reaction without modifying the overall standard Gibbs energy change in the reaction.

Catalyst: a substance that alters the rate of a chemical reaction and may be recovered essentially unaltered in form or amount at the end of the reaction.

Cation exchange: the interchange between a cation in solution and another cation in the boundary layer between the solution and surface of negatively charged material such as clay or organic matter.

Cation exchange capacity (CEC): the sum of the exchangeable bases plus total soil acidity at a specific pH, usually 7.0 or 8.0. When acidity is expressed as salt extractable acidity, the cation exchange capacity is called the effective cation exchange capacity (ECEC), because this is considered to be the CEC of the exchanger at the native pH value. It is usually expressed in centimoles of charge per kilogram of exchanger (cmol/kg) or millimoles of charge per kilogram of exchanger.

CERCLA: Comprehensive Environmental Response, Compensation, and Liability Act. This law created a tax on the chemical and petroleum industries, and provided broad federal authority to respond directly to releases or threatened releases of hazardous substances that may endanger public health or the environment.

Check standard: an analyte with a well-characterized property of interest, e.g., concentration, density, and other properties, that is used to verify method, instrument, and operator performance during regular operation; *check standards* may be obtained from a certified supplier, may be a pure substance with properties obtained from the literature, or may be developed in-house.

Chemical bond: the forces acting among two atoms or groups of atoms that lead to the formation of an aggregate with sufficient stability to be considered as an independent molecular species.

Chemical dispersion: in relation to oil spills, this term refers to the creation of oil-in-water *emulsions* by the use of chemical dispersants made for this purpose.

Chemical induction (coupling): when one reaction accelerates another in a chemical system, there is said to be chemical induction or coupling. Coupling is caused by an intermediate or by-product of the inducing reaction that participates in a second reaction; chemical induction is often *observed* in oxidation-reduction reactions.

Chemical reaction: a process that results in the interconversion of chemical species.

Chlorinated solvent: a volatile organic compound containing chlorine; common solvents are trichloroethylene, tetrachloroethylene, and carbon tetrachloride.

Cis trans isomers: the difference in the positions of atoms (or groups of atoms) relative to a reference plane in an organic molecule; in a *cis-isomer,* the atoms are on the same side of the molecule, but are on opposite sides in the *trans*-isomer; sometimes called stereoisomers; these arrangements are common in alkenes and cycloalkanes.

Clean Water Act: the Clean Water Act establishes the basic structure for regulating discharges of pollutants into the waters of the United States.

It gives the EPA the authority to implement pollution control programs, such as setting wastewater standards for industry; also continued requirements to set water quality standards for all contaminants in surface waters, and makes it unlawful for any person to discharge any pollutant from a point source into navigable waters, unless a permit was obtained under its provisions.

Co-metabolism: the process by which compounds in petroleum may be enzymatically attacked by microorganisms without furnishing carbon for cell growth and division; a variation on biodegradation, in which microbes transform a contaminant even though the contaminant cannot serve as the primary energy source for the organisms. To degrade the contaminant, the microbes require the presence of other compounds (primary substrates) that can support their growth.

Complex modulus: a measure of the overall resistance of a material to flow under an applied stress, in units of force per unit area. It combines *viscosity* and elasticity elements to provide a measure of "stiffness," or resistance to flow. The *complex modulus* is more useful than *viscosity* for assessing the physical behavior of very non-Newtonian materials, such as *emulsions*.

Concentration: composition of a mixture characterized in terms of mass, amount, volume, or number concentration with respect to the volume of the mixture.

Conservative constituent or compound: one that does not degrade, is unreactive, and its *movement* is not retarded within a given environment (aquifer, stream, contaminant plume).

Constituent: an essential part or component of a system or group (that is, an ingredient of a chemical mixture); for example, benzene is one constituent of gasoline.

Covalent bond: a region of *relatively* high electron density between atomic nuclei that results from sharing of electrons and that *gives* rise to an attractive force and a characteristic internuclear distance; carbon-hydrogen bonds are covalent bonds.

Culture: the growth of cells or microorganisms in a controlled artificial environment.

Cycloalkanes (naphthene, cycloparaffin): A saturated, cyclic compound containing only carbon and hydrogen. One of the simplest *cycloalkanes* is cyclohexane (C_6H_{12}). Steranes and triterpanes are branched naphthenes consisting of multiple condensed five- or six-carbon rings.

Daughter product: a compound that results directly from the degradation of another chemical.

Degradation: the breakdown or transformation of a compound into byproducts and/or end products.

Dehydrohalogenation: removal of hydrogen and halide ions from an alkane, resulting in the formation of an alkene.

Denitrification: bacterial reduction of nitrate to nitrite to gaseous nitrogen or nitrous oxides under anaerobic conditions.

Density: the mass per unit volume of a substance. *Density* is temperature-dependent, generally decreasing with temperature. The density of oil relative to water, its specific gravity, governs whether a particular oil will float on water. Most fresh crude oils and fuels will float on water. Bitumen and certain residual fuel oils, however, may have densities greater than water at some temperature ranges, and may submerge in water. The density of a spilled oil will also increase with time, as components are lost due to weathering.

Detection limit (in analysis): the minimum single result that, with a stated probability, can be distinguished from a representative blank value during the laboratory analysis of substances such as water, soil, air, rock, and biota.

Dichloroelimination: removal of two chlorine atoms from an alkane compound and the formation of an alkene compound within a reducing environment.

Dihaloelimination: removal of two halide atoms from an alkane compound and the formation of an alkene compound within a reducing environment.

Diols: chemical compounds that contain two hydroxy (–OH) groups, generally assumed to be, but not necessarily, alcoholic; aliphatic diols are also called glycols.

Downgradient: in the direction of decreasing static hydraulic head.

Dispersant (chemical dispersant): a chemical that reduces the surface tension between water and a hydrophobic substance, such as oil. In the case of an oil spill, dispersants facilitate the breakup and dispersal of an oil slick throughout the water column in the form of an oil-in-water emulsion; chemical dispersants can only be used in areas where biological damage will not occur, and must be approved for use by government regulatory agencies.

Electron acceptor: the compound that receives electrons (and therefore is reduced) in the energy-producing oxidation-reduction reactions that are essential for the growth of microorganisms and bioremediation; common electron acceptors in bioremediation are oxygen, nitrate, sulfate, and iron.

Electron donor: the compound that donates electrons (and therefore is oxidized). In bioremediation, the organic contaminant often serves as an electron donor.

Electronegativity: the power of an atom to attract electrons to itself.

Elimination: a reaction where two groups, such as chlorine and hydrogen, are lost from adjacent carbon atoms, and a double bond is formed in their place.

Emulsan: a polyanionic heteropolysaccharide bioemulsifier produced by *Acinetobacter calcoaceticus* RAG-1; used to stabilize oil-in-water emulsions.

Emulsion: a stable mixture of two immiscible liquids, consisting of a continuous phase and a dispersed phase. Oil and water can form both oil-in-water and water-in-oil emulsions. The former is termed a dispersion, while *emulsion* implies the latter. Water-in-oil emulsions formed from petroleum and brine can be grouped into four stability classes: stable, a formal emulsion that will persist indefinitely; meso-stable, which gradually degrades over time due to a lack of one or more stabilizing factors; entrained water, a mechanical mixture characterized by high viscosity of the petroleum component which impedes separation of the two phases; and unstable, which are mixtures that rapidly separate into immiscible layers.

Emulsion stability: generally accompanied by a marked increase in *viscosity* and elasticity over that of the parent oil which significantly changes behavior. Coupled with the increased volume due to the introduction of brine, emulsion formation has a large effect on the choice of countermeasures employed to combat a spill.

Emulsification: the process of *emulsion* formation, typically by mechanical mixing. In the environment, *emulsions* are most often formed as a result of wave action. Chemical agents can be used to prevent the formation of *emulsions* or to "break" the *emulsions* to their component oil and water phases.

Endergonic reaction: a chemical reaction that requires energy to proceed. A chemical reaction is endergonic when the change in free energy is positive.

Engineered bioremediation: a type of remediation that increases the growth and degradative activity of microorganisms by using engineered systems that supply nutrients, electron acceptors, and/or other growth-stimulating materials.

Enhanced bioremediation: a process which involves the addition of microorganisms (e.g., fungi, bacteria, and other microbes) or nutrients (e.g. oxygen, nitrates) to the subsurface environment to accelerate the natural biodegradation process.

Enzyme: a macromolecule, mostly proteins or conjugated proteins produced by living organisms, that facilitates the degradation of a chemical compound (catalyst); in general, an enzyme catalyzes only one reaction type (reaction specificity) and operates on only one type of substrate (substrate specificity); any of a group of catalytic proteins that are produced by cells and that mediate or promote the chemical processes of life without themselves being altered or destroyed.

Epoxidation: a reaction wherein an oxygen molecule is inserted in a carbon-carbon double bond, and an epoxide is formed.

Epoxides: a subclass of epoxy compounds containing a saturated three-membered cyclic ether. See *Epoxy compounds*.

Epoxy compounds: compounds in which an oxygen atom is directly attached to two adjacent or nonadjacent carbon atoms in a carbon chain or ring system; thus, cyclic ethers.

Equipment blank: a sample of analyte-free media that has been used to rinse the sampling equipment. It is collected after completion of decontamination and prior to sampling. This blank is useful in documenting and controlling the preparation of the sampling and laboratory equipment.

Ex-situ bioremediation: a process which involves removing the contaminated soil or water to another location before treatment.

Facultative anaerobes: microorganisms that use (and prefer) oxygen when it is available, but can also use alternate electron acceptors, such as nitrate, under anaerobic conditions when necessary.

Fermentation: the process whereby microorganisms use an organic compound as both electron donor and electron acceptor, converting the compound to fermentation products such as organic acids, alcohols, hydrogen, and carbon dioxide; microbial metabolism in which a particular compound is used both as an electron donor and an electron acceptor, resulting in the production of oxidized and reduced daughter products.

Field capacity or in situ (field water capacity): the water content, on a mass or volume basis, remaining in soil two or three days after having been wetted with water and after free drainage is negligible.

Fingerprint: a chromatographic signature of relative intensities used in oil-oil or oil-source rock correlations. Mass chromatograms of *steranes* or *terpanes* are examples of *fingerprints* that can be used for qualitative or quantitative comparison of oils.

Flash point: the temperature at which the vapor over a liquid will ignite when exposed to an ignition source. A liquid is considered to be flammable if its *flash point* is less than 60°C. *Flash point* is an extremely important factor in relation to the safety of spill cleanup operations. Gasoline and other light fuels can ignite under most ambient conditions, and therefore are a serious hazard when spilled. Many freshly spilled crude oils also have low *flash points* until the lighter components have evaporated or dispersed.

Fraction: one of the portions of a chemical mixture separated by chemical or physical means from the remainder.

Functional group: an atom or a group of atoms attached to the base structure of a compound that has similar chemical properties irrespective of the compound to which it is a part; a means of defining the characteristic physical and chemical properties of families of organic compounds.

Gas chromatography (GC): a separation technique involving passage of a gaseous moving phase through a column containing a fixed liquid phase; it is used principally as a quantitative analytical technique for compounds that are volatile or can be converted to volatile forms.

Gaseous nutrient injection: a process in which nutrients are fed to contaminated groundwater and soil via wells to encourage and feed naturally occurring microorganisms: the most common added gas is air. In the presence of sufficient oxygen, microorganisms convert many organic contaminants to carbon dioxide, water, and microbial cell mass. In the absence of oxygen, organic contaminants are metabolized to methane, limited amounts of carbon dioxide, and trace amounts of hydrogen gas. Another gas that is added is methane. It enhances degradation by co-metabolism in which as bacteria consume the methane; they produce enzymes that react with the organic contaminant, and degrade it to harmless minerals.

GC-MS: gas chromatography-mass spectrometry.

GC-TPH: GC detectable total petroleum hydrocarbons, that is, the sum of all GC-resolved and unresolved hydrocarbons. The resolvable hydrocarbons appear as peaks, and the unresolvable hydrocarbons appear as the area between the lower baseline and the curve defining the base of resolvable peaks.

Gravimetric analysis: a technique of quantitative analytical chemistry in which a desired constituent is efficiently recovered and weighed.

Half-life: the time required to reduce the concentration of a chemical to 50 percent of its initial concentration; units are typically in hours or days.

Halide: an element from the halogen group, which include fluorine, chlorine, bromine, iodine, and astatine.

Halogen: Group 17 in the periodic table of the elements; these elements are the reactive nonmetals, and are electronegative.

Henry's law: the relation between the partial pressure of a compound and the equilibrium concentration in the liquid through a proportionality constant known as the Henry's law constant.

Henry's law constant: the concentration ratio between a compound in air (or vapor) and the concentration of the compound in water under equilibrium conditions.

Heterogeneous: varying in structure or composition at different locations in space.

Heterotroph: an organism that cannot synthesize its own food, and is dependent on complex organic substances for nutrition.

Heterotrophic bacteria: bacteria that utilize organic carbon as a source of energy; organisms that derive carbon from organic matter for cell growth.

Hopane: a pentacyclic *hydrocarbon* of the *triterpane* group believed to be derived primarily from bacteriohopanoids in bacterial membranes.

Homogeneous: having uniform structure or composition at all locations in space.

Hydration: the addition of a water molecule to a compound within an aerobic degradation pathway.

Hydrocarbon: one of a very large and diverse group of chemical compounds composed only of carbon and hydrogen; the largest source of hydrocarbons is petroleum crude oil; the principal constituents of crude oils and refined petroleum products.

Hydrogen bond: a form of association between an electronegative atom and a hydrogen atom attached to a second, relatively electronegative atom; best considered as an electrostatic interaction, heightened by the small size of hydrogen, which permits close proximity of the interacting dipoles or charges.

Hydrogenation: a process whereby an enzyme in certain microorganisms catalyzes the hydrolysis or reduction of a substrate by molecular hydrogen.

Hydrogenolysis: a reductive reaction in which a carbon-halogen bond is broken, and hydrogen replaces the halogen substituent.

Hydrolysis: a chemical transformation process in which a chemical reacts with water. In the process, a new carbon-oxygen bond is formed with oxygen derived from the water molecule, and a bond is cleaved within the chemical between carbon and some functional group.

Hydroxylation: addition of a hydroxyl group to a chlorinated aliphatic hydrocarbon.

Infiltration rate: the time required for water at a given depth to soak into the ground.

Inhibition: the decrease in rate of reaction brought about by the addition of a substance (inhibitor), by virtue of its effect on the concentration of a reactant, catalyst, or reaction intermediate.

Inoculum: a small amount of material (either liquid or solid) containing bacteria removed from a culture in order to start a new culture.

Inorganic: pertaining to, or composed of, chemical compounds that are not organic, that is, contain no carbon-hydrogen bonds; examples include chemicals with no carbon, and those with carbon in non-hydrogen-linked forms.

In situ: in its original place; unmoved; unexcavated; remaining in the subsurface.

In-situ bioremediation: a process which treats the contaminated water or soil where it was found.

Interfacial tension: the net energy per unit area at the interface of two substances, such as oil and water or oil and air. The air/liquid interfacial tension is often referred to as surface tension. The SI units for *interfacial tension* are milli-Newtons per meter (mN/m). The higher the *interfacial tension*, the less attractive the two surfaces are to each other, and the more size of the interface will be minimized. Low surface tensions can drive the spreading of one fluid on another. The surface tension of an oil, together with its viscosity, affects the rate at which spilled oil will spread over a water surface or into the ground.

Internal Standard (IS): a pure analyte added to a sample extract in a known amount, which is used to measure the relative responses of other analytes and surrogates that are components of the same solution. The *internal standard* must be an analyte that is not a sample component.

Intrinsic bioremediation: a type of bioremediation that manages the innate capabilities of naturally occurring microbes to degrade contaminants without taking any engineering steps to enhance the process.

IUPAC: International Union of Pure and Applied Chemistry.

Lag phase: the growth interval (adaption phase) between microbial inoculation and the start of the exponential growth phase, during which there is little or no microbial growth.

Loading rate: the amount of material that can be absorbed per volume of soil.

LTU: Land Treatment Unit; a physically delimited area where contaminated land is treated to remove/minimize contaminants, and where parameters such as moisture, pH, salinity, temperature and nutrient content can be controlled.

Measurement: a description of a property of a system by means of a set of specified rules, that maps the property on to a scale of specified values, by direct or mathematical comparison with specified references.

Metabolic by-product: a product of the reaction between an electron donor and an electron acceptor; metabolic by-products include volatile fatty acids, daughter products of chlorinated aliphatic hydrocarbons, methane, and chloride.

Metabolism: the physical and chemical processes by which foodstuffs are synthesized into complex elements, complex substances are transformed into simple ones, and energy is made available for use by an organism; thus, all biochemical reactions of a cell or tissue, both synthetic and degradative, are included; the sum of all of the enzyme-catalyzed reactions in living cells that transform organic molecules into simpler compounds used in biosynthesis of cellular components or in extraction of energy used in cellular processes.

Metabolize: a product of metabolism.

Methanogens: strictly anaerobic archaebacteria, able to use only a very limited spectrum of substrates (for example, molecular hydrogen, formate, methanol, methylamine, carbon monoxide, or acetate) as electron donors for the reduction of carbon dioxide to methane.

Methanogenic: the formation of methane by certain anaerobic bacteria (methanogens) during the process of anaerobic fermentation.

Microcosm: a diminutive, representative system analogous to a larger system in composition, development, or configuration.

Microorganism (micro-organism): an organism of microscopic size that is capable of growth and reproduction through biodegradation of food sources, which can include hazardous contaminants; microscopic organisms, including bacteria, yeasts, filamentous fungi, algae, and protozoa; a living organism too small to be seen with the naked eye; includes bacteria, fungi, protozoans, microscopic algae, and viruses.

Microbe: the shortened term for microorganism.

Mineralization: the biological process of complete breakdown of organic compounds, whereby organic materials are converted to inorganic products (e.g., the conversion of hydrocarbons to carbon dioxide and water); the release of inorganic chemicals from organic matter in the process of aerobic or anaerobic decay.

Monoaromatic: aromatic hydrocarbons containing a single benzene ring.

MTBE (Methyl Tertiary Butyl Ether): a fuel additive that has been used in the United States since 1979. Its use began as a replacement for lead in gasoline because of health hazards associated with lead. MTBE has distinctive physical properties that result in it being highly soluble, persistent in the environment, and able to migrate through the ground. Environmental regulations have required the monitoring and cleanup of MTBE at petroleum contaminated sites since February 1990; the program continues to monitor studies focusing on the potential health effects of MTBE and other fuel additives.

NCP: National Contingency Plan, also called the National Oil and Hazardous Substances Pollution Contingency Plan; provides a comprehensive system of accident reporting, spill containment, and cleanup, and established response headquarters (National Response Team and Regional Response Teams).

Nitrate enhancement: a process in which a solution of nitrate is sometimes added to groundwater to enhance anaerobic biodegradation.

Nucleophile: a chemical reagent that reacts by forming covalent bonds with electronegative atoms and compounds.

Nutrients: major elements (for example, nitrogen and phosphorus) and trace elements (including sulfur, potassium, calcium, and magnesium) that are essential for the growth of organisms.

Octanol-water partition coefficient (K_{ow}): the equilibrium ratio of a chemical's concentration in octanol (an alcoholic compound) to its concentration in the aqueous phase of a two-phase octanol-water system, typically expressed in log units (log K_{ow}); K_{ow} provides an indication of a chemical's solubility in fats (lipophilicity), its tendency to bioconcentrate in aquatic organisms, or sorb to soil or sediment.

Oleophilic: oil seeking or oil loving (e.g., nutrients that stick to or dissolve in oil).

Order of reaction: a chemical rate process occurring in systems for which concentration changes (and hence the rate of reaction) are not themselves measurable, provided it is possible to measure a chemical flux.

Organic: chemical compounds based on carbon that also contain hydrogen, with or without oxygen, nitrogen, and other elements.

Organic carbon (soil) partition coefficient (K_{oc}): the proportion of a chemical sorbed to the solid phase, at equilibrium in a two-phase, water/soil or water/sediment system expressed on an organic carbon basis. Chemicals with higher Koc values are more strongly sorbed to organic carbon and, therefore, tend to be less mobile in the environment.

Organic liquid nutrient injection: an enhanced bioremediation process in which an organic liquid, which can be naturally degraded and fermented in the subsurface, is injected to result in the generation of hydrogen. The most commonly added for enhanced anaerobic bioremediation include lactate, molasses, hydrogen release compounds (HRCs®), and vegetable oils.

Osmotic potential: expressed as a negative value (or zero), indicates the ability of the soil to dissolve salts and organic molecules; the reduction of soil water osmotic potential is caused by the presence of dissolved solutes.

OPA: Oil Pollution Act of 1990; an act that addresses oil pollution and establishes liability for the discharge and substantial threat of a discharge of oil to navigable waters and shorelines of the United States.

Oven dry: the weight of a soil after all water has been removed by heating in an oven at a specified temperature (usually in excess of 100°C, 212°F) for water; temperatures will vary if other solvents have been used.

Oxidation: the transfer of electrons away from a compound, such as an organic contaminant; the coupling of oxidation to reduction (see below) usually supplies energy that microorganisms use for growth and reproduction. Often (but not always), oxidation results in the addition of an oxygen atom and/or the loss of a hydrogen atom.

Oxygen enhancement with hydrogen peroxide: an alternative process to pumping oxygen gas into groundwater, involves injecting a dilute solution of hydrogen peroxide. Its chemical formula is H_2O_2, and it easily releases the extra oxygen atom to form water and free oxygen. This circulates through the contaminated groundwater zone to enhance the rate of aerobic biodegradation of organic contaminants by naturally occurring microbes. A solid peroxide product [e.g., oxygen releasing compound (ORC®)] can also be used to increase the rate of biodegradation.

PAHs: polycyclic aromatic hydrocarbons. Alkylated *PAHs* are *alkyl group* derivatives of the parent *PAHs*. The five target alkylated *PAHs* referred to in this report are the alkylated naphthalene, phenanthrene, dibenzothiophene, fluorene, and chrysene series.

Pathogen: an organism that causes disease (e.g., some bacteria or viruses).

Permeability: the capability of the soil to allow water or air movement through it. The quality of the soil that enables water to move downward through the profile, measured as the number of inches per hour that water moves downward through the saturated soil.

Phytodegradation: the process in which some plant species can metabolize VOC contaminants. The resulting metabolic products include trichloroethanol, trichloroacetic acid, and dichloracetic acid. Mineralization products are probably incorporated into insoluble products, such as components of plant cell walls.

Phytovolatilization: the process in which VOCs are taken up by plants and discharged into the atmosphere during transpiration.

Polar compound: an organic compound with distinct regions of positive and negative charge. *Polar compounds* include alcohols, such as sterols, and some *aromatics*, such as monoaromatic-steroids. Because of their polarity, these compounds are more soluble in polar solvents, including water, compared to non-polar compounds of similar molecular structure.

Pour point: the lowest temperature at which an oil will appear to flow under ambient pressure over a period of five seconds. The *pour point* of crude oils generally varies from –60°C to 30°C. Lighter oils with low *viscosities* generally have lower *pour points*.

Primary substrates: The electron donor and electron acceptor that are essential to ensure the growth of microorganisms. These compounds can be viewed as analogous to the food and oxygen that are required for human growth and reproduction.

Propagule: any part of a plant (e.g., bud) that facilitates dispersal of the species, and from which a new plant may form.

Rate: a derived quantity in which time is a denominator quantity, so that the progress of a reaction is measured with time.

Rate constant, k: see *Order of reaction*.

Rate-controlling step (rate-limiting step, rate-determining step): the elementary reaction having the largest control factor exerts the strongest influence on the rate; a step having a control factor much larger than any other step is said to be rate-controlling.

Recalcitrant: unreactive, non-degradable, refractory.

Redox (reduction-oxidation reactions): oxidation and reduction occur simultaneously; in general, the oxidizing agent gains electrons in the process (and is reduced), while the reducing agent donates electrons (and is oxidized).

Reduction: the transfer of electrons to a compound, such as oxygen, that occurs when another compound is oxidized.

Reductive dehalogenation: a variation on biodegradation, in which microbially catalyzed reactions cause the replacement of a halogen atom on an organic compound with a hydrogen atom. The reactions result in the net addition of two electrons to the organic compound.

Resins: the name given to a large group of *polar compounds* in oil. These include hetero-substituted *aromatics*, acids, ketones, alcohols, and monoaromatic steroids. Because of their polarity, these compounds are more soluble in *polar* solvents, including water, than the non-polar compounds, such as *waxes* and *aromatics*, of similar molecular weight. They are largely responsible for oil *adhesion*.

Respiration: the process of coupling oxidation of organic compounds with the reduction of inorganic compounds, such as oxygen, nitrate, iron (III), manganese (IV), and sulfate.

Rhizodegradation: the process whereby plants modify the environment of the root zone soil by releasing root exudates and secondary plant metabolites. Root exudates are typically photosynthetic carbon, low molecular weight molecules, and high molecular weight organic acids. This complex mixture modifies and promotes the development of a microbial community in the rhizosphere. These secondary metabolites have a potential role in the development of naturally occurring contaminant-degrading enzymes.

Rhizosphere: the soil environment encompassing the root zone of the plant.

RRF: relative response factor.

Saturated hydrocarbon: a saturated carbon-hydrogen compound with all carbon bonds filled; that is, there are no double or triple bonds, as in olefins or acetylenes.

Saturation: the maximum amount of solute that can be dissolved or absorbed under prescribed conditions.

SIM (Selecting Ion Monitoring): mass spectrometric monitoring of a specific mass/charge (m/z) ratio. The *SIM* mode offers better sensitivity than can be obtained using the full scan mode.

Solubility: The amount of a substance (solute) that dissolves in a given amount of another substance (solvent). Particularly relevant to oil spill cleanup is the measure of how much and the composition of oil that will dissolve in the water column. This is important as the soluble fractions of the oil are often toxic to aquatic life, especially at high concentrations. The *solubility* of oil in water is very low, generally less than 1 part per million (ppm).

Soluble: capable of being dissolved in a solvent.

Solvolysis: generally, a reaction with a solvent, involving the rupture of one or more bonds in the reacting solute: more specifically, the term is used for substitution, elimination, or fragmentation reactions in which a solvent species is the nucleophile; hydrolysis, if the solvent is water, or alcoholysis, if the solvent is an alcohol.

Stable: as applied to chemical species, the term expresses a thermodynamic property, which is quantitatively measured by relative molar standard Gibbs energies; a chemical species A is more stable than its isomer B under the same standard conditions.

Substrate component: in a nutrient medium, supplying microorganisms with carbon (C-substrate) and nitrogen (N-substrate) as food needed to grow.

Surface-active agent: a compound that reduces the surface tension of liquids, or reduces interfacial tension between two liquids, or a liquid and a solid; also known as surfactant, wetting agent, or detergent.

Sustainable enhancement: an intervention action that continues until such time that the enhancement is no longer required to reduce contaminant concentrations or fluxes.

Steranes: a class of tetracyclic, saturated biomarkers constructed from six isoprene subunits (~C_{30}). *Steranes* are derived from sterols, which are important membrane and hormone components in eukaryotic organisms. Most commonly used *steranes* are in the range of C_{26} to C_{30} and are detected using m/z 217 mass chromatograms.

Surrogate analyte: a pure analyte that is extremely unlikely to be found in any sample, which is added to a sample aliquot in a known amount, and is measured with the same procedures used to measure other components. The purpose of a *surrogate analyte* is to monitor the method performance with each sample.

Terminal electron acceptor (TEA): a compound or molecule that accepts an electron (is reduced) during metabolism (oxidation) of a carbon source; under aerobic conditions, molecular oxygen is the terminal electron acceptor; under anaerobic conditions, a variety of terminal electron acceptors may be used. In order of decreasing redox potential, these terminal electron acceptors include nitrate, manganic manganese, ferric iron, sulfate, and carbon dioxide; microorganisms preferentially utilize electron acceptors that provide the maximum free energy during respiration; of the common terminal electron acceptors listed above, oxygen has the highest redox potential, and provides the most free energy during electron transfer.

Terpanes: a class of branched, cyclic alkane biomarkers, including *hopanes* and tricyclic compounds. They are commonly monitored using m/z 191 mass chromatograms.

Total n-alkanes: the sum of all resolved *n-alkanes* (from C_8 to C_{40} plus pristane and phytane).

Total 5 alkylated PAH homologs: the sum of the 5 target PAHs (naphthalene, phenanthrene, dibenzothiophene, fluorene, chrysene) and their alkylated (C_1 to C_4) homologues, as determined by GCMS. These 5 target alkylated PAH homologous series are oil-characteristic aromatic compounds.

Total aromatics: the sum of all resolved and unresolved aromatic hydrocarbons, including the total of BTEX and other alkyl benzene compounds, total 5 target alkylated PAH homologues, and other EPA priority PAHs.

Total saturates: the sum of all resolved and unresolved aliphatic hydrocarbons, including the total n-alkanes, branched alkanes, and cyclic saturates.

TPH: total petroleum hydrocarbons; the total measurable amount of petroleum-based hydrocarbons present in a medium, as determined by gravimetric or chromatographic means.

Triterpanes: a class of cyclic saturated *biomarkers* constructed from six isoprene subunits. Cyclic *terpane* compounds containing two, four, and six isoprene subunits are called monoterpane (C_{10}), diterpane (C_{20}), and *triterpane* (C_{30}), respectively.

UCM: unresolved complex mixture of hydrocarbons on, for example, a gas chromatographic tracing; the UCM appear as the *envelope* or *hump area* between the solvent baseline and the curve defining the base of resolvable peaks.

Underground storage tank: a storage tank that is partially or completely buried in the earth.

Unsaturated zone: the zone between land surface and the capillary fringe within which the moisture content is less than saturation and pressure is less than atmospheric; soil pore spaces also typically contain air or other gases; the capillary fringe is not included in the unsaturated zone (See *Vadose zone*).

Upgradient: in the direction of increasing potentiometric (piezometric) head. See also *Downgradient*.

US EPA: United States Environmental Protection Agency.

USGS: United States Geological Survey.

Vadose zone: the zone between land surface and the water table within which the moisture content is less than saturation (except in the capillary fringe) and pressure is less than atmospheric; soil pore spaces also typically contain air or other gases; the capillary fringe is included in the vadose zone.

Vapor pressure: a measure of how oil partitions between the liquid and gas phases, or the partial pressure of a vapor above a liquid oil at a fixed temperature; the force per unit area exerted by a vapor in an equilibrium state with its pure solid, liquid, or solution at a given temperature.

Viscosity: the resistance of a fluid to shear, movement, or flow. The viscosity of an oil is a function of its composition. In general, the greater the fraction of *saturates* and *aromatics* and the lower the amount of *asphaltenes* and *resins*, the lower the viscosity. As oil weathers, the evaporation of the lighter components leads to increased viscosity.

Viscosity also increases with decreased temperature, and decreases with increased temperature.

The viscosity of an ideal, non-interacting fluid does not change with shear rate. Such fluids are called Newtonian. Most crude oils and oil products are Newtonian. The viscosity of non-Newtonian materials may vary with shear rate, as well as duration of shear. Oils with high *wax* content are often non-Newtonian, and stable water-in-oil *emulsions* are always non-Newtonian. A material that exhibits a decrease in viscosity with shear stress is termed pseudoplastic, while those that exhibit a decrease in viscosity with time of applied shear force are referred to as thixotropic. Both effects are caused by internal interactions of the molecules and larger structures in the fluid, which change with the movement of the material under applied stress. Generally, non-Newtonian oils are pseudoplastic, while *emulsions* may be either thixotropic or pseudoplastic.

In terms of oil spill cleanup, viscous oils do not spread rapidly, do not penetrate soils as rapidly, and affect the ability of pumps and skimmers to handle the oil.

Volatile: readily dissipating by evaporation.

Volatile organic compounds (VOC): organic compounds with high *vapor pressures* at normal temperatures. *VOCs* include light *saturates* and *aromatics*, such as pentane, hexane, *BTEX*, and other lighter substituted benzene compounds, which can make up to a few percent of the total mass of some crude oils.

Water solubility: the maximum amount of a chemical that can be dissolved in a given amount of pure water at standard conditions of temperature and pressure; typical units are milligrams per liter (mg/L), gallons per liter (g/L), or pounds per gallon (lbs/gall).

Waxes: waxes are predominately straight-chain *saturates* with melting points above 20°C (generally, the *n*-alkanes C_{18} and higher molecular weight).

Weathering: processes related to the physical and chemical actions of air, water, and organisms after oil spill. The major weathering processes include evaporation, dissolution, dispersion, photochemical oxidation, water-in-oil *emulsification*, microbial degradation, adsorption onto suspended particulate materials, interaction with mineral fines, sinking, sedimentation, and formation of tar balls.

Wilting point: the largest water content of a soil at which indicator plants, growing in that soil, wilt and fail to recover when placed in a humid chamber.

Conversion Factors

1. Concentration Conversions

1 part per million (1 ppm) = 1 microgram per liter (1 μg/L)
1 microgram per liter (1 μg/L) = 1 milligram per kilogram (1 mg/kg)
1 microgram per liter (μg/L) × 6.243 × 10^8 = 1 lb per cubic foot (1 lb/ft^3)
1 microgram per liter (1 μg/L) × 10^{-3} = 1 milligram per liter (1 mg/L)
1 milligram per liter (1 mg/L) × 6.243 × 10^5 = 1 pound per cubic foot (1 lb/ft^3)
I gram mole per cubic meter (1 g mol/m^3) × 6.243 × 10^5 = 1 pound per cubic foot (1 lb/ft^3)
10,000 ppm = 1% w/w
1 ppm hydrocarbon in soil × 0.002 = 1 lb of hydrocarbons per ton of contaminated soil

2. Sludge Conversions

1,700 lbs wet sludge = 1 yd^3 wet sludge
1 yd^3 sludge = wet tons/0.85
Wet tons sludge × 240 = gallons sludge
1 wet ton sludge × % dry solids/100 = 1 dry ton of sludge

3. Weight Conversion

1 ounce (1 oz) = 28.3495 grams (18.2495 g)
1 pound (1 lb) = 0.454 kilogram
1 pound (1 lb) = 454 grams (454 g)
1 kilogram (1 kg) = 2.20462 pounds (2.20462 lb)
1 stone (English, 1 st) = 14 pounds (14 lb)

1 ton (US; 1 short ton) = 2,000 lbs
1 ton (English; 1 long ton) = 2,240 lbs
1 metric ton = 2204.62262 pounds
1 tonne = 2204.62262 pounds

4. Temperature Conversions

°F = (°C × 1.8) + 32
°C = (°F – 32)/1.8
(°F – 32) × 0.555 = °C
Absolute zero = –273.15°C
Absolute zero = –459.67°F

5. Area

1 square centimeter (1 cm^2) = 0.1550 square inches
1 square meter 1 (m^2) = 1.1960 square yards
1 hectare = 2.4711 acres
1 square kilometer (1 km^2) = 0.3861 square miles
1 square inch (1 $inch^2$) = 6.4516 square centimeters
1 square foot (1 ft^2) = 0.0929 square meters
1 square yard (1 yd^2) = 0.8361 square meters
1 acre = 4046.9 square meters
1 square mile (1 mi^2) = 2.59 square kilometers

6. Nutrient Conversion Factor

1 pound phosphorus × 2.3 (1 lb P × 2.3) = 1 pound phosphorous pentoxide (1 lb P_2O_5)
1 pound potassium × 1.2 (1 lb K × 1.2) = 1 pound potassium oxide (1 lb K_2O)

7. Other Approximations

14.7 pounds per square inch (14.7 psi) – 1 atmosphere (1 atmos)
1 kiloPascal (kPa) × 9.8692 × 10^{-3} = 14.7 pounds per square inch (14.7 psi)
1 yd^3 = 27 ft^3
1 US gallon of water = 8.34 lbs
1 imperial gallon of water – 10 lbs

1 yd^3 = 0.765 m^3
1 acre-inch of liquid = 27,150 gallons = 3.630 ft^3
1 ft depth in 1 acre (in-situ) = 1,613 × (20 to 25 % excavation factor) = ~2,000 yd^3
1 yd^3 (clayey soils-excavated) = 1.1 to 1.2 tons (US)
1 yd^3 (sandy soils-excavated) = 1.2 to 1.3 tons (US)

Index

Accuracy 242
Acid rain 123
Activated sludge 375
Adhesion 18, 263
Adsorption chromatography 66
Advantages/disadvantages 9, 25, 524
Aerobic biodegradation 15
Air sparging 26, 368
Air stripping 369
Alkanes 50, 307
Alkenes 50
Anaerobes 10
Anaerobic biodegradation 15, 21
Anaerobic sparging 27
Analysis 134
  Accuracy 242
  Measurement 240
  Precision 243
Analytical methods 259, 289
  Standards organizations 45
API gravity 64, 266
Aromatic hydrocarbons 7, 309
  Multi-ring 51
  Single ring 50
Asphalt 104, 467, 470, 473
  Acid number 187
  Asphaltene content 188
  Biodegradation 469, 471
  Carbon disulfide insolubles 189
  Composition 190
  Density 192
  Elemental analysis 195
  Environmental impact 197
  Float test 196
  Production 465, 467
  Softening point 196
  Viscosity 197
  Weathering 197
Asphalt manufacture 186
Asphaltene constituents 56, 338
Associated gas 121
Atmospheric distillation 53, 85, 185
Autoignition temperature 269
Automotive gasoline 92
Aviation fuel 93
Aviation gasoline 93
Aviation turbine fuel 93

Beavon process 132
Beta-oxidation process 18
Bioaugmentation 12, 22, 25, 336
Bioavailability 8
Biochemical oxygen demand 30
Biodegradation 3, 10, 305, 322, 501
  Aerobic 15
  Alkanes 307
  Anaerobic 15
  Aromatic hydrocarbons 309
  Biodegradation of petroleum 305
  Chemical history 306
  Chlorinated compounds 319
  Definition 3
  Effect of dispersants 329, 338

Effect of pH 497
Effect of salinity 498
Effect of temperature 4, 327, 496
Factors relating to 79, 323, 490
Inherent 79
Kinetics 19
Mechanism 15
Mineral oil 31
Phenolic compounds 316
Petroleum 319
Polynuclear aromatic hydrocarbons 310
Rate 329
Site factors 3, 490
Spent caustic 317
Spills 24, 333
Test methods 30
Wastewater 318
Biodegradation of kerosene and diesel 385, 389
Biodegradation of fuel oil 413, 421
Biodegradation of lubricating oil 431, 438
Petroleum-based oils 440
Synthetic lubricating oil 433, 442
Used lubricating oil 443
Vegetable oils 448
Biodegradation of naphtha and gasoline 361
Dual-phase vacuum extraction 371
Hydrogen peroxide/Fenton's reagent 371
In-situ air stripping 369
In situ chemical oxidation 371
Permanganates 372
Pump-and-treat systems 370
Remediation 363
Single-phase vacuum extraction 371
Site remediation 364
Surfactant enhanced aquifer redemption 371
Vacuum enhanced recovery 370
Biodegradation of residua and asphalt 463, 469, 471
Asphaltene constituents 473
Biodegradation of vegetable oil 448
Bioengineering in bioremediation 29
Biological oxygen demand 263
Bioremediation 1, 3, 10, 502
Advantages 9, 524
Challenges 452
Chemical reactions 15
Control and optimization 8
Definition 3
Disadvantages 524
Effect of deep-sea environments 507
Effect of dispersants 329, 338, 506
Effect of flowing water 506
Effect of nutrients 325, 505
Effect of oxygen 505
Effect of petroleum characteristics 506
Effect of pH 497
Effect of prior exposure 506
Effect of salt 20, 498
Effect of temperature 496, 503
Factors relating to 4, 79, 323, 490
Future 515
Kinetic aspects 19
Method parameters 23
Methods 483
Natural 10
Soil characteristics 492
Extreme environments 522
Biopiling 23
Bioremediation in extreme environments 522
Bioremediation of waterways 9
Bioremediation methods 22, 483
Application 336
Conventional 11, 23, 517
Effect of pH 497
Effect of salinity 20, 498

Effect of temperature 4, 496
Land ecosystems 483
Parameters 23
Site evaluation 486
Soil evaluation 334, 489
Status 517
Traditional 11
Water ecosystems 498
Bio-slurping 27
Bioslurry reactor 11, 23
Biosparging 26
Biostimulation 12, 22, 25, 336
Biosurfactants 28, 521
Bioventing 26, 27, 368
BOD 30
Boiling fractions of petroleum 83
Boiling point
Distribution 264
Molecular weight 75
Boiling ranges of petroleum fraction 265
BTEX 9, 309, 377
Bunker C oil 97

Calorific value 136
Carbenes 56
Carboids 56
Cation exchange capacity 492
Challenges for bioremediation 452
Chaos theory 214
Chemical dispersibility 265
Chemo-autotrophic aerobes 16
Chlorinated hydrocarbons 7, 319
Claus tail gas 132
Coal tar 466
Coke 105, 199
Ash 201
Composition 202
Delayed 200
Density 204
Dust control 205
Environmental impact 207
Fluid 201
Hardness 205
Metals 206
Sulfur 207
Combined cycle generation units 124
Co-metabolism 15
Composting 11, 23
Condensate releases
Sampling 227
Conditions for biodegradation 323
Containment 2
Conventional bioremediation 11, 23, 517
Conversion processes 83
Crankcase oil 99
Cycloalkanes 50

Deep-sea environments 507
Delayed coke 200
Delayed coking 87
Demet 113
Density 141, 158, 167, 266
Dew point 119
Dielectric oil 100
Diesel fuel 94, 98, 387, 391
Bioremediation 389
Component streams 388
Diesel fuel additives 401
Direct oxygen injection 376
Dissolved gas 121
Dissolved oil 178
Distillate fuel oil 95, 162
Distillation 52, 185
Atmospheric 53, 85
Vacuum 54, 85
Doctor test method 170
Domestic fuel oil 95, 162
Dual-phase vacuum extraction 371

Effects of biodegradation 322
Effect of dispersants 329, 338
Effect of nutrients 325
Effect of temperature 4, 327, 496
Effect of weathering 332
Electrospray ionization 62

Emulsified oil 178
Emulsion formation 40, 267
Enhanced bioremediation 12, 519
Environmental aspects of lubricating oil use 438
Environmental effects 161, 175, 197, 207
  Refining 129
EPA quality control methods 246, 247, 248
Essential factors for bioremediation 4, 490
Evaporation 158, 268
Ex situ bioremediation 13, 24
Ex situ waste treatment 2
Extract concentration 237
Extraction methods 224

Factors relating to biodegradability 79, 490
Fenton's reagent 371
Fertilization 5
Fire point 74, 269
Flame emission spectroscopy 61
Flame photometry 61
Flash point 74, 159, 269
Fluid catalytic cracking 88
Fluid coke 201
Fluid coking 88
Fractionation 270
Free oil 178
Fuel gas 91
Fuel oil 95, 162, 413, 421
  Asphaltene content 164
  Biodegradation 421
  Composition 165
  Density (specific gravity) 167
  Elemental analysis 168
  Environmental impact 175
  Flash point 172
  Grades 414
  Metals content 172
  Pour point 173
  Precipitation number 165
  Production 417
  Properties 96, 419
  Stability 174
  Viscosity 173
  Weathering 424
Fugitive emissions 117

Gas chromatographic methods 286
Gas chromatography 285, 290
Gas chromatography-mass spectrometry 288
Gaseous products
  Composition and properties 115
Gasoline 92, 363
  Component streams 86
  Ex situ destruction 367
  Reformulated 93
Gel permeation chromatography 66
Gravimetry 296
Grease 101
Group analyses 279

Heat of combustion 136
Heavy fuel oil 96
Heteroatom systems 40, 45
High performance liquid chromatography 66, 287
Hydrocarbon degraders 338
Hydrocarbon gases 130
Hydrocarbons
  Number of isomers 56
Hydrocracking 89
Hydrofluoric acid 109
Hydrogen peroxide/Fenton's reagent 371
Hydrotreating 90

Ignition temperature 269
Immunoassay 284, 297
Incineration 2
Indigenous bacteria 10
Infiltration rate 491
Infrared spectroscopy 58, 294

Inherent biodegradability 79
Inorganic gases 130
Insecticides 100
In-situ air stripping 369
In situ bioremediation 13, 24
In situ chemical oxidation 371
In situ-enhanced bioremediation 372
Insulating oil 100
Interfacial tension 69, 273
Intrinsic aerobic biodegradation 5
Intrinsic bioremediation 5
Ion exchange chromatography 66

Jet fuel 93, 98
  Bioremediation 402

Kerosene 94, 388, 389
  Bioremediation 389

Laboratory tests and full-scale operations 7
Land ecosystems 483
Land farming 11, 23
Leachability 270
Lighter non-aqueous phase liquids 28
Liquefaction 67
Liquefied petroleum gas 118
  Properties 116
  Stenching 119
Liquid chromatography 62, 66
Liquid products
  Composition 147, 158
  Properties 147
Litho-trophic aerobes 16
LNAPLs 28
LPG 90, 116
Lower explosive limit 269
Lubricating oil 97
  Biodegradation 30, 438
  Classification 435
  Composition 436
  Environmental impact 435, 438
  Petroleum-based 440
  Production 434
  Properties 436
  SAE standards 99
  Synthetic 433, 442
  Thermal stability 437
  Used 443
  Viscosity 99

Maltene fraction 56
Mass spectrometry 59
Measurement 240
Mechanism of biodegradation 15
Metals content 271
Methanogenic bacteria 17
Method detection limit 256
Method validation 245
Methyl tert-butyl ether 380
Met-x 113
Microbial inocula 13
Middle distillate 81
Mineralization 2, 15, 306
Mineral oil 431
Mineral oil biodegradability 31
Missing oil 40
Molecular weight
  Boiling point 75
Monitored natural attenuation 25, 376
Motor oil 99
MTBE 379, 380
Multi-ring aromatics 51

Naphtha 148
  Boiling range 362
  Composition 153
  Density 158
  Environmental impact 161
  Evaporation rate 158
  Flash point 159
  Odor and color 159
  Terminology 362
  Volatility 160
NAPLs 8, 27

Natural attenuation 3, 22, 25
Natural bioremediation 10
Natural gas 91, 121
Non-aqueous phase liquids 8, 27
Non-point sources of pollution 39, 117, 337
Non-volatile compounds 215
  Sampling 228
Non-volatile organic compounds 215
Nuclear magnetic resonance spectroscopy 58
Nutrient enrichment 5

Odor and color 159
Oil spill bioremediation 24

Particulate matter 133
Permanganates 372
Petrochemicals 90
Petroleum
  History 41
Petroleum-based lubricating oil 440
Petroleum biodegradation 305, 319
  Conditions for biodegradation 323
  Effects of biodegradation 322
  Effect of dispersants 329, 338
  Effect of nutrients 325
  Effect of temperature 327, 496
  Effect of weathering 332
  Rates of oil biodegradation 329
  Specific constituents 307
Petroleum chemicals 215
Petroleum composition 19, 43, 46
  Chemical composition 50
  Chromatographic fractionation 65
  Composition by volatility 52, 83
  Composition by fractionation 55
  Composition by spectroscopy 57
  Elemental composition 47
  Heteroatom systems 40, 45
  Metals content 68
  Variation 44
Petroleum gas 118
Petroleum group analyses 261, 279
Petroleum products
  History 41
Petroleum properties 19, 39, 42, 62, 262
  Density and specific gravity 63
  Elemental analysis 64
  Liquefaction and solidification 67
  Surface and interfacial tension 69
  Ultimate analysis 64
  Variation 44
  Viscosity 71
  Volatility 74
Petroleum solvents 94
Phenolic compounds 316
Phytoextraction 6
Phytoremediation 5
Phytostabilization 6
Phytotransformation 6
Phytovolatilization 6
Pipestill gas 91
Pitch 466
Point sources of pollution 39, 117
Polynuclear aromatic systems 40, 309
Pour point 272
Precipitation number 165
Precision 243
Primary refining process 52
Principles of bioremediation 7
Propane deasphalting 186
Pump-and-treat systems 2, 370

Quality assurance 252, 255
Quality control 246, 247, 248, 252, 253

Rates of oil biodegradation 329
Reduced crude 185

Refinery 82
Conversion processes 83
Distillation 185
Environmental effects 129
Separation processes 83
Refinery chemicals 106, 215
Demet 113
Hydrofluoric acid 109
Met-x 113
Spent acid 107
Spent catalysts 109
Spent caustic 106
Sulfuric acid sludge 109
Refinery gas 125
Analysis 134
Calorific value 136
Composition 126, 136
Density 141
Heat of combustion 136
Hydrocarbon gases 130
Inorganic gases 130
Relative density 141
Sulfur content 142
Vapor density 141
Vapor pressure and volatility 143
Refinery products 79, 81
Asphalt 104
Coke 105
Diesel fuel 94
Environmental impact 161
Fuel oil 95
Gases 115, 118
Gasoline 92
Grease 101
Insecticides 100
Kerosene 94
Liquefied petroleum gas 90
Lubricating oil 98
Naphtha 92, 148
Gasoline 92
Wax 102
White oil, insulating oil,
Refinery by-products 79
Sludge 208
Refinery gas 91
Reformulated gasoline 93
Regulated compounds 215
Remediation 363
Relative density 141
Residua 184, 467, 471
Acid number 187
Asphaltene content 188
Biodegradation 469
Carbon disulfide insolubles 189
Composition 48, 49, 54, 190
Density 192
Elemental analysis 195
Environmental impact 197
Float test 196
Production 463
Properties 464
Softening point 196
Viscosity 197
Weathering 197, 332
Residual fuel oil 95, 162
Resin constituents 56, 338
Rhizofiltration 6
Rhizosphere bioremediation 6, 28
Rhizovolatilization 6

Sample cleanup 240
Sample collection 213, 218, 222
Condensate releases 227
Extraction methods 224
Semi-volatile and non-volatile compounds 228
Solids 233
Volatile compounds 225
Water samples 236
Sample 213
Sample cleanup 240
Sample collection 218
Sample preparation 218
Sampling from an aqueous medium 221
Sampling log 217
Sampling protocols 219
SARA analysis 55, 166

SCOT process 132
Sediment 232
Seeding 5
Semi-volatile compounds 215
  Sampling 228
Separation processes 83
Sewage treatment 2
Simdis 75
Simulated distillation 75
Single-phase vacuum extraction 371
Single ring aromatics 50
Site evaluation 486
Site factors for biodegradation 3, 490
Site remediation 364
Sludge 208
Soil classification 490
Soil contamination 1, 334
Soil evaluation 489
Soil sampling 220
Soil vapor extraction 2, 26
Solidification 67
Solid products 183
  Acid number 187
  Asphalt 184
  Asphaltene constituents 188, 473
  Carbon disulfide insoluble constituents 189
  Composition 190
  Density 192
  Elemental analysis 195
  Environmental impact 197
  Float test 196
  Residua 48, 49, 184
  Sampling 233
  Softening point 196
  Specific gravity 192
  Viscosity 197
  Weathering 197, 332
Solubility in aqueous media 272
Specific gravity 63, 158, 266
Spent acid 107
Spent catalysts 109
Spent caustic 106, 317
Spills 39, 333
  Missing oil 40
Standards organizations for analytical methods 45
Stenching 119
Stoddard solvent 94
Stove oil 96, 163
Straight-run fractions 52
Sulfur content 273
Sulfuric acid sludge 109
Supercritical fluid 67
Surface tension 69, 273
  Selected hydrocarbons 70
Surfactant enhanced aquifer redemption 371
Suspended particulate matter 133
Synthetic lubricating oil 433, 442

Tar 466
TCLP 271
Thin layer chromatography 283
Total petroleum hydrocarbons 261, 274, 289
Total suspended particulate matter 133
Toxic chemicals 2
Toxicity 270
Toxicity characteristic leaching procedure 271
TPH 261, 274, 289
Traditional bioremediation 11
Transformer oil 100
Trichloroethylene 7

Ultimate analysis 64
Unassociated gas 121
Unregulated compounds 215
Upper explosive limit 269
Used lubricating oil 443
Use of biosurfactants 28

Vacuum distillation 54, 85, 185
Vacuum enhanced recovery 370

Vapor density 141
Vapor pressure 75, 143
Vegetable oils 448
Visbreaking 87
Viscosity 71, 277
Viscosity index 73
Volatile compounds
  Sampling 225
Volatile organic compounds 215, 278
Volatility 74, 143, 160

Wastewater 177, 318
Water content 279
Water ecosystems 499
Water samples 236
Wax 102
Weathering 197, 332, 500
Wellman-Lord process 132
White distillate 100
White oil 100

Xenobiotic chemical 520

Also of Interest

Check out these other related titles from Scrivener Publishing

From the Same Author

An Introduction to Petroleum Technology, Economics, and Politics, by James Speight, ISBN 9781118012994. The perfect primer for anyone wishing to learn about the petroleum industry, for the layperson or the engineer. *NOW AVAILABLE!*

Ethics in Engineering, by James Speight and Russell Foote, ISBN 9780470626023. Covers the most thought-provoking ethical questions in engineering. *NOW AVAILABLE!*

Other Related Titles from Scrivener Publishing

Fundamentals of the Petrophysics of Oil and Gas Reservoirs, by Buryakovsky, Chilingar, Rieke, and Shin. ISBN 9781118344477. The most comprehensive book ever written on the basics of petrophysics for oil and gas reservoirs. *NOW AVAILABLE!*

Petroleum Accumulation Zones on Continental Margins, by Grigorenko, Chilingar, Sobolev, Andiyeva, and Zhukova. Coming in September 2012, ISBN 9781118385074. Some of the best-known petroleum engineers in the world have come together to produce one of the first comprehensive publications on the detailed (zonal) forecast of offshore petroleum potential, a must-have for any petroleum engineer or engineering student.

Mechanics of Fluid Flow, by Basniev, Dmitriev, and Chilingar. Coming in September 2012, ISBN 9781118385067. The mechanics of fluid flow is one of the most important fundamental engineering disciplines explaining both natural phenomena and human-induced processes. A group of some of the best-known petroleum engineers in the world give a thorough understanding of this important discipline, central to the operations of the oil and gas industry.

Zero-Waste Engineering, by Rafiqul Islam, ISBN 9780470626047. In this controversial new volume, the author explores the question of zero-waste engineering and how it can be done, efficiently and profitably. *NOW AVAILABLE!*

Sustainable Energy Pricing, by Gary Zatzman, ISBN 9780470901632. In this controversial new volume, the author explores a new science of energy pricing and how it can be done in a way that is sustainable for the world's economy and environment. *NOW AVAILABLE!*

Formulas and Calculations for Drilling Engineers, by Robello Samuel, ISBN 9780470625996. The most comprehensive coverage of solutions for daily drilling problems ever published. *NOW AVAILABLE!*

Emergency Response Management for Offshore Oil Spills, by Nicholas P. Cheremisinoff, PhD, and Anton Davletshin, ISBN 9780470927120. The first book to examine the Deepwater Horizon disaster and offer processes for safety and environmental protection. *NOW AVAILABLE!*

Advanced Petroleum Reservoir Simulation, by M.R. Islam, S.H. Mousavizadegan, Shabbir Mustafiz, and Jamal H. Abou-Kassem, ISBN 9780470625811. The state of the art in petroleum reservoir simulation. *NOW AVAILABLE!*

Energy Storage: A New Approach, by Ralph Zito, ISBN 9780470625910. Exploring the potential of reversible concentrations cells, the author of this groundbreaking volume reveals new technologies to solve the global crisis of energy storage. *NOW AVAILABLE!*